Springer Proceedings in Mathematics & Statistics

Volume 390

This book series features volumes composed of selected contributions from workshops and conferences in all areas of current research in mathematics and statistics, including data science, operations research and optimization. In addition to an overall evaluation of the interest, scientific quality, and timeliness of each proposal at the hands of the publisher, individual contributions are all refereed to the high quality standards of leading journals in the field. Thus, this series provides the research community with well-edited, authoritative reports on developments in the most exciting areas of mathematical and statistical research today.

Luigi Accardi · Farrukh Mukhamedov ·
Ahmed Al Rawashdeh
Editors

Infinite Dimensional Analysis, Quantum Probability and Applications

QP41 Conference, Al Ain, UAE, March 28–April 1, 2021

Springer

Editors
Luigi Accardi
Vito Volterra Interdepartamental Centre
Universita degli Studi di Roma
Rome, Italy

Ahmed Al Rawashdeh
Department of Mathematical Sciences
College of Science
United Arab Emirates University
Al Ain, Abu Dhabi, United Arab Emirates

Farrukh Mukhamedov
Department of Mathematical Sciences
College of Science
United Arab Emirates University
Al Ain, Abu Dhabi, United Arab Emirates

ISSN 2194-1009 ISSN 2194-1017 (electronic)
Springer Proceedings in Mathematics & Statistics
ISBN 978-3-031-06169-1 ISBN 978-3-031-06170-7 (eBook)
https://doi.org/10.1007/978-3-031-06170-7

Mathematics Subject Classification: 37C45, 37N25, 47Axx, 47B90, 47B93, 47D07, 60D05, 60H40, 81Pxx, 81Sxx

This Springer imprint is published by the registered company Springer Nature Switzerland AG
The registered company address is: Gewerbestrasse 11, 6330 Cham, Switzerland

Organization

QP41 2021 is organized by the Department of Mathematical Sciences, United Arab Emirates University (UAEU), and AQPIDA (The Association of Quantum Probability, Infinite Dimensional Analysis). We acknowledge the following UAEU grants: UPAR Grant No. G00003447 and UPAR-2019 (Fund Number 31S397).

Executive Committee

Conference Chair
Prof. Farrukh Mukhamedov, UAE University, UAE

Conference Co-Chair
Prof. Luigi Accardi, University of Rome "Tor Vergata", Italy
Prof. Ahmed Al Rawashdeh, UAE University, UAE

Organizing Committee

Dr. Muhammad Imran, UAE University, UAE
Dr. Humberto Rafeiro, UAE University, UAE
Dr. Ho Hon Leung, UAE University, UAE
Dr. Youssef El Khatib, UAE University, UAE
Ms. Fouzia Shaheen, UAE University, UAE

International Scientific Committee

Preface

The fact that Quantum Theory (QT) is not only a new mechanics but also a new probability was clear from the very beginning of this discipline, in the first quarter of the twentieth century. While the systematic comparison between classical and quantum mechanics has always been present in QT, the systematic investigation of the probabilistic aspects of quantum theory began in the early 1970s. From that time to the present day, the field of quantum probability (QP) has grown enormously.

At the beginning, the QP conferences were a meeting point for a small group of people coming from different countries who had a clear intuition of the potentialities of this new field and were united by the goal of promoting its development. Since the interactions between probability and QT arise in a multiplicity of scientific levels, from philosophical problems concerning the foundations of probability and physics, to the solution of specific problems arising from physics and the development of new mathematical techniques to deal with them, the multidisciplinarity of the field was one of the characteristics of QP since its early developments. Practically, no other mathematical discipline, after the Second World War, has given so many conceptually deep and innovative contributions to so many branches of science. From the mathematical explanation of the old quantum paradoxes, which the philosopher Karl Popper labeled *the great quantum muddle*, to the development of the first deductive (as opposed to phenomenological) theory of quantum transport in open systems, to the emerging applications of non-Kolmogorovian models in sciences different from physics, to the discovery of the natural emergence, from the combination of classical probability with the classical theory of orthogonal polynomials, of *general quantization rules* which explain the probabilistic roots of Heisenberg commutation relations and explains why future *interacting quantum theories* cannot be based on such relations. Parallel to this challenge to physics, the same discovery poses an optimistic challenge to classical probability and statistics: to learn how to interpret and put to use the quantum aspects of classical random variables, processes, and fields which are ubiquitous outside quantum physics.

If, from deep conceptual innovations, we pass to relevant technical progresses (which often are the basis of conceptual innovations), then the list of contributions of QP becomes considerably longer. From the emergence of quantum Markov chains, on

which nowadays the most sophisticated numerical technique for the approximation of ground states of quantum Hamiltonians is based, to quantum stochastic calculus, without which no deductive theory of quantum transport could exist, to white noise hamiltonian calculus, unifying the stochastic differential equation approach used in mathematics with the distribution (white noise) approach used in physics, to new objects in operator theory like the canonical conditional expectation of a state, to the introduction of local equilibrium fields, to the universality of the non-crossing diagrams in the stochastic limit of strongly non-linear interactions (versus the universality of the gaussian diagrams in usual quantum field models), to the connections between random matrices and free independence or between the adjacency matrices of graphs and various forms of independence (monotone, boolean, ...), to the deep connections between interacting Fock spaces and classical orthogonal polynomials, to the first deduction of the natural emergence of Hilbert modules in QT (notice that the great mathematical development of Hilbert's module theory in the last three decades occurred after this discovery). The above list is far from exhaustive, but it is sufficient to show that QP abundantly passes the only criterium that distinguishes deep innovations from individual, although important, results, namely, the capacity to give essential contributions to different fields of science, and in particular of mathematics.

Clearly, given the width of the scientific front, very few people could contribute to all these developments. However, quite a number of people gave relevant contributions to several of them and this fact, on one side, played a relevant role in keeping up the initial multidisciplinary inspiration, on the other side, confirms the historical fact that every deep innovation in science is a collective phenomenon, involving several groups of scientists in several different countries.

Now, in almost 50 years of life, QP has reached maturity and begins to face the problems intrinsic to every mature mathematical discipline, namely, the conflict between specialization and multidisciplinarity. This induced a gradual change in the purpose of the QP conferences: from meeting points of a few researchers with strong interactions among themselves to moment of interaction and mutual confrontation among groups dealing with problems that, although stemming from a common root, employ different and sometimes very mathematical techniques for different purposes. With the explosion of Quantum Technology, the promotion of applications to industrial problems is going to become one of the main goals of the conferences.

The 41st International Conference on Quantum Probability and Related Topics, was held at the UAE University, Al Ain city-UAE from March 28 to April 1, 2021. The conference was dedicated to the memory of Robin Hudson, one of the pioneers of quantum probability, who passed away on January 12, 2021. Unfortunately, just a few months before the opening of the conference, two other pioneers of quantum probability passed away: Andrzej Kossakowski, famous for his contributions to the theory of open quantum systems, and Wilhelm von Waldenfels author of the first universal quantum central limit theorems and who introduced co-algebra techniques in QP long before the emergence of quantum groups. During the conference, a commemorative speech on Kossakowski who passed away on February 1, 2021 was delivered by Dariusz Chruscinski and one in memory of Wilhelm von Waldenfels was given

by Luigi Accardi. The latter will appear in a special issue of IDAQP 2022 dedicated to both Robin Hudson and Wilhelm von Waldenfels, good friends in life and united by an almost simultaneous death.

As usual, this conference covered recent developments in Quantum Probability and Infinite Dimensional Analysis, with applications to Mathematical Physics, Quantum Information Theory, and other related fields.

The multidisciplinary nature of Quantum Probability is well reflected by the structure of the volume, which is divided into six parts:

I. Quantum Probability Methods
II. Quantum Information Methods
III. Quantum Dynamical Systems
IV. Infinite Dimensional Analysis
V. Operator Algebras
VI. Stochastic Operators

This wide spectrum of topics shows that the original goal of the organizers was fully realized. Several new results, some of which are of great depth, were presented at the conference and show the vitality of the field and its continual propensity to interact with different disciplines and to cover a wide spectrum of interests.

The 42nd QP conference will take place in the last week of January 2022 in Bangalore thus continuing the tradition and confirming the international reality of the QP community (the past ten QP conferences have taken place always in different nations, covering four continents). Readers of the volume are anticipated to be graduate students and research mathematicians interested in functional analysis, classical and quantum probability, operator algebras, Lie algebras, mathematical physics, and quantum information, quantum technology.

Rome, Italy — Luigi Accardi
Al Ain, United Arab Emirates — Farrukh Mukhamedov
Al Ain, United Arab Emirates — Ahmed Al Rawashdeh
November 2021

Contents

Part I
Quantum Probability Methods

The Non–linear and Quadratic Quantization Programs

Luigi Accardi, Andreas Boukas, Yun-Gang Lu, and Alexander Teretenkov

Abstract After a short outline of the notion of canonical quantum decomposition of a classical random field and of its connection with the program of non–linear quantization, we concentrate our attention on quadratic quantization. We introduce the $*$–Lie algebra of homogeneous quadratic Boson fields denoted $\text{heis}_{2,d}(\mathbb{C})$. Then we recall the main notions related to the complex symplectic Lie algebra $sp(2d, \mathbb{C})$ and how it is possible to define a natural involution $\diamond$ on it, thus obtaining a $*$–Lie algebra $sp_{\diamond}(2d, \mathbb{C})$. We prove that, with this involution, there is a $*$–isomorphism between $\text{heis}_{2,d,cls}(\mathbb{C})$ and $sp_{\diamond}(2d, \mathbb{C})$. We show that that the real Lie sub–algebra $sp_{\diamond,skew}(2d, \mathbb{C})$ of $sp_{\diamond}(2d, \mathbb{C})$ consisting of its skew–adjoint elements is isomorphic, as a real $*$–Lie algebra to $sp_{-}(2d, \mathbb{R})$, but the involution allowing this isomorphism is not the restriction to $sp(2d, \mathbb{R})$ of the $\diamond$–involution. We recall the expressions of the vacuum characteristic functions of quadratic Weyl operators, i.e. exponentials of elements of $sp_{\diamond,skew}(2d, \mathbb{C})$. We describe the Lie groups associated with the symplectic algebra. Finally we discuss the problems of diagonalizability and vacuum factorizability for quadratic fields, i.e. elements of $sp_{\diamond}(2d, \mathbb{C})$, and we give a necessary and sufficient condition for diagonalizability.

L. Accardi (✉)
Centro Vito Volterra, Università di Roma Tor Vergata, via Columbia 2, 00133 Roma, Italy
e-mail: accardi@volterra.mat.uniroma2.it

A. Boukas
Centro Vito Volterra and Hellenic Open University, Graduate School of Mathematics, Patras, Greece
e-mail: boukas.andreas@ac.eap.gr

Y. G. Lu
Dipartimento di Matematica, Università di Bari, n.4, Via E. Orabona, 70125 Bari, Italy
e-mail: yungang.lu@uniba.it

A. Teretenkov
Department of Mathematical Methods for Quantum Technologies, Steklov Mathematical Institute of Russian Academy of Sciences, 8 Gubkina St., Moscow 119991, Russia
e-mail: taemsu@mail.ru

L. Accardi et al. (eds.), *Infinite Dimensional Analysis, Quantum Probability and Applications*, Springer Proceedings in Mathematics & Statistics 390,
https://doi.org/10.1007/978-3-031-06170-7_1

Keywords Quantum decomposition of a classical random field · Non–linear quantization · Symplectic Lie algebra · Quadratic quantization

1 Introduction

In this Introduction we will try to explain, in a short and qualitative way, without entering into details, the main ideas of the **program of non–linear quantization**. Later on we will give precise definitions of all the objects discussed here.

The starting point is the fact that every classical random field $X \equiv (X_j)_{j \in D}$ (D being a finite or infinite set) has a **canonical quantum decomposition**:

$$X_j = a_j^+ + a_j^0 + a_j$$

where, in the right hand side, the following **CAP operators** appear:

$$\begin{array}{ll} a_j^+ & \textbf{creation} \\ a_j^0 & \textbf{preservation} \\ a_j & \textbf{annihilation} \end{array}$$

One can think of the X_j as components of a classical vector field along linearly independent directions of a vector space and the index j varies in a set that can be finite or infinite.

The CAP operators satisfy **two types of commutation relations**, in other words: a **classical random field** X is canonically associated to a **large set of mathematical structures**. In particular:

- a **∗–Lie algebra** $\mathcal{L}_X$;
- a **quantum probability** space, namely the pair

$$(\mathcal{P}_X(CAP\ operators), \varphi_X) \quad (\text{∗–algebra generated by the CAP operators})$$

where φ_X, is called **vacuum state** and is uniquely determined by the probability distribution of the classical random field X;

- a **semi–finite trace** on $\mathcal{P}_X(CAP\ operators)$ (so one can speak of **density operators**);
- classical dynamical systems are lifted **algebraically** to quantum dynamical systems;
- **quantum Hamiltonian systems**, given for example by the free or harmonic oscillator Hamiltonian (plus possible potentials) find their natural analogues in this approach;
- on $\mathcal{P}_X(CAP\ operators)$ there is a natural notion of **normal order** that extends, in a non–trivial way, the usual (bosonic or fermionic) notion.

In conclusion:
Any classical random field X is canonically associated to its own quantum theory.
The following question then naturally arises:
To which classical random field X is usual quantum theory canonically associated?
The answer is given by the following Theorem.
Theorem A

$$\mathcal{L}_X = \text{Heisenberg algebra} \iff X \in \text{gaussian class}$$

In other words: **usual quantum theory is the quantum theory canonically associated to classical gaussian random fields.**
From this a second natural question arises:

What about the quantum theories canonically associated to **other classes** of classical random fields?

Is the study of these theories **interesting**? Or it is a purely **formal exercise**?

There are **many reasons** why the study of these theories is interesting, for example:

- To learn the **meaning of quantum correlations in classical frameworks**: physicists have learned to interpret quantum correlations. Now that we know that all classical random variables have their quantum correlations, we have to understand their meaning in each specific context (biology, economy, engineering, ...).
- Several open problems of classical probability **were solved using the quantum decomposition of classical random variables** (for example the problem to express the Jacobi sequences, of (general) powers of a real valued random variable X, in terms of Jacobi sequences of X, see [10]).
- **To go beyond the gaussian framework.**

Gaussian random fields are fundamental. However a gaussian measure is equivalent, up to a linear transformation in the supporting space, to a **product measure.**

It is a known, universal phenomenon, that **product measure $\iff$ absence of correlations** i.e., **absence of interaction.**

The new approach to quantization, shows that: **correlations of a classical random field X are coded into the commutation relations** canonically associated to it, i.e. into $\mathcal{L}_X$. In fact the following theorem holds.
Theorem B A classical random field X is of product type **iff** there exists a parametrization of its CAP operators satisfying the following commutation relations

$$j \neq k \Rightarrow [a_j^+, a_k] = [a_j^+, a_k^0] = 0 \tag{1}$$

i.e. CAP operators associated to independent degrees of freedom commute.

In usual (boson) quantum mechanics one can always find a parametrization of the CAP operators such that (1) holds.

A consequence of the above theorem is that: **a truly interacting quantum theory cannot be based** on the usual boson commutation relations described in Sect. 2.1.

The program of non–linear quantization is to develop such a truly interacting quantum theory and to deduce the associated commutation relations. This program can be realized in the following two steps:

(1) Look for **the simplest** classical random fields whose Lie algebras $\mathcal{L}_X$ **do not satisfy** (1) (**interacting Lie algebras**).

(2) Construct a Hilbert space representation (typically one takes the analogue of the Fock representation) for (at least a sub–class) of these algebras.

1.1 Quadratic Quantization

The main success of this program up to now has been **quadratic quantization** of systems with 1 **degree of freedom** (or equivalently for factorizable systems with infinitely many degrees of freedom): a theory which, although far from complete, is sufficiently developed and is based on an extremely rich mathematical structure which allows explicit, highly non–trivial, calculations of the physically relevant quantities. This theory has been the basic model for all successive steps in the non–linear quantization program. In one degree of freedom it is known that, if a^+ and a are boson operators ($[a, a^+] = 1$), then

$$\{a^{+2}, a^2, a^+a, 1\} \tag{2}$$

generate a $*$–Lie algebra isomorphic to the 1–dimensional central extension of $sl(2, \mathbb{R})$. For this reason one speaks of **quadratic quantization**. It has been shown in [9] that the vacuum distributions of the self–adjoint elements of this algebra coincide with the class of non–standard Meixner classes (Gammas, negative binomials and proper Meixner). The standard Meixner classes (Gauss and Poisson) appear in usual quantization as vacuum distributions of the extended Heisenberg algebra (whose elements are the skew–hermitean generators of the 1–parameter groups of the Weyl representation of the full Euclidean group of the 1–particle space). Thus quadratic quantization provides the first example of non–gaussian, i.e. non–linear, quantization.

The limit of the present version of this theory is that it still leads to independent increment processes, i.e. it is still inside the class of product (or factorizable) measures. In other words the (quadratic) non–linearity manifests itself **at the level of each single degree of freedom**, while independent degrees of freedom still satisfy (1).

The problem is then: **to extend quadratic quantization to the multi–dimensional case** abandoning the factorization requirement.

From what said above, it follows that there are two natural ways to try and extend the quadratic quantization to the multi–dimensional case:
(1) **The quantum decomposition approach** starts from a classical multi–dimensional random field, finds its canonical quantum decomposition and deduces from it the commutation relations.
(2) **The quadratic boson approach** considers, in analogy with the 1–dimensional case, the natural multi–dimensional extensions of the products (2), i.e. the homogeneous, normally ordered, quadratic expressions in creation and annihilation operators of a system with $d(\in \mathbb{N})$ degrees of freedom (**homogeneous quadratic boson fields (HQBF)**) (which, together with the central element **1**, are a $*$–Lie algebra), finds some $*$–representation of this $*$–Lie algebra, cyclic with respect to a unit vector Φ (e.g. the Fock one), and looks for a maximal abelian $*$–Lie sub–algebra (**Cartan sub–algebra**) such that the Φ–distributions of its hermitean generators are non–factorizable measures.

Remark Notice that the two approaches are, in some sense, **symmetric**.

The quantum decomposition approach starts from a commutative $*$–algebra (the polynomial algebra in the components of the classical fields) and deduces, from the commutation relations, a $*$–Lie algebra and its Fock representation.

The quadratic boson approach starts from a $*$–Lie algebra and a representation of it (not necessarily the Fock one) and from it constructs a classical multi–dimensional random field.

Remark Usually in physics, quadratic expressions in Boson fields are related to dynamics and one speaks of **quadratic Hamiltonians**.

One of the first fall–outs of the quadratic quantization program, confirmed a posteriori by the new developments in the theory of orthogonal polynomials is that quadratic expressions in Boson fields have a natural **kinematical interpretation** as generalized field operators and for this reason in the following we call them **quadratic fields**. With this new interpretation, the positivity of these quadratic expressions is not required.
The problem with Approach (1) is that there exist in the literature several different definitions of *multi–dimensional Meixner distributions* whose equivalence is by far not clear. They share the property of giving the usual Meixner classes in 1 dimension, but in some cases little less. Moreover a well defined class of densities, or of explicit expressions for the probabilities, is not yet emerged from any of these definitions.
With Approach (2) there is a similar problem concerning explicit formulas for the characteristic functions of homogeneous quadratic boson fields. In the present paper we review some recent progresses concerning Approach (2). The motivation for this paper arose from the remark in [3] that the $*$–Lie algebra of HQBF has a structure that strongly resembles a *non-commutative extension* of $\mathrm{sl}(2, \mathbb{C})$ (more precisely of its unique 1–dimensional central extension). On the other hand, in the huge physics and mathematics literature devoted to HQBF starting from the late 1940's (see the bibliography below) the algebra generated by these fields was identified, **as a Lie algebra**, with a symplectic algebra of appropriate dimensions (see further). Since

both algebras describe the same object, i.e. the HQBF, it was natural to conjecture that the resulting Lie algebras should coincide. In Sect. 3.2 we prove that this conjecture is correct: the centerless version of sl(2, $\mathbb{C}$) is Lie–algebra isomorphic to the Lie algebra of HQBF: the difference between the two is just in the way of choosing the linear generators and of writing the commutation relations. In particular, the symplectic algebra is a matrix representation of both. Moreover, noting that the algebra of HQBF has a natural involution (given by the adjoint) which makes it a $*$–Lie algebra, following [25], we introduce a (non–trivial) involution on the symplectic algebra which also makes it a $*$–Lie algebra. Then we prove the isomorphism of the two **as $*$–Lie algebras**. More explicitly, we identify the centerless $*$–Lie sub–algebra of the $*$–Lie algebra of HQBF with the symplectic $*$–Lie algebra (see Sect. 4.2).

This **extends, to the $*$–Lie algebra level**, the above mentioned result on the Lie algebra isomorphism between the symplectic algebra and the algebra of HQBF. This extension is non–trivial because it is known (see [6]) that even on sl(2, $\mathbb{C}$) there exist non–isomorphic $*$–Lie algebra structures. From this, we deduce an identification of the group generated by the exponentials of the skew–adjoint with the metaplectic group (see Sect. 7).

On the other hand, in the paper [24] it is proved that the quadratic Heisenberg group is isomorphic to PSU(1, 1). It is known that SU(1, 1) is isomorphic to Sp(2, $\mathbb{R}$) (see [26], pages 170-171) where an explicit isomorphism between these two groups can be found. It is also known that the Lie algebras, su(1, 1), sp(2, $\mathbb{R}$) and sl(2, $\mathbb{R}$) are isomorphic. Therefore also the Lie algebras psu(1, 1), sl(2, $\mathbb{R}$), and $sp(2, \mathbb{R})$ are isomorphic. As the Lie algebra of a Lie group is uniquely determined by the connected component of the identity, the Lie algebras of PSU(1, 1) and of $Sp(2, \mathbb{R})$ are isomorphic. However it is known that the same Lie algebra can give raise to different Lie groups. The standard choice, that we follow, for the Lie group associated to $sp(2, \mathbb{R})$ is the metaplectic group and it is known that this group is not isomorphic to PSU(1, 1). Therefore the natural problem arise to explain why these two groups naturally emerge from the same Lie algebra. This problem will be discussed in a future paper [5].

Finally, in Sect. 8, we discuss the connections between the diagonalizability and the vacuum–factorizability problems for their vacuum distributions. We prove a necessary and sufficient condition for diagonalizability and we use this result to prove, with some counterexamples, that factorizability is a strictly stronger notion that diagonalizability.

2 Some Properties of $*$–Lie Algebras

Since the literature on $*$–Lie algebras is very small, in this section, we recall some of the main notions used in the following. In the following, **unless explicitly specified, all vector spaces** (in particular all $*$–algebras and Lie algebras) are **complex**. A **real structure** on a vector space V is given by a sub–space $V_0 \subset V$ such that

$$V = V_0 \dot{+} i V_0$$

Since, in what follows, we will meet several different involutions, we introduce the following definition.

Definition 1 Let V be a vector space. A **vector space involution** on V is a map $\iota : V \to V$ with the following properties:

$$\iota^2 = id_V, \quad \textbf{(involution property)} \tag{3}$$

$$\iota(xv) = \overline{x}\,\iota(v), \quad x \in \mathbb{C}\,,\ v \in V \ \textbf{(anti – linearity)} \tag{4}$$

If V is a **real** vector space, (4) is replaced by real linearity.

Maps satisfying (3), but not (4) will be called **involutive (or idempotent) maps**. If V is an **associative algebra**, it is additionally required that

$$\iota(uv) = \iota(v)\,\iota(u), \quad u, v \in V \ \textbf{(algebraanti – homomorphism)} \tag{5}$$

and, in this case, ι is simply called an **algebra involution** on V.

If V is a **Lie algebra**, it is additionally required that

$$\iota([u, v]) = [\iota(v), \iota(u)], \quad u, v \in V \ \textbf{(Liealgebraanti – homomorphism)} \tag{6}$$

and, in this case, ι is called an **Lie algebra involution** on V.

Proposition 1 *Let $\mathcal{L}$ be a $*$–Lie algebra with involution denoted by $*$. Then:*
(i) **Every linear basis** *B of $\mathcal{L}$ is an hermitean set, i.e.:*

$$B = B^* := \{b^* : b \in B\} \tag{7}$$

In particular every such a linear basis B has a **partition** *of the form*

$$B = B_+ \cup B_0 \cup B_- \tag{8}$$

where

$$a_0 \in B_0 \iff a_0 \in B \ \text{and}\ a_0 = a_0^* \tag{9}$$

In particular B_0 is the same for all partitions (8) of B.
(ii) The **skew–hermitean** *elements of $\mathcal{L}$ are a* **real $*$–Lie sub–algebra** *of $\mathcal{L}$, denoted $\mathcal{L}_{skew}(\mathbb{R})$, on which the involution acts as sign change $X^* = -X$. $\mathcal{L}_{skew}(\mathbb{R})$ is a* **real form** *of $\mathcal{L}$, i.e. $\mathcal{L}$ is the* **complexification of** *$\mathcal{L}_{skew}(\mathbb{R})$:*

$$\mathcal{L} = \mathcal{L}_{skew}(\mathbb{R}) \dot{+} i\mathcal{L}_{skew}(\mathbb{R}) \tag{10}$$

where $\dot{+}$ means **vector space direct sum.**
(iii) Every $*$–map $T : \mathcal{L} \to \mathcal{L}$ maps $\mathcal{L}_{skew}(\mathbb{R})$ into itself.

Proof (i). Define

$$B_H := \{b \in B : b^* = b\}; \quad B_{NH} := \{b \in B : b^* \neq b\}$$

Suppose, by contradiction, that there exists $b_0 \in B$ such that $b_0^* \notin B$. Then $b_0 \in B_{NH}$. Since B is a linear basis of $\mathcal{L}$, there exist $\{c_b, d_b \in \mathbb{C} : b \in B\}$ such that

$$b_0^* - b_0 = \sum_{b \in B} d_b b; \qquad b_0^* = \sum_{b \in B} c_b b \tag{11}$$

Therefore

$$b_0 = b_0^* - (b_0^* - b_0) = \sum_{b \in B} c_b b - \sum_{b \in B} d_b b = \sum_{b \in B} (c_b - d_b) b$$

Hence the linear independence of the elements of B implies that

$$c_b = d_b \text{if} b \neq b_0$$
$$c_{b_0} - d_{b_0} = 1$$

Therefore (11) becomes

$$b_0^* - b_0 = d_{b_0} b_0 + \sum_{b \in B, b \neq b_0} d_b b; \quad b_0^* = (1 + d_{b_0}) b_0 + \sum_{b \in B, b \neq b_0} d_b b \tag{12}$$

and this implies that

$$b_0^* - 2b_0 = (b_0^* - b_0) - b_0 = d_{b_0} b_0 - (1 + d_{b_0}) b_0 = -b_0 \iff b_0^* = b_0$$

contradicting the assumption that $b_0 \in B_{NH}$. Define B_0 by (9). If $B \setminus B_0 = \emptyset$, then $B = B_0$ and (8), (9) hold with $B_+ = \emptyset$. If $B \setminus B_0 \neq \emptyset$, then there exists $a_+ \in B \setminus B_0$ and, since $a_+ \neq a_+^*$, $\{a_+\} \cap \{a_+^*\} = \emptyset$. Therefore the family of sub–sets

$$\{C \subset B \setminus B_0 : C \cap C^* = \emptyset\}$$

is not empty. By Zorn Lemma, such a family has a maximal element B_+. Define $B_- := B_+^*$. Then, by construction $B_+ \cap B_- = \emptyset$. Suppose by contradiction that $B_+ \cup B_- \neq B \setminus B_0$ and let $b \in (B \setminus B_0) \setminus (B_+ \cup B_-)$. Then, since $B \setminus B_0$ is an hermitean set because B is such, $b^* \in (B \setminus B_0)$. Moreover b^* cannot belong to $B_+ \cup B_-$ because, if $b^* \in B_+$, then $b \in B_-$, against the assumption that $b \in (B \setminus B_0) \setminus (B_+ \cup B_-)$ and a similar argument works if $b^* \in B_-$. In conclusion $\{b, b^*\} \subseteq (B \setminus B_0) \setminus (B_+ \cup B_-)$. But then $(B_+ \cup \{b\}) \cap (B_- \cup \{b^*\}) = \emptyset$, contradicting maximality. This proves (i).
(ii). The first statement of (ii) is clear. To prove (10), notice that every element $X \in \mathcal{L}$ can be written as

$$X = \frac{1}{2i} i(X + X^*) + i\frac{1}{2}(X - X^*)$$

and $i(X + X^*), X - X^* \in \mathcal{L}_{skew}(\mathbb{R})$. Therefore $\mathcal{L} = \mathcal{L}_{skew}(\mathbb{R}) + i\mathcal{L}_{skew}(\mathbb{R})$. Moreover the sum is direct because elements of $\mathcal{L}_{skew}(\mathbb{R})$ are skew–hermitean and elements of $i\mathcal{L}_{skew}(\mathbb{R})$ are hermitean and only 0 can have both properties. This proves (ii).

(iii) holds because $\mathcal{L}_{skew}(\mathbb{R})$ is the set of all the skew–hermitean elements of $\mathcal{L}$. Hence, if $l \in \mathcal{L}_{skew}(\mathbb{R})$, then

$$T(l)^* = T(l^*) = T(-l) = -T(l)(\iff T(l) \in \mathcal{L}_{skew}(\mathbb{R})) \qquad \square$$

2.1 *The Complex d–Dimensional Heisenberg Algebra: $heis_{1,d}(\mathbb{C})$*

Definition 2 For $d \in \mathbb{N}$, the complex $*$–**Lie algebra** with generators

$$\mathbf{1} \text{ (central element)}, a_j^+, a_k, \quad j, k \in \{1, \ldots, d\} \tag{13}$$

commutation relations

$$[a_i, a_j^+] = \delta_{ij}; \quad [a_i, a_j] = [a_i^+, a_j^+] = 0, \quad \forall i, j \in \{1, \ldots, d\} \tag{14}$$

and involution given by

$$a_j^* = a_j^+; \quad \left(a_j^+\right)^* = a_j; \quad 1^* = 1 \tag{15}$$

is called the **complex d–dimensional Heisenberg algebra** and denoted $\mathrm{heis}_{1,d}(\mathbb{C})$.

The ***d*–dimensional Heisenberg algebra**, denoted $\mathrm{heis}_{1,d,skew}(\mathbb{C})$, is the real $*$–Lie sub–algebra of $\mathrm{heis}_{1,d}(\mathbb{C})$ consisting of its **skew–hermitean elements**.

The associated Lie groups are denoted respectively $\mathrm{Heis}_{1,d}(\mathbb{C})$ and $\mathrm{Heis}_{1,d,un}(\mathbb{C})$.

We identify $\mathrm{heis}_{1,d}(\mathbb{C})$ to a sub–algebra of its **universal enveloping** $*$–algebra **denoted** $UE(\mathrm{heis}_{1,d}(\mathbb{C}))$. It is known that the algebra involution on $UE(\mathrm{heis}_{1,d}(\mathbb{C}))$ is **uniquely determined** by the Lie algebra involution on $\mathrm{heis}_{1,d}(\mathbb{C})$.

Remark 1 In the case of $*$–Lie algebras or groups, one should also introduce in the notations the involution implementing the $*$–structure. We will do this only in the cases when there might be **several inequivalent involutions**.

Remark 2 It is known that the generators (13) are linearly independent. Thus, fixing the canonical linear basis (e_j) in $\mathbb{C}^d$ the linear map

$$f = (f_1, \ldots, f_d) \in \mathbb{C}^d \mapsto a_f^+ := \sum_{i=1}^{d} f_i a_i^+ \in \mathrm{heis}_{1,d}(\mathbb{C})$$

is **injective** and the vector space generated by the a_f, a_g^+ ($f, g \in \mathbb{C}^d$), which is isomorphic to $\mathbb{C}^{2d}$ has a natural **real structure** which makes it isomorphic to $\mathbb{C}^{2d}$ considered as a **real vector space** for its natural real structure.

For each $k \in \mathbb{N}$, denote

$$C_k : f = (f_1, \ldots, f_k) \in \mathbb{C}^k \to C_k f := \bar{f} := f = (\overline{f_1}, \ldots, \overline{f_k}) \in \mathbb{C}^k \qquad (16)$$

the **conjugation operator** on $\mathbb{C}^k$. The **conjugate** of a matrix $A \equiv (A_{ij}) \in M_k(\mathbb{C})$ (the $k \times k$–matrices with entries in $\mathbb{C}$), **with respect to** C_k, is defined by

$$\overline{A} := C_k A C_k \equiv (\overline{A_{ij}}) \qquad (17)$$

and the **transpose** of A, by

$$A^T := (C_k A C_k)^* \equiv (A_{ji}) \qquad (18)$$

2.2 *The Complex d–Dimensional Quadratic Heisenberg Algebra* $heis_{2,d}(\mathbb{C})$

Denote $M_d(\mathbb{C})$ (resp. $M_{d,sym}(\mathbb{C})$) the algebra of $d \times d$ complex matrices (resp. symmetric matrices). For $M \equiv (M_{j,k}) \in M_d(\mathbb{C})$, the transpose, conjugate and adjoint of M with respect to the involution (16) are given respectively by:

$$(M^T)_{j,k} := M_{k,j}; \ (\overline{M})_{j,k} := \overline{M_{j,k}}; \ (M^*)_{j,k} := (\overline{M})^T_{j,k} = \overline{(M)^T}_{j,k} = \overline{M_{k,j}}$$

We identify $\text{heis}_{1,d}(\mathbb{C})$ to a sub–algebra of its universal enveloping $*$–algebra $UE(\text{heis}_{1,d}(\mathbb{C}))$ and consider the space of all **homogeneous normally ordered quadratic expressions** in the generators (13):

$$a^+ N a^+ := \sum_{j,k=1}^{d} N_{j,k} a_j^+ a_k^+ \qquad (19)$$

$$a^+ L a := \sum_{j,k=1}^{d} L_{j,k} a_j^+ a_k \qquad (20)$$

$$a M a := \sum_{j,k=1}^{d} M_{j,k} a_j a_k \qquad (21)$$

Notice that, since creators (resp. annihilators) mutually commute, one has

$$a^+Na^+ = \sum_{j,k=1}^{d} \frac{1}{2}\left(N_{j,k}+N_{k,j}\right)a_j^+a_k^+ ; \quad aMa = \sum_{j,k=1}^{d} \frac{1}{2}\left(M_{j,k}+M_{k,j}\right)a_ja_k \tag{22}$$

i.e. the expressions a^+Na^+ and aMa are parametrized by **symmetric matrices**, $N^T = N$ and $M^T = M$. Moreover

$$(a^+Na^+)^* = \sum_{j,k=1}^{d} \overline{N_{j,k}}a_ka_j = \sum_{j,k=1}^{d} \overline{N_{k,j}}a_ka_j = a\overline{N}a \tag{23}$$

$$(a^+La)^* = \sum_{j,k=1}^{d} \overline{L_{j,k}}a_k^+a_j = \sum_{j,k=1}^{d} \overline{L_{k,j}}a_j^+a_k = a^+L^*a \tag{24}$$

$$(aMa)^* = a^+\overline{M}a^+ = a^+M^*a^+ \tag{25}$$

where we used (23).

Definition 3 Denote $\mathrm{heis}_{2,d}(\mathbb{C})$ the $*$–vector space:

$$\mathrm{heis}_{2,d}(\mathbb{C}) := \mathbb{C}\cdot\mathbf{1} \oplus \mathbb{C}\text{–linear span of}$$

$$\left\{a^+Na^+,\ a^+La,\ aMa : L \in M_d(\mathbb{C}),\ N, M \in M_{d,sym}(\mathbb{C})\right\}$$

$$= \mathbb{C}\text{–linear span of } \{\mathbf{1},\ a_j^+a_k^+,\ a_j^+a_k,\ a_ja_k,\ j,k \in \{1\dots,d\}\}; \quad \mathbf{1}^* = \mathbf{1}$$

with involution given by (23), (24), (25).

Denote $\mathrm{heis}_{2,d,skew}(\mathbb{C})$ the $*$–vector sub–space of $\mathrm{heis}_{2,d}(\mathbb{C})$ defined by:

$$\mathrm{heis}_{2,d,skew}(\mathbb{C}) := \{X \in \mathrm{heis}_{2,d}(\mathbb{C}) : X^* = -X\} \tag{26}$$

Remark Because of the linear independence of the set $\{\mathbf{1},\ a_j^+a_k^+,\ a_j^+a_k,\ a_ja_k\}$, $j,k \in \{1\dots,d\}$, $\mathrm{heis}_{2,d}(\mathbb{C})$ is the range of the vector space isomorphism

$$\begin{aligned}(c1_{M_d}, N, L, M) &\in \mathbb{C}\cdot 1_{M_d} \times M_{d,sym}(\mathbb{C}) \times M_d(\mathbb{C}) \times M_{d,sym}(\mathbb{C}) \\ &\mapsto c\mathbf{1} + a^+Na^+ + a^+La + aMa \in \mathrm{heis}_{2,d,skew}(\mathbb{C})\end{aligned} \tag{27}$$

and $\mathrm{heis}_{2,d,skew}(\mathbb{C})$ is the range of the vector space isomorphism

$$\begin{aligned}(c1_{M_d}, N, L) &\in \mathbb{R}\cdot 1_{M_d} \times M_{d,sym}(\mathbb{C}) \times M_{d,herm}(\mathbb{C}) \\ &\mapsto i\left(c\mathbf{1} + a^+Na^+ + a^+La + a\overline{N}a\right) \in \mathrm{heis}_{2,d,skew}(\mathbb{C})\end{aligned} \tag{28}$$

where $M_{d,herm}(\mathbb{C})$ denotes the space of **hermitean complex** $d \times d$**–matrices**.

In the following we will use the notation

$$\mathbf{c1} \equiv c; \qquad c \in \mathbb{C}$$

The involution (15) on $\text{heis}_{1,d}(\mathbb{C})$ induces an involution on $\text{heis}_{2,d}(\mathbb{C})$ given by

$$(a^+Na^+)^* = aN^*a, \quad (a^+La)^* = a^+L^*a, \quad (aMa)^* = a^+M^*a^+$$

With this involution (27) is a $*$–isomorphism of $*$–vector spaces.

Lemma 1 $\text{heis}_{2,d}(\mathbb{C})$ is a $*$–Lie algebra with involution given by (15), central element $\mathbf{1}$ and with the following commutation relations:

$$[aMa, a^+Na^+] = 2\,\text{Tr}(NM) + 4\,a^+NMa; \;\; M, N \in M_{d,sym}(\mathbb{C}) \tag{29}$$

$$[aMa, a^+Na] = a\left(MN + (MN)^T\right)a; \;\; M \in M_{d,sym}(\mathbb{C}), \;\; N \in M_d(\mathbb{C}) \tag{30}$$

$$[a^+Ma, a^+Na] = a^+[M, N]a; \;\; M, N \in M_d(\mathbb{C}) \tag{31}$$

$$[a^+Ma^+, a^+Na^+] = [aMa, aNa] = 0; \;\; M, N \in M_{d,sym}(\mathbb{C}) \tag{32}$$

and

$$[a^+Ma, a^+Na^+] = a^+\left(MN + (MN)^T\right)a^+; \;\; N \in M_{d,sym}(\mathbb{C}), \;\; M \in M_d(\mathbb{C}) \tag{33}$$

Proof See [3].

Remark It is clear that, for $d = 1$, the above commutation relations reduce to the usual commutation relations of $sl(2, \mathbb{R})$. This explains in what sense $\text{heis}_{2,d}(\mathbb{C})$ can be considered as a **non–commutative extension of** $\text{heis}_{2,d}(\mathbb{C})$.

Definition 4 The **complex d–dimensional quadratic Boson algebra**, still denoted $\text{heis}_{2,d}(\mathbb{C})$, is the $*$–Lie algebra with generators

$$\left\{\mathbf{1},\; B^+(N),\; B^0(L),\; B(M) : \; N, M \in M_{d,sym}(\mathbb{C}),\; L \in M_d(\mathbb{C})\right\} \tag{34}$$

central element $\mathbf{1}$, involution given by

$$B^+(N)^* = B(N^*); \;\; B^0(L)^* = B^0(L^*); \;\; B(M)^* = B^+(M^*) \tag{35}$$

and Lie brackets given by:

$$[B(M), B^+(N)] = 2\,\text{Tr}(NM) + 4\,B^0(NM) \tag{36}$$

$$[B^0(L_1), B^0(L_2)] = B^0([L_1, L_2]) \tag{37}$$

$$[B(M), B^0(L)] = B\left(ML + (ML)^T\right) \tag{38}$$

$$[B^+(M), B^0(L)] = -B^+\left(LM + (LM)^T\right) \tag{39}$$

$$[B^+(M), B^+(N)] = [B(M), B(N)] = 0 \tag{40}$$

The **skew–hermitean** elements of $\mathrm{heis}_{2,d}(\mathbb{C})$, are a **real** $*$–Lie sub–algebra denoted $\mathrm{heis}_{2,d,skew}(\mathbb{C})$ and called the **d–dimensional quadratic Heisenberg algebra**. The local Lie groups associated to $\mathrm{heis}_{2,d}(\mathbb{C})$ and $\mathrm{heis}_{2,d,skew}(\mathbb{C})$ are denoted respectively $\mathrm{Heis}_{2;\mathbb{C}}(d)$ and $\mathrm{Heis}_{2,d,skew}(\mathbb{C})$.

The following simple remark will play a crucial role in what follows.

Corollary 1 *The set*

$$\left\{a^+Na^+ + a^+2La + aMa + Tr(L) \ : \ L \in M_d(\mathbb{C}),\ N, M \in M_{d,sym}(\mathbb{C})\right\} \quad (41)$$

is also a set of linear generators of $\mathrm{heis}_{2,d}(\mathbb{C})$.

Proof Denote $\mathcal{L}$ the $*$–Lie algebra generated by the set (41). By construction $\mathcal{L} \subseteq \mathrm{heis}_{2,d}(\mathbb{C})$. So one has to prove the converse inclusion. Since $L \in M_d(\mathbb{C})$ and N, $M \in M_{d,sym}(\mathbb{C})$ are arbitrary, the multiples of the identity belong to $\mathcal{L}$. With a similar argument, one concludes that all the generators of $\mathrm{heis}_{2,d}(\mathbb{C})$ belong to $\mathcal{L}$. Hence $\mathcal{L} = \mathrm{heis}_{2,d}(\mathbb{C})$. □

Remark In order to simplify notations we use the **same symbol** $\mathrm{heis}_{2,d}(\mathbb{C})$ for the abstract $*$–Lie algebra and for its Boson realization and the same for their central element. The identifications

$$B^+(N) \equiv a^+Na^+, \quad B^0(L) \equiv a^+La, \quad B(M) \equiv aMa$$

are clear from the commutation relations and in the following we will use them constantly because the Boson notation are more intuitive. These identifications should not create confusion provided they are handled carefully. For example the following identities, where summation on repeated indices is understood, make sense in a Boson context:

$$\begin{aligned} aNa^+ &= a_i(N)_{ij}a_j^+ = a_iN_{ij}a_j^+ = N_{ij}([a_i, a_j^+] + a_j^+a_i) \qquad (42)\\ &= N_{ij}(\delta_{ij} + a_j^+a_i) = N_{ij}\delta_{ij} + N_{ij}a_j^+a_i = N_{ii}\delta_{ij} + a_j^+(N^T)_{ji}a_i \\ &= \mathrm{Tr}(N) + a_j^+((N)^T)_{ji}a_i = \mathrm{Tr}(N) + a^+N^Ta \end{aligned}$$

but, while expressions of the form aNa^+ do not make sense in the quadratic Boson algebra, both terms in the sum appearing in the last identity in (42) are in the quadratic Boson algebra.

Remark In the physics literature, instead of (35), one uses the involution

$$B^0(L)^* = B^0(L^*); \ \ B(M)^* = B^+(M),$$

where annihilators depend **anti–linearly** on their **test matrices**. This has the advantage that annihilators are simply defined as adjoints of creators and one is not obliged to define an involution on the test function space. However, since most test function spaces concretely used have a natural involution, the choice (35) seems to be more natural.

3 The Symplectic Approach to Homogeneous Quadratic Boson Fields

In the symplectic approach to quadratic Boson fields, one introduces a single **creation–annihilation** map

$$\mathbf{a} : f \equiv (\overline{f}_1, f_2) \in \mathbb{C}^{2d} \equiv (\mathbb{C}^d)^* \oplus \mathbb{C}^d \to \mathbf{a}_f \in heis_{2,d}(\mathbb{C})$$
$$\mathbf{a}_f := a_{\overline{f_1}} + a^+_{f_2} = \sum_{i=1}^{d} (f_{1;i} a_i + f_{2;i} a_i^+) \tag{43}$$

where $(\mathbb{C}^d)^*$ denotes the dual of $(\mathbb{C}^d)$.

Remark Notice that **a** is a **complex linear** map and that

$$\mathbf{a}^+_f := a_{f_2} + a^+_{\overline{f_1}} = \mathbf{a}^+_{\tilde{C}_{2d} f} \tag{44}$$

where, using (16),

$$\tilde{C}_{2d} f = \tilde{C}_{2d}(\overline{f}_1, f_2) := (\overline{f}_2, \overline{f}_1) = (C_d f_2, C_d f_1) \tag{45}$$

It is convenient to introduce the canonical coordinates also on $\mathbb{C}^{2d}$, i.e. $f \equiv (f_1, \dots, f_{2d})$ and, in these coordinates,

$$\mathbf{a} \equiv \left(a_1, \cdots, a_d, a_1^+, \cdots, a_d^+\right)^T : f \in \mathbb{C}^{2d} \to \mathbf{a}_f := \sum_{i=1}^{d} (f_i a_i + f_{i+d} a_i^+) \in heis_{2,d}(\mathbb{C})$$

Lemma 2 *With these notations, the* **canonical commutation relations** *(14) take the coordinate independent form*

$$[\mathbf{a}_f, \mathbf{a}_g] = -f^T J g, \quad \forall g, f \in \mathbb{C}^{2d} \tag{46}$$

where

$$J = \begin{pmatrix} 0 & -1_d \\ 1_d & 0 \end{pmatrix} \tag{47}$$

$J \equiv J_d$ is the standard symplectic matrix of order 2d and 1_d is the identity matrix in the space $M_d(\mathbb{C})$ of complex $d \times d$ matrices.

Proof Using (14) one finds

$$[\mathbf{a}_f, \mathbf{a}_g] = \left[\sum_{i=1}^{d} (f_i a_i + f_{i+d} a_i^+), \sum_{j=1}^{d} (g_j a_j + g_{j+d} a_j^+) \right] =$$

$$\sum_{i=1}^{d}\sum_{j=1}^{d}(f_i g_j[a_i, a_j] + f_i g_{j+d}[a_i, a_j^+] + f_{i+d} g_j[a_i^+, a_j] + f_{i+d} g_{j+d}[a_i^+, a_j^+]) =$$

$$\sum_{i=1}^{d}\sum_{j=1}^{d}(f_i g_{j+d}\delta_{ij} - f_{i+d} g_j \delta_{ij}) = -f^T Jg \qquad \square$$

Remark The identity (46) is the complexified version of the starting point of the **symplectic approach to quantization**. In it, the starting point is a pair (V, ω), where V is a real vector space and ω a non–degenerate symplectic form on it. In this approach, the CCR are formulated **without a priori distinction** between creators and annihilators (or position and momentum). This distinction is introduced fixing a ω–**Lagrangian sub–space** Q of V (ω is identically zero on Q and Q is maximal for this property) and defining P as the ω–orthogonal complement of Q. Then $V \equiv Q \oplus P$ and this decomposition allows to define a complex structure on V. From this point of view, the decomposition $\mathbb{C}^d \oplus \mathbb{C}^d$ in (43) represents a *doubling of the phase space*. This is not justified in usual quantization, but we will see that is essential for quadratic quantization because in that case the natural choice for V is $V \equiv M_{d,sym}(\mathbb{C})$.

3.1 The ∗–Lie Algebra of Homogeneous Quadratic Boson Fields

Definition 5 Denote

$$M_{d,sym}(\mathbb{C}) := \{K \in M_d(\mathbb{C}) : K = K^T\}$$

the space of **symmetric** complex $d \times d$ matrices. To any $K \in M_{2d,sym}(\mathbb{C})$ one associates the **homogeneous quadratic Boson field (HQBF)**,

$$\mathbf{a}K\mathbf{a} := \sum_{i=1}^{d}\sum_{j=1}^{d}(K_{i,j}a_i a_j + K_{i+d,j}a_i^+ a_j + K_{i,j+d}a_i a_j^+ + K_{i+d,j+d}a_i^+ a_j^+) \quad (48)$$

Lemma 3 *$K \in M_{2d,sym}(\mathbb{C})$ iff it can be written in the form*

$$K = \begin{pmatrix} M & L^T \\ L & N \end{pmatrix} \quad (49)$$

where $M, N \in M_{d,sym}(\mathbb{C})$ and $L \in M_d(\mathbb{C})$.

Proof If $K \in M_{2d,sym}(\mathbb{C})$, define

$$M := (K_{i,j})_{i,j=1}^d; \quad N := (K_{d+i,d+j})_{i,j=1}^d; \quad L \equiv (K_{i,j+d})_{i,j=1}^d$$

Since K is symmetric, such are M and N. Moreover $K_{i+d,j} = K_{j,i+d}$. Therefore K has the form (49). Conversely, it is clear that the right hand side of (49) is a symmetric matrix whenever $M, N \in M_{d,sym}(\mathbb{C})$ and $L \in M_d(\mathbb{C})$. □

Remark With the notations (49), (19), (20), (21), one can write (48) in the form

$$\begin{aligned}\mathbf{a}K\mathbf{a} &= \sum_{i=1}^{d}\sum_{j=1}^{d}(M_{i,j}a_ia_j + (L)^T_{i,j}a_i^+a_j + L_{i,j}a_ia_j^+ + N_{i,j}a_i^+a_j^+) \\ &= aMa + a^+(L)^Ta + aLa^+ + a^+Na^+ \end{aligned} \tag{50}$$

Lemma 4 *With the same notations used in (50), the normally ordered form of the HQBF (48) is:*

$$\mathbf{a}K\mathbf{a} = :\mathbf{a}K\mathbf{a}: + Tr(L) = :\mathbf{a}K\mathbf{a}: + \frac{1}{2}Tr(EK), \quad L \in M_{2d}(\mathbb{C}) \tag{51}$$

where K has the form (49) and

$$:\mathbf{a}K\mathbf{a}: \;:=\; aMa + a^+2(L)^Ta + a^+Na^+ \tag{52}$$

$$E := \begin{pmatrix} 0 & 1_d \\ 1_d & 0 \end{pmatrix} \tag{53}$$

Proof Using (50) and with the convention of summation over repeated indices, the normally ordered form of $\mathbf{a}K\mathbf{a}$ is:

$$\begin{aligned}\mathbf{a}K\mathbf{a} &= M_{i,j}a_ia_j + (L)^T_{i,j}a_i^+a_j + L_{i,j}a_ia_j^+ + N_{i,j}a_i^+a_j^+ \\ &= M_{i,j}a_ia_j + (L)^T_{i,j}a_i^+a_j + L_{i,j}[a_i, a_j^+] + L_{i,j}a_j^+a_i + N_{i,j}a_i^+a_j^+ \\ &= M_{i,j}a_ia_j + (L)^T_{i,j}a_i^+a_j + L_{j,i}a_i^+a_j + N_{j,i}a_j^+a_i^+ + \mathrm{Tr}(L) \\ &= M_{i,j}a_ia_j + 2(L)^T_{i,j}a_i^+a_j + N_{j,i}a_j^+a_i^+ + \mathrm{Tr}(L) \\ &= aMa + a^+2(L)^Ta + a^+Na^+ + \mathrm{Tr}(L)\end{aligned}$$

□

3.2 *Identification of the ∗–Lie Algebra of Homogeneous Quadratic Boson Fields with* $\mathbf{heis}_{2,d}(\mathbb{C})$

Theorem 1 *The ∗–Lie algebra of HQBF and* $\mathrm{heis}_{2,d}(\mathbb{C})$ *coincide as ∗–Lie algebras.*

Proof Both the HQBF ∗–Lie algebra and $\mathrm{heis}_{2,d}(\mathbb{C})$ are sub–∗–Lie algebras of the universal enveloping algebra $UE(\mathrm{heis}_{1;\mathbb{C}}(d))$ of $\mathrm{heis}_{1;\mathbb{C}}(d)$. Thus, to prove that they coincide as ∗–Lie algebras, it is sufficient to prove that their intersection contains a set which linearly generates both algebras. In fact the involution and commutation relations on this set are uniquely determined, by restriction, from those on $UE(\text{heis}_{1;\mathbb{C}}(d))$. Notice that, since in (52) $L \in M_{2d}(\mathbb{C})$ is arbitrary, one can replace in it L by $(L)^T$. Thus, according to (51) and (52), a set of linear generators of the $*$–Lie algebra of HQBF is given by

$$\left\{a^+Na^+ + a^+2(L)^T a + aMa + \text{Tr}(L) : L \in M_d(\mathbb{C}),\ N, M \in M_{d,sym}(\mathbb{C})\right\} \quad (54)$$

But, according to Corollary 1, this is also a set of generators of $\text{heis}_{1;\mathbb{C}}(d)$. Therefore the thesis follows from the argument above. □

3.3 *Central Decomposition of $\text{heis}_{2,d}(\mathbb{C})$: $\text{heis}_{2,d,cls}(\mathbb{C})$*

Recall the following definition.

Definition 6 Let $\mathcal{L}$ be a $*$–Lie algebra and $\mathcal{L}_0 \subseteq \mathcal{L}$ a $*$–Lie sub–algebra. $\mathcal{L}$ is called a trivial 1–**dimensional central extension** of $\mathcal{L}_0$ if there is an **hermitean** central element $\mathbf{1} \in \mathcal{L}$ such that

$$\mathcal{L} = \mathcal{L}_0 \dot{+} \mathbb{C}\mathbf{1}, \quad [L_0 \dot{+} c\mathbf{1}, M_0 \dot{+} d\mathbf{1}] := [L_0, M_0], \quad L_0, M_0 \in \mathcal{L}_0, \quad c, d \in \mathbb{C} \quad (55)$$

where $\dot{+}$ denotes direct sum of vector spaces.

Remark It is clear that, if $\mathbf{1} \in \mathcal{L}$ is the unique, up to real multiples, hermitean central element of $\mathcal{L}$, then, if $\mathcal{L} = \mathcal{L}_1 \dot{+} \mathbb{C}E$ is another decomposition of $\mathcal{L}$ as a trivial 1–dimensional central extension of $\mathcal{L}_1$, then one must have $E = c\mathbf{1}$ with $c \in \mathbb{R}$ and $\mathcal{L}_1$ and $\mathcal{L}_0$ **are isomorphic as $*$–Lie algebras**. The isomorphism is established by considering two decompositions of a generic element $L \in \mathcal{L}$ as

$$L = L_0 \dot{+} c_0\mathbf{1} = L_1 \dot{+} c_1\mathbf{1}$$

In fact the map $L_0 \mapsto L_1$, $c_0\mathbf{1} \mapsto c_1\mathbf{1}$ is clearly a $*$–map and satisfies

$$[L, L'] = [L_0 \dot{+} c_0\mathbf{1}, L_0' \dot{+} c_0'\mathbf{1}] = [L_1 \dot{+} c_1\mathbf{1}, L_1' \dot{+} c_1'\mathbf{1}]$$
$$\iff [L, L'] = [L_0, L_0'] = [L_1, L_1']$$

therefore it is a $*$–Lie algebra isomorphism.

Moreover, if (55) holds, any representation π_0 of $\mathcal{L}_0$ on a vector space V can be extended to a representation π of $\mathcal{L}$ on V defining

$$\pi(\mathbf{1}) := 1_V; \quad \pi(X_0) := \pi_0(X_0); \quad \forall X_0 \in \mathcal{L}_0$$

Thus the problem to find representations of $\mathcal{L}$ is reduced to that of finding representations of $\mathcal{L}_0$.

Theorem 2 *The ∗–Lie sub–algebra of* $heis_{2,d}(\mathbb{C})$ *with generators*

$$\hat{B}^0(L) := a^+La + Tr(L) \tag{56}$$

$$\hat{B}(M) := aMa \tag{57}$$

$$\hat{B}^+(N) := a^+Na^+ \tag{58}$$

with $L \in M_d(\mathbb{C})$ *and* $N, M \in M_{d,sym}(\mathbb{C})$*, denoted* $heis_{2,d,cls}(\mathbb{C})$ *(**centerless**), is such that* $heis_{2,d}(\mathbb{C})$ *is the trivial 1–dimensional central extension of* $heis_{2,d,cls}(\mathbb{C})$*:*

$$heis_{2,d}(\mathbb{C}) = heis_{2,d,cls}(\mathbb{C}) \dot{+} \mathbb{C}\mathbf{1} \tag{59}$$

Moreover $heis_{2,d,cls}(\mathbb{C})$ *is the unique, up to isomorphisms, ∗–Lie sub–algebra of* $heis_{2,d}(\mathbb{C})$ *with this property. Finally the Lie brackets of the generators* $\hat{B}^0(L)$, $\hat{B}(M)$, $\hat{B}^+(N)$ *are given by:*

$$[\hat{B}(M), \hat{B}^+(N)] = 4\,\hat{B}^0(NM) \tag{60}$$

$$[\hat{B}^0(L_1), \hat{B}^0(L_2)] = \hat{B}^0([L_1, L_2]) \tag{61}$$

$$[\hat{B}(M), \hat{B}^0(L)] = \hat{B}\left(ML + (ML)^T\right) \tag{62}$$

$$[\hat{B}^+(M), \hat{B}^0(L)] = -\hat{B}^+\left(LM + (LM)^T\right) \tag{63}$$

$$[\hat{B}^+(M), \hat{B}^+(N)] = [\hat{B}(M), \hat{B}(N)] = 0 \tag{64}$$

and the involution by

$$\hat{B}^+(N)^* = \hat{B}(N^*), \quad \hat{B}^0(L)^* = \hat{B}^0(L^*), \quad \hat{B}(M)^* = \hat{B}^+(M^*) \tag{65}$$

Proof The commutation relations from (60) to (64) follow immediately from the commutation relations (36), ..., (40). Similarly, (65) follows from (35). Thus $\text{heis}_{2,d,cls}(\mathbb{C})$ is a ∗–Lie sub–algebra of $\text{heis}_{2,d}(\mathbb{C})$. The decomposition (59) follows from the linear independence of $\mathbf{1}$ from the generators $\hat{B}^0(L)$, $\hat{B}(M)$, $\hat{B}^+(N)$. $\mathbf{1} \in \mathcal{L}$ is the unique, up to real multiples, hermitean central element of $\mathcal{L}$. Therefore, by the Remark after Definition 6, $\text{heis}_{2,d,cls}(\mathbb{C})$ is the unique, up to isomorphisms, ∗–Lie sub–algebra of $\text{heis}_{2,d}(\mathbb{C})$ satisfying (59). □

The **the skew–hermitean** elements of $\text{heis}_{2,d,cls}(\mathbb{C})$, are a **real** ∗–Lie sub–algebra denoted $\text{heis}_{2,d,cls,skew}(\mathbb{C})$, called the **centerless *d*–dimensional quadratic Heisen-**

berg algebra. The local Lie groups associated to $\mathrm{heis}_{2,d,cls}(\mathbb{C})$ and $\mathrm{heis}_{2,d,cls,skew}(\mathbb{C})$ are denoted respectively $\mathrm{Heis}_{2,d,cls}(\mathbb{C})$ and $\mathrm{Heis}_{2,d,cls,skew}(\mathbb{C})$.

4 The Complex Symplectic ∗–Lie Algebra $sp(2d, \mathbb{C})$

The following Proposition summarizes some more or less known facts.

Proposition 2 *Let J be the standard symplectic matrix given by (47). The space*

$$\{X \in M_{2d}(\mathbb{C}) : JXJ = X^T\} \tag{66}$$

is a Lie algebra, closed under transposition, called the **complex symplectic algebra** *and denoted $sp(2d, \mathbb{C})$. Moreover*

$$sp(2d, \mathbb{C}) = \{X \in M_{2d}(\mathbb{C}) : \exists K \in M_{2d,sym}(\mathbb{C}),\ X = KJ\} \tag{67}$$

$$= \{X \in M_{2d}(\mathbb{C}) : \exists H \in M_{2d,sym}(\mathbb{C}),\ X = JH\}$$

In particular

$$sp(2d, \mathbb{C}) =$$

$$= \left\{X \in M_{2d}(\mathbb{C}) : \exists L \in M_d(\mathbb{C}),\ \exists N, M \in M_{d,sym}(\mathbb{C}),\ X = \begin{pmatrix} L^T & -M \\ N & -L \end{pmatrix}\right\} \tag{68}$$

Proof If $X, Y \in sp(2d, \mathbb{C})$, then

$$J[X, Y]J = JXYJ - JYXJ = -(JXJ)(JYJ) + (JYJ)(JXJ) = -X^T Y^T + Y^T X^T$$

$$= [Y^T, X^T] = [X, Y]^T$$

Therefore $sp(2d, \mathbb{C})$ is a Lie algebra. Since

$$JX^T J = (JXJ)^T = (X^T)^T = X$$

$sp(2d, \mathbb{C})$ is closed under transposition. If $K \in M_{2d,sym}(\mathbb{C})$, then

$$J(KJ)J = -JK = (KJ)^T \tag{69}$$

Therefore the right hand side of the first identity in (67) is contained in $sp(2d, \mathbb{C})$. To prove the converse inclusion, let $X \in sp(2d, \mathbb{C})$. Then $X = (-XJ)J$ and $(-XJ) \in M_{2d,sym}(\mathbb{C})$ because

$$(-XJ)^T = JX^T = J(JXJ) = -XJ$$

A similar argument shows that the second identity in (67) holds.

Notice that, by Lemma (3), any $K \in M_{2d,sym}(\mathbb{C})$ has the form

$$K = \begin{pmatrix} M & L^T \\ L & N \end{pmatrix}, \quad L \in M_d(\mathbb{C}), N, M \in M_{d,sym}(\mathbb{C}) \tag{70}$$

Therefore, if $X = KJ \in sp(2d, \mathbb{C})$, it will have the form

$$X = KJ = \begin{pmatrix} M & L^T \\ L & N \end{pmatrix} \begin{pmatrix} 0 & -1_d \\ 1_d & 0 \end{pmatrix} = \begin{pmatrix} L^T & -M \\ N & -L \end{pmatrix} \tag{71}$$

which is (68). □

Remark Using the second identity in (67), one can obtain an equivalent parametrization of $sp(2d, \mathbb{C})$.

Remark Elements of $sp(2d, \mathbb{C})$ are called (complex) Hamiltonian matrices (with respect to the symplectic form J). (See e.g. [18].)

Proposition 3 *The commutation relations among the generators (50) of $heis_{2,d}(\mathbb{C})$ are the following*

$$\left[\frac{1}{2}\mathbf{a}K\mathbf{a}, \frac{1}{2}\mathbf{a}M\mathbf{a}\right] = \frac{1}{2}\mathbf{a}[KJ, MJ]J\mathbf{a}, \quad \forall K, M \in M_{2d,sym}(\mathbb{C}) \tag{72}$$

Proof For any $K, M \in M_{2d,sym}(\mathbb{C})$, one has (see [25], Lemma 3)

$$\begin{aligned} \left[\frac{1}{2}\mathbf{a}K\mathbf{a}, \frac{1}{2}\mathbf{a}M\mathbf{a}\right] &= \frac{1}{2}\mathbf{a}(MJK - KJM)\mathbf{a} = \frac{1}{2}\mathbf{a}(KJMJ - MJKJ)J\mathbf{a} \\ &= \frac{1}{2}\mathbf{a}[KJ, MJ]J\mathbf{a} \end{aligned} \tag{73}$$

Since $sp(2d, \mathbb{C})$ is a Lie algebra and (67) holds, there exists $H \in M_{2d,sym}(\mathbb{C})$ such that $[KJ, MJ] = HJ$. Therefore $[KJ, MJ]J = (HJ)J = -H$ which is symmetric. □

Theorem 3 *The linear map h, from the Lie-algebra of homogeneous Boson quadratic fields to $sp(2d, \mathbb{C})$, defined by*

$$h : \frac{1}{2}\mathbf{a}K\mathbf{a} \in heis_{2,d}(\mathbb{C}) \to KJ \in sp(2d, \mathbb{C}), \quad K \in M_{2d,sym}(\mathbb{C}) \tag{74}$$

is a Lie-algebra isomorphism.

Proof The map (74) is clearly linear and onto $sp(2d, \mathbb{C})$ because of (67). It is injective because the map $K \in M_{2d,sym}(\mathbb{C}) \mapsto (\mathbf{a}K\mathbf{a})$ is injective. Therefore it is a vector space isomorphism. To complete the proof we have to prove that

$$h\left(\left[\frac{1}{2}\mathbf{a}K\mathbf{a},\frac{1}{2}\mathbf{a}M\mathbf{a}\right]\right)=\left[h(\frac{1}{2}\mathbf{a}K\mathbf{a}),h(\frac{1}{2}\mathbf{a}M\mathbf{a})\right]$$

This follows from the fact that, for all $K, M \in M_{2d,sym}(\mathbb{C})$, using (74) one has, using (72)

$$h\left(\left[\frac{1}{2}\mathbf{a}K\mathbf{a},\frac{1}{2}\mathbf{a}M\mathbf{a}\right]\right)=-h\left(\frac{1}{2}\mathbf{a}[KJ,MJ]J\mathbf{a}\right)=-([KJ,MJ]J)J=[KJ,MJ]$$

and

$$\left[\frac{1}{2}h(\mathbf{a}K\mathbf{a}),h(\frac{1}{2}\mathbf{a}M\mathbf{a})\right]=[KJ,MJ]$$

□

4.1 The ⋄ Involution on $sp(2d,\mathbb{C})$: $sp_\diamond(2d,\mathbb{C})$

Lemma 5 *The map*

$$K \in M_{2d}(\mathbb{C}) \mapsto K^\diamond := -E\overline{K}E \tag{75}$$

where E is defined by (53), is a **Lie algebra anti-linear involution** *called the ⋄–involution. Moreover one has*

$$M_{2d,sym}(\mathbb{C})^\diamond = M_{2d,sym}(\mathbb{C});\quad sp(2d,\mathbb{C})^\diamond = sp(2d,\mathbb{C}) \tag{76}$$

i.e. ⋄ induces an involution on $sp(2d,\mathbb{C})$ denoted by the same symbol.

Proof ⋄ is an anti-linear involution on $M_{2d}(\mathbb{C})$ because $E^2 = 1$ and it is a Lie algebra involution because

$$[X,Y]^\diamond=(XY-YX)^\diamond=-E\overline{(XY-YX)}E=-E\overline{XY}E+E\overline{YX}E$$

$$=-E\overline{X}\,\overline{Y}E+E\overline{Y}\,\overline{X}E=-E\overline{X}EE\overline{Y}E+E\overline{Y}EE\overline{X}E=[Y^\diamond,X^\diamond]$$

The first identity in (76) is clear because E is symmetric and conjugation preserves symmetry. To prove the second identity in (76), recall that the generic element of $sp(2d,\mathbb{C})$ has the form $X = KJ$ (see (67). Therefore, using (75), one has

$$J(KJ)^\diamond J=-JE\overline{K}JEJ$$

Using

$$JEJ=\begin{pmatrix}0&-1_d\\1_d&0\end{pmatrix}\begin{pmatrix}0&1_d\\1_d&0\end{pmatrix}\begin{pmatrix}0&-1_d\\1_d&0\end{pmatrix}=\begin{pmatrix}-1_d&0\\0&1_d\end{pmatrix}\begin{pmatrix}0&-1_d\\1_d&0\end{pmatrix}=\begin{pmatrix}0&1_d\\1_d&0\end{pmatrix}=E$$

one obtains

$$J(KJ)^{\diamond}J = -JE\overline{K}E$$

Therefore, using the first identity in (76), one obtains

$$J(KJ)^{\diamond}J = -J(E\overline{K}E)^T = J^T(E\overline{K}E)^T = (KJ)^T$$

Since, according to (66), the above identity characterizes elements in $sp(2d, \mathbb{C})$, it follows that $sp(2d, \mathbb{C})$ is invariant under the $\diamond$**–involution.** □

Remark The – sign in (75) is essential. In fact the map

$$K \in M_{2d}(\mathbb{C}) \mapsto K^{\sim} := E\overline{K}E \tag{77}$$

is not a $*$–Lie algebra involution because from a $*$–Lie algebra involution it is required that

$$[X, Y]^* = [Y^*, X^*]$$

however

$$[Y^{\sim}, X^{\sim}] = E\overline{Y}EE\overline{X}E - E\overline{X}EE\overline{Y}E = E(\overline{Y}\overline{X} - \overline{X}\overline{Y})E$$

$$= E\overline{(YX - XY)}E = E\overline{[Y, X]}E = [Y, X]^{\sim} = -[X, Y]^{\sim}$$

i.e. $\sim$ is an **anti–linear Lie algebra isomorphism** on $\mathcal{L}$, but not **a $*$–Lie algebra involution.**

Definition 7 With the $*$–involution $\diamond$ the Lie algebra $sp(2d, \mathbb{C})$ becomes a $*$–Lie algebra **denoted** $sp_{\diamond}(2d, \mathbb{C})$.

4.2 *$*$–Isomorphism Between* $\mathbf{heis}_{2,d,cls}(\mathbb{C})$ *and* $\mathbf{sp}_{\diamond}(2d, \mathbb{C})$

Because of Theorem 1, one can identify the *centerless* $*$–Lie sub–algebra of $\text{heis}_{2,d,cls}(\mathbb{C})$ with the $*$–Lie algebra of HQBF (see Sect. 3.2). In this section, we deal with the HQBF realization.

Theorem 4 *One has, for h given by (74),*

$$(\mathbf{a}K\mathbf{a})^+ = -\mathbf{a}(K^{\diamond})\mathbf{a}, \quad K \in M_{2d,sym}(\mathbb{C}) \tag{78}$$

$$h((\mathbf{a}K\mathbf{a})^+) = (h(\mathbf{a}K\mathbf{a}))^{\diamond}$$

In particular the Lie-algebra isomorphism h, defined in Theorem 3, i.e.

$$h\left(\frac{1}{2}\mathbf{a}K\mathbf{a}\right) = KJ, \quad K \in M_{2d,sym}(\mathbb{C}) \tag{79}$$

is a $*$**–Lie algebra isomorphism** *from* $heis_{2,d}(\mathbb{C})$ *to* $sp_{\diamond}(2d, \mathbb{C})$.

Proof Writing $K \in M_{2d,sym}(\mathbb{C})$ in the form (70), i.e.

$$K = \begin{pmatrix} M & L^T \\ L & N \end{pmatrix}, \quad L \in M_d(\mathbb{C}), N, M \in M_{d,sym}(\mathbb{C}) \tag{80}$$

Equation (48) becomes

$$\begin{aligned}\mathbf{a}K\mathbf{a} &= \sum_{i=1}^{d}\sum_{j=1}^{d}(K_{i,j}a_ia_j + K_{i+d,j}a_i^+a_j + K_{i,j+d}a_ia_j^+ + K_{i+d,j+d}a_i^+a_j^+) \\ &= \sum_{i=1}^{d}\sum_{j=1}^{d}(M_{i,j}a_ia_j + L_{i,j}a_i^+a_j + L^T_{i,j}a_ia_j^+ + N_{i,j}a_i^+a_j^+)\end{aligned} \tag{81}$$

Therefore

$$\begin{aligned}(\mathbf{a}K\mathbf{a})^+ &= \sum_{i=1}^{d}\sum_{j=1}^{d}(\overline{M_{i,j}}a_j^+a_i^+ + \overline{L_{i,j}}a_j^+a_i + \overline{L^T_{i,j}}a_ja_i^+ + \overline{N_{i,j}}a_ja_i) \\ &= \sum_{i=1}^{d}\sum_{j=1}^{d}(\overline{M_{i,j}}a_i^+a_j^+ + \overline{L_{i,j}}a_j^+a_i + \overline{L^T_{i,j}}a_ja_i^+ + \overline{N_{i,j}}a_ia_j) \\ &= \sum_{i=1}^{d}\sum_{j=1}^{d}(\overline{M_{i,j}}a_i^+a_j^+ + \overline{L_{j,i}}a_i^+a_j + \overline{L^T_{j,i}}a_ia_j^+ + \overline{N_{i,j}}a_ia_j) \\ &= \sum_{i=1}^{d}\sum_{j=1}^{d}(\overline{M_{i,j}}a_i^+a_j^+ + \overline{L^T_{i,j}}a_i^+a_j + \overline{L_{i,j}}a_ia_j^+ + \overline{N_{i,j}}a_ia_j) \\ &= \sum_{i=1}^{d}\sum_{j=1}^{d}(\overline{N_{i,j}}a_ia_j + \overline{L_{i,j}}^T a_i^+a_j + \overline{L_{i,j}}a_ia_j^+ + \overline{M_{i,j}}a_i^+a_j^+) \\ &= \mathbf{a}\begin{pmatrix} \overline{N} & \overline{L} \\ \overline{L}^T & \overline{M} \end{pmatrix}\mathbf{a} =: \mathbf{a}\widehat{K}\mathbf{a}\end{aligned} \tag{82}$$

On the other hand

$$\widehat{K} \equiv \begin{pmatrix} \overline{N} & \overline{L} \\ \overline{L}^T & \overline{M} \end{pmatrix} = \begin{pmatrix} 0 & I_n \\ I_n & 0 \end{pmatrix}\begin{pmatrix} \overline{M} & \overline{L}^T \\ \overline{L} & \overline{N} \end{pmatrix}\begin{pmatrix} 0 & I_n \\ I_n & 0 \end{pmatrix} = E\overline{K}E = -K^{\diamond}$$

and this implies (78). From this and the identity

$$JE = \begin{pmatrix} 0 & -1_d \\ 1_d & 0 \end{pmatrix}\begin{pmatrix} 0 & 1_d \\ 1_d & 0 \end{pmatrix} = \begin{pmatrix} -1_d & 0 \\ 0 & 1_d \end{pmatrix} = -\begin{pmatrix} 1_d & 0 \\ 0 & -1_d \end{pmatrix}$$

$$= -\begin{pmatrix} 0 & 1_d \\ 1_d & 0 \end{pmatrix}\begin{pmatrix} 0 & -1_d \\ 1_d & 0 \end{pmatrix} = -EJ$$

one deduces that

$$h(\frac{1}{2}(\mathbf{a}K\mathbf{a})^+) = h(\frac{1}{2}\mathbf{a}\widehat{K}\mathbf{a}) = \widehat{K}J = E\overline{K}EJ = -E\overline{K}JE = -E\overline{KJ}E = (KJ)^\diamond = (h(\frac{1}{2}\mathbf{a}K\mathbf{a}))^\diamond$$

Hence h is a $*$–isomorphism. □

4.3 *The Isomorphism Between* $heis_{2,d}(\mathbb{C})$ *and* $sp(2d, \mathbb{C})$*: Direct Proof*

A direct proof of the isomorphism between $heis_{2,d}(\mathbb{C})$ and $sp(2d, \mathbb{C})$, i.e. one based directly on the normally ordered commutation relations in Lemma (1) rather than the non–normally ordered one (50) can be obtained as follows. The commutation relations in Lemma (1), in centerless form, are:

$$[aMa, a^+Na^+] = 4\left(a^+NMa + \frac{1}{2}\mathrm{Tr}MN\right) \tag{83}$$

$$[a^+Ma^+, a^+Na^+] = 0 \tag{84}$$

$$[aMa, aNa] = 0 \tag{85}$$

$$\left[aMa, a^+La + \frac{1}{2}\mathrm{Tr}L\right] = a(ML + (ML)^T)a \tag{86}$$

$$\left[a^+Ma^+, a^+La + \frac{1}{2}\mathrm{Tr}L\right] = -a^+(LM + (LM)^T)a^+ \tag{87}$$

where $L \in M_d(\mathbb{C})$ and $N, M \in M_{d,sym}(\mathbb{C})$. The substitutions

$$aMa \mapsto B_M; \quad a^+Na^+ \mapsto B_N^+; \quad a^+La + \frac{1}{2}\mathrm{Tr}L \mapsto B_L^0$$

lead to the $*$–Lie algebra characterized by:

$$[B_M, B_N^+] = 4B_{NM}^0 \tag{88}$$

$$[B_M^+, B_N^+] = 0 \tag{89}$$

$$[B_M, B_N] = 0 \tag{90}$$

$$\left[B_M, B_L^0\right] = B_{ML+(ML)^T} \tag{91}$$

$$\left[B^+_M, B^0_L\right] = -B^+_{LM+(LM)^T} \tag{92}$$

Proposition 4 *The unique linear extension of the map*

$$h : B_M \to \begin{pmatrix} 0 & 2M \\ 0 & 0 \end{pmatrix}, \quad h : B^+_N \to \begin{pmatrix} 0 & 0 \\ -2N & 0 \end{pmatrix}, \quad h : B^0_L \to \begin{pmatrix} -L^T & 0 \\ 0 & L \end{pmatrix} \tag{93}$$

(still denoted h) is a Lie algebra isomorphism from $heis_{2,n}(\mathbb{C})$) to $sp(2d, \mathbb{C})$.

Proof The two Lie algebras are isomorphic as vector spaces over $\mathbb{C}$ because they have the same dimension. To prove that h preserves the Lie brackets it is sufficient to show that it preserves the commutation relation among the generators. One has:

$$[h(B_M), h(B^+_N)] = \left[\begin{pmatrix} 0 & 2M \\ 0 & 0 \end{pmatrix}, \begin{pmatrix} 0 & 0 \\ -2N & 0 \end{pmatrix}\right]$$

$$\begin{pmatrix} -4MN & 0 \\ 0 & 0 \end{pmatrix} - \begin{pmatrix} 0 & 0 \\ 0 & -4NM \end{pmatrix} = \begin{pmatrix} -4MN & 0 \\ 0 & 4NM \end{pmatrix}$$

$$= 4\begin{pmatrix} -(NM)^T & 0 \\ 0 & NM \end{pmatrix} = 4h\left(B^0_{NM}\right) \tag{94}$$

Thus (88) is preserved.

$$[h(B^+_M), h(B^+_N)] = \left[\begin{pmatrix} 0 & 0 \\ -2M & 0 \end{pmatrix}, \begin{pmatrix} 0 & 0 \\ -2N & 0 \end{pmatrix}\right] = 0$$

Thus (89) is preserved.

$$[h(B_M), h(B_N)] = \left[\begin{pmatrix} 0 & 2M \\ 0 & 0 \end{pmatrix}, \begin{pmatrix} 0 & 2N \\ 0 & 0 \end{pmatrix}\right] = 0$$

Thus (90) is preserved.

$$\left[h(B_M), h\left(B^0_L\right)\right] = \left[\begin{pmatrix} 0 & 2M \\ 0 & 0 \end{pmatrix}, \begin{pmatrix} -L^T & 0 \\ 0 & L \end{pmatrix}\right] = \begin{pmatrix} 0 & 2ML + 2L^T M \\ 0 & 0 \end{pmatrix}$$

$$= \begin{pmatrix} 0 & 2(ML + (ML)^T) \\ 0 & 0 \end{pmatrix} = h(B_{ML+(ML)^T})$$

Thus (91) is preserved.

$$[h(B_M^+), h(B_L^0)] = \left[\begin{pmatrix} 0 & 0 \\ -2M & 0 \end{pmatrix}, \begin{pmatrix} -L^T & 0 \\ 0 & L \end{pmatrix}\right] = \begin{pmatrix} 0 & 0 \\ 2ML^T + 2LM & 0 \end{pmatrix}$$

$$= -\begin{pmatrix} 0 & 0 \\ -2(LM + (LM)^T) & 0 \end{pmatrix} = -h(B^0_{LM+(LM)^T})$$

Thus (92) is preserved. This proves the Lie algebra isomorphism. That this is a $*$–Lie algebra isomorphism is proved exactly as in the second part of Theorem 4. □

5 Real Lie Sub–algebras of $\mathbf{heis}_{2,d}(\mathbb{C})$ and $sp_{skew,\diamond}(2d, \mathbb{C})$

Proposition 5 *An homogeneous Boson quadratic field is* **skew–hermitean** *with respect to the* $^+$*-involution, iff it has the form*

$$i\frac{1}{2}\mathbf{a}K\mathbf{a}, \quad K = K^T = -K^\diamond \tag{95}$$

The set of these fields is a **real $*$–Lie sub–algebra** *of* $heis_{2,d}(\mathbb{C})$ *with respect to the involution* $+$.

Proof One has, using (78),

$$\left(i\frac{1}{2}\mathbf{a}K\mathbf{a}\right)^+ = -i\frac{1}{2}(\mathbf{a}K\mathbf{a})^+ = i\frac{1}{2}\mathbf{a}(K^\diamond)\mathbf{a}$$

So $i\frac{1}{2}\mathbf{a}K\mathbf{a}$ is skew–hermitean with respect to the $^+$-involution iff

$$i\frac{1}{2}\mathbf{a}(K^\diamond)\mathbf{a} = -i\frac{1}{2}\mathbf{a}K\mathbf{a}$$

i.e. iff $K = -K^\diamond$, which is (95). □

Definition 8 The set of matrices corresponding to skew–hermitean fields is a $*$–Lie sub–algebra of $sp(2d, \mathbb{R})$ denoted

$$sp_{skew,\diamond}(2d, \mathbb{C}) := \{iKJ \in sp(2d, \mathbb{C}) : K \in M_{2d}(\mathbb{C}), \ K = K^T = -K^\diamond\} \tag{96}$$

Remark Notice that $sp_{skew,\diamond}(2d, \mathbb{C})$ is a **real** $*$–Lie algebra, but the associated matrices are **complex**. In Theorem 5 below, we will prove that $sp_{skew,\diamond}(2d, \mathbb{C})$ is isomorphic to $sp(2d, \mathbb{R})$ as a real $*$–Lie algebra with an appropriate $*$-Lie algebra involution (different from the one induced by $\diamond$).

Lemma 6 *An homogeneous Boson quadratic boson field* $i\frac{1}{2}\mathbf{a}K\mathbf{a}$ *is* **skew–hermitean** *with respect to the* $^+$*–involution, iff* K *has the form:*

$$K = \begin{pmatrix} M & \overline{L} \\ L & M^* \end{pmatrix} = \begin{pmatrix} M & \overline{L} \\ L & \overline{M} \end{pmatrix}, \quad M \in M_{d,sym}(\mathbb{C}), \ L \in M_{d,herm}(\mathbb{C}) \tag{97}$$

Proof Since $K \in M_{d,sym}(\mathbb{C})$, by Lemma 3, it can be written in the form

$$K = \begin{pmatrix} M & L^T \\ L & N \end{pmatrix}$$

with $M, N \in M_{d,sym}(\mathbb{C}), \ L \in M_d(\mathbb{C})$. Then,

$$K^\diamond = -\begin{pmatrix} 0 & 1_d \\ 1_d & 0 \end{pmatrix}\begin{pmatrix} \overline{M} & \overline{L}^T \\ \overline{L} & \overline{N} \end{pmatrix}\begin{pmatrix} 0 & 1_d \\ 1_d & 0 \end{pmatrix} = -\begin{pmatrix} \overline{L} & \overline{N} \\ \overline{M} & \overline{L}^T \end{pmatrix}\begin{pmatrix} 0 & 1_d \\ 1_d & 0 \end{pmatrix}$$

$$= -\begin{pmatrix} \overline{N} & \overline{L} \\ \overline{L}^T & \overline{M} \end{pmatrix} = -\begin{pmatrix} \overline{N} & \overline{L} \\ L^* & \overline{M} \end{pmatrix}$$

Therefore

$$K = -K^\diamond \iff \begin{pmatrix} M & L^T \\ L & N \end{pmatrix} = \begin{pmatrix} \overline{N} & \overline{L} \\ L^* & \overline{M} \end{pmatrix} \iff \begin{cases} M = \overline{N} \iff N = \overline{M} \\ L = L^* \iff L^T = \overline{L} \end{cases} \tag{98}$$

and this is equivalent to (97). □

The set of skew–hermitean fields is a **real** $*$**–Lie sub-algebra** of $sp(2d, \mathbb{R})$ with respect to the involution $+$. The corresponding set of matrices is **denoted**

$$sp_{skew,\diamond}(2d, \mathbb{C}) := \{iKJ \, : \, K \in M_{2d}(\mathbb{C}), \ K = K^T = -K^\diamond\} \tag{99}$$

5.1 *Real* $*$*–Lie Algebra–Isomorphism Between* $sp_{skew,\diamond}(2d, \mathbb{C})$ *and* $sp_-(2d, \mathbb{R})$

Notice that $sp_{skew,\diamond}(2d, \mathbb{C})$ is a **real** $*$–Lie algebra, but the associated matrices are **complex**.

Definition 9 Define

$$sp(2d, \mathbb{R}) := sp(2d, \mathbb{C}) \cap M_{2d}(\mathbb{R}) \tag{100}$$

$sp(2d, \mathbb{R})$ inherits the Lie algebra structure from $sp(2d, \mathbb{C})$.

Definition 10 Introduce the **involution on** $sp(2d, \mathbb{R})$, given by **sign change**

$$X \mapsto -X, \quad \forall X \in sp(2d, \mathbb{R}) \tag{101}$$

With the involution (101), $sp(2d, \mathbb{R})$ becomes a **real *–Lie algebra** denoted $sp_-(2d, \mathbb{R})$.

Remark Notice that this involution **is not the restriction** to $sp(2d, \mathbb{R})$ of the involution on $sp_\diamond(2d, \mathbb{C})$. In fact, for $K = \begin{pmatrix} M & L^T \\ L & N \end{pmatrix} \in sp(2d, \mathbb{R})$, this restriction gives the map

$$\begin{pmatrix} M & L^T \\ L & N \end{pmatrix} \mapsto -\begin{pmatrix} N & L \\ L^T & M \end{pmatrix} \neq -\begin{pmatrix} M & L^T \\ L & N \end{pmatrix}$$

Lemma 7 *Defining the unitary matrix*

$$W := \frac{1}{\sqrt{2}} \begin{pmatrix} 1_d & i1_d \\ 1_d & -i1_d \end{pmatrix} \tag{102}$$

if $-K^\diamond = K$ *then*

$$W^T i K J (W^T)^{-1} \in sp(2d, \mathbb{R})$$

Proof From Lemma 6 we know that

$$K = \begin{pmatrix} M & \overline{L} \\ L & \overline{M} \end{pmatrix}$$

with $M \in M_{d,sym}(\mathbb{C})$ and $L \in M_{d,herm}(\mathbb{C})$. Therefore, using (102)

$$W^T i K J (W^T)^{-1} = i\frac{1}{\sqrt{2}} \begin{pmatrix} 1_d & 1_d \\ i1_d & -i1_d \end{pmatrix} \begin{pmatrix} \overline{L} & -M \\ \overline{M} & -L \end{pmatrix} \frac{1}{\sqrt{2}} \begin{pmatrix} 1_d & -i1_d \\ 1_d & i1_d \end{pmatrix} = \tag{103}$$

$$\frac{1}{2}\begin{pmatrix} -i(L - \overline{L} + M - \overline{M}) & L + \overline{L} + M + \overline{M} \\ M + \overline{M} - (L + \overline{L}) & -i(L - \overline{L} - (M - \overline{M})) \end{pmatrix} = \begin{pmatrix} \mathrm{Im}(L + M) & \mathrm{Re}(L + M) \\ \mathrm{Re}(M - L) & \mathrm{Im}(L - M) \end{pmatrix}$$

So $W^T i K J (W^T)^{-1}$ is a matrix with real entries. Notice that one has:

$$W^T i K J (W^T)^{-1} = \begin{pmatrix} -\mathrm{Re}(L + M) & \mathrm{Im}(L + M) \\ -\mathrm{Im}(L - M) & \mathrm{Re}(M - L) \end{pmatrix} \begin{pmatrix} 0 & -1_d \\ 1_d & 0 \end{pmatrix} \tag{104}$$

and both $-\mathrm{Re}(L + M)$ and $\mathrm{Re}(M - L)$ are symmetric and

$$(\mathrm{Im}(L + M))^T = \mathrm{Im}(L^T + M^T) = \mathrm{Im}(\overline{L} + M) = \mathrm{Im}(-L + M) = -\mathrm{Im}(L - M)$$

then the matrix

$$R_K := \begin{pmatrix} -\mathrm{Re}(L+M) & \mathrm{Im}(L+M) \\ -\mathrm{Im}(L-M) & \mathrm{Re}(M-L) \end{pmatrix} = \begin{pmatrix} -\mathrm{Re}(L+M) & \mathrm{Im}(L+M) \\ (\mathrm{Im}(L+M))^T & \mathrm{Re}(M-L) \end{pmatrix}$$

is symmetric, hence $W^T iKJ(W^T)^{-1} = R_K J \in sp(2d, \mathbb{R})$. Thus $W^T iKJ(W^T)^{-1} \in sp(2d, \mathbb{R})$. □

Remark W transforms the annihilation–creation generators into **coordinate–momentum generators.**

Theorem 5 *The map* $W^T(\cdot)(W^T)^{-1}$ *is a* $*$*–Lie algebra isomorphism from* $sp_{skew,\diamond}(2d, \mathbb{C})$ *with the* $\diamond$*–involution to* $sp_-(2d, \mathbb{R})$ *(see Definition 9).*

Proof The map $W^T(\cdot)(W^T)^{-1}$ is invertible as map from $M_d(\mathbb{C})$ to $M_d(\mathbb{C})$. Namely, its inverse is the map $(W^T)^{-1}(\cdot)W^T$, so due to Lemma 7, it defines an isomorphism between $sp_{skew,\diamond}(2d, \mathbb{C})$ and $sp_-(2d, \mathbb{R})$ as linear spaces. Because of $W^T[X, Y](W^T)^{-1} = [W^T X(W^T)^{-1}, W^T Y(W^T)^{-1}]$ this map is a Lie algebra isomorphism. Finally, to prove that the map $W^T(\cdot)(W^T)^{-1}$ is a $*$–map, notice that

$$(iKJ)^\diamond = -(-i)\begin{pmatrix} 0 & 1_d \\ 1_d & 0 \end{pmatrix}\begin{pmatrix} L & -\overline{M} \\ M & -\overline{L} \end{pmatrix}\begin{pmatrix} 0 & 1_d \\ 1_d & 0 \end{pmatrix} = i\begin{pmatrix} -\overline{L} & M \\ -\overline{M} & L \end{pmatrix}$$

So, in (103), $\diamond$ maps $L \to -L$ and $M \to -M$, thus

$$W^T(iKJ)^\diamond(W^T)^{-1} = \begin{pmatrix} -\mathrm{Im}(L+M) & -\mathrm{Re}(L+M) \\ -\mathrm{Re}(M-L) & -\mathrm{Im}(L-M) \end{pmatrix} = -W^T iKJ(W^T)^{-1}$$

which agrees with the involution in $sp_-(2d, \mathbb{R})$. □

6 Vacuum Averages

Let us recall in a bit different notation the results of [12].

Theorem 6 *The normally ordered form of* $e^{it\frac{1}{2}\mathbf{a}K\mathbf{a}}$*, where* $t \in \mathbb{R}$ *is:*

$$e^{it\frac{1}{2}\mathbf{a}K\mathbf{a}} = \tag{105}$$

$$= \frac{1}{\sqrt{\det\left(P_- e^{iKJt} + P_+\right)}} : \exp\left(-\frac{1}{2}\mathbf{a}(e^{iKJt} - I)(P_- e^{iKJt} - P_+)^{-1}E\mathbf{a}\right) :$$

where

$$P_+ = \begin{pmatrix} 1_d & 0 \\ 0 & 0 \end{pmatrix}, \quad P_- = \begin{pmatrix} 0 & 0 \\ 0 & 1_d \end{pmatrix}$$

and the square root in (105) is continuous root, i.e. the branch of it is chosen depending on t *in such a way that it is continuous in* t*. In particular,*

$$\langle \Phi, e^{it\frac{1}{2}\mathbf{a}K\mathbf{a}}\Phi \rangle = \frac{1}{\sqrt{\det\left(P_-e^{iKJt} + P_+\right)}}, \tag{106}$$

which could be written in the form

$$\frac{1}{\sqrt{\det\left(P_-e^{iKJt} + P_+\right)}} = \exp\left(-\frac{1}{2}\int_0^t d\tau Tr P_- iKJe^{iKJ\tau}\left(P_-e^{iKJ\tau} + P_+\right)^{-1}\right) \tag{107}$$

Proposition 6 *The vacuum* **cumulants** *of the random variable* $\frac{1}{2}\mathbf{a}K\mathbf{a}$ *defined by*

$$\langle \Phi, e^{it\frac{1}{2}\mathbf{a}K\mathbf{a}}\Phi \rangle = e^{\sum_{k=1}^{\infty}\kappa_k \frac{(it)^k}{k!}}$$

are given by:

$$\kappa_k = -\frac{1}{2}\sum_{k_1+\cdots+k_q=k-1}\frac{(k-1)!}{k_1!\cdots k_q!}Tr P_- KJP_+(KJ)^{k_1}\dots P_+(KJ)^{k_q}$$

Proof (idea) This is just a direct expansion of the right hand side of (107). □

7 Lie Groups Associated with the Symplectic Algebra

In this section we identify the Lie group formed by the unitarities of the form $e^{i\frac{1}{2}\mathbf{a}K\mathbf{a}}$ with $K = -K^{\diamond}$, i.e. with self-adjoint $\frac{1}{2}\mathbf{a}K\mathbf{a}$.

It is known that the correspondence between Lie algebras and Lie groups is not one–to–one.

We will see (Theorem 10 below) that in our case the correct group is not $Sp(2d, \mathbb{R})$ but the metaplectic group $Mp(2d, \mathbb{R})$, which is the double cover of $Sp(2d, \mathbb{R})$ (the double cover is unique, see Theorem 8). It is also sometimes called the **oscillator representation** of the symplectic group.

Let us recall the following definitions:

$$Sp(2d, \mathbb{C}) = \{S \in M_{2d}(\mathbb{C}) : SJS^T = J\}$$

$$Sp(2d, \mathbb{R}) = \{S \in M_{2d}(\mathbb{R}) : SJS^T = J\}$$

$$Sp_{skew,\diamond}(2d, \mathbb{C}) = \{S : SJS^T = J, S = -S^{\diamond}\}$$

They are generated by the corresponding algebras.

7.1 The Siegel Unit Disk

Theorem 7 *(i) The elements $S \in Sp(2d, \mathbb{R})$ can be represented in the block form*

$$S = \begin{pmatrix} U & V \\ \overline{V} & \overline{U} \end{pmatrix} \tag{108}$$

where $U, V \in M_d(\mathbb{C})$, which satisfy (see [11], p. 90, Eq. (4.5), (4.8)).

$$UU^* - VV^* = 1_d, \qquad \overline{U}V^* = \overline{V}U^* \tag{109}$$

or equivalently

$$U^*U - V^T\overline{V} = 1_d, \quad U^T\overline{V} = V^*U \tag{110}$$

(ii) Defining the **Siegel unit disk** *by:*

$$\mathbf{D} = \{Z \in M_d(\mathbb{C}) : 1_d - Z^*Z > 0\} \tag{111}$$

$Sp_{skew,\diamond}(2d, \mathbb{C})$ acts on $\mathbf{D}$ as follows:

$$Z \in \mathbf{D} \mapsto S \circ Z = (UZ + V)(\overline{U} + \overline{V}Z)^{-1} \in \mathbf{D}$$

(iii) For each $Z \in \mathbf{D}$

$$m(S, Z) := \det(\overline{U} + \overline{V}Z)$$

is a **multiplicative 1-cocycle** *for this action, i.e.*

$$m(S_1S_2, Z) = m(S_1, S_2 \circ Z)m(S_2, Z) \tag{112}$$

Proof (i) Let us represent $S \in M_{2d}(\mathbb{C})$ in block form

$$S = \begin{pmatrix} U & V \\ V_1 & U_1 \end{pmatrix}$$

where $U, V, U_1, V_1 \in M_d(\mathbb{C})$. Now let us take into account $S = -S^\diamond$ from the definition of $Sp_{skew,\diamond}(2d, \mathbb{C})$, then

$$\begin{pmatrix} U & V \\ V_1 & U_1 \end{pmatrix} = \begin{pmatrix} 0 & 1_d \\ 1_d & 0 \end{pmatrix} \begin{pmatrix} \overline{U} & \overline{V} \\ \overline{V_1} & \overline{U_1} \end{pmatrix} \begin{pmatrix} 0 & 1_d \\ 1_d & 0 \end{pmatrix} = \begin{pmatrix} \overline{U_1} & \overline{V_1} \\ \overline{V} & \overline{U} \end{pmatrix}$$

which leads to $V_1 = \overline{V}$ and $U_1 = \overline{V}$, i.e. to (108). Now let us calculate

$$SJS^T = \begin{pmatrix} U & V \\ \overline{V} & \overline{U} \end{pmatrix} \begin{pmatrix} 0 & -1_d \\ 1_d & 0 \end{pmatrix} \begin{pmatrix} U^T & V^* \\ V^T & U^* \end{pmatrix} = \begin{pmatrix} VU^T - UV^T & VV^* - UU^* \\ \overline{U}U^T - \overline{V}V^T & \overline{U}V^* - \overline{V}U^* \end{pmatrix}$$

Thus, $SJS^T = J$ is equivalent to (109). As for square matrices the left and right inverse are equal, then $SJS^T = J$ is equivalent to $S^T JS = J$. So let us calculate

$$S^T JS = \begin{pmatrix} U^T & V^* \\ V^T & U^* \end{pmatrix} \begin{pmatrix} 0 & -1_d \\ 1_d & 0 \end{pmatrix} \begin{pmatrix} U & V \\ \overline{V} & \overline{U} \end{pmatrix} = \begin{pmatrix} V^*U - U^T\overline{V} & V^*V - U^T\overline{U} \\ U^*U - V^T\overline{V} & U^*V - V^T\overline{U} \end{pmatrix}$$

Thus, $S^T JS = J$ is equivalent to (110).
(ii) Let us calculate

$$((U^T + Z^*V^T)(\overline{U} + \overline{V}Z) - (Z^*U^* + V^*)(UZ + V)) =$$

$$(U^T\overline{U} + Z^*V^T\overline{U} + U^T\overline{V}Z + Z^*V^T\overline{V}Z) - (Z^*U^*UZ + Z^*U^*V + V^*UZ + V^*V) =$$

$$((U^T\overline{U} - V^*V) + Z^*(V^T\overline{U} - U^*V) + (U^T\overline{V} - V^*U)Z + Z^*(V^T\overline{V} - U^\dagger U)Z) =$$

$$1_d - Z^*Z > 0$$

Hence,

$$(\overline{U} + \overline{V}Z)^*(\overline{U} + \overline{V}Z) = (U^T + Z^*V^T)(\overline{U} + \overline{V}Z) > (Z^*U^* + V^*)(UZ + V) =$$

$$(UZ + V)^*(UZ + V) \geq 0$$

so

$$(\overline{U} + \overline{V}Z)^*(\overline{U} + \overline{V}Z) > 0$$

Thus, $\overline{U} + \overline{V}Z$ has no zero eigenvalues and $Y := (UZ + V)(\overline{U} + \overline{V}Z)^{-1}$ is well-defined. Moreover, due to the above calculations we have

$$1_d - Y^*Y = 1_d - (U^T + Z^*V^T)^{-1}(Z^*U^* + V^*)(UZ + V)(\overline{U} + \overline{V}Z)^{-1} =$$

$$(U^T + Z^TV^*)^{-1}((U^T + Z^*V^T)(\overline{U} + \overline{V}Z) - (Z^*U^* + V^*)(UZ + V))(\overline{U} + \overline{V}Z)^{-1} =$$

$$((\overline{U} + \overline{V}Z)^{-1})^*(1_d - Z^*Z)(\overline{U} + \overline{V}Z)^{-1} > 0$$

Thus, $Y \in \mathbf{D}$. Now, let us show that $\circ$ is a (left) group action. Let us take

$$S_j = \begin{pmatrix} U_j & V_j \\ \overline{V}_j & \overline{U}_j \end{pmatrix} \in Sp_{skew,\diamond}(2d, \mathbb{C}), \quad j = 1, 2$$

then

$$S_1 \circ (S_2 \circ Z) = S_1 \circ ((U_2Z + V_2)(\overline{U}_2 + \overline{V}_2Z)^{-1}) =$$

$$(U_1(U_2Z + V_2)(\overline{U}_2 + \overline{V}_2Z)^{-1} + V_1)(\overline{U}_1 + \overline{V}_1(U_2Z + V_2)(\overline{U}_2 + \overline{V}_2Z)^{-1})^{-1} =$$

$$(U_1(U_2Z + V_2) + V_1(\overline{U}_2 + \overline{V}_2Z))(\overline{U}_1(\overline{U}_2 + \overline{V}_2Z) + \overline{V}_1(U_2Z + V_2))^{-1} =$$

$$((U_1U_2+V_1\overline{V}_2)Z+U_1V_2+V_1\overline{U}_2)(\overline{U}_1\overline{U}_2+\overline{V}_1V_2+(\overline{U}_1\overline{V}_2+\overline{V}_1U_2)Z)^{-1}=$$

$$\begin{pmatrix} U_1U_2+V_1\overline{V}_2 & U_1V_2+V_1\overline{U}_2 \\ \overline{U}_1\overline{U}_2+\overline{V}_1V_2 & \overline{U}_1\overline{V}_2+\overline{V}_1U_2 \end{pmatrix}\circ Z=\begin{pmatrix} U_1 & V_1 \\ \overline{V}_1 & \overline{U}_1 \end{pmatrix}\begin{pmatrix} U_2 & V_2 \\ \overline{V}_2 & \overline{U}_2 \end{pmatrix}\circ Z=(S_1S_2)\circ Z$$

and $1_{2d}\circ Z=Z$.
(iii) From the previous item we have

$$\begin{aligned} m(S_1S_2,Z)&=\det(\overline{U}_1\overline{U}_2+\overline{V}_1V_2+(\overline{U}_1\overline{V}_2+\overline{V}_1U_2)Z)\\ m(S_1,S_2\circ Z)&=m(S_1,(U_2Z+V_2)(\overline{U}_2+\overline{V}_2Z)^{-1})=\\ &\det(\overline{U}_1+\overline{V}_1(U_2Z+V_2)(\overline{U}_2+\overline{V}_2Z)^{-1})\\ m(S_2,Z)&=\det(\overline{U}_2+\overline{V}_2Z) \end{aligned}$$

So we obtain

$$m(S_1,S_2\circ Z)m(S_2,Z)=\det(\overline{U}_1+\overline{V}_1(U_2Z+V_2)(\overline{U}_2+\overline{V}_2Z)^{-1})\det(\overline{U}_2+\overline{V}_2Z)=$$

$$\det((\overline{U}_1+\overline{V}_1(U_2Z+V_2)(\overline{U}_2+\overline{V}_2Z)^{-1})(\overline{U}_2+\overline{V}_2Z))=$$

$$\det(\overline{U}_1(\overline{U}_2+\overline{V}_2Z)+\overline{V}_1(U_2Z+V_2))=\det(\overline{U}_1\overline{U}_2+\overline{V}_1V_2+(\overline{U}_1\overline{V}_2+\overline{V}_1U_2)Z)=$$

$$m(S_1S_2,Z),$$

which is (112). □

Remark Actually, the Siegel disk is expected here due to the following fact. The following subset of $Sp_{skew,\diamond}(2d,\mathbb{C})$

$$U_{sp,c}(d)=\left\{\begin{pmatrix} U & 0 \\ 0 & \overline{U} \end{pmatrix} : U\in U(d)\right\}$$

is a maximal compact sub–group (isomorphic to $U(d)$) of $Sp_{skew,\diamond}(2d,\mathbb{C})$. The Siegel disk can be identified with the homogeneous space $Sp_{skew,\diamond}(2d,\mathbb{C})/U_{sp,c}(d)$. (see [22]).

Proposition 7 *The fundamental group of $Sp_{skew,\diamond}(2d,\mathbb{C})$ is isomorphic to $\mathbb{Z}$.*

The statement of this proposition for $Sp(2d,\mathbb{R})\sim Sp_{skew,\diamond}(2d,\mathbb{C})$ can be found in ([15], p. 16), but its proof is spread over the text. So let us summarize the idea of it. As the fundamental group $\pi_1(U_{sp,c}(d))\sim\pi_1(U(d))\sim\mathbb{Z}$ and the Siegel disk is contractible (hence, $Sp_{skew,\diamond}(2d,\mathbb{C})$ and $U_{sp,c}(d)$ are homotopic), then the fundamental group $\pi_1(Sp_{skew,\diamond}(2d,\mathbb{C}))\sim\mathbb{Z}$.

Theorem 8 *The connected m-covering group of $Sp_{skew,\diamond}(2d,\mathbb{C})$ is unique up to isomorphism.*

Proof There is a one-to-one correspondence between conjugacy classes of sub–groups of the fundamental group of path-connected, locally path-connected, and semi–locally simply-connected topological spaces and the isomorphism classes of path-connected covering spaces of it (see [16], Theorem 1.38). Any Lie group (in particular, $Sp_{skew,\diamond}(2d, \mathbb{C})$) is a manifold. Moreover, $Sp_{skew,\diamond}(2d, \mathbb{C}) \sim Sp(2d, \mathbb{R})$ is connected (see [14], p. 28). Any connected manifold is path-connected, locally path-connected and semi–locally simply-connected. By definition the covering group is a covering space which is a Lie group at the same time, so every connected covering group is path connected. Thus, there is one-to-one correspondence between conjugacy classes of sub–groups of the fundamental group $\pi_1(Sp_{skew,\diamond}(2d, \mathbb{C}))$ and isomorphism classes of covering groups of $Sp(2d, \mathbb{R})$.

All the sub–groups of $\mathbb{Z}$ are $m\mathbb{Z}$, and there is one-to-one correspondence between them and conjugacy classes due to normality. (All sub–groups of an Abelian group are normal. And all the normal sub–groups are self-conjugate.) The indexes of such sub–groups, i.e. the numbers of elements in the factor groups $\mathbb{Z}/m\mathbb{Z} = \mathbb{Z}_m$ are exactly m. Due to (see [16] Proposition 1.32) the number of sheets of a covering space equals the index of the corresponding sub–group, hence there is unique (up to isomorphism) with exactly m sheets, thus m-th covering group is unique up to isomorphism. □

7.2 The Metaplectic Group

Theorem 9 *Let $\varepsilon(S, Z)$ be a holomorphic function of Z on* **D** *such that*

$$(\varepsilon(S, Z))^2 = m(S, Z) \tag{113}$$

for some $S \in Sp_{skew,\diamond}(2d, \mathbb{C})$. Then the set of pairs

$$(S, \varepsilon(S, Z))$$

is a group for the multiplication law

$$(S_1, \varepsilon_1(S_1, Z)) \cdot (S_2, \varepsilon_2(S_2, Z)) = (S_1 S_2, \varepsilon_1(S_1, S_2 \circ Z)\varepsilon_2(S_2, Z)) \tag{114}$$

This group is called **the metaplectic group**. *It is the unique double cover up to isomorphism.*

Proof To prove the group property we should show that $S_1 S_2$ is an element of $Sp_{skew,\diamond}(2d, \mathbb{C})$, which follows from $S_1, S_2 \in Sp_{skew,\diamond}(\mathbb{C}, 2d)$ and the fact that $Sp_{skew,\diamond}(2d, \mathbb{C})$ is a group, and that $\varepsilon_1(S_1, S_2 \circ Z)\varepsilon_2(S_2, Z)$ is such that $(\varepsilon_1(S_1, S_2 \circ Z)\varepsilon_2(S_2, Z))^2 = m(S_1 S_2, Z)$, which follows from (?? and the fact that $m(S_2, Z)$ is a 1-cocycle:

$$(\varepsilon_1(S_1, S_2 \circ Z)\varepsilon_2(S_2, Z))^2 = (\varepsilon_1(S_1, S_2 \circ Z))^2(\varepsilon_2(S_2, Z))^2 = \\ = m(S_1, S_2 \circ Z)m(S_2, Z) = m(S_1 S_2, Z)$$

There are exactly two such functions $\varepsilon(S, Z)$ for a given S, namely $\pm e^{\frac{1}{2}Log m(S,Z)}$, where Log is the principal branch of the logarithm, so there are exactly two sheets of the covering space, i.e. this is a double cover. □

Remark The multiplication law (114) defines the semidirect product of $Sp_{skew,\diamond}(2d, \mathbb{C})$ and the group consisting of two roots of $m(S, Z)$, which is isomorphic to $\mathbb{Z}_2$.

Lemma 8 *The vacuum average is defined by the* 1*-cocycle* $m(S, Z)$*, through the identity*

$$\langle \Phi, e^{it\frac{1}{2}\mathbf{a}K\mathbf{a}}\Phi \rangle = \frac{1}{\sqrt{m(e^{iKJt}, 0)}} \tag{115}$$

Proof If one defines $U_t, V_t \in M_d(\mathbb{C})$ as blocks of the matrix

$$S_t = e^{iKJt} = \begin{pmatrix} U_t & V_t \\ \overline{V}_t & \overline{U}_t \end{pmatrix}$$

(it is possible for every $S \in Sp_{skew,\diamond}(2d, \mathbb{C})$ due to Theorem 7), then $\det(P_- e^{iKJt} + P_+) = \det \overline{U}_t = m(S_t, 0)$. Thus, formula (106) takes the form (115). □

Theorem 10 *The metaplectic group is isomorphic to the group generated by* $heis_{2;d}(\mathbb{C})$.

Proof (idea). The detailed proof could be found in [13, Chap. 4, Sect. 2]. The basic idea consists it the fact $e^{it\frac{1}{2}\mathbf{a}K\mathbf{a}}$ is fully defined by $e^{iKJt} \in Sp_{skew,\diamond}(2d, \mathbb{C})$ and the choice of the branch of square root in (115), which depends on path as the continuous square root is used. The two possibilities of this choice identify $e^{it\frac{1}{2}\mathbf{a}K\mathbf{a}}$ with elements of the double cover, i.e. the metaplectic group. □

7.3 The Abstract Symplectic Algebra and Its Lie Groups

The CCR (46)

$$[\mathbf{a}_f, \mathbf{a}_g] = -f^T J g$$

imply the following commutation relations among HQBF:

$$\left[\frac{1}{2}\mathbf{a}K\mathbf{a}, \frac{1}{2}\mathbf{a}M\mathbf{a}\right] = \frac{1}{2}\mathbf{a}[KJ, MJ]J\mathbf{a} \tag{116}$$

This suggests to introduce an abstract Lie algebra with generators s_K ($K \in M_{2d,sym}(\mathbb{C})$), involution given by

$$s_K^* := s_{K^\diamond}$$

and commutation relations

$$[s_K, s_M] = s_{[KJ,MJ]J} \tag{117}$$

The correspondence between Lie algebras and Lie groups is not one-to-one. Namely, the following proposition can be found in [19], Sect. 3.5.

Proposition 8 *For any given real Lie algebra, there is one and only one (up to isomorphism) simply connected Lie group, whose algebra coincides with given algebra. This group is the universal cover as it is a covering group of any Lie group, whose algebra coincides with given algebra.*

If we consider $sp_{skew,\diamond}(2d, \mathbb{C})$, we have an ambiguity because $Sp_{skew,\diamond}(2d, \mathbb{C})$ is not simply connected. In [8] is actually the simply connected Lie group was considered, so it is isomorphic to the universal cover, which is different from the metaplectic group. In the case of $sp(2d, \mathbb{C})$ the corresponding Lie group is unique (because $Sp(2d, \mathbb{C})$ is simply connected).

But for the representation based on the CCR, not all the exponentials of algebra elements are defined, so this representation does not generate a Lie group at all.

Universal Cover

Proposition 9 *Let $\varphi(S, Z)$ be a holomorphic function $\varepsilon(Z)$ on $\mathbf{D}$ such that $e^{\varphi(S,Z)} = m(S, Z)$ for some $S \in Sp_{skew,\diamond}(2d, \mathbb{C})$.*

For each $S \in Sp_{skew,\diamond}(2d, \mathbb{C})$, let $\varphi(S, Z)$ be a holomorphic function on $\mathbf{D}$ such that $e^{\varphi(S,Z)} = m(S, Z)$.

The set of pairs

$$(S, \varphi(S, Z))$$

with multiplcation

$$(S_1, \varphi_1(S_1, Z)) \cdot (S_2, \varphi_2(S_2, Z)) = (S_1 S_2, \varphi_1(S_1, S_2 \circ Z) + \varphi_2(S_2, Z)) \tag{118}$$

is a group called **universal cover** *of $Sp_{skew,\diamond}(2d, \mathbb{C})$.*

Remark The possible choices of the logarithm of $m(S, Z)$ are parametrized by $\mathbb{Z}$.

The proof could be found in [20, p. 80] for $d = 1$ and [23] for the general case.

In [20] the addition is used as group operation for the second component of $(S, \varphi(S, Z))$. To make the comparison with our definition (see Theorem 9) of the metaplectic group more explicit, let us define $\varepsilon(S, Z) = e^{\nu\varphi(S,Z)} = (m(S, Z))^\nu$, $\nu \geq 0$ and the multiplication

$$(S_1, \varepsilon_1(S_1, Z)) \cdot (S_2, \varepsilon_2(S_2, Z)) = (S_1 S_2, \varepsilon_1(S_1, S_2 \circ Z)\varepsilon_2(S_2, Z)) \tag{119}$$

then for $\nu = 1$ we recover the initial group (as there are one-to-one correspondence between the analytic function $\varepsilon(S, Z) = m(S, Z)$ of Z and $S \in Sp_{skew,\diamond}(2d, \mathbb{C})$), for $\nu = 1/2$ we recover the metaplectic case (as there exactly two analytic functions of Z such that $(\varepsilon(S, Z))^2 = m(S, Z)$ for given $S \in Sp_{skew,\diamond}(2d, \mathbb{C})$), and similarly we have k-covers for $\nu = 1/k$, $k \in \mathbb{N}$ (as there are k analytic functions of Z such that $(\varepsilon(S, Z))^k = m(S, Z)$ for given $S \in Sp_{skew,\diamond}(2d, \mathbb{C})$). In general, all possible functions $\varepsilon(S, Z)$ have the form

$$\varepsilon(S, Z) = e^{\nu(Log(m(S,Z))+2\pi q)}, \qquad q \in \mathbb{Z}$$

where Log is the principal branch of the logarithm. For rational ν some of them are coincide and (119) defines the k-cover, but for irrational ν they are different, so (119) defines the group isomorphic to the universal cover.

8 The Problems of Diagonalizability and Vacuum Factorizability

Definition 11 A real valued classical random variable X is called **factorizable** if it can be expressed in the form

$$X = X_1 + X_2 \tag{120}$$

where X_1, X_2 are neither zero nor with atomic distribution concentrated on a single point and satisfy

$$E_X\left(e^{it(X_1+X_2)}\right) = E_X\left(e^{itX_1}\right) E_X\left(e^{itX_2}\right) \tag{121}$$

where E_X denotes expectation with respect to the probability measure of the process $X \equiv (X_1, X_2)$. In this case also the distribution of X is called factorizable (or **decomposable**) and the distributions of the X_j are called its **factors** (see [21], p. 103).

For classical **random fields** with values in a real finite dimensional vector space V a **weaker notion** of factorizability is important for applications.

Recall that a V–**valued classical random field** is a pair (X, μ), where μ is a probability measure on V^* (the dual of V) and $X : v \in V \to X_v$ is a linear map from V to the real valued random variables defined on the probability space (V^*, μ) defined by:

$$X_v(u^*) := \langle u^*, v\rangle, \quad v \in V,\ u^* \in V^* \ \text{(coordinate functions)}$$

Definition 12 A V–valued random field (X, μ) is called **factorizable** if there exists a linear basis $e = (e_j)_{j\in D}$ of V such that the classical random variables X_{e_j} are independent, i.e.

$$E_X\left(e^{\sum_{j\in D} t_j X_{e_j}}\right) = \prod_{j\in D} E_X\left(e^{t_j X_{e_j}}\right), \quad \forall t_j \in \mathbb{R},\ j \in D \tag{122}$$

8.1 Diagonalizability and Factorizability of Quadratic Fields

If Q is a self–adjoint operator acting on a Hilbert space $\mathcal{H}$ and $\Phi \in \mathcal{H}$ is a unit vector, the spectral measure of Q in Φ, defined by

$$\mu_{Q,\Phi}(I) := \langle \Phi, \chi_I(Q)\Phi \rangle, \quad I \in \text{Borel}(\mathbb{R}) \tag{123}$$

where $\chi_I(Q)$ is defined by the spectral theorem, is a probability measure, hence it defines, up to stochastic equivalence, a classical random variable X_Q. Conversely, any classical real valued random variable X acts, as multiplication operator, on the L^2–space of its probability distribution. The linear operator Q corresponding to this action is self–adjoint with spectral measure $I \in \text{Borel}(\mathbb{R}) \mapsto$ Multiplication by χ_I. In this sense **every self–adjoint linear operator on a Hilbert space defines a classical real valued random variable and conversely**.

In the paper [4], it is proved that the quadratic boson fields are essentially self–adjoint. Therefore for any quadratic field $F_{A,C}$, denoting still $F_{A,C}$ its unique self–adjoint extension, the pair $(F_{A,C}, \Phi)$, where Φ is the Fock vacuum, can be identified to a real valued random variable.

In this section we fix $D := \{1, \ldots, d\}$ with $d \in \mathbb{N}^*$ where $d := \dim(V)$ and we suppose that V is a semi–Hilbert space so that one can speak of **adjoint** of a linear operator on V (see [7]). In this case, for any linear basis $e \equiv (e_j)_{j \in D}$ of V any homogeneous quadratic boson field X can be written as

$$F_{A,C} = \sum_{i,j \in D} \left(A_{ij}^{(e)} a_{e_i}^+ a_{e_j}^+ + \bar{A}_{ji}^{(e)} a_{e_j} a_{e_i} + C_{ij}^{(e)} a_{e_i}^+ a_{e_j} \right) \tag{124}$$

where $A^{(e)} := (A_{ij}^{(e)}) \in M_{d,sym}(\mathbb{C})$ and $C^{(e)} := (C_{ij}^{(e)}) \in M_{d,herm}(\mathbb{C})$ are the matrices of two linear operators A, C on V.

Under a unitary change of basis in V, the field $F_{A,C}$ remains the same, while the pair of matrices (i.e. the coefficients of the CAP operators in the given basis) changes. Linear changes of bases are parametrized by the group $Un(V)$ of unitary operators of V. Thus these changes of bases induce an action of $Un(V)$ on the pairs $(A^{(e)}, C^{(e)})$ where $A^{(e)}$ is symmetric and $C^{(e)}$ hermitean. The following Lemma determines the form of this action.

Remark Since the field $F_{A,C}$ is uniquely determined by the pair (A, C) we often **identify** the two.

Lemma 9 *Let $F_{A,C}$ be the homogeneous quadratic boson field given by (124). If $f \equiv (f_j)$ is another linear basis and $U := (U_{jk})$ is the coefficient matrix of e_k in the (f_j)–basis*

$$e_k := \sum_j U_{jk} f_j = U f_k; \qquad U := (U_{jk}) \in GL(\mathbb{C}^d)$$

Then the standard form of $F_{A,C}$ with respect to the linear basis $f \equiv (f_j)$ is characterized by the pair $(A^{(f)}, C^{(f)})$ given by:

$$C^{(f)} = UC^{(e)}U^* \tag{125}$$

$$A^{(f)} = UA^{(e)}U^T \tag{126}$$

Proof See [3].

Definition 13 A quadratic field $F_{A,C}$ is called **diagonalizable** if there exists $U \in Un(V)$, such that $M := UAU^T \in M_d(\mathbb{C})$ and $L := UCU^* \in M_d(\mathbb{R})$ are diagonal matrices.

$F_{A,C}$ is called **vacuum–factorizable** if, $U \in Un(V), n \in D, n \geq 2$, and vacuum–independent quadratic fields $F_{A_1,C_1}, \ldots, F_{A_n,C_n}$ with $(A_j, C_j) \neq (0, 0)$ for all $j \in \{1, \ldots, n\}$ such that:

$$\langle \Phi, e^{itF_{A,C}}\Phi \rangle = \langle \Phi, e^{it(F_{A_1,C_1}+\cdots+F_{A_n,C_n})}\Phi \rangle = \prod_{j\in\{1,\ldots,n\}} \langle \Phi, e^{itF_{A_j,C_j}}\Phi \rangle \tag{127}$$

Remark Diagonalizability implies the additive decomposition (129) for $F_{A,C}$. Conditions (129) (resp. (127) are, a priori, **stronger** than (120) (resp. (121) because of the **additional condition** that the additive decomposition of $F_{A,C}$ must be obtained through other HQBF and not through arbitrary random variables. However a proof of their inequivalence is at the moment not available.

8.2 *Diagonalizability Implies Vacuum–Factorizability*

Remark The following result shows that diagonalizability implies vacuum–factorizability.

Theorem 11 *If the field $F_{A,C}$ is diagonalizable then the vacuum random variable*

$$F_{A,C} = \sum_{i,j=1}^{d} \left(A_{ij}a_i^+a_j^+ + \bar{A}_{ij}a_ia_j + C_{ij}a_i^+a_j \right) \tag{128}$$

takes the form

$$F_{A,C} = \sum_{i=1}^{d} \left(m_{ii}\,(b_i^+)^2 + \bar{m}_{ii}\,(b_i)^2 + l_{ii}b_i^+b_i \right) \tag{129}$$

where the l_{ii} are real numbers and, for $i \in \{1, 2, \ldots, d\}$,

$$b_i^+ = \sum_{k=1}^{d} \overline{u_{ik}} a_k^+; \qquad b_i = \sum_{k=1}^{d} u_{ik} a_k \tag{130}$$

and the matrices

$$U = \left(u_{ij}\right), \ L = \left(l_{ij}\right) \ M = \left(m_{ij}\right)$$

are as in Lemma 9. Moreover, for $i, j \in \{1, 2, ..., d\}$,

$$[b_i, b_j^+] = \delta_{i,j}, \ [b_i^+, b_j^+] = [b_i, b_j] = 0$$

Proof We notice that

$$F_{A,C} = a^+ A a^+ + a \bar{A} a + a^+ C a$$

where

$$a^+ = (a_1^+, \cdots, a_d^+); \quad a = (a_1, \cdots, a_d)$$

In the notations of Lemma 9, one has

$$A = U^* M \bar{U}, \ C = U^* L U$$

so that

$$F_{A,C} = a^+ U^* M \bar{U} a^+ + a \overline{U^* M \bar{U}} a + a^+ U^* L U a$$

$$= a^+ U^* M \bar{U} a^+ + a U^T \bar{M} U a + a^+ U^* L U a = b^+ M b^+ + b \bar{M} b + b^+ L b$$

where

$$b^+ = a^+ U^*, \ b = a U^T$$

i.e.,

$$b^+ = \left(b_1^+ \ldots b_d^+\right), \ b = \left(b_1 \ldots b_d\right)$$

where the $b_j^{\pm}$ are as in (130). Since L, M are diagonal, we obtain

$$F_{A,C} = \sum_{i=1}^{d} \left(m_{ii} \left(b_i^+\right)^2 + \bar{m}_{ii} \left(b_i\right)^2 + l_{ii} b_i^+ b_i\right)$$

Moreover

$$[b_i, b_j^+] = \sum_{p,q=1}^{d} u_{ip}\overline{u_{jq}}[a_p, a_q^+] = \sum_{p,q=1}^{d} u_{ip}\overline{u_{jq}}\delta_{p,q} = \sum_{p=1}^{d} u_{ip}\overline{u_{jp}} = \left(UU^*\right)_{ij} = \delta_{i,j}$$

and

$$[a_p^+, a_q^+] = [a_p, a_q] = 0 \Rightarrow [b_i^+, b_j^+] = [b_i, b_j] = 0$$

□

Theorem 12 *Let Φ be a Fock vacuum unit vector, i.e., $\|\Phi\| = 1$ and $a_j\Phi = 0$ for all $j \in \{1, 2, ..., d\}$, let*

$$F_{A,C} = \sum_{i,j=1}^{d} \left(A_{ij}a_i^+a_j^+ + \bar{A}_{ij}a_ia_j + C_{ij}a_i^+a_j\right) \tag{131}$$

be diagonalizable and let, for each $j = 1, 2, ..., d$,

$$X_j = m_{jj}\left(b_j^+\right)^2 + \bar{m}_{jj}\left(b_j\right)^2 + l_{jj}b_j^+b_j \tag{132}$$

Then, for suitably small $s \in \mathbb{R}$ and $i^2 = -1$, the vacuum characteristic function of the operator random variable $F_{A,C}$ is

$$\langle \Phi, e^{i\,s\,F_{A,C}}\,\Phi\rangle = \prod_{j=1}^{d} \sqrt{\frac{\cos L_j}{\cos(iK_js + L_j)}} \tag{133}$$

where, for each $j = 1, 2, ..., d$,

$$K_j = \sqrt{4\,|m_{jj}|^2 - l_{jj}^2};\; L_j = \arctan\left(\frac{l_{jj}}{K_j}\right)$$

Proof By Lemma 3.3, Proposition 3.4 and Remark 3.6 of [1] (for, in the notation of [1], $a = 0, b = m_{jj}, \lambda = -\frac{l_{jj}}{2}, \nu = l_{jj}$), for each $j = 1, 2, ..., d$:

$$e^{i\,s\,X_j}\,\Phi = p_j(s)e^{q_j(s)(b_j^+)^2}\,\Phi$$

and

$$\langle \Phi, e^{i\,s\,X_j}\,\Phi\rangle = p_j(s)$$

where

$$p_j(s) = e^{-\frac{il_{jj}s}{2}} \sqrt{\frac{\cos L_j}{\cos(iK_j s + L_j)}}, \; q_j(s) = \frac{K_j \tan(iK_j s + L_j) - l_{jj}}{4\bar{m}_{jj}}$$

$$K_j = \sqrt{4\,|m_{jj}|^2 - l_{jj}^2}; \; L_j = \arctan\left(\frac{l_{jj}}{K_j}\right)$$

By Theorem 13, $F_{A,C} = \sum_{j=1}^{d} X_j$ where $[X_j, X_k] = 0$ for $j \neq k$. Thus

$$\begin{aligned}
\langle \Phi, e^{i\,s\,F_{A,C}}\,\Phi\rangle &= \langle \Phi, e^{i\,s\,\sum_{j=1}^{d} X_j}\,\Phi\rangle = \langle \Phi, \prod_{j=1}^{d} e^{i\,s\,X_j}\,\Phi\rangle \\
&= \langle e^{-i\,s\,X_1}\Phi, \prod_{j=2}^{d} e^{i\,s\,X_j}\,\Phi\rangle = p_1(s)\langle e^{q_1(s)(b_1^+)^2}\Phi, \prod_{j=2}^{d} e^{i\,s\,X_j}\,\Phi\rangle \\
&= p_1(s)\langle \Phi, e^{\overline{q_1(s)}(b_1)^2} \prod_{j=2}^{d} e^{i\,s\,X_j}\,\Phi\rangle = p_1(s)\langle \Phi, \prod_{j=2}^{d} e^{i\,s\,X_j} e^{\overline{q_1(s)}(b_1)^2}\,\Phi\rangle \\
&= p_1(s)\langle \Phi, \prod_{j=2}^{d} e^{i\,s\,X_j}\,\Phi\rangle = p_1(s)\langle e^{-i\,s\,X_2}\Phi, \prod_{j=3}^{d} e^{i\,s\,X_j}\,\Phi\rangle \\
&= p_1(s)p_2(s)\langle \Phi, \prod_{j=3}^{d} e^{i\,s\,X_j}\,\Phi\rangle = ... = \prod_{j=1}^{d} p_j(s)\langle \Phi, \Phi\rangle = \prod_{j=1}^{d} p_j(s) \\
&= \prod_{j=1}^{d} \langle \Phi, e^{i\,s\,X_j}\,\Phi\rangle = \prod_{j=1}^{d} \sqrt{\frac{\cos L_j}{\cos(iK_j s + L_j)}}
\end{aligned}$$

□

Remark Notice that, in (132), m_{jj} and l_{jj} are arbitrary. Therefore one can replace them by $t_j m_{jj}$ and $t_j l_{jj}$ respectively. With this change, the characteristic function $\langle \Phi, e^{i\,s\,F_{A,C}}\,\Phi\rangle$ in (133) becomes

$$\langle \Phi, e^{i\,s\,F_{A,C}}\,\Phi\rangle = \langle \Phi, e^{i\,\sum_{j=1}^{d} s\,t_j X_j}\,\Phi\rangle = \langle \Phi, e^{i\,\sum_{j=1}^{d} s\,t_j \left(m_{jj}\left(b_j^+\right)^2 + \bar{m}_{jj}(b_j)^2 + l_{jj} b_j^+ b_j\right)}\,\Phi\rangle$$

and, since both s and the t_j are arbitrary, such are the products st_j. Therefore Theorem 12 implies that the probability measure $\mu_{F_{A,C}}$ associated with the operator random variable $F_{A,C}$ is factorizable in the form $\mu_{F_{A,C}} = \mu_{X_1} \times \cdots \times \mu_{X_d}$, where μ_{X_j} is the probability measure associated with the operator random variable X_j, $j \in \{1, 2, ..., d\}$.

Remark For $m_{jj} = 1$ and $l_{jj} = 0$, $j \in \{1, 2, ..., d\}$,

$$\langle \Phi, e^{i\,s\,X_j}\,\Phi\rangle = (\operatorname{sech} 2s)^{1/2}$$

with corresponding, via Bochner's theorem, probability density function

$$p(\lambda) = \frac{1-i}{8\pi}\left(e^{-\frac{\pi\lambda}{4}}\, B\left(-1; \frac{1-i\lambda}{4}, \frac{1}{2}\right) + e^{\frac{\pi\lambda}{4}}\, B\left(-1; \frac{1+i\lambda}{4}, \frac{1}{2}\right)\right)$$

where

$$B(z; a, b) = \int_0^z t^{a-1}(1-t)^{b-1}\, dt = z^a \sum_{d=0}^{\infty} \frac{(1-b)^d}{d!\,(a+n)} z^n$$

is the **incomplete Beta function** and, for $x \in \mathbb{R}$, $(x)^n = x(x+1)(x+2)\cdots(x+n-1)$ is the increasing partial factorial.

A characterization of diagonalizable HQBF is given as follows.

Theorem 13 *A field $F_{A,C}$ is diagonalizable if and only if CA is symmetric, i.e. if and only if $CA = A\overline{C}$.*

Proof See [5].

The truncated Virasoro fields is an example of a non diagonalizable Quadratic Hamiltonian: For $n \in \mathbb{N}$, the **Virasoro fields truncated at** n (see [3]) is defined by:

$$\begin{aligned} Y_n(w) = \alpha_1 \mathbf{I} + \sum_{j=1}^{n} \alpha_2^j a_j^+ + \sum_{j=1}^{n} \alpha_3^j a_j + \frac{1}{2}\sum_{j,k=1}^{n} \alpha_4^{j,k} a_j^+ a_k^+ \\ + \frac{1}{2}\sum_{j,k=1}^{n} \alpha_5^{j,k}\left(a_j^+ a_k + a_k a_j^+\right) + \frac{1}{2}\sum_{j,k=1}^{n} \alpha_6^{j,k} a_j a_k \end{aligned} \tag{134}$$

where $n \geq 1$,

$$\alpha_1 = \frac{w_0}{2}\left(\mu^2 - \frac{n(n+1)}{2}\right); \quad \alpha_2^j = \alpha_3^j = \mu\, w_j \sqrt{j}$$

$$\alpha_4^{j,k} = \alpha_6^{j,k} = \chi_{n]}(j+k) w_{j+k}\sqrt{jk}$$

$$\alpha_5^{j,k} = \chi_{n]}(j+k) w_{|k-j|}\sqrt{jk}$$

where

$$\chi_{n]}(a) = \begin{cases} 1 & \text{if } a \leq n \\ 0 & \text{otherwise} \end{cases}$$

and $w_m \in \mathbb{R}$ so that the coefficient $\alpha_5^{j,k}$ is common for $a_j^+ a_k$ and $a_k a_j^+$.

The quadratic part of $Y_n(w)$ is

$$Q_n(w) = \frac{1}{2}\sum_{j,k=1}^{n} \alpha_4^{j,k} a_j^+ a_k^+ + \frac{1}{2}\sum_{j,k=1}^{n} \alpha_5^{j,k}\left(a_j^+ a_k + a_k a_j^+\right) + \frac{1}{2}\sum_{j,k=1}^{n} \alpha_6^{j,k} a_j a_k$$

Using

$$[a_k, a_j^+] = \delta_{k,j}$$

we see that the homogeneous quadratic part of $Y_n(w)$ is

$$X_n(w) = \frac{1}{2}\sum_{j,k=1}^{n} \alpha_4^{j,k} a_j^+ a_k^+ + \sum_{j,k=1}^{n} \alpha_5^{j,k} a_j^+ a_k + \frac{1}{2}\sum_{j,k=1}^{n} \alpha_6^{j,k} a_j a_k ,$$

so

$$A = (A_{j,k}) = \left(\frac{1}{2}\alpha_4^{j,k}\right) , \quad C = (C_{j,k}) = \left(\alpha_5^{j,k}\right)$$

Since both A, C are real, the condition on the diagonalizability of the pair (A, C) reduces to the commutativity of A and C, i.e. to $[A, C] = 0$ which is equivalent to $[\alpha_4, \alpha_5] = 0$. In general, the matrices α_4, α_5 do not commute. For example, for $n = 4$

$$\alpha_4 = \begin{pmatrix} w_2 & \sqrt{2}w_3 & \sqrt{3}w_4 & 0 \\ \sqrt{2}w_3 & 2w_4 & 0 & 0 \\ \sqrt{3}w_4 & 0 & 0 & 0 \\ 0 & 0 & 0 & 0 \end{pmatrix} , \quad \alpha_5 = \begin{pmatrix} w_0 & \sqrt{2}w_1 & \sqrt{3}w_2 & 0 \\ \sqrt{2}w_1 & 2w_0 & 0 & 0 \\ \sqrt{3}w_2 & 0 & 0 & 0 \\ 0 & 0 & 0 & 0 \end{pmatrix}$$

and $[\alpha_4, \alpha_5]$ is

$$\begin{pmatrix} 0 & -\sqrt{2}(2w_1w_4 - w_1w_2 - w_0w_3) & -\sqrt{3}(w_0w_4 - w_2^2) & 0 \\ \sqrt{2}(2w_1w_4 - w_1w_2 - w_0w_3) & 0 & -\sqrt{6}(w_1w_4 - w_2w_3) & 0 \\ \sqrt{3}(w_0w_4 - w_2^2) & \sqrt{6}(w_1w_4 - w_2w_3) & 0 & 0 \\ 0 & 0 & 0 & 0 \end{pmatrix}$$

However they do commute, in fact they are equal, when w_m is constant for all m. **Non-diagonalizable Does Not Imply Non-factorizable.** We consider the quadratic field (128), i.e.

$$F_{A,C} = \sum_{i,j=1}^{2} \left(A_{ij}a_i^+a_j^+ + \bar{A}_{ij}a_ia_j + C_{ij}a_i^+a_j\right)$$

where

$$A = (A_{ij}) = \begin{pmatrix} a & b \\ b & c \end{pmatrix} , \quad C = (C_{ij}) = \begin{pmatrix} k & l \\ \bar{l} & m \end{pmatrix} , \quad a, b, c, l \in \mathbb{C} , \quad k, m \in \mathbb{R}$$

The diagonalizability condition $CA = A\overline{C}$ is equivalent to the equation

$$kb + lc = a\bar{l} + bm \tag{135}$$

Choosing $a = c = 1$, $b = 0$, $k = m = 0$, and $l = i$ where $i^2 = -1$, we have the matrices

$$A = I = \begin{pmatrix} 1 & 0 \\ 0 & 1 \end{pmatrix}, \quad C = iS = i\begin{pmatrix} 0 & 1 \\ -1 & 0 \end{pmatrix}$$

corresponding to the non-diagonalizable quadratic field

$$Z = \left(a_1^+\right)^2 + \left(a_2^+\right)^2 + a_1^2 + a_2^2 + i\left(a_1^+ a_2 - a_2^+ a_1\right) \tag{136}$$

On the real plane, S is the matrix of the transformation describing symmetry with respect to the line $y = x$, followed by reflection across the x-axis.

Theorem 14 *For real t sufficiently close to zero, the vacuum characteristic function of Z is*

$$\langle \Phi, e^{itZ} \Phi \rangle = \text{sech } 2\text{t}$$

Proof In the notation of Theorem 2 of [2], with the identification $a_i = D_i$, $a_i^+ = x_i$, we have:

$$\alpha_1 = \alpha_2 = \alpha_3 = 0, \alpha_4 = 2A = 2I, \alpha_6 = 2\bar{A} = 2I, \alpha_5 = C = iS$$

Therefore

$$v = \begin{pmatrix} \alpha_5 & \alpha_4 \\ -\alpha_6 & -\alpha_5^T \end{pmatrix} = \begin{pmatrix} C & 2A \\ -2\bar{A} & -C^T \end{pmatrix} = \left(\begin{array}{cc|cc} 0 & i & 2 & 0 \\ -i & 0 & 0 & 2 \\ \hline -2 & 0 & 0 & i \\ 0 & -2 & -i & 0 \end{array}\right)$$

and

$$e^{tv} = \begin{pmatrix} P(t) & Q(t) \\ -R(t) & S(t) \end{pmatrix} = \left(\begin{array}{cc|cc} \cosh t \cos 2t & i \cos 2t \sinh t & \cosh t \sin 2t & i \sin 2t \sinh t \\ -i \cos 2t \sinh t & \cos 2t \cosh t & -i \sin 2t \sinh t & \cosh t \sin 2t \\ \hline -\cosh t \sin 2t & -i \sin 2t \sinh t & \cos 2t \cosh t & i \cos 2t \sinh t \\ i \sin 2t \sinh t & -\cosh t \sin 2t & -i \cos 2t \sinh t & \cos 2t \cosh t \end{array}\right)$$

Thus,

$$A_1(t) = A_2(t) = A_3(t) = 0$$

and

$$A_4(t) = Q(t)S(t)^{-1}, \; A_5(t) = -\log\left(S(t)^T\right), \; A_6(t) = S(t)^{-1}R(t)$$

Thus, by Theorem 2 of [2] and the matrix identity

$$e^{\text{Tr}(\log \text{A})} = \det A$$

for non-singular matrices A, we have

$$\langle \Phi, e^{itZ}\Phi\rangle = e^{\frac{1}{2}\sum_{j=1}^{2} A_5^{j,j}(it)} = e^{\frac{1}{2}\mathrm{TrA_5(it)}}$$
$$= \left(e^{\mathrm{Tr}\log(\mathrm{S(it)^T})}\right)^{-1/2} = (\det S(it))^{-1/2} = \left(\cosh^2 2t\right)^{-1/2} = \text{sech 2t} \qquad \square$$

Corollary 2 *The vacuum characteristic function of Z is factorizable.*

Proof Let

$$Z_j = \left(a_j^+\right)^2 + a_j^2, \quad j = 1, 2$$

Then, Z_j is diagonalizable for $j = 1, 2$ and, as in the proof of Theorem 14,

$$v_1 = \left(\begin{array}{cc|cc} 0 & 0 & 2 & 0 \\ 0 & 0 & 0 & 0 \\ \hline -2 & 0 & 0 & 0 \\ 0 & 0 & 0 & 0 \end{array}\right), \quad e^{tv_1} = \begin{pmatrix} P_1(t) & Q_1(t) \\ -R_1(t) & S_1(t) \end{pmatrix} = \left(\begin{array}{cc|cc} \cos 2t & 0 & \sin 2t & 0 \\ 0 & 1 & 0 & 0 \\ \hline -\sin 2t & 0 & \cos 2t & 0 \\ 0 & 0 & 0 & 1 \end{array}\right)$$

and

$$v_2 = \left(\begin{array}{cc|cc} 0 & 0 & 0 & 0 \\ 0 & 0 & 0 & 2 \\ \hline 0 & 0 & 0 & 0 \\ 0 & -2 & 0 & 0 \end{array}\right), \quad e^{tv_2} = \begin{pmatrix} P_2(t) & Q_2(t) \\ -R_2(t) & S_2(t) \end{pmatrix} = \left(\begin{array}{cc|cc} 1 & 0 & 0 & 0 \\ 0 & \cos 2t & 0 & \sin 2t \\ \hline 0 & 0 & 1 & 0 \\ 0 & -\sin 2t & 0 & \cos 2t \end{array}\right)$$

Thus, for $j = 1, 2$,

$$\langle \Phi, e^{itZ_j}\Phi\rangle = \left(\det S_j(it)\right)^{-1/2} = (\cosh 2t)^{-1/2} = (\text{sech 2t})^{1/2}$$

Therefore

$$\langle \Phi, e^{itZ}\Phi\rangle = \langle \Phi, e^{itZ_1}\Phi\rangle\langle \Phi, e^{itZ_2}\Phi\rangle$$

so Z is factorizable. $\square$

Recall that

$$Z = \left(a_1^+\right)^2 + \left(a_2^+\right)^2 + a_1^2 + a_2^2 + i\left(a_1^+ a_2 - a_2^+ a_1\right) \tag{137}$$

Lemma 10

$$[\left(a_1^+\right)^2 + \left(a_2^+\right)^2 + a_1^2 + a_2^2, i\left(a_1^+ a_2 - a_2^+ a_1\right)] = 0 \tag{138}$$

Proof

$$[a_1, a_1^+ a_2 - a_2^+ a_1] = a_2; \qquad [a_2, a_1^+ a_2 - a_2^+ a_1] = -a_2$$

$$[a_1^2, a_1^+ a_2 - a_2^+ a_1] = a_1[a_1, a_1^+ a_2 - a_2^+ a_1] + [a_1, a_1^+ a_2 - a_2^+ a_1]a_1$$

$$= a_1 a_2 + a_2 a_1 = 2a_1 a_2$$

$$[a_2^2, a_1^+ a_2 - a_2^+ a_1] = a_2[a_2, a_1^+ a_2 - a_2^+ a_1] + [a_2, a_1^+ a_2 - a_2^+ a_1]a_2$$
$$= -a_2 a_1 - a_1 a_2 = -2a_1 a_2$$

Summing the last two identities, one has

$$[a_1^2 + a_2^2, a_1^+ a_2 - a_2^+ a_1] = 2a_1 a_2 - 2a_1 a_2 = 0$$

Consequently

$$0 = [a_1^2 + a_2^2, a_1^+ a_2 - a_2^+ a_1]^* = [a_1 a_2^+ - a_2 a_1^+, (a_1^+)^2 + (a_2^+)^2]$$
$$= -[a_1^+ a_2 - a_1 a_2^+, (a_1^+)^2 + (a_2^+)^2] = [(a_1^+)^2 + (a_2^+)^2, a_1^+ a_2 - a_1 a_2^+]$$

Summing the last two identities, one finds

$$[a_1^2 + a_2^2 + (a_1^+)^2 + (a_2^+)^2, a_1^+ a_2 - a_2^+ a_1] = 0$$

which is equivalent to (138). □

Corollary 3 *For t sufficiently small, one has:*

$$e^{itZ} = e^{it\left((a_1^+)^2 + (a_2^+)^2 + a_1^2 + a_2^2\right)} e^{-t(a_1^+ a_2 - a_2^+ a_1)} \tag{139}$$

i.e. the vacuum characteristic function of Z is factorizable.

Proof Equation (139) follows from (138). □

Remark Corollary 3 means that the decomposition

$$Z = \left((a_1^+)^2 + (a_2^+)^2 + a_1^2 + a_2^2\right) + i\left(a_1^+ a_2 - a_2^+ a_1\right)$$

expresses Z as a sum of two vacuum independent fields, identified to classical random variables.

Remark The result of Corollary 3 does not contradicts Corollary 2 for the following reasons: due to (139)

$$e^{itZ} = e^{it\left((a_1^+)^2 + (a_2^+)^2 + a_1^2 + a_2^2\right)} e^{-t(a_1^+ a_2 - a_2^+ a_1)} \tag{140}$$
$$= e^{it((a_1^+)^2 + a_1^2)} e^{it((a_2^+)^2 + a_2^2)} e^{-t(a_1^+ a_2 - a_2^+ a_1)}$$

Now notice that

$$e^{-t(a_1^+ a_2 - a_2^+ a_1)}\Phi = \Phi$$

Therefore

$$\langle \Phi, e^{-t(a_1^+ a_2 - a_2^+ a_1)} \Phi \rangle = 1$$

This means that the vacuum distribution of the quantum random variable $i(a_1^+ a_2 - a_2^+ a_1)$ is the δ–measure at 0.

Theorem 15 *Let*

$$F_{A,C} = X_1 + X_2 + X_3$$

where

$$\begin{aligned} X_1 &= a\,(a_1^+)^2 + \bar{a}\,a_1^2 + k\,a_1^+ a_1 \\ X_2 &= c\,(a_2^+)^2 + \bar{c}\,a_2^2 + m\,a_2^+ a_2 \\ X_3 &= 2\,b\,a_1^+\,a_2^+ + 2\bar{b}\,a_1 a_2 + l\,a_1^+\,a_2 + \bar{l}\,a_2^+ a_1 \end{aligned}$$

where $a, b, c, l \in \mathbb{C}\ ,\ k, m \in \mathbb{R}$, *i.e.,* $F_{A,C}$ *is of the form*

$$F_{A,C} = \sum_{i,j=1}^{2} \left(A_{ij} a_i^+ a_j^+ + \bar{A}_{ij} a_i a_j + C_{ij} a_i^+ a_j \right)$$

where

$$A = (A_{ij}) = \begin{pmatrix} a & b \\ b & c \end{pmatrix}\ ,\ C = (C_{ij}) = \begin{pmatrix} k & l \\ \bar{l} & m \end{pmatrix}$$

Then,

$$[X_1, X_2] = 0[X_1, X_3] = 2(kb - a\bar{l})a_1^+ a_2^+ + 2(\bar{a}l - k\bar{b})a_1 a_2 \tag{141}$$

$$+ (kl - 4a\bar{b})a_1^+ a_2 + (4\bar{a}b - k\bar{l})a_2^+ a_1 \tag{142}$$

$$[X_2, X_3] = 2(mb - cl)a_1^+ a_2^+ + 2(\bar{c}\bar{l} - m\bar{b})a_1 a_2 + (4b\bar{c} - ml)a_1^+ a_2 + (m\bar{l} - 4\bar{b}c)a_2^+ a_1 \tag{143}$$

Proof Equation (141) follows from the commutativity of everything that has index 1 with everything that has index 2. Using commutation relations (14) to put terms in **normal order**, i.e., a^+'s on the left and a's on the right, we obtain the commutation relations:

$$\begin{aligned} [(a_1^+)^2, a_1^+\,a_2^+] = [(a_1^+)^2, a_1^+\,a_2] = [a_1^2, a_1 a_2] = [a_1^2, a_2^+ a_1] = 0 \\ [(a_1^+)^2, a_1 a_2] = -2a_1^+ a_2\ ,\ [(a_1^+)^2, a_2^+ a_1] = -2a_1^+ a_2^+ \\ [a_1^2, a_1^+ a_2^+] = 2a_2^+ a_1\ ,\ [a_1^2, a_1^+ a_2] = 2a_1 a_2 \\ [a_1^+ a_1, a_1^+ a_2^+] = a_1^+ a_2^+\ ,\ [a_1^+ a_1, a_1 a_2] \end{aligned}$$

$$= -a_1 a_2 [a_1^+ a_1, a_1^+ a_2] = a_1^+ a_2\ ,\ [a_1^+ a_1, a_2^+ a_1] = -a_2^+ a_1$$

and

$$[(a_2^+)^2, a_1^+a_2^+] = [(a_2^+)^2, a_2^+a_1] = [a_2^2, a_1a_2] = [a_2^2, a_1^+a_2] = 0$$

$$[(a_2^+)^2, a_1a_2] = -2a_2^+a_1 \, , [(a_2^+)^2, a_1^+a_2] = -2a_1^+a_2^+[a_2^2, a_1^+ a_2^+] = 2a_1^+a_2 \, , [a_2^2, a_2^+ a_1]$$

$$= 2a_1a_2[a_2^+a_2, a_1^+a_2^+] = a_1^+a_2^+ \, , [a_2^+a_2, a_1a_2]$$

$$= -a_1a_2[a_2^+a_2, a_1^+a_2] = -a_1^+a_2 \, , [a_2^+a_2, a_2^+a_1] = a_2^+a_1$$

Using the above commutation relations and the bilinearity of the Lie bracket, we obtain (142) and (143). □

Corollary 4 *In the notation of Theorem 15,* $[X_1 + X_2, X_3] = 0$ *if and only if*

$$(kb - a\bar{l}) + (mb - cl) = 0 \tag{144}$$

$$(kl - 4a\bar{b}) + (4b\bar{c} - ml) = 0 \tag{145}$$

In particular, if $[X_1 + X_2, X_3] = 0$ *then* $F_{A,C}$ *is factorizable.*

Proof Using (142), (143), and the linear independence of $a_1^+a_2^+, a_1a_2, a_1^+a_2$ and $a_1a_2^+$, we find that $[X_1 + X_2, X_3] = 0$ if and only if

$$2(kb - a\bar{l}) + 2(mb - cl) = 0 \, ,$$

$$2(\bar{a}l - k\bar{b}) + 2(\bar{c}\bar{l} - m\bar{b}) = 0,$$

$$(kl - 4a\bar{b}) + (4b\bar{c} - ml) = 0, \quad (4\bar{a}b - k\bar{l}) + (m\bar{l} - 4\bar{b}c) = 0$$

from which, since the second and fourth equations are the conjugates of the first and third equation respectively, we obtain (144) and (145). Since $[X_1 + X_2, X_3] = 0$ and $[X_1, X_2] = 0$,

$$\langle \Phi, e^{i\,s\,F_{A,C}}\,\Phi\rangle = \langle \Phi, e^{i\,s\,(X_1+X_2)}e^{i\,s\,X_3}\,\Phi\rangle = \langle \Phi, e^{i\,s\,X_1}e^{i\,s\,X_2}e^{i\,s\,X_3}\,\Phi\rangle$$

$$= \langle \Phi, e^{i\,s\,X_1}e^{i\,s\,X_2}\,\Phi\rangle = \langle \Phi, e^{i\,s\,X_1}\,\Phi\rangle\langle \Phi, e^{i\,s\,X_2}\,\Phi\rangle$$

as in the proof of Theorem 12, so $F_{A,C}$ is factorizable. □

Theorem 16 *Let*

$$X = Y_1 + Y_2 + Y_3$$

where

$$Y_1 = a\,(a_1^+)^2 + \bar{a}\,a_1^2$$
$$Y_2 = c\,(a_2^+)^2 + \bar{c}\,a_2^2$$
$$Y_3 = k\,a_1^+ a_1 + m\,a_2^+ a_2 + 2\,b\,a_1^+\,a_2^+ + 2\bar{b}\,a_1 a_2 + l\,a_1^+\,a_2 + \bar{l}\,a_2^+ a_1$$

where $a, b, c, l \in \mathbb{C}$, $k, m \in \mathbb{R}$, *i.e.,* $F_{A,C}$ *is of the form*

$$F_{A,C} = \sum_{i,j=1}^{2} \left(A_{ij} a_i^+ a_j^+ + \bar{A}_{ij} a_i a_j + C_{ij} a_i^+ a_j \right)$$

where

$$A = (A_{ij}) = \begin{pmatrix} a & b \\ b & c \end{pmatrix} , \quad C = (C_{ij}) = \begin{pmatrix} k & l \\ \bar{l} & m \end{pmatrix}$$

Then,

$$[Y_1, Y_2] = 0 \tag{146}$$

$$\begin{aligned} [Y_1, Y_3] &= -2ak\,(a_1^+)^2 - 4a\bar{b}\,a_1^+\,a_2 - 2a\bar{l}\,a_1^+\,a_2^+ \\ &\quad + 2\bar{a}k\,a_1^2 + 4\bar{a}b\,a_2^+\,a_1 + 2\bar{a}l\,a_1 a_2 \end{aligned} \tag{147}$$

$$\begin{aligned} [Y_2, Y_3] &= -2cm\,(a_2^+)^2 - 4c\bar{b}\,a_2^+\,a_1 - 2c\bar{l}\,a_1^+\,a_2^+ \\ &\quad + 2\bar{c}m\,a_2^2 + 4\bar{c}b\,a_1^+\,a_2 + 2\bar{c}\bar{l}\,a_1 a_2 \end{aligned} \tag{148}$$

Proof The proof is similar to that of theorem (15), using the additional commutation relations:

$$[a_2^2, a_1^+\,a_1] = [(a_2^+)^2, a_1^+\,a_1] = 0$$
$$[(a_1^+)^2, a_1^+ a_1] = -2a_1^+)^2 \ , [(a_2^+)^2, a_2^+ a_2] = -2(a_2^+)^2$$
$$[a_1^2, a_1^+\,a_1] = 2a_1^2 \ , [a_2^2, a_2^+\,a_2] = 2a_2^2$$

□

Corollary 5 *In the notation of Theorem 16,* $[Y_1 + Y_2, Y_3] = 0$ *if and only if*

$$ak = cm = (a + c)l = \bar{c}b - a\bar{b} = 0\,. \tag{149}$$

In particular, if $[X_1 + X_2, X_3] = 0$ *then* $F_{A,C}$ *is factorizable.*

Proof Using (147), (148), and the linear independence of $(a_1^+)^2$, $(a_2^+)^2, a_1^2, a_2^2$, $a_1^+ a_2^+$, $a_1 a_2$, $a_1^+ a_2$ and $a_2^+ a_1$, we find that $[Y_1 + Y_2, Y_3] = 0$ if and only if (149) is true. The factorizability of $F_{A,C}$ follows as in the Proof of Corollary 4. □

References

1. Accardi, L., Boukas, A.: On the characteristic function of random variables associated with Boson Lie algebras. Commun. Stoch. Anal. **4**(4), 493–504 (2010)
2. Accardi, L., Boukas, A.: Fourier transform of random variables associated with the multi-dimensional Heisenberg Lie algebra. Proc. Am. Math. Soc. **143**, 4095–4101 (2015)
3. Accardi, L., Boukas, A.: Canonical forms of quadratic Boson fields: Proceedings of the Conference QBIC 2014, dedicated to Prof. Ohya, M., Accardi, L., Freudenberg, W., Watanabe, N., (eds.) (2020). http://www.worldscientific.com/worldscibooks/
4. Accardi, L., Boukas, A., Lu, Y.G.: The n-dimensional quadratic Heisenberg algebra as a "non–commutative" sl(2, $\mathbb{C}$). To appear in: J. Stoch. Anal. (JOSA) (2021)
5. Accardi, L., Boukas, A., Lu, Y.G., Teretenkov, A.E.: In preparation
6. Accardi, L., Aref'eva, I.Y., Lu, Y.G., Volovich, I.V.: Two non–$*$–isomorphic $*$–Lie algebra structures on $sl(2, \mathbb{R})$ and their physical origins, Submitted to IDAQP
7. Accardi, L., Barhoumi, A., Dhahri, A.: Identification of the theory of orthogonal polynomials in d-indeterminates with the theory of 3-diagonal symmetric interacting Fock spaces on $\mathbb{C}R^d$. Infin. Dimens. Anal. Quantum. Probab. Relat. Top. **20**(1), 1750004–1–55 (2017)
8. Accardi, L., Ouerdiane, H., Rebei, H.: On the quadratic Heisenberg group. Infin. Dimens. Anal. Quantum Probab. Relat. Top. **13**(04), 551–587 (2010)
9. Accardi, L., Franz, U., Skeide, M.: Squared white noise and other non-Gaussian Noises as Lévy processes on real lie algebras. Commun. Math. Phys. **228**, 123–150 (2002) Preprint Volterra, N. 423 (2000)
10. Accardi, L., Barhoumi, A., Rhaima, M.: Jacobi sequences of powers of random variables. Infin. Dimens. Anal. Quantum Probab. Relat. Top. (IDA-QP) **21**(1), 1850001-1-22 (2018)
11. Berezin, F.A.: The Method of Second Quantization. Pure and Applied Physics, vol. 24, 1st edn. Academic, New York (1966)
12. Chebotarev, A.M., Tlyachev, T.V.: Normal forms, inner products, and Maslov indices of general multimode squeezing. Math. Notes **95**(5–6), 721–37 (2014)
13. Folland, G.B.: Harmonic Analysis in Phase Space, vol. 122. Princeton University Press, Princeton (1989)
14. de Gosson, M.A.: Symplectic Methods in Harmonic Analysis and in Mathematical Physics. Springer, Berlin (2011)
15. Hall, B.: Lie Groups, Lie Algebras, and Representations: An Elementary Introduction, 1st ed. 2003. Corr. 2nd printing 2004 edition. Springer, Berlin (2004)
16. Hatcher, A.: Algebraic Topology, 1st edn. Cambridge University Press, Cambridge (2001)
17. Horn, R.A., Johnson, C.R.: Matrix Analysis, 2nd edn. Cambridge University Press, Cambridge (2013)
18. Ikramov, K.D.: Hamiltonian square roots of skew-Hamiltonian matrices revisited. Linear Algebra Appl. **325**(1–3), 101–107 (2001)
19. Isaev, A.P., Rubakov, V.A.: Theory of Groups and Symmetries: Finite Groups, Lie Groups, and Lie Algebras. World Scientific, Singapore (2018)
20. Lion, G., Vergne, M.: The Weil Representation, Maslov Index and Theta Series. Progress in Mathematics. Birkauser, Boston (1980)
21. Eugene Lukacs Characteristic functions Griffin (1970)
22. Perelomov, A.: Generalized Coherent States and Their Applications, p. 103. Springer Science & Business Media (2012)
23. Rawnsley, J.: On the universal covering group of the real symplectic group. J. Geom. Phys. **62**(10), 2044–2058 (2012)
24. Rebei, H., Rguigui, H., Anis, R., Alhussain, Z.A.: Identification of the one-mode quadratic Heisenberg group with the projective group $PSU(1, 1)$ and holomorphic representation. Infin. Dimens. Anal. Quantum Probab. Relat. Top. **23**(04), 2050023 (2020)
25. Teretenkov, A.E.: Irreversible quantum evolution with quadratic generator. Infin. Dimens. Anal. Quantum Probab. Relat. Top. (IDAQP) **22**(4), 1930001 (2019)
26. Wolf, K.B.: Geometric Optics on Phase Space. Theoretical and Mathematical Physics. Springer, Berlin (2004)

A Pedagogical Note on the Computation of Relative Entropy of Two n-Mode Gaussian States

K. R. Parthasarathy

In memory of Robin L Hudson.

Abstract A formula for the relative entropy $S(\rho||\sigma) = \text{Tr}\,\rho\,(\log\,\rho - \log\,\sigma)$ of two gaussian states ρ, σ in the boson Fock space $\Gamma(\mathbb{C}^n)$ is presented. It is shown that the relative entropy has a classical and a quantum part: The classical part consists of a weighted linear combination of relative Shannon entropies of n pairs of Bernouli trials arising from the thermal state composition of the gaussian states ρ and σ. The quantum part has a sum of n terms, that are functions of the annihilation means and the covariance matrices of 1-mode marginals of the gaussian state ρ', which is equivalent to ρ under a disentangling unitary gaussian symmetry operation of the state σ. A generalized formula for the Petz-Rényi relative entropy $S_\alpha(\rho||\sigma) = -\frac{1}{\alpha-1}\log \text{Tr}\,\rho^\alpha\sigma^{1-\alpha},\ 0 < \alpha < 1$ for gaussian states ρ, σ is also presented. Furthermore it is shown that $S_\alpha(\rho||\sigma)$ converges to the limit $S(\rho||\sigma)$ as α increases to 1.

Keywords Boson Fock space · Gaussian state · Weyl operator · Symplectic group · Relative entropy

1 Introduction

Every gaussian state in $\Gamma(\mathbb{C}^n)$ is completely determined by its annihilation mean and covariance matrix. A finer analysis of the covariance matrix reveals that such a guassian state is composed of a product of n thermal states (including the vacuum

K. R. Parthasarathy (✉)
Indian Statistical Institute, Theoretical Statistics and Mathematics Unit,
Delhi Centre, 7 S. J. S. Sansanwal Marg, New Delhi 110 016, India
e-mail: krp@isid.ac.in

L. Accardi et al. (eds.), *Infinite Dimensional Analysis, Quantum Probability and Applications*, Springer Proceedings in Mathematics & Statistics 390,
https://doi.org/10.1007/978-3-031-06170-7_2

state) and an entangling unitary gaussian symmetry operator arising from the group generated by phase translations and symplectic transformations. Every thermal state, when viewed in a product basis, gives rise to a natural Bernoulli (binomial) trial with success and failure probabilities corresponding to detection or no-detection of a particle.

In this paper we present a formula for the relative entropy $S(\rho||\sigma) = \mathrm{Tr}\rho(\log\rho - \log\sigma)$ of two gaussian states $\rho,\ \sigma$ in $\Gamma(\mathbb{C}^n)$. It is shown that the relative entropy $S(\rho||\sigma)$ contains two distinguished components: the first part is a weighted linear combination of classical Shannon relative entropies $H(p_{1j} : p_{2j}),\ j = 1, 2, \ldots, n$ of n pairs of Bernouli trials arising from the thermal components of the gaussian states ρ and σ. The second part is purely quantum and it contains a sum of n terms, each of which is arising from the annihilation mean $\alpha_k \in \mathbb{C}$ and the 2×2 covariance matrices T_k of 1-mode marginal states $\rho(\alpha_k, T_k)$ of a gaussian state $\rho' = U^\dagger \rho U$ where $U = W(\boldsymbol{\ell})\,\Gamma(L)$ corresponds to the disentangling gaussian symmetry of the state σ.

Extending the above concepts we derive an explicit formula for the Petz-Rényi relative entropy $S_\alpha(\rho||\sigma) = -\frac{1}{\alpha-1}\log \mathrm{Tr}\,\rho^\alpha\sigma^{1-\alpha},\ 0 < \alpha < 1$ for two gaussian states $\rho,\ \sigma$ in $\Gamma(\mathbb{C}^n)$. We show that in the limit $\alpha \to 1$ the Petz-Rényi relative entropy $S_\alpha(\rho||\sigma)$ reduces to the relative entropy $S(\rho||\sigma)$ as expected.

2 Preliminary Concepts on Gaussian States

We focus our attention on gaussian states in the complex Hilbert space $L^2(\mathbb{R}^n)$ of a quantum system with n degrees of freedom, which constitute a natural extension of the concept of normal (gaussian) distributions in classical probability [1–5].

We shall use the Dirac notation $\langle u|v\rangle$ for the scalar product of any two elements u and v of a complex Hilbert space. The scalar product $\langle u|v\rangle$ is assumed to be linear in v and conjugate linear in u. For any operator A and elements $u,\ v$ in the Hilbert space we write $\langle u|A|v\rangle = \langle A^\dagger u|v\rangle$ whenever v is in the domain of A and u is in the domain of $A^\dagger$, provided they are well-defined.

Consider the Hilbert space $L^2(\mathbb{R}^n)$, or equivalently, the boson Fock space $\mathcal{H} = \Gamma(\mathbb{C}^n)$, over the complex Hilbert space $\mathbb{C}^n$ of finite dimension n. Fix a canonical orthonormal basis $\{e_k,\ 1 \leq k \leq n\}$ in $\mathbb{C}^n$ with

$$e_k = (0, 0, \ldots, 0, 1, 0, \ldots, 0)^T$$

where '1' appears in the kth position.

At every element $\mathbf{u} \in \mathbb{C}^n$ we associate a pair of operators $a(\mathbf{u}),\ a^\dagger(\mathbf{u})$, called annihilation, creation operators, respectively in the boson Fock space $\mathcal{H}$ and

$$W(\mathbf{u}) = e^{a^\dagger(\mathbf{u}) - a(\mathbf{u})} \tag{1}$$

denotes the Weyl displacement operator or simply Weyl operator in $\mathcal{H}$. For any state ρ in $\mathcal{H}$ the complex-valued function on $\mathbb{C}^n$ given by

$$\hat{\rho}(\mathbf{u}) = \operatorname{Tr} W(\mathbf{u})\, \rho, \quad \mathbf{u} \in \mathbb{C}^n \tag{2}$$

is the quantum characteristic function of ρ at $\mathbf{u}$. The state ρ is said to have finite second moments if

$$\langle a^\dagger(\mathbf{u})\, a(\mathbf{u})\rangle_\rho = \operatorname{Tr} a^\dagger(\mathbf{u})\, a(\mathbf{u})\, \rho < \infty, \quad \forall\, \mathbf{u} \in \mathbb{C}^n. \tag{3}$$

The annihilation mean, or simply the mean $\mathbf{m} \in \mathbb{C}^n$ in the state $\rho \in \mathcal{H}$ is defined through

$$\langle a(\mathbf{u})\rangle_\rho = \operatorname{Tr} a(\mathbf{u})\, \rho = \langle \mathbf{u}|\mathbf{m}\rangle, \quad \mathbf{u} \in \mathbb{C}^n. \tag{4}$$

Then

$$\langle a^\dagger(\mathbf{u})\rangle_\rho = \langle \mathbf{m}|\mathbf{u}\rangle. \tag{5}$$

Define the observables

$$q(\mathbf{u}) = \frac{a(\mathbf{u}) + a^\dagger(\mathbf{u})}{\sqrt{2}} \tag{6}$$

$$p(\mathbf{u}) = \frac{a(\mathbf{u}) - a^\dagger(\mathbf{u})}{i\,\sqrt{2}}. \tag{7}$$

Then the quantum characteristic function of ρ at $\mathbf{u}$ can be expressed as

$$\hat{\rho}(\mathbf{u}) = \operatorname{Tr} e^{-i\,\sqrt{2}\, p(\mathbf{u})}\, \rho. \tag{8}$$

Let $\mathbf{u} = \mathbf{x} + i\,\mathbf{y},\ \mathbf{x},\ \mathbf{y} \in \mathbb{R}^n$. Then $\mathbf{x} \to q(\mathbf{x}), \mathbf{y} \to p(\mathbf{y})$ are position and momentum fields obeying the commutation relations

$$\left[q(\mathbf{x}),\ p(\mathbf{y})\right] = i\,\mathbf{x}^T\mathbf{y}. \tag{9}$$

Then

$$q_k = q(e_k),\ p_k = p(e_k),\ \ k = 1, 2, \ldots, n$$

yield the canonical commutation relations (CCR) [1]

$$\left[q_j,\ q_k\right] = 0,\ \left[p_j,\ p_k\right] = 0,\ \left[q_j,\ p_k\right] = i\,\delta_{jk}. \tag{10}$$

Variance of $p(\mathbf{u})$ in the state ρ yields a quadratic form in $(x_1, x_2, \ldots, x_n;\, y_1, y_2, \ldots, y_n)^T \in \mathbb{R}^{2n}$ so that

$$\mathrm{Var}_\rho\, p(\mathbf{x}+i\,\mathbf{y}) = (\mathbf{x},\mathbf{y})^T\, C \begin{pmatrix}\mathbf{x}\\ \mathbf{y}\end{pmatrix}. \tag{11}$$

Here C is a real $2n \times 2n$ positive definite matrix, satisfying the matrix inequality

$$C + \frac{i}{2}\, J \geq 0 \tag{12}$$

with

$$J = \begin{pmatrix} 0_n & I_n \\ -I_n & 0_n \end{pmatrix} \tag{13}$$

being the canonical symplectic matrix in the real symplectic matrix group of order $2n$:

$$\mathrm{Sp}(2n,\mathbb{R}) = \left\{L \mid L^T\, J\, L = J\right\}.$$

In (13) the right hand side is expressed in the block notation, with 0_n, I_n denoting $n \times n$ null and identity matrices respectively. The correspondence

$$t \mapsto W(t\,\mathbf{u}) = e^{-i\,t\,\sqrt{2}\,p(\mathbf{u})} \tag{14}$$

is a strongly continuous one-parameter group of unitary operators in $t \in \mathbb{R}$, and $\hat{\rho}(t,\mathbf{u}) = \mathrm{Tr}\, W(t\,\mathbf{u})\,\rho$ is the characteristic function of the observable $p(\mathbf{u})$, which has the normal or gaussian distribution with mean value

$$\frac{\langle \mathbf{m}|\mathbf{u}\rangle - \langle \mathbf{u}|\mathbf{m}\rangle}{i\,\sqrt{2}} \tag{15}$$

and variance

$$(\mathbf{x}^T,\mathbf{y}^T)\; C \begin{pmatrix}\mathbf{x}\\ \mathbf{y}\end{pmatrix}, \quad \mathbf{x}+i\,\mathbf{y} = \mathbf{u}. \tag{16}$$

Then

$$\widehat{\rho}\,(t\,(\mathbf{x}+i\,\mathbf{y})) = \exp\left[t\,\left(\langle \mathbf{m}\,|\mathbf{u}\rangle - \langle \mathbf{u}|\mathbf{m}\rangle\right) - t^2\,(\mathbf{x}^T,\mathbf{y}^T)\; C \begin{pmatrix}\mathbf{x}\\ \mathbf{y}\end{pmatrix}\right], \tag{17}$$

for all $t \in \mathbb{R}, \mathbf{u} \in \mathbb{C}^n$ with $\mathbf{u} = \mathbf{x} + i\,\mathbf{y}$. Thus ρ is a quantum gaussian state with mean $\mathbf{m}$ and covariance matrix C if and only if

$$\begin{aligned}\widehat{\rho}\,(\mathbf{x}+i\,\mathbf{y}) &= \exp\left[-2\,i\,\mathrm{Im}(\mathbf{x}-i\,\mathbf{y})^T\,\mathbf{m} - (\mathbf{x}^T,\mathbf{y}^T)\; C \begin{pmatrix}\mathbf{x}\\ \mathbf{y}\end{pmatrix}\right] \\ &= \exp\left[-2\,i\,\left(\mathbf{x}^T\,\mathrm{Im}\,\mathbf{m} - \mathbf{y}^T\,\mathrm{Re}\,\mathbf{m}\right) - (\mathbf{x}^T,\mathbf{y}^T)\; C \begin{pmatrix}\mathbf{x}\\ \mathbf{y}\end{pmatrix}\right] \end{aligned} \tag{18}$$

for all $\mathbf{x},\ \mathbf{y} \in \mathbb{R}^n$. We write

$$\rho(\mathbf{m}, C) = n\text{-mode gaussian state with mean } \mathbf{m} \text{ and covariance matrix } C.$$

The covariance matrix may be expressed in the $n \times n$ block matrix notation as

$$C = \begin{pmatrix} C_{11} & C_{12} \\ C_{21} & C_{22} \end{pmatrix} \tag{19}$$

where

$$\begin{aligned} C_{11} &= \mathrm{Cov}\,(p_1, p_2, \ldots\ p_n) \\ C_{22} &= \mathrm{Cov}\,(q_1, q_2, \ldots\ q_n) \\ C_{12} &= \big(-\mathrm{Cov}\,(p_j, q_k)\big) = C_{21}^T. \end{aligned} \tag{20}$$

Note that

$$W(\mathbf{z})\,\rho(\mathbf{m}, C)\,W(\mathbf{z})^\dagger = \rho(\mathbf{m} + \mathbf{z}, C), \quad \mathbf{z} \in \mathbb{C}^n. \tag{21}$$

3 Williamson's Theorem Applied to n-Mode Gaussian Covariance Matrix

Let C be a covariance matrix of an n-mode gaussian state ρ as described above. Then there exist $\nu_1 \geq \nu_2 \geq \cdots \geq \nu_n \geq 1/2$ and a sympletic matrix $L \in \mathrm{Sp}(2n, \mathbb{R})$ such that

$$C = L^T \begin{pmatrix} \mathrm{diag}\,(\nu_1, \nu_2, \ldots, \nu_n) & 0_n \\ 0_n & \mathrm{diag}\,(\nu_1, \nu_2, \ldots, \nu_n) \end{pmatrix} L \tag{22}$$

Theorem 1 *To every $L \in \mathrm{Sp}(2n, \mathbb{R})$, there exists a unitary operator $\Gamma(L)$ in the boson Fock space $\mathcal{H} = \Gamma(\mathbb{C}^n)$ satisfying*

$$\Gamma(L)\,W(\mathbf{u})\,\Gamma(L)^{-1} = W(L \circ \mathbf{u}), \ \mathbf{u} \in \mathbb{C}^n \tag{23}$$

where

$$\begin{aligned} L \circ \mathbf{u} &= \big(I_n, i\,I_n\big)\,L\begin{pmatrix} \mathbf{x} \\ \mathbf{y} \end{pmatrix} \\ &= (A_{11}\,\mathbf{x} + A_{12}\,\mathbf{y}) + i\ (A_{21}\,\mathbf{x} + A_{22}\,\mathbf{y}), \end{aligned} \tag{24}$$

with $A_{11}, A_{12}, A_{21}, A_{22}$ denoting $n \times n$ real matrices constituting L as

$$L = \begin{pmatrix} A_{11} & A_{12} \\ A_{21} & A_{22} \end{pmatrix}, \tag{25}$$

and $\mathbf{u} = \mathbf{x} + i\,\mathbf{y},\ \mathbf{x}, \mathbf{y} \in \mathbb{R}^n$. *The unitary operator* $\Gamma(L)$ *is unique up to a scalar multiple of modulus unity.*

Proof Follows as a consequence of Stone-von Neumann theorem on Weyl operator [1, 2].

4 Structure Theorem for n-Mode Gaussian States

For $0 < s \leq \infty$, we may associate a single mode gaussian thermal state $\rho(s)$ in $\Gamma(\mathbb{C})$ by

$$\rho(s) = (1 - e^{-s})\, e^{-s\, a^\dagger a} \tag{26}$$

where s denotes inverse temperature and $\rho(\infty) = |\Omega\rangle\langle\Omega|$, the vacuum state or the thermal state at zero temperature. Every thermal state $\rho(s)$ yields a natural binomial distribution with probability for success equal to e^{-s}, where success stands for the event that the number of particles detected is greater than or equal to 1. Then, failure is equivalent to no particle count with probability $1 - e^{-s}$.

The von Neumann entropy of the thermal state $\rho(s)$ is given by

$$\begin{aligned} S\left(\rho(s)\right) &= -\mathrm{Tr}\rho(s)\ \log\ \rho(s) \\ &= \frac{H(e^{-s})}{1 - e^{-s}}. \end{aligned} \tag{27}$$

Here $H(p) = -p\ \log\ p - (1-p)\ \log\,(1-p),\ \ 0 \leq p \leq 1$ is the Shannon function. Note that $\rho(s) \to \rho(\infty) = |\Omega\rangle\langle\Omega|$ as $s \to \infty$. The state $\rho(\infty)$ being pure, $S\left(\rho(\infty)\right) = 0$.

Consider any n-mode gaussian state $\rho(\mathbf{m}, C)$ with mean $\mathbf{m} \in \mathbb{C}^n$ and real symmetric positive definite $2n \times 2n$ covariance matrix C. Then, there exist $0 < s_1 \leq s_2 \leq \cdots \leq s_n \leq \infty$, and a symplectic matrix $L \in \mathrm{Sp}(2n, \mathbb{R})$ such that

$$\rho(\mathbf{m}, C) = W(\mathbf{m})\, \Gamma(L)\, \rho(s_1) \otimes \rho(s_2) \otimes \cdots \otimes \rho(s_n)\, \Gamma(L)^{-1}\, W(\mathbf{m})^{-1}. \tag{28}$$

The sequence $s_1, s_2, \ldots, s_n$ is unique. The covariance matrix C is given by

$$C = \left(L^{-1}\right)^T\ D(\mathbf{s})L^{-1},\ \ D(\mathbf{s}) = \begin{pmatrix} D_0(\mathbf{s}) & 0_n \\ 0_n & D_0(\mathbf{s}) \end{pmatrix}, \tag{29}$$

$$D_0(\mathbf{s}) = \mathrm{diag}\left[\frac{1}{2}\coth\left(\frac{s_k}{2}\right), k = 1, 2, \ldots, n\right]. \tag{30}$$

Thus the von Neumann entropy $S(\rho)$ is given by

$$S(\rho) = \sum_{k:\, s_k < \infty} \frac{H\left(e^{-s_k}\right)}{1 - e^{-s_k}}. \tag{31}$$

5 Relative Entropy $S(\rho||\sigma)$ of Two Gaussian States

To set the stage ready for the computation of relative entropy $S(\rho||\sigma) = \mathrm{Tr}\,\rho\,(\log\rho - \log\sigma)$ of two gaussian states ρ, σ we present the following lemmas.

Lemma 1 *Let* $|\psi\rangle \in \mathcal{H}$, $||\,\psi\,|| = 1$, *and* ρ *be a state in* $\mathcal{H}$ *such that* $\langle\psi|\rho\,|\psi\rangle < 1$. *Then*

$$S(\,\rho\,||\,|\psi\rangle\langle\psi|\,) := \infty. \tag{32}$$

Proof Write $|\psi_0\rangle = |\psi\rangle,\ |\psi_1\rangle,\ |\psi_2\rangle, \ldots$ such that $\{|\psi_k\rangle,\ k = 0, 1, \ldots\}$ forms an orthonormal basis (ONB) in $\mathcal{H}$. Then

$$\sum_{k=0}^{\infty}\langle\psi_k|\rho\,|\psi_k\rangle = 1. \tag{33}$$

For all $k_0 : k_0 \geq 1$ such that $\langle\psi_{k_0}|\rho\,|\psi_{k_0}\rangle > 0$, $|\psi_{k_0}\rangle$ is an eigenvector of $-\log\,|\psi_0\rangle\langle\psi_0|$ with eigenvalue ∞. Thus

$$-\mathrm{Tr}\,\rho\log\,|\psi_0\rangle\langle\psi_0| \geq \langle\psi_{k_0}|\rho\,|\psi_{k_0}\rangle\ \times\infty = \infty.$$

Then

$$S(\,\rho\,||\,|\psi_0\rangle\langle\psi_0|\,) = \mathrm{Tr}\,\rho\,\log\,\rho - \mathrm{Tr}\,\rho\,\log\,|\psi_0\rangle\langle\psi_0| = \infty. \tag{34}$$

Lemma 2 *Consider Hilbert spaces* $\mathcal{H}_k,\ k = 1, 2, \ldots, n$ *and states* σ_k *in* $\mathcal{H}_k$. *Relative entropy of any arbitrary state* ρ *with respect to the state* $\sigma = \sigma_1 \otimes \sigma_2 \otimes \cdots \otimes \sigma_n$, *in* $\mathcal{H}_1 \otimes \mathcal{H}_2 \otimes \cdots \otimes \mathcal{H}_n$ *is given by*

$$S\,(\,\rho\,||\,\sigma\,) = -S(\rho) - \sum_{k=1}^{n}\mathrm{Tr}\,\rho_k\,\log\sigma_k \tag{35}$$

where ρ_k *is the* k*-th marginal of the state* ρ *in* $\mathcal{H}_k$.

Proof Write $\widetilde{\sigma}_k = I \otimes I \otimes \cdots \otimes I \otimes \sigma_k \otimes I \otimes \cdots \otimes I$, where σ_k appears at the k-th position. Then

$$\log\,\sigma = \sum_k \log\,\widetilde{\sigma}_k \tag{36}$$

and

$$\begin{aligned}\mathrm{Tr}\,\rho\,\log\,\sigma &= \sum_k \mathrm{Tr}\,\rho\,\log\,\widetilde{\sigma}_k \\ &= \sum_k \mathrm{Tr}\,\rho_k\,\log\,\sigma_k \end{aligned} \tag{37}$$

So

$$\begin{aligned} S(\rho \,||\, \sigma) &= \operatorname{Tr} \rho \, \log \rho - \operatorname{Tr} \rho \, \log \sigma \\ &= -S(\rho) - \sum_k \operatorname{Tr} \rho_k \, \log \sigma_k . \end{aligned} \tag{38}$$

Lemma 3 *Let $\rho(m, T)$ be a 1-mode gaussian state with annihilation mean $m \in \mathbb{C}$ and 2×2 covariance matrix T. Consider a thermal state $\rho(t) = (1 - e^{-t})\, e^{-t\, a^\dagger a}$ with a, $a^\dagger$ denoting 1-mode annihilation and creation operators respectively. Then*

$$\operatorname{Tr} \rho(m, T) \, \log \rho(t) = \log (1 - e^{-t}) - \frac{t}{2} \left(\operatorname{Tr} T + 2\,|m|^2 - 1\right), \quad \text{if } 0 < t < \infty. \tag{39}$$

Proof It is readily seen that

$$\begin{aligned} \operatorname{Tr} \rho(m, T) \, \log \rho(t) &= \operatorname{Tr} \rho(m, T) \, \left\{\log (1 - e^{-t}) - t\, a^\dagger a\right\} \\ &= \log (1 - e^{-t}) - t \left\langle a^\dagger a\right\rangle_{\rho(m,T)} . \end{aligned} \tag{40}$$

Expressing

$$\begin{aligned} \left\langle a^\dagger a\right\rangle_{\rho(m,T)} &= \frac{1}{2} \left\langle (p^2 + q^2 - 1) \right\rangle_{\rho(m,T)} \\ &= \frac{1}{2} \left(\operatorname{Var}(p)_{\rho(m,T)} + \operatorname{Var}(q)_{\rho(m,T)} + \langle p \rangle^2_{\rho(m,T)} + \langle q \rangle^2_{\rho(m,T)} - 1\right) \\ &= \frac{1}{2} \left(\operatorname{Tr} T + 2 \left| \left\langle \frac{q + ip}{\sqrt{2}} \right\rangle_{\rho(m,T)} \right|^2 - 1\right) \\ &= \frac{1}{2} \left(\operatorname{Tr} T + 2\,|m|^2 - 1\right) \end{aligned} \tag{41}$$

we obtain (39).

Lemma 4 *Let $\rho = \rho(\mathbf{m}, C)$, σ be gaussian states in $\Gamma(\mathbb{C}^n)$, where $\mathbf{m}$ is the annihilation mean i.e., $\langle a(\mathbf{u})\rangle_\rho = \langle \mathbf{u}|\mathbf{m}\rangle$ and C is the covariance matrix of ρ. Let the standard structure of the gaussian state σ be*

$$\sigma = W(\boldsymbol{\ell})\, \Gamma(L)\, \rho(t_1) \otimes \rho(t_2) \otimes \cdots \otimes \rho(t_n)\, \Gamma(L)^{-1}\, W(\boldsymbol{\ell})^{-1} \tag{42}$$

so that $\boldsymbol{\ell}$ is the annihilation mean of σ. Here $L \in \mathrm{Sp}(2n, \mathbb{R})$ with

$$\Gamma(L)\, W(\mathbf{u})\, \Gamma(L)^{-1} = W(L \circ \mathbf{u})$$

for all $\mathbf{u} \in \mathbb{C}^n$ and $0 < t_1 \le t_2 \le \cdots \le t_n \le \infty$ as before in the structure theorem. Then

$$S(\rho || \sigma) = S(\rho' || \rho(t_1) \otimes \rho(t_2) \otimes \cdots \otimes \rho(t_n)) \tag{43}$$

where $\rho' \equiv \rho'\left(\mathbf{m}', C'\right)$ is the gaussian state characterized by the annihilation mean

$$\mathbf{m}' = L^{-1} \circ (\mathbf{m} - \boldsymbol{\ell}) \tag{44}$$

and the covariance matrix

$$C' = L^T \, C \, L. \tag{45}$$

Proof For any unitary U in $\Gamma(\mathbb{C}^n)$ and for $\rho' = U^\dagger \rho \, U, \;\; \sigma' = U^\dagger \, \sigma \, U$ we have

$$S(\rho' || \sigma') = S(\rho \, || \, \sigma), \tag{46}$$

Choose $U = W(\boldsymbol{\ell}) \, \Gamma(L)$ to be the gaussian symmetry leading to the standard form (42) of σ i.e.,

$$\sigma' = (W(\boldsymbol{\ell}) \, \Gamma(L))^{-1} \; \sigma \; W(\boldsymbol{\ell}) \, \Gamma(L) = \rho(t_1) \otimes \rho(t_2) \otimes \cdots \otimes \rho(t_n).$$

Noting that

$$\begin{aligned} W(\mathbf{z})^{-1} \, \rho(\mathbf{m}, \, C) \, W(\mathbf{z}) &= \rho(\mathbf{m} - \mathbf{z}, \, C), \;\; \mathbf{z} \in \mathbb{C}^n \\ \Gamma(M)^{-1} \, \rho(\mathbf{m}, \, C) \, \Gamma(M) &= \rho\left(M^{-1} \circ \mathbf{m}, \; M^T \, C \, M\right), \;\; M \in \mathrm{Sp}(2n, \mathbb{R}) \end{aligned}$$

we recover (44), (45) respectively for the annihilation mean and the covariance matrix of the transformed state $\rho'(\mathbf{m}', C') = (W(\boldsymbol{\ell}) \, \Gamma(L))^{-1} \;\; \rho \; (W(\boldsymbol{\ell}) \, \Gamma(L))$. Hence (43) follows.

Let us denote

$$\begin{aligned} \boldsymbol{m}_\rho &= L^{-1} \circ (\mathbf{m} - \boldsymbol{\ell}) = ((m_\rho)_1, (m_\rho)_2, \ldots, (m_\rho)_n)^T \\ T_k &= \text{ the } k\text{-th 1-mode covariance matrix of order 2 in } L^T \, C \, L. \end{aligned}$$

The 1-mode covariance matrix T_k is made up of the (kk)-th, $(k, k+n)$-th, $(k+n, k)$-th, $(k+n, k+n)$-th elements of the transformed $2n \times 2n$ covariance matrix $C' = L^T \, C \, L$.

Following observations are useful for computing relative entropy $S(\rho || \sigma)$ of ρ with respect to σ.

Observation 1: The k-th mode marginal of the transformed state

$$\rho'(\mathbf{m}', C') = (W(\boldsymbol{\ell})\, \Gamma(L))^{-1}\ \rho\, W(\boldsymbol{\ell})\, \Gamma(L)$$

is the 1-mode gaussian state denoted by $\rho((m_\rho)_k,\ T_k)$.

Observation 2: Using Lemma 4 we have

$$S(\rho||\sigma) = S(\rho'||\rho(t_1) \otimes \rho(t_2) \otimes \cdots \otimes \rho(t_n)).$$

Since ρ' and ρ are unitarily equivalent it follows that $S(\rho') = S(\rho)$.

Observation 3: By Lemma 2 it follows that

$$S(\rho||\sigma) = -S(\rho) - \sum_k \operatorname{Tr}\rho((m_\rho)_k,\ T_k) \log \rho(t_k).$$

If $t_k = \infty$, then $\rho(t_k) = |\Omega\,\rangle\langle\Omega\,|$ (vacuum state). If, in addition, $\rho((m_\rho)_k,\ T_k) \neq |\Omega\,\rangle\langle\Omega\,|$, then by Lemma 1 we have

$$-\operatorname{Tr}\rho((m_\rho)_k,\ T_k)\ \log\ \rho(t_k) = \infty.$$

Since $-\operatorname{Tr}\rho((m_\rho)_k,\ T_k)\ \log\ \rho(t_k) > 0\ \forall\, k$, it follows that $S(\rho||\sigma) = \infty$.

Observation 4: Since $\rho(t_k) = |\Omega\,\rangle\langle\Omega\,|$ whenever $t_k = \infty$, and by Lemma 3

$$\operatorname{Tr}\rho((m_\rho)_k,\ T_k)\ \log\ \rho(t_k)\ = \log(1 - e^{-t_k}) - \frac{t_k}{2}\left(\operatorname{Tr} T_k + 2\,|(m_\rho)_k|^2 - 1\right), \quad \text{if } t_k < \infty.$$

we have

$$S(\rho||\sigma) = -S(\rho) + \sum_{k:\, t_k<\infty} \left\{-\log(1 - e^{-t_k}) + \frac{t_k}{2}\left[\operatorname{Tr} T_k + 2\,|(m_\rho)_k|^2 - 1\right]\right\}.$$

Thus we obtain

$$S(\rho||\sigma) = \begin{cases} \infty, & \text{if } \exists\, k \text{ such that } t_k = \infty \text{ and } \rho((m_\rho)_k,\ T_k) \neq |\Omega\rangle\langle\Omega|, \\ -S(\rho) + \displaystyle\sum_{k:t_k<\infty} \left(-\log(1 - e^{-t_k}) + \frac{t_k}{2}\left(\operatorname{Tr}T_k + 2|(m_\rho)_k|^2 - 1\right)\right), & \text{otherwise.} \end{cases} \tag{47}$$

Let $0<s_1\leq s_2\leq\cdots\leq s_n\leq\infty$, $0 < t_1 \leq t_2 \leq \cdots \leq t_n \leq \infty$ and $S(\rho||\sigma)\ <\ \infty$. It may be noted that $t_k = \infty$ implies $\rho\left((m_\rho)_k,\ T_k\right) = |\Omega\,\rangle\langle\Omega\,|$ and therefore

$(m_\rho)_k = 0$, $\mathrm{Tr}\, T_k = 1$. We interpret $t_k \left(\mathrm{Tr}\, T_k + 2|(m_\rho)_k|^2 - 1\right)$ to be 0. With this understanding we express

$$\begin{aligned} S(\rho||\sigma) &= -S(\rho) + \sum_{k=1}^{n} \left\{ -\log\left(1 - e^{-t_k}\right) + \frac{t_k}{2}\left[\mathrm{Tr}\, T_k + 2\,|(m_\rho)_k|^2 - 1\right] \right\} \\ &= \sum_{k=1}^{n} \left\{ \frac{-H(e^{-s_k})}{(1 - e^{-s_k})} - \log\left(1 - e^{-t_k}\right) + \frac{t_k}{2}\left[\mathrm{Tr}\, T_k + 2\,|(m_\rho)_k|^2 - 1\right] \right\}. \end{aligned} \tag{48}$$

Drop the suffix k for the k-th term of the summation in (48) for convenience of computation and express it as

$$\frac{-H(e^{-s})}{(1 - e^{-s})} - \log\left(1 - e^{-t}\right) + \frac{t}{2}\left(\mathrm{Tr}\, T + 2\,|m_\rho|^2 - 1\right). \tag{49}$$

The first two terms of (49) can be expressed as

$$\frac{-H(e^{-s})}{(1 - e^{-s})} - \log\left(1 - e^{-t}\right) = \frac{H(p_1 : p_2)}{(1 - e^{-s})} - \frac{t\, e^{-s}}{(1 - e^{-s})} \tag{50}$$

where

$$\begin{aligned} H(p_1 : p_2) &= p_1 \log p_1 + (1 - p_1)\log(1 - p_1) - p_1 \log p_2 - (1 - p_1)\log(1 - p_2), \\ p_1 &= e^{-s},\ p_2 = e^{-t} \end{aligned} \tag{51}$$

is the classical Shannon relative entropy of a binomial trial with probability of success p_1 with respect to another with probability of success p_2. Thus (49) takes the form

$$\begin{aligned} \frac{H(e^{-s} : e^{-t})}{(1 - e^{-s})} + \frac{t}{2}\left[\mathrm{Tr}\, T - \left(1 + \frac{2\, e^{-s}}{1 - e^{-s}}\right) + 2\,|m_\rho|^2\right] &= \frac{H(e^{-s} : e^{-t})}{(1 - e^{-s})} \\ &+ \frac{t}{2}\left[\mathrm{Tr}\, T - \coth\left(\frac{s}{2}\right) + 2\,|m_\rho|^2\right]. \end{aligned} \tag{52}$$

Substituting (52) in (48) leads to

$$S(\rho||\sigma) = \sum_{k=1}^{n} \left\{ \frac{H(e^{-s_k} : e^{-t_k})}{(1 - e^{-s_k})} + \frac{t_k}{2}\left[\mathrm{Tr}\, T_k - \coth\left(\frac{s_k}{2}\right) + 2\,|(m_\rho)_k|^2\right] \right\} \tag{53}$$

whenever $S(\rho||\sigma) < \infty$.

Remark 1 Relative entropy $S(\rho||\sigma)$ of two gaussian states ρ and σ consists of a classical part

$$\sum_{j=1}^{n} \frac{H(e^{-s_k} : e^{-t_k})}{(1 - e^{-s_k})}$$

and a quantum part

$$\sum_{k=1}^{n} \frac{t_k}{2} \left[\operatorname{Tr} T_k - \coth\left(\frac{s_k}{2}\right) + 2\,|(m_\rho)_k|^2 \right].$$

Remark 2 Note that $\coth\left(\frac{s_k}{2}\right)$ is equal to the trace of the 2×2 covariance matrix of the single mode gaussian state $\rho(s_k)$.

Remark 3 Note that

$$\begin{aligned} e^{-s_k} &= \text{Success probability} \\ &= Pr(\#\ \text{particles} \geq 1 \text{ in the thermal state } \rho(s_k)), \\ e^{-t_k} &= \text{Success probability} \\ &= Pr(\#\ \text{particles} \geq 1 \text{ in the thermal state } \rho(t_k)) \end{aligned}$$

whereas $1 - e^{-s_k}$ and $1 - e^{-t_k}$ are respectively the probabilities of the "no particle count" in the thermal states $\rho(s_k)$ and $\rho(t_k)$.

6 Petz-Rényi Relative Entropy $S_\alpha(\rho||\sigma)$, $0 < \alpha < 1$ of Two Gaussian States

Consider the α-dependent Petz-Rényi relative entropy [6, 7] between two states ρ, σ:

$$S_\alpha(\rho||\sigma) \equiv \frac{1}{\alpha - 1} \log \operatorname{Tr} \rho^\alpha \sigma^{1-\alpha} \tag{54}$$

for $0 < \alpha < 1$. Let ρ and σ be two n-mode gaussian states with their respective standard forms given by

$$\begin{aligned} \rho &= W(\boldsymbol{\ell})\,\Gamma(L)\,\rho(s_1) \otimes \rho(s_2) \otimes \cdots \otimes \rho(s_n)\ (W(\boldsymbol{\ell})\,\Gamma(L))^{-1} \\ \sigma &= W(\mathbf{m})\,\Gamma(M)\,\rho(t_1) \otimes \rho(t_2) \otimes \cdots \otimes \rho(t_n)\ (W(\mathbf{m})\,\Gamma(M))^{-1} \end{aligned} \tag{55}$$

where $\boldsymbol{\ell}, \mathbf{m} \in \mathbb{C}^n$; L, $M \in$ Sp$(2n, \mathbb{R})$; $0 < s_1 \leq s_2 \leq \ldots s_n \leq \infty$; $0 < t_1 \leq t_2 \leq \ldots t_n \leq \infty$. Let $U = W(\mathbf{m})\,\Gamma(M)$. Then $\rho' = U^{-1} \rho\, U$ and $\sigma' = U^{-1} \sigma\, U$ are gaussian states whose standard forms are given by

$$\begin{aligned} \rho' &= W(M^{-1} \circ (\boldsymbol{\ell} - \mathbf{m}))\,\Gamma(M^{-1} L)\,\rho(s_1) \otimes \rho(s_2) \otimes \cdots \otimes \rho(s_n) \\ &\quad \times \left(W(M^{-1} \circ (\boldsymbol{\ell} - \mathbf{m}))\,\Gamma(M^{-1} L)\right)^{-1} \\ \sigma' &= \rho(t_1) \otimes \rho(t_2) \otimes \cdots \otimes \rho(t_n). \end{aligned} \tag{56}$$

and

$$S_\alpha(\rho||\sigma) = S_\alpha(\rho'||\sigma'). \tag{57}$$

Thus, in the computation of $S_\alpha(\rho||\sigma)$ we may and do assume that ρ and σ have their standard forms given by

$$\begin{aligned}\rho &= W(\mathbf{m}_\rho)\,\Gamma(L_\rho)\,\rho(s_1)\otimes\rho(s_2)\otimes\cdots\otimes\rho(s_n)\,\big(W(\mathbf{m}_\rho)\,\Gamma(L_\rho)\big)^{-1}\\ \sigma &= \rho(t_1)\otimes\rho(t_2)\otimes\cdots\otimes\rho(t_n).\end{aligned} \tag{58}$$

where the covariance matrix

$$\begin{aligned}C_\rho &= L_\rho^T\,D(\mathbf{s})\,L_\rho,\quad L_\rho\in\mathrm{Sp}(2n,\,\mathbb{R})\\ D(\mathbf{s}) &= \begin{pmatrix} D_0(\mathbf{s}) & 0_n\\ 0_n & D_0(\mathbf{s})\end{pmatrix}\\ D_0(\mathbf{s}) &= \mathrm{diag}\left[\frac{1}{2}\coth\left(\frac{s_k}{2}\right),\,k=1,2,\ldots,n\right]\end{aligned} \tag{59}$$

and

$$\begin{aligned}C_\sigma &= D(\mathbf{t}) = \begin{pmatrix} D_0(\mathbf{t}) & 0_n\\ 0_n & D_0(\mathbf{t})\end{pmatrix},\\ D_0(\mathbf{t}) &= \mathrm{diag}\left[\frac{1}{2}\coth\left(\frac{t_k}{2}\right),\,k=1,2,\ldots,n\right]\end{aligned} \tag{60}$$

We denote

$$\begin{aligned}\rho &= \rho(\mathbf{m}_\rho,\,L_\rho^T\,D\,(\mathbf{s})\,L_\rho)\\ \sigma &= \rho(\mathbf{0},\,D(\mathbf{t})).\end{aligned} \tag{61}$$

Let us assume that $s_k<\infty$ and $t_k\le\infty$ for all $k=1,2,\ldots,n$. For $0<\alpha<1$ we have

$$\rho^\alpha = \frac{p^\alpha(\mathbf{s})}{p(\alpha\,\mathbf{s})}\,\rho(\mathbf{m}_\rho,\,L_\rho^T\,D\,(\alpha\,\mathbf{s})\,L_\rho) \tag{62}$$

$$\sigma^{1-\alpha} = \frac{p^{1-\alpha}(\mathbf{t})}{p((1-\alpha)\,\mathbf{t})}\,\rho(\mathbf{0},\,D((1-\alpha)\,\mathbf{t})) \tag{63}$$

where $p(\mathbf{r})=\prod_{k=1}^{n}(1-e^{-r_k})$. Note that $\rho^\alpha,\sigma^{1-\alpha}$ are also gaussian states

$$\rho\left(\mathbf{m}_\rho,\,L_\rho^T\,D\,(\alpha\,\mathbf{s})\,L_\rho\right),\ \rho(\mathbf{0},\,D((1-\alpha)\,\mathbf{t}))$$

up to multiplication by scalar factors $\frac{p^\alpha(\mathbf{s})}{p(\alpha\,\mathbf{s})},\ \frac{p^{1-\alpha}(\mathbf{t})}{p((1-\alpha)\,\mathbf{t})}$ respectively.

By Wigner's theorem [4, 5] and gaussian integral formula we obtain

$$\mathrm{Tr}\rho^\alpha\,\sigma^{1-\alpha} = \frac{1}{\pi^n}\int_{\mathbb{R}^{2n}} d\mathbf{x}\,d\mathbf{y}\,\widehat{(\rho^\alpha)}(\mathbf{x}+i\mathbf{y})\,\widehat{(\sigma)^{1-\alpha}}(\mathbf{x}+i\mathbf{y})$$

$$= \left(\frac{p^{\alpha}(\mathbf{s})}{p(\alpha\,\mathbf{s})}\right)\left(\frac{p^{1-\alpha}(\mathbf{t})}{p((1-\alpha)\,\mathbf{t})}\right)\frac{\exp\left[-\widetilde{\mathbf{m}}_{\rho}^{T}\left(L_{\rho}^{T}\,D(\alpha\,\mathbf{s})\,L_{\rho}+D((1-\alpha)\,\mathbf{t})\right)^{-1}\widetilde{\mathbf{m}}_{\rho}\right]}{\sqrt{\det\left[L_{\rho}^{T}\,D(\alpha\mathbf{s})\,L_{\rho}+D((1-\alpha)\mathbf{t})\right]}} \tag{64}$$

where $\widetilde{\mathbf{m}}_{\rho} = (-\operatorname{Im}\mathbf{m}_{\rho},\ \operatorname{Re}\mathbf{m}_{\rho})^{T}$. We express the Petz-Rényi relative entropy $S_{\alpha}(\rho||\sigma)$, $0 < \alpha < 1$ as

$$S_{\alpha}(\rho||\sigma) = \mathcal{R}_{\alpha}^{I} + \mathcal{R}_{\alpha}^{II} + \mathcal{R}_{\alpha}^{III} + \mathcal{R}_{\alpha}^{IV} \tag{65}$$

where

$$\mathcal{R}_{\alpha}^{I} = -\frac{1}{1-\alpha}\sum_{k=1}^{n}\left[\alpha\,\log\left(1-e^{-s_k}\right)-\log\left(1-e^{-\alpha\,s_k}\right)\right] \tag{66}$$

$$\mathcal{R}_{\alpha}^{II} = \sum_{k=1}^{n}\left[-\log\left(1-e^{-t_k}\right)+\frac{1}{1-\alpha}\,\log\left(1-e^{-t_k(1-\alpha)}\right)\right] \tag{67}$$

$$\mathcal{R}_{\alpha}^{III} = \frac{1}{1-\alpha}\left[\widetilde{\mathbf{m}}_{\rho}^{T}\left\{L_{\rho}^{T}\,D\,(\alpha\,\mathbf{s})\,L_{\rho}+D((1-\alpha)\,\mathbf{t})\right\}^{-1}\widetilde{\mathbf{m}}_{\rho}\right] \tag{68}$$

$$\mathcal{R}_{\alpha}^{IV} = \frac{1}{2(1-\alpha)}\,\log\det\left[L_{\rho}^{T}\,D\,(\alpha\,\mathbf{s})\,L_{\rho}+D((1-\alpha)\mathbf{t})\right] \tag{69}$$

The Petz-Rényi relative entropy $S_{\alpha}(\rho||\sigma)$ given by Eqs. (65)–(69) converges to $S(\rho||\sigma)$ in the limit $\alpha \to 1$.

By L'Hospital rule we obtain

$$\begin{aligned}\lim_{\alpha\to 1} R_{\alpha}^{I} &= \sum_{k=1}^{n}\lim_{\alpha\to 1}\left[-\frac{\alpha\,\log\left(1-e^{-s_k}\right)}{1-\alpha}+\frac{\log\left(1-e^{-\alpha\,s_k}\right)}{1-\alpha}\right]\\ &= \sum_{k=1}^{n}\left[\log\left(1-e^{-s_k}\right)-\frac{s_k\,e^{-s_k}}{1-e^{-s_k}}\right]\\ &= -\sum_{k=1}^{n}\frac{H(e^{-s_k})}{1-e^{-s_k}}\end{aligned} \tag{70}$$

and

$$\begin{aligned}\lim_{\alpha\to 1} R_{\alpha}^{II} &= \sum_{k}\left[-\log\left(1-e^{-t_k}\right)+\lim_{\alpha\to 1}\frac{1}{1-\alpha}\,\log\left(1-e^{-t_k(1-\alpha)}\right)\right]\\ &= \sum_{k}\left[-\log\left(1-e^{-t_k}\right)+\lim_{\alpha\to 1}\frac{t_k\,e^{-t_k(1-\alpha)}}{(1-e^{-t_k(1-\alpha)})}\right]\\ &= \sum_{k}\left[-\log\left(1-e^{-t_k}\right)+\lim_{\alpha\to 1}\frac{t_k^2\,e^{-t_k(1-\alpha)}}{(-t_k\,e^{-t_k(1-\alpha)})}\right]\end{aligned}$$

$$= \sum_k \left[-\log\left(1 - e^{-t_k}\right) - t_k \right]. \tag{71}$$

In order to compute $\lim_{\alpha \to 1} R_\alpha^{III}$ we take the factor $\frac{1}{1-\alpha}$ inside the bracket $\{\cdots\}^{-1}$ in (68) to obtain

$$\lim_{\alpha \to 1} R_\alpha^{III} = \left[\widetilde{\mathbf{m}}_\rho^T \lim_{\alpha \to 1} \left\{ (1-\alpha)\, L_\rho^T\, D\,(\alpha\, \mathbf{s})\; L_\rho + (1-\alpha) D((1-\alpha)\,\mathbf{t}) \right\}^{-1} \widetilde{\mathbf{m}}_\rho \right]. \tag{72}$$

Note that any entry in the $2n \times 2n$ matrix $D(\alpha\, \mathbf{s})$ has the form (see (59)) $\frac{1}{2}\coth\left(\frac{\alpha\, s_k}{2}\right)$, which converges to $\frac{1}{2}\coth\left(\frac{s_k}{2}\right)$ as α increases to 1. Thus, the first term inside the bracket $\{\cdots\}^{-1}$ in (72) is zero. Any entry in the second term $(1-\alpha)\; D((1-\alpha)\,\mathbf{t})$ is of the form

$$\left(\frac{1-\alpha}{2}\right) \coth\left(\frac{(1-\alpha)\, t_k}{2}\right) = \left(\frac{1-\alpha}{2}\right) \frac{1 + e^{t_k(1-\alpha)}}{1 - e^{t_k(1-\alpha)}}.$$

Expanding $e^{t_k(1-\alpha)}$ in the denominator and applying L'Hospital rule we get

$$\lim_{\alpha \to 1} \left(\frac{1-\alpha}{2}\right) \frac{1 + e^{-t_k(1-\alpha)}}{1 - e^{-t_k(1-\alpha)}} = \frac{1}{t_k}.$$

Thus,

$$\begin{aligned} \lim_{\alpha \to 1} R_\alpha^{III} &= \widetilde{\mathbf{m}}_\rho^T \,\mathrm{diag}\,(t_1,\, t_2,\, \ldots,\, t_n;\, t_1,\, t_2,\, \ldots,\, t_n)\; \widetilde{\mathbf{m}}_\rho \\ &= \sum_{k=1}^{n} t_k\, |(m_\rho)_k\,|^2 \end{aligned} \tag{73}$$

To determine the limit $\alpha \to 1$ of the term $\mathcal{R}_\alpha^{IV}$ (see (69)) we pull out the diagonal $2n \times 2n$ matrix $D((1-\alpha)\mathbf{t})$ from the determinant to obtain

$$R_\alpha^{IV} = \frac{1}{2(1-\alpha)} \left\{ \log\, \det(D((1-\alpha)\mathbf{t})) + \log\, \det\left(I_{2n} + B_\alpha\, L_\rho^T\, D\,(\alpha\, \mathbf{s})\; L_\rho\; B_\alpha \right) \right\} \tag{74}$$

where

$$B_\alpha = D^{-\frac{1}{2}}((1-\alpha)\mathbf{t}). \tag{75}$$

Substituting $\det(D((1-\alpha)\mathbf{t})) = 2 \sum_{k=1}^{n} \log\left[\frac{1}{2}\coth\left(\frac{(1-\alpha)\, t_k}{2}\right)\right]$ in (74) and simplifying, we obtain

$$R_\alpha^{IV} = \frac{1}{2(1-\alpha)} \left\{ \sum_{k=1}^{n} 2 \left[\log\left(\frac{1 + e^{-t_k(1-\alpha)}}{2}\right) - \log\left(1 - e^{-t_k(1-\alpha)}\right) \right] \right.$$

$$+\log\det\left(I_{2n}+B_\alpha\,L_\rho^T\,D\,(\alpha\,\mathbf{s})\,L_\rho\,B_\alpha\right)\Big\}. \tag{76}$$

Applying L'Hospital rule we obtain

$$\lim_{\alpha\to 1}\frac{1}{(1-\alpha)}\log\left(\frac{1+e^{-t_k(1-\alpha)}}{2}\right)=\lim_{\alpha\to 1}\frac{-\,t_k\,e^{-t_k(1-\alpha)}}{(1+e^{-t_k(1-\alpha)})}=-\frac{t_k}{2} \tag{77}$$

$$\lim_{\alpha\to 1}\frac{-1}{(1-\alpha)}\log\left(1-e^{-t_k(1-\alpha)}\right)=t_k. \tag{78}$$

To facilitate the computation of $\lim_{\alpha\to 1}\frac{1}{2(1-\alpha)}\log\det\left(I_{2n}+B_\alpha\,L_\rho^T\,D\,(\alpha\,\mathbf{s})\,L_\rho\,B_\alpha\right)$ we present the following Lemma.

Lemma 5 *Let* $0<\theta<1$ *be a parameter and* F_θ *be positive matrices of finite order such that*

$$\lim_{\theta\to 0}F_\theta=F. \tag{79}$$

Then

$$\lim_{\theta\to 0}\frac{1}{\theta}\log\det\,(I+\theta\,F_\theta)=\mathrm{Tr}\,F. \tag{80}$$

Proof The positive matrices F_θ admit eigen decomposition

$$F_\theta=\sum_j\lambda_j(\theta)\,|\psi_j(\theta)\rangle\langle\psi_j(\theta)| \tag{81}$$

using which we express

$$\frac{1}{\theta}\log\det\,(I+\theta\,F_\theta)=\frac{1}{\theta}\log\prod_j\left(1+\theta\,\lambda_j(\theta)\right) \tag{82}$$

$$=\sum_j\frac{1}{\theta}\log\left(1+\theta\,\lambda_j(\theta)\right). \tag{83}$$

As $F_\theta\to F$ in the limit $\theta\to 0$, the eigenvalues $\lambda_j(\theta)$ are expected to remain positive and bounded. Expanding $\log(1+\theta\,\lambda_j(\theta))$ in the small θ limit, up to second order in θ i.e.,

$$\log\left(1+\theta\,\lambda_j(\theta)\right)\approx\theta\,\lambda_j(\theta)+\mathcal{O}(\theta^2) \tag{84}$$

we get

$$\lim_{\theta \to 0} \sum_j \frac{1}{\theta} \log\left(1 + \theta\, \lambda_j(\theta)\right) = \lim_{\theta \to 0} \sum_j \lambda_j(\theta)$$
$$= \operatorname{Tr} F. \tag{85}$$

We apply Lemma 5 by substituting $\theta{=}(1-\alpha)$ and $F_\theta{=}\frac{1}{\theta}\, B_\theta\, L_\rho^T\, D\,((1{-}\theta)\,\mathbf{s})\, L_\rho\, B_\theta$, $B_\theta = D^{-\frac{1}{2}}(\theta\, \mathbf{t})$ to obtain

$$\lim_{\alpha \to 1} \frac{\log \det\left(I_{2n} + B_\alpha\, L_\rho^T\, D\,(\alpha\, \mathbf{s})\, L_\rho\, B_\alpha\right)}{2(1-\alpha)} = \lim_{\alpha \to 1} \frac{\operatorname{Tr} D^{-1}((1-\alpha)\,\mathbf{t})\, L_\rho^T\, D\,(\alpha\, \mathbf{s})\, L_\rho}{2(1-\alpha)}. \tag{86}$$

The $2n \times 2n$ diagonal matrix $D((1-\alpha)\,\mathbf{t})$ has entries of the form $\frac{1}{2}\coth\left(\frac{1-\alpha}{2}\, t_k\right)$ and we know that

$$\lim_{\alpha \to 1} \left(\frac{1-\alpha}{2}\right) \coth\left(\frac{1-\alpha}{2}\, t_k\right) = \lim_{\alpha \to 1} \left(\frac{1-\alpha}{2}\right) \frac{1 + e^{-t_k(1-\alpha)}}{1 - e^{-t_k(1-\alpha)}} = \frac{1}{t_k}.$$

Thus,

$$\lim_{\alpha \to 1} \frac{D^{-1}((1-\alpha)\,\mathbf{t})}{(1-\alpha)} = \operatorname{diag}\,(t_1, t_2, \ldots, t_n; t_1, t_2, \ldots, t_n) \tag{87}$$

Substituting (87) in (86) we get

$$\lim_{\alpha \to 1} \frac{\operatorname{Tr} D^{-1}((1-\alpha)\,\mathbf{t})\, L_\rho^T\, D(\alpha\, \mathbf{s})\, L_\rho}{2(1-\alpha)} = \frac{1}{2} \operatorname{Tr} \operatorname{diag}\,(t_1, t_2, \ldots, t_n; t_1, t_2, \ldots, t_n)\, C_\rho$$
$$= \sum_{k=1}^{n} t_k \operatorname{Tr} T_k \tag{88}$$

where $C_\rho = L^\rho\, D\,(\mathbf{s})\, L_\rho^T$ is the covariance matrix of the n-mode gaussian state ρ and T_k is the 2×2 single mode covariance matrix of the k-th mode marginal system of ρ. Thus,

$$\lim_{\alpha \to 1} R_\alpha^{IV} = \sum_{k=1}^{n} \frac{t_k}{2}\,(1 + \operatorname{Tr} T_k) \tag{89}$$

Putting together (70), (71), (73) and (89) we obtain

$$\lim_{\alpha \to 1} S_\alpha(\rho||\sigma) = \sum_{k=1}^{n} \left\{ \frac{-H(e^{-s_k})}{(1 - e^{-s_k})} - \log\,(1 - e^{-t_k}) + \frac{t_k}{2}\left[\operatorname{Tr} T_k + 2\,|(m_\rho)_k|^2 - 1\right]\right\}$$
$$= S(\rho||\sigma)$$

as expected.

For alternate approaches on the computation of relative entropy between two gaussian states see Refs. [8–11].

Acknowledgements My thanks to A. R. Usha Devi for making a readable manuscript out of my scribbled algebra. I thank Mark M. Wilde for pointing out a misprint in (39) of the first version of our manuscript and Ajit Iqbal Singh for going through the manuscript meticulously.

References

1. Parthasarathy, K.R.: What is a gaussian state? Commun. Stoch. Anal. **4**, 143–160 (2010). MR 2662722
2. Parthasarathy, K.R.: The symmetry group of gaussian states in $L^2(\mathbb{R}^n)$. In: Shiryaev, A.N., Varadhan, S.R.S., Presman, E.L. (eds.) Prokhorov and Contemporary Probability Theory. Springer Proceedings in Mathematics and Statistics, vol. 33, pp. 349–369. Springer, Heidelberg (2013). https://doi.org/10.1007/978-3-642-33549-5. MR 3070484
3. Parthasarathy, K.R., Sengupta, R.: From particle counting to gaussian tomography. Infin. Dimens. Anal., Quantum Probab. Relat. Top. **18**, 1550023 (2015). https://doi.org/10.1142/S021902571550023X
4. John, T.C., Parthasarathy, K.R.: Klauder-Bargmann Integral Representation of Gaussian Symmetries and Generating Functions of Gaussian States. arXiv:1911.06555v3 [quant-ph]
5. John, T.C., Parthasarathy, K.R.: A common parametrization for finite mode Gaussian states, their symmetries, and associated contractions with some applications. J. Math. Phys. **62**, 022102 (2021). https://doi.org/10.1063/5.0019413
6. Rényi, A.: On measures of entropy and information. In: Neyman, J. (ed.) Proceedings of the 4th Berkeley Symposium on Mathematics, Statistics and Probability, vol. 1, pp. 547–561. University of California Press, Berkeley (1961)
7. Petz, D.: Quasi-entropies for finite quantum systems. Rep. Math. Phys. **23**, 57–65 (1986). https://doi.org/10.1016/0034-4877(86)90067-4
8. Scheel, S., Welsch, D.-G.: Entanglement generation and degradation by passive optical devices. Phys. Rev. A. **64**, 063811 (2001). https://doi.org/10.1103/PhysRevA.64.063811
9. Chen, X.Y.: Gaussian relative entropy of entanglement. Phys. Rev. A. **71**, 062320 (2005). https://doi.org/10.1103/PhysRevA.71.062320
10. Pirandola, S., Laurenza, R., Ottaviani, C., Banchi, L.: Fundamental limits of repeaterless quantum communications. Nat. Commun. **8**, 15043 (2017). https://doi.org/10.1038/ncomms15043
11. Seshadreesan, K.P., Lami, L., Wilde, M.M.: Rényi relative entropies of quantum Gaussian states. J. Math. Phys. **59**, 072204 (2018). https://doi.org/10.1063/1.5007167

Quantum Operators of the Semicircle Distributions

Gabriela Popa and Aurel I. Stan

Dedicated to Professor R. L. Hudson

Abstract The decomposition of the quantum operators, of the semicircle distributions, in terms of powers of the 0-derivative operator will be presented.

Keywords Quantum operators · Position operator · 0-momentum operator

1 Introduction

From a classic probabilistic point of view, to know a real valued random variable X means to understand its distribution, which requires to possess a formula for calculating the probability $P(a \leq X < b)$ for all real numbers $a < b$. From a quantum probabilistic point of view, identifying a real valued random variable X, having finite moments of all orders, with the multiplication operator that it generates, to know a random variable X means to have precise formulas for each of its three quantum operators: creation, preservation, and annihilation operators. In [8] the so called "position-momentum" decomposition of a linear operator mapping the space of polynomial random variables of X into itself was introduced, where X is the multiplication operator generated by a random variable having finite moments of all orders. That means, if $T : F \to F$ is a linear operator, where F denotes the space of

G. Popa (✉)
Ohio University, College of Arts and Science, 240 Elton Hall, Zanesville 43701, USA
e-mail: popag@ohio.edu

A. I. Stan
The Ohio State University at Marion, Mount Vernon Avenue, Marion 43302, USA
e-mail: stan.7@osu.edu

L. Accardi et al. (eds.), *Infinite Dimensional Analysis, Quantum Probability and Applications*, Springer Proceedings in Mathematics & Statistics 390,
https://doi.org/10.1007/978-3-031-06170-7_3

all random variables of the form $P(X)$, for P a polynomial function, then T can be written uniquely as:

$$T = \sum_{n=0}^{\infty} A_n(X) D^n, \tag{1}$$

where $D = d/dx$ denotes the formal differentiation operator and, for all $n \in \mathbb{N} \cup \{0\}$, $A_n(X)$ is a polynomial of the multiplication operator X. X is understood as being "the position" operator, while D as the "momentum operator". In this decomposition, it must be noted that all the position factors $A_n(X)$ are placed to the left of the momentum factors D^n.

In this paper, we will replace the classic differentiation operator D, by the 0-differentiation operator D_0, and find a similar decomposition for the quantum operators of the semicircle distributions.

2 Background

Let X be a real valued random variable, defined on the probability space $(\Omega, \mathcal{F}, P)$, having finite moments of all orders. That means, for all $p > 0$,

$$E\left[|X|^p\right] < \infty, \tag{2}$$

where E denotes the expectation. Let us define the spaces:

$$F := \{f(X) \mid f \in \mathbb{C}[x]\}, \tag{3}$$

where $\mathbb{C}[x]$ denotes the space of all polynomial functions in one variable with complex coefficients, and for all $n \in \mathbb{N} \cup \{0\}$, let us consider the spaces:

$$F_n := \{f(X) \mid f \in \mathbb{C}[x],\ \deg(f) \leq n\}, \tag{4}$$

where $\deg(f)$ denotes the degree of the polynomial function f.
Because X has finite moments of all orders, we have:

$$\mathbb{C} \equiv F_0 \subseteq F_1 \subseteq F_2 \subseteq \cdots \subseteq L^2(\Omega, P). \tag{5}$$

Since, for each $n \in \mathbb{N} \cup \{0\}$, F_n is a finite dimensional subspace of $L^2(\Omega, P)$, being spanned by the random variables $1, X, X^2, \dots, X^n$, the space F_n is a closed subspace of $L^2(\Omega, P)$. Thus, we can orthogonalize the spaces $\{F_n\}_{n\geq 0}$, defining the spaces $G_0 := F_0 \equiv \mathbb{C}$, and for all $n \geq 1$,

$$G_n := F_n \ominus F_{n-1}, \tag{6}$$

that means G_n is the orthogonal complement of F_{n-1} into F_n.

For each nonnegative integer n, we call G_n the n-th homogenous chaos space generated by$X$, and refer to each random variable $f(X) \in G_n$, as being a homogenous polynomial random variable of degree n.

From now on we identify the random variable X with the multiplication operator $X : F \to F$, that maps:

$$f(X) \mapsto Xf(X).$$

The following lemma was proven in [2, 4], not only in the one-dimensional case, but also in the multi-dimensional case:

Lemma 1 *If we view X as the multiplication operator that it generates, then for all non-negative integers n and k, such that $k \notin \{n-1, n, n+1\}$, the space XG_n is orthogonal to the space G_k.*

Due to this lemma, for all nonnegative integers n, we have:

$$XG_n \subseteq G_{n-1} \oplus G_n \oplus G_{n+1}, \tag{7}$$

where for $n = 0$, $G_{n-1} = G_{-1} = \{0\}$ is the null space.

On the other hand, since for all $n \geq 0$, $F_n = F_{n-1} + \mathbb{C}X^n$, we can see that the dimension of the space G_n, which is equal to the co-dimension of F_{n-1} into F_n, is at most equal to 1. Thus, if G_n is not the null space, then there exists and is unique a polynomial random variable $f_n(X)$ in G_n having the leading coefficient equal to 1. We call this polynomial random variable the n-th monic orthogonal polynomial generated by$X$. In this way, the three-term inclusion formula (7) becomes the classic three-term recursive formula:

$$Xf_n(X) = f_{n+1}(X) + \alpha_n f_n(X) + \omega_n f_{n-1}(X), \tag{8}$$

for some real numbers α_n and ω_n, with the coefficient of $f_{n+1}(X)$ in the right side of this formula being equal to 1 to agree with the fact that the coefficient of X^{n+1} in the left side of (8) is equal to 1. For $n = 0$, since $f_{-1}(X) = 0$, the coefficient ω_0 can be chosen arbitrarily. The numbers $\{\alpha_n\}_{n \geq 0}$ and $\{\omega_n\}_{n \geq 1}$ are called the Szegö-Jacobi parameters of the random variable X. It is also known that, for all $n \in \mathbb{N}$, the square of the L^2-norm of $f_n(X)$ is:

$$\| f_n(X) \|_2^2 = \omega_1 \omega_2 \cdots \omega_n. \tag{9}$$

For all $n \geq 0$, we define the operators:

$$D_n^+ : G_n \to G_{n+1}, \tag{10}$$

$$D_n^+ (cf_n(X)) := cf_{n+1}(X), \tag{11}$$

and call it a creation operator since it increases the degree of a homogenous polynomial random variable by one unit,

$$D_n^0 : G_n \to G_n, \tag{12}$$

$$D_n^0 (cf_n(X)) := c\alpha_n f_n(X), \tag{13}$$

and call it a preservation operator since it does not change the degree of a homogenous polynomial random variable, and

$$D_n^- : G_n \to G_{n-1}, \tag{14}$$

$$D_n^- (cf_n(X)) := c\omega_n f_{n-1}(X), \tag{15}$$

and call it an annihilation operator since it decreases the degree of a homogenous polynomial random variable by one unit, where c is any complex number.

We extend the definition of the creation, preservation, and annihilation operators to the space F of all polynomial random variables, in a linear way. That means, for a fixed polynomial random variable:

$$f(X) = \sum_{k=0}^{n} c_k f_k(X), \tag{16}$$

we define:

$$a^+ f(x) = \sum_{k=0}^{n} c_k D_k^+ f_k(X) \tag{17}$$

$$= \sum_{k=0}^{n} c_k f_{k+1}(X), \tag{18}$$

and call a^+ the creation operator,

$$a^0 f(x) = \sum_{k=0}^{n} c_k D_k^0 f_k(X) \tag{19}$$

$$= \sum_{k=0}^{n} c_k \alpha_k f_k(X), \tag{20}$$

and call a^0 the preservation operator, and

$$a^- f(x) = \sum_{k=0}^{n} c_k D_k^- f_k(X) \tag{21}$$

$$= \sum_{k=1}^{n} c_k \omega_k f_{k-1}(X), \tag{22}$$

and call a^- the annihilation operator generated by X, where $n \in \mathbb{N} \cup \{0\}$, and $c_0, c_1, \ldots, c_n$ are arbitrary complex numbers. Finally, we refer to the operators a^+, a^0, and a^- as being the quantum operators generated by the random variable X. In this way, the three-term recursive relation (8) becomes:

$$X f_n(X) = a^+ f_n(X) + a^0 f_n(X) + a^- f_n(X), \tag{23}$$

for all nonnegative integers n. That means we have the following quantum decomposition formula of the multiplication operator generated by X:

Theorem 1 *The multiplication operator X can be written as the sum of the three quantum operators, that means:*

$$X = a^+ + a^0 + a^-. \tag{24}$$

The domain of all four operators: X, a^+, a^0, and a^-, involved in this equality, is considered to be the space F of all polynomial random variables in X.

It is also known, see [2], that $a^0 = 0$ if and only if X is polynomially symmetric, that means, $E[X^{2n+1}] = 0$, for all nonnegative integers n.

3 Position-(0-Momentum) Decomposition of the Semicircle Distributions

It was shown in [8] that each linear operator, $T : F \to F$, can be written uniquely as

$$T = \sum_{n=0}^{\infty} A_n(X) D^n, \tag{25}$$

where for each $n \geq 0$, $A_n(X)$ is a polynomial of the multiplication operator by X, and $D : F \to F$ is the formal differentiation operator mapping $X^k \mapsto kX^{k-1}$, for all non-negative integers k. Interpreting X as the position operator and D as the momentum operator, formula (25) was called the position-momentum decomposition of T. This formula must be interpreted with care, in the case when the probability distribution of X has a finite support, since the momentum operator is not well defined in that case, due to the fact that we may have two different polynomials $f(x)$ and $g(x)$, such that $f(X) = g(X)$, P-almost surely, but $f'(X) \neq g'(X)$.

For every real number $q \neq 1$, we can replace the classic derivative operator D by the q-derivative operator D_q, where for any polynomial function $f(x)$, we define:

$$D_q f(x) = \frac{f(qx) - f(x)}{qx - x}. \tag{26}$$

It is not hard to see that since for any three polynomials $f(x)$, $g(x)$, and $h(x)$, the polynomial $f(g(x)) - f(h(x))$ is divisible by the polynomial $g(x) - h(x)$, D_q maps the space $\mathbb{C}[x]$ into $\mathbb{C}[x]$. We will call D_q the q-momentum operator, with the observation that when $q = 1$, the 1-momentum operator, which is defined as the point-wise limit of the q-momentum operator, as q approaches 1, becomes the classic momentum operator D. In the same way that the position-momentum decomposition was proven in [8], we can prove the following:

Theorem 2 *Let $T : F \to F$ be a linear operator. Then, for any real number $q \neq 1$, there exists and is unique a sequence of polynomial multiplication operators $\{A_n(X)\}_{n \geq 0}$, such that for all polynomial random variables $f(X)$ in F, we have:*

$$Tf(X) = \sum_{n=0}^{\infty} A_n(X) D_q^n f(X). \tag{27}$$

Formula (27) is written operatorially as:

$$T = \sum_{n=0}^{\infty} A_n(X) D_q^n. \tag{28}$$

This formula must be interpreted with care, in the case when the probability distribution of X has a finite support. In that case, for a given polynomial random variable Y, the set of polynomial functions $\mathcal{P}_Y := \{f \in \mathbb{C}[x] \mid f(X) = Y\}$ is infinite, and we must impose the condition that once a polynomial function f from $\mathcal{P}_Y$ is chosen, then the same polynomial function must be used for all the terms $A_n(X) D_q^n f(X)$, $n \geq 0$, of the sum from the right side of (27). We call formula (28) the position-(q-momentum) decomposition of T.

For the remaining part of this paper, we will choose $q = 0$, and study the position-(0-momentum) decomposition of the quantum operators generated by the semicircle distributions.

A random variable X, having finite moments of all orders, is said to have a semicircle distribution, if its monic orthogonal polynomials, $\{f_n\}_{n \geq 0}$, satisfy the recursive relation, for all $n \geq 0$:

$$f_n(X) = f_{n+1}(X) + cf_{n-1}(X), \tag{29}$$

where c is a positive constant. That means, its Szegö-Jacobi parameters are $\alpha_n = 0$ and $\omega_n = c$, for all $n \geq 1$. Therefore, its preservation operator is $a^0 = 0$.

Since, for all $n \geq 0$, $a^- f_n(X) = cf_{n-1}(X)$, we can see that a^- has a continuous extension to the Hilbert space

$$H := \sum_{n=0}^{\infty} \mathbb{C} f_n(X), \tag{30}$$

which is the closure of the space of polynomial random variables F in $L^2(\Omega, P)$, defined as follows. For all $f(x) = \sum_{n=0}^{\infty} c_n f_n(X) \in H$, which means $\| f(X) \|_2^2 = \sum_{n=0}^{\infty} |c_n|^2 \| f_n(X) \|_2^2 < \infty$, where $\| \cdot \|_2$ denotes the L^2-norm, we define:

$$a^- f(X) = \sum_{n=1}^{\infty} c_n c f_{n-1}(X). \tag{31}$$

The operator a^- is bounded since:

$$\begin{aligned}
\| a^- f(X) \|_2^2 &= \sum_{n=1}^{\infty} |c_n c|^2 \| f_{n-1}(X) \|^2 \\
&= c^2 \sum_{n=1}^{\infty} |c_n|^2 (\omega_1 \omega_2 \cdots \omega_{n-1}) \\
&= c^2 \sum_{n=1}^{\infty} |c_n|^2 c^{n-1} \\
&= c \sum_{n=1}^{\infty} |c_n|^2 c^n \\
&= c \sum_{n=1}^{\infty} |c_n|^2 \| f_n(X) \|^2 \\
&= c \left(\| f(X) \|_2^2 - |c_0|^2 \| f_0(X) \|_2^2 \right) \\
&\leq c \| f(X) \|_2^2 .
\end{aligned} \tag{32}$$

Let us compute the commutator between the annihilation operator a^- and creation operator a^+, $[a^-, a^+]$, by applying it to each monic orthogonal polynomial random variable $f_n(X)$, for $n \geq 0$.

For $n = 0$, we have:

$$\begin{aligned}\left[a^-, a^+\right] f_0(X) &= a^- a^+ f_0(X) - a^+ a^- f_0(X)\\ &= a^- f_1(X) - a^+ 0\\ &= c f_0(X).\end{aligned} \tag{33}$$

For $n \geq 1$, we have:

$$\begin{aligned}\left[a^-, a^+\right] f_n(X) &= a^- a^+ f_n(X) - a^+ a^- f_n(X)\\ &= a^- f_{n+1}(X) - a^+ \left(c f_{n-1}(X)\right)\\ &= c f_n(X) - c f_n(X)\\ &= 0.\end{aligned} \tag{34}$$

Thus, if for all $n \geq 1$, we denote by $\overline{P}_n$ the orthogonal projection of $L^2(\Omega, P)$ onto the closed subspace G_n, and by P_n the restriction of $\overline{P}_n$ to F, then we conclude that:

$$\left[a^-, a^+\right] = c P_0. \tag{35}$$

It follows from here that:

$$\begin{aligned}\left[a^-, X\right] &= \left[a^-, a^+ + a^-\right]\\ &= \left[a^-, a^+\right]\\ &= c P_0.\end{aligned} \tag{36}$$

For all natural numbers n, the random variable X^n can be obtained by repeatedly applying the multiplication operator by X, n times, to the constant polynomial random variable 1. We have:

$$a^- X^n = a^- X^n 1. \tag{37}$$

Using now Leibniz commutator rule:

$$[A, B_1 B_2 \cdots B_n] = \sum_{i=1}^{n} B_1 \cdots B_{i-1} [A, B_i] B_{i+1} \cdots B_n, \tag{38}$$

for all operators $A, B_1, B_2, \ldots, B_n$, we obtain:

$$\begin{aligned}a^- X^n &= a^- X^n 1\\ &= X^n a^- 1 + \left[a^-, X^n\right] 1\\ &= 0 + \sum_{i=1}^{n} X^{i-1} [a^-, X] X^{n-i} 1\\ &= \sum_{i=1}^{n} X^{i-1} c P_0 X^{n-i},\end{aligned} \tag{39}$$

where when we write $X^{n-i}1$, we interpret it as the composition of the multiplication operator by X with itself, $n - i$ times, applied to the constant polynomial random variable 1, while when we write X^{n-i} (without 1 to its right), we interpret it as the polynomial random variable X^{n-i}.

If $\langle \cdot, \cdot \rangle$ denotes the inner product of the Hilbert space $L^2(\Omega, P)$, then since P_0 is the projection onto $G_0 = \mathbb{C}1$, we have:

$$\begin{aligned} a^- X^n &= c \sum_{i=1}^{n} X^{i-1} P_0 X^{n-i} \\ &= c \sum_{i=1}^{n} X^{i-1} \langle X^{n-i}, 1 \rangle 1 \\ &= c \sum_{i=1}^{n} E\left[X^{n-i}\right] X^{i-1}. \end{aligned} \tag{40}$$

This formula can be used first to recover the moments of the random variable X, in the following way. Since $a^0 = 0$, all the odd moments are zero. The even moments can be computed as follows. For all non-negative integer k, we have:

$$\begin{aligned} E\left[X^{2k+2}\right] &= \langle X X^{2k+1} 1, 1 \rangle \\ &= \langle \left(a^+ + a^-\right) X^{2k+1} 1, 1 \rangle \\ &= \langle a^- X^{2k+1} 1, 1 \rangle + \langle X^{2k+1} 1, a^- 1 \rangle, \end{aligned} \tag{41}$$

since a^- is the polynomial dual of a^+. Because $a^- 1 = 0$, we obtain:

$$\begin{aligned} E\left[X^{2k+2}\right] &= \langle a^- X^{2k+1} 1, 1 \rangle \\ &= E\left[a^- X^{2k+1}\right]. \end{aligned} \tag{42}$$

Using formula (40) and the fact that all odd moments are zero, we obtain:

$$\begin{aligned} E\left[X^{2k+2}\right] &= E\left[a^- X^{2k+1}\right] \\ &= E\left[c \sum_{i=1}^{2k+1} E\left[X^{2k+1-i}\right] X^{i-1}\right] \\ &= c \sum_{j=0}^{k} E\left[X^{2(k-j)}\right] E\left[X^{2j}\right]. \end{aligned} \tag{43}$$

We prove first, by induction, that given any number $p > 1$, there exists a constant d (depending on p), such that for all nonnegative integers n, we have:

$$E\left[X^{2n}\right] \leq \frac{d^n}{(n+1)^p}. \tag{44}$$

Indeed, since $p > 1$, the series $\sum_{m \geq 1} 1/m^p$ is convergent. Let

$$d := 2^p c \sum_{m=1}^{\infty} \frac{1}{m^p}. \tag{45}$$

Then for $n = 0$, we have:

$$\begin{aligned} E\left[X^{2(0)}\right] &= 1 \\ &= \frac{d^0}{(0+1)^p}. \end{aligned} \tag{46}$$

So, the induction basis is checked.

Now, let us suppose that inequality (44) is true for all nonnegative integers $n \leq k$, and prove that it remains true for $n = k + 1$, where k is an arbitrary nonnegative integer. Using the recursive formula (43) and the induction hypothesis, we have:

$$\begin{aligned} E\left[X^{2(k+1)}\right] &= c \sum_{j=0}^{k} E\left[X^{2(k-j)}\right] E\left[X^{2j}\right] \\ &\leq c \sum_{j=0}^{k} \frac{d^{k-j}}{(k-j+1)^p} \cdot \frac{d^j}{(j+1)^p} \\ &= c \frac{d^k}{(k+2)^p} \sum_{j=0}^{k} \frac{(k+2)^p}{(k-j+1)^p (j+1)^p} \\ &= c \frac{d^k}{(k+2)^p} \sum_{j=0}^{k} \left[\frac{(k-j+1)+(j+1)}{(k-j+1)(j+1)}\right]^p \\ &= c \frac{d^k}{(k+2)^p} \sum_{j=0}^{k} \left(\frac{1}{k-j+1} + \frac{1}{j+1}\right)^p. \end{aligned} \tag{47}$$

Since $p > 1$, the function $x \mapsto x^p$ is convex on $[0, \infty)$, and applying Jensen inequality we obtain:

$$(x+y)^p \leq 2^{p-1}\left(x^p + y^p\right), \tag{48}$$

for all nonnegative real numbers x and y. Using this inequality, he have:

$$\begin{aligned}
E\left[X^{2(k+1)}\right] &\leq c\frac{d^k}{(k+2)^p}\sum_{j=0}^{k}\left(\frac{1}{k-j+1}+\frac{1}{j+1}\right)^p \\
&\leq c\frac{d^k}{(k+2)^p}\sum_{j=0}^{k}2^{p-1}\left[\left(\frac{1}{k-j+1}\right)^p+\left(\frac{1}{j+1}\right)^p\right] \\
&= d^k\frac{2^{p-1}c}{(k+2)^p}\cdot 2\sum_{m=1}^{k+1}\frac{1}{m^p} \\
&< d^k\frac{2^p c}{(k+2)^p}\sum_{m=1}^{\infty}\frac{1}{m^p} \\
&= \frac{d^{k+1}}{(k+2)^p}.
\end{aligned} \tag{49}$$

Hence, inequality (44) is proven by the principle of mathematical induction.

Because of inequality (44), the moment generating function:

$$\varphi(t) = E\left[\frac{1}{1-tX}\right] \tag{50}$$

$$= E\left[\sum_{n=0}^{\infty}t^nX^n\right] \tag{51}$$

is well-defined on a neighborhood V of 0, and using the Lebesgue dominated convergence theorem, we can interchange the expectation and sum, obtaining (since the odd moments vanish):

$$\begin{aligned}
\varphi(t) &= 1+\sum_{k=0}^{\infty}t^{2(k+1)}E\left[X^{2(k+1)}\right] \\
&= 1+ct^2\sum_{k=0}^{\infty}t^{2k}\sum_{j=0}^{k}E\left[X^{2(k-j)}\right]E\left[X^{2j}\right] \\
&= 1+ct^2\left(\sum_{j=0}^{\infty}t^{2j}E\left[X^{2j}\right]\right)^2 \\
&= 1+ct^2\left[\varphi(t)\right]^2.
\end{aligned} \tag{52}$$

Thus, we obtain that $\varphi(t)$ is a solution of the quadratic equation:

$$ct^2\left[\varphi(t)\right]^2-\varphi(t)+1=0. \tag{53}$$

Since $\varphi(0) = 1$ and φ is continuous on a neighborhood of 0, using the quadratic formula, we conclude that:

$$E\left[\frac{1}{1-tX}\right] = \frac{1-\sqrt{1-4ct^2}}{2ct^2}, \tag{54}$$

for all $t \in V$.

It is not hard to see that the random variable which has this moment generating function is given by the density function:

$$f_c(x) = \frac{1}{2\pi c}\sqrt{1-4cx^2}, \quad \text{for} \quad -2\sqrt{c} < x < 2\sqrt{c}, \tag{55}$$

and the space of polynomial random variables F is dense in $L^2(\Omega, \sigma(X), P)$, where $\sigma(X)$ is the sigma-algebra generated by X. Thus, the closure H of F in $L^2(\Omega, P)$ is equal to $L^2(\Omega, \sigma(X), P)$.

We are interested in finding a position-(0-momentum) decomposition of the quantum operators a^- and a^+ generated by X. For this, we go back to formula (40). Multiplying both sides of this formula by t^n, and summing up from $n = 1$ to ∞, we obtain:

$$\sum_{n=1}^{\infty} t^n a^- X^n = \sum_{n=1}^{\infty} t^n c \sum_{i=1}^{n} E\left[X^{n-i}\right] X^{i-1}. \tag{56}$$

Since a^- is a bounded operator from $H = L^2(\Omega, \sigma(X), P)$ to itself, and it is not hard to see that for t in a neighborhood V of 0, the random variable $1/(1-tX)$ belongs to $L^2(\Omega, \sigma(X), P)$ and $\sum_{k=0}^{n} t^k X^k \to 1/(1-tX)$ in $L^2(\Omega, \sigma(X), P)$, as $n \to \infty$, we can interchange $\sum_{n=0}^{\infty}$ and a^- in the last equation and obtain:

$$\begin{aligned} a^-\left[\sum_{n=1}^{\infty} t^n X^n\right] &= c\sum_{n=1}^{\infty} t^n \sum_{i=1}^{n} E\left[X^{n-i}\right] X^{i-1} \\ &= ct \sum_{n=1}^{\infty}\sum_{i=1}^{n} t^{n-i} E\left[X^{n-i}\right] t^{i-1} X^{i-1}. \end{aligned} \tag{57}$$

Making the change of index of summation $j = i - 1$ in the second sum, and using the fact that $a^-(t^0x^0) = a^-1 = 0$, we obtain:

$$a^-\left[\sum_{n=0}^{\infty} t^n X^n\right] = ct \sum_{n=1}^{\infty}\sum_{j=0}^{n-1} t^{n-1-j} E\left[X^{n-1-j}\right] t^j X^j. \tag{58}$$

Making now the change of the index of summation $m = n - 1$ in the first sum from the right side of the last formula, we obtain:

$$a^{-}\left[\sum_{n=0}^{\infty} t^n X^n\right] = ct \sum_{m=0}^{\infty}\sum_{j=0}^{m} t^{m-j} E\left[X^{m-j}\right] t^j X^j$$
$$= ct \left(\sum_{p=0}^{\infty} t^p E\left[X^p\right]\right)\left(\sum_{q=0}^{\infty} t^q X^q\right). \tag{59}$$

This is equivalent to:

$$a^{-}\left(\frac{1}{1-tX}\right) = ctE\left[\frac{1}{1-tX}\right]\frac{1}{1-tX}. \tag{60}$$

Using now formula (54), we obtain:

$$a^{-}\left(\frac{1}{1-tX}\right) = ct \cdot \frac{1-\sqrt{1-4ct^2}}{2ct^2} \cdot \frac{1}{1-tX}$$
$$= \frac{1-\sqrt{1-4ct^2}}{2t} \cdot \frac{1}{1-tX}. \tag{61}$$

This proves that the random variables $1/(1-tX)$ are eigenfunctions of the annihilation operator a^- corresponding to the eigenvalues $\lambda_t := (1-\sqrt{1-4ct^2})/(2t)$, for t in a neighborhood V of 0. This fact links the annihilation operator a^- to the 0-differentiation operator because of the following result.

Proposition 1 *For any t in a neighborhood of 0, the L^2-random variable*

$$\psi_t(X) = \frac{1}{1-tX} \tag{62}$$

is an eigenfunction of the 0-differentiation (0-momentum) operator, D_0, corresponding to the eigenvalue t.

Proof Indeed, for all t in a neighborhood of 0, we have:

$$D_0\psi_t(X) = \frac{\psi_t(0\cdot X) - \psi_t(X)}{0\cdot X - X}$$
$$= \frac{\psi_t(X) - \psi_t(0)}{X}$$
$$= \frac{1/(1-tX) - 1/(1-t\cdot 0)}{X}$$
$$= \frac{tX/(1-tX)}{X}$$
$$= t\cdot\frac{1}{1-tX}$$
$$= t\psi_t(X). \tag{63}$$

Using an idea from Riesz-Dunford Functional Calculus, if f is a holomorphic function, then $\psi_t(X)$ is an eigenfunction of $f(D_0)$ corresponding to the eigenvalue $f(t)$. We can take f to be the function:

$$\begin{aligned} f(z) &= \frac{1-\sqrt{1-4cz^2}}{2z} \\ &= \frac{1}{2z}\cdot\left[1-\left(1-\sum_{n=1}^{\infty}\frac{2^n(2n-3)!!c^n}{n!}z^{2n}\right)\right] \\ &= \sum_{n=1}^{\infty}\frac{2^{n-1}(2n-3)!!c^n}{n!}z^{2n-1}, \end{aligned} \tag{64}$$

which is holomorphic on a neighborhood of 0, and conclude that $\psi_t(X)$ is an eigenfunction of $f(D_0)$ corresponding to the eigenvalue $f(t)$, for all t in a neighborhood of 0. On the other hand, we have seen that $\psi_t(X)$ is also an eigenfunction of a^- corresponding to the same eigenvalue $f(t)$, for all t in a neighborhood V of 0. Since the vector space spanned by the random variables $\{\psi_t(X)\}_{t\in V}$ is dense in $L^2(\Omega, \sigma(X), P)$, we conclude that:

$$\begin{aligned} a^- &= f(D_0) \\ &= \sum_{n=1}^{\infty}\frac{2^{n-1}(2n-3)!!c^n}{n!}D_0^{2n-1}. \end{aligned} \tag{65}$$

Formula (65) represents the position-(0-momentum) decomposition of the annihilation operator, a^-. There is no problem with the convergence of the series from this formula, since the "classic" domain of a^- is the space F of polynomial random variables, $f(X)$. Thus if k denotes the degree of the polynomial f, then $D_0^n f(X) = 0$, for all $n > k$, and so:

$$\begin{aligned} a^- f(X) &= \sum_{n=1}^{\infty}\frac{2^{n-1}(2n-3)!!c^n}{n!}D_0^{2n-1}f(X) \\ &= \sum_{n=1}^{\lfloor (k+1)/2\rfloor}\frac{2^{n-1}(2n-3)!!c^n}{n!}D_0^{2n-1}f(X). \end{aligned} \tag{66}$$

Finally, having the position-(0-momentum) decomposition of a^-, we can easily find the position-(0-momentum) decomposition of the creation operator a^+ as:

$$\begin{aligned} a^+ &= X - a^- \\ &= X - \sum_{n=1}^{\infty}\frac{2^{n-1}(2n-3)!!c^n}{n!}D_0^{2n-1} \end{aligned} \tag{67}$$

$$= \sum_{n=0}^{\infty}A_n(X)D_0^n, \tag{68}$$

where:

$$A_n(X) = \begin{cases} X & \text{if} \quad n = 0 \\ 0 & \text{if} \quad n = 2k \quad \text{with } k \geq 1 \\ -\frac{2^{k-1}(2k-3)!!c^k}{k!} & \text{if } n = 2k-1 \text{ with } k \geq 1 \end{cases}, \tag{69}$$

where $(-1)!! := 1$.

Final Observation. For any real number q, if $T : F \to F$ is a linear operator, such that in its position-(q-momentum) decomposition:

$$T = \sum_{n=0}^{\infty} A_n(X) D_q^n,$$

all the position coefficients, $A_n(X) = a_n$, are constants (not depending on the position operator X), for $n \geq 0$, then we may call T a purely q-momentous operator. Then, using this definition, we can see a similarity between, what may be called, via the Central Limit Theorem, the most important random variables in the Classical Probability Theory, namely the gaussians, X_g, and the most important random variables in the Free Probability Theory, namely the semicircle random variables, X_s, by comparing their annihilation operators. For a gaussian random variable, its annihilation operator is:

$$a^- = \sigma D, \tag{70}$$

where σ is a positive constant and $D = D_1$ is the classic differentiation (momentum) operator, while for a semicircle distribution, its annihilation operator is:

$$a^- = \sum_{n=1}^{\infty} \frac{2^{n-1}(2n-3)!!c^n}{n!} D_0^{2n-1}.$$

Even though the decomposition of a^- in the semicircle case is much more complicated than in the gaussian case, the annihilation operator for the gaussian and semicircle random variables is purely 1-momentous and 0-momentous, respectively.

References

1. Acardi, L., Bożejko, M.: Interacting Fock spaces and Gaussianization of probability measures. Infin. Dimesn. Anal. Quantum Probab. Relat. Top. **1**(4), 663–670 (1998)
2. Accardi, L., Kuo, H.–H., Stan, A.I.: Characterization of probability measures through the canonically associated interacting Fock spaces. Infin. Dimens. Anal. Quantum Probab. Relat. Top. **7**(4), 485–505 (2004)
3. Accardi, L., Kuo, H.–H., Stan, A.I.: Moments and commutators of probability measures. Infin. Dimens. Anal. Quantum Probab. Relat. Top. **10**(4), 591–612 (2007)

4. Accardi, L., Nahni, M.: Interacting Fock spaces and orthogonal polynomials in several variables. In: Non-Commutativity, Infinite-Dimensionality and Probability at the Crossroads, vol. 16, pp. 192–205 (2002)
5. Asai, N., Kubo, I., Kuo, H.-H.: Generating functions of orthogonal polynomials and Szegö-Jacobi parameters. Prob. Math. Stat. **23**, 273–291 (2003)
6. Chihara, T.S.: An Introduction to Orthogonal Polynomials. Gordon & Breach, New York (1978)
7. Stan, A.I.: PAC commutators and the R-transform. Commun. Stoch. Anal. **3**(1), 15–31 (2009)
8. Stan, A.I., Popa, G., Dutta, R.: Position-momentum decomposition of linear operators defined on algebras of polynomials. J. Math. Phys. **62**(1) (2021). https://doi.org/10.1063/5.0008155
9. Szegö, M.: Orthogonal Polynomials, vol. 23. Colloquium Publications, American Mathematical Society (1975)
10. Voiculescu, D.V., Dykema, K.J., Nica, A.: Free Random Variables, vol. 1. American Mathematical Society, Providence, Rhode Island, USA, CRM Monograph Series (1992)

Quantum Probability for Modeling Cognition, Decision Making, and Artificial Intelligence

Andrei Khrennikov

Abstract This is the brief review on quantum-like modeling, quantum probability versus classical probability in cognition, psychology, decision making, and artificial intelligence. We describe the problems induced by using classical probability (CP) for mathematical formalization of decision making—paradoxes and probability fallacies. These problems are related to the violation of the formula of total probability (FTP) derived in CP: either violation of additive of probability or Bayes formula for conditional probability. Quantum probability (QP) generates FTP with additional interference term, the latter provides the right balance of experimental probabilities obtained in cognitive psychology, decision making, and artificial intelligence. We consider application of QP to model violation of the Savage Sure Thing Principle (the axiom of agents' rationality in CP-based decision making) and the order and response replicability effects. We make comparison CP versus QP rationality and briefly consider coupling with social laser theory.

Keywords Quantum-like modeling · Quantum probability · Classical probability · Cognition · Psychology · Decision making · Artificial intelligence · Order effect · Response replicability effect

1 Introduction

This brief review is not about mathematical modeling of the "quantum brain" in the spirit of Penrose and Hameroff. We do not try to reduce information processing by cognitive systems to the quantum physical effects in them. On the other hand, we do not criticize the quantum brain project, although its difficulties are well known: the brain is too hot and big, the scales of neurons' operating are too rough comparing with

---

A. Khrennikov (✉)
Linnaeus University, International Center for Mathematical Modeling in Physics and Cognitive Sciences, SE-351 95, Växjö, Sweden
e-mail: Andrei.Khrennikov@lnu.se

L. Accardi et al. (eds.), *Infinite Dimensional Analysis, Quantum Probability and Applications*, Springer Proceedings in Mathematics & Statistics 390,
https://doi.org/10.1007/978-3-031-06170-7_4

the quantum physical scales. For our approach, it is simply non-important whether genuine quantum physical processes in brain's cells contribute to cognition or not.

In our modeling, the brain is a black box which information processing cannot be described by classical probability theory (CP). We point out that a plenty of such nonclassical probabilistic data can be found in cognitive psychology, game theory, decision making, social science, economics, finance, and politics. In decision theory, such data was coupled to probability fallacies and irrational behavior of agents [1–4].

Our suggestion [5–7] is to model cognition and decision making with the most well developed non-classical theory of probability, namely, based on the mathematical formalism of QM—quantum probability (QP). One may say that the appeal to QP (and consequently to quantum information) to model humans' behavior and decision making too exotic. QP was developed to model the behavior of micro-systems, atoms, electrons, photons. It seems that the domain of its application is the micro-world. Why do we apply this theory to humanities?

We recall that as early as the 1970s, Tversky (one of the most cited psychologists of all time) and Kahneman (Nobel prize in economics in 2002, for prospect theory, which he co-developed with Tversky) have been demonstrating cases where classical probability prescription and human behavior persistently diverge [1–4]. Kahneman pointed out that today, we are at the theoretical cross-roads, with huge divisions across conflicting, entrenched theoretical positions:

Should we continue relying on CP as the basis for descriptive and normative predictions in decision making (and perhaps ascribe inconsistencies to methodological idiosyncrasies)?

Should we abandon probability theory completely and instead pursue explanations based on heuristics, as Tversky and Kahneman proposed?

However, the use of the probabilistic and statistical methods is really the cornerstone of the modern scientific methodology, both in natural and social sciences. Thus, although the heuristic approach to decision making cannot be discarded completely, it seems more natural not reject completely the probabilistic approach to decision making, but to search for novel probabilistic models. And QP is a good candidate for an alternative to CP.

Mathematical models which are based on QP and applied outside of physics are know as *quantum-like* (the terminology was invented by the author to distinguish such studies from genuine quantum physics). Recently quantum-like modeling became fashionable in humanities, decision making, economics and finance (see, e.g., [7–14]).

2 Classical Versus Quantum Probability

CP was mathematically formalized by Kolmogorov [15] (1933). This is the calculus of probability measures, non-negative weight $p(A)$ is assigned to any event A. The main property of CP is its additivity: if two events A_1, A_2 are disjoint, then the probability of disjunction of these events equals to the sum of probabilities:

$$P(A_1 \vee A_2) = P(A_1) + P(A_2).$$

QP is the calculus of complex amplitudes or in the abstract formalism complex vectors. Thus, instead of operations on probability measures one operates with vectors. Each complex amplitude ψ gives the probability by the Born's rule [16]:

Probability is the square of the absolute value of the complex amplitude:

$$p = |\psi|^2.$$

2.1 Interference of Probabilities

By operating with complex probability amplitudes, instead of the direct operation with probabilities, one can violate the basic laws of CP, in particular, additivity of probability. One can get that, for disjoint events, the probability of disjunction is strictly smaller or larger than the sum of probabilities:

$$P(A_1 \vee A_2) < P(A_1) + P(A_2)$$

or

$$P(A_1 \vee A_2) > P(A_1) + P(A_2),$$

since QP calculus leads to the formula [17] (see also [18, 19] for discussion at the physical level of rigorousness)

$$P(A_1 \vee A_2) = P(A_1) + P(A_2) + 2\cos\theta\sqrt{P(A_1)P(A_2)}. \quad (1)$$

The additional term is known as the interference term.

We recall that interference is the basic feature of waves, so often one speaks about the probability waves. These are not usual waves in physical space, these are fluctuations of probability which can lead to its amplification (constructive interference for $\cos\theta > 0$) or diminishing (constructive interference for $\cos\theta < 0$).

2.2 Bayesian Versus Non-Bayesian Inference

We recall the *Bayes formula* for conditional probability in the CP-framework: $\mathbf{P}(B|A) = \frac{\mathbf{P}(B\cap A)}{\mathbf{P}(A)}$, $\mathbf{P}(A) > 0$. The *formula of total probability* (FTP) is a simple consequence of the Bayes formula. Consider a pair a and b of discrete random variables CP-representing observables. Then

$$\mathbf{P}(b = \beta) = \sum_{\alpha} \mathbf{P}(a = \alpha)\mathbf{P}(b = \beta|a = \alpha). \quad (2)$$

By using Bayesian inference, one can find the posterior probability with the aid of the Bayes theorem (a trivial consequence of the Bayes formula for conditional probability):

$$P(H \mid E) = \frac{P(E \mid H) \cdot P(H)}{P(E)} \tag{3}$$

where H stands for a hypothesis which validity is checked. Typically decision maker considers a few competing hypotheses (H_i), and the task is to determine which is the most probable; $P(H)$ is the prior probability, it gives the estimate of hypothesis' probability before the observational data E is obtained. The output of inference is the posterior probability P$(H \mid E)$; this is the probability after data E is observed. Probability $P(E)$ is determined by FTP:

$$P(E) = \sum_j P(H_j)P(E|H_j),$$

where (H_j) is complete set of hypothesis under consideration. Hence,

$$P(H_i \mid E) = \frac{P(E \mid H_i) \cdot P(H_i)}{\sum_j P(H_j)P(E|H_j)} \tag{4}$$

To make a decision, one compares posterior probabilities $(P(H_i \mid E))$ and finds the maximal of them. We remark that Bayesian approach to conditional probability plays the crucial role in CP modeling of cognition, machine learning, artificial intelligence, especially applications of Bayesian networks.

QP provides the possibility to relax some constraints on statistical data posed by CP. One of such constraints is classical FTP which is a consequence of additivity of CP and the Bayes formula in definition of conditional probability. In terms of quantum conditional probabilities formula (1) can be written for dichotomous quantum observables as

$$\mathbf{P}(b = \beta) = \mathbf{P}(a = \alpha_1)\mathbf{P}(b = \beta|a = \alpha_1) + \mathbf{P}(a = \alpha_2)\mathbf{P}(b = \beta|a = \alpha_2) \tag{5}$$

$$+2\cos\theta\sqrt{\mathbf{P}(a = \alpha_1)\mathbf{P}(b = \beta|a = \alpha_1)\mathbf{P}(a = \alpha_2)\mathbf{P}(b = \beta|a = \alpha_2)}$$

This is the quantum version of classical FTP–FTP with interference term. This formula was derived in author's book [17] and then widely used in various applications outside of physics [20–22].

More generally, QP extends the calculus of conditional probabilities and the operation of the probability update is generalized via the quantum operation of the state update. The latter is based on the projection postulate or quantum instrument theory [23–27]. Thus the QP decision making project can be considered as extension of boundaries of Bayesian probability inference, with wide applications to decision making [28].

3 Quantum-Like Paradigm

The following paradigm can be used to motivate the applications of QP outside of physics.

Quantum-like paradigm (Khrennikov 1999): *The mathematical formalism of quantum information and probability theories can be used to model behavior not only of genuine quantum physical systems, but all context-sensitive systems, e.g., humans.*

Contextual information processing cannot be based on complete resolution of ambiguity. It is meaningless to do this for the concrete context, if tomorrow context will be totally different. Therefore such systems process ambiguities, represented by superpositions of alternatives.

We illustrate such information processing with the following example. Ivan has girl-friend Sara; he is not sure whether he love her or not. His belief-state can be mathematically represented as superposition

$$|\psi\rangle(t) = c_0(t)|not - love\rangle + c_1(t)|love\rangle, \tag{6}$$

where $|c_0(t)|^2 + |c_1(t)|^2 = 1$. These complex amplitudes determine intensities of his feelings. They fluctuate in the process of his belief-state evolution.

He can process information in two ways:

- classically: at each instant of time t, Ivan resolves uncertainty and determines whether he loves Sara and he updates his behavior on the basis of this information;
- quantumly: Ivan does not try to resolve uncertainty and proceeds with superposition $|\psi\rangle(t)$.

The latter processing saves a lot of computational and psychic resources. Ivan being in love-superposition can use these resources to solve other life-problems, e.g., getting education. Of course, even "quantum-like Ivan" would soon or latter resolve this ambiguity—to perform the self-measurement, but he may as well to proceed with this love-superposition for years and even for the whole life. Self-measurement can happen as the result of some external circumstances—spontaneous collapse of the mental wave function.

4 Paradoxes of CP Decision Theory and Their QP Resolution

The use QP, instead of CP, can resolve some paradoxes of CP-based theory of decision making, economics, and game theory; e.g., the Allais (1953), Ellsberg (1961), and Machina (1987) paradoxes. Typically a paradox (as a sign of irrational behavior) is probabilistically expressed as the violation of the law of additivity of probability (or FTP), i.e., that for disjoint events

$$P(A_1 \vee A_2) \neq P(A_1) + P(A_2).$$

QP resolves all known paradoxes of classical decision theory

The number of paradoxes generated by the classical decision making theory is really amazing. The authors of the recent review (Erev and Ert 2015) counted 35 basic paradoxes. During many years DM-theory was developed through creation of paradoxes and resolving them through modifications of the theory, e.g., from expected utility theory to the prospect theory. But any modified theory suffered of new paradoxes. The use of QP can resolve all such paradoxes, at least this was shown in paper [12].

QP was also successfully applied to model a variety of psychological effects, e.g., the order, disjunction, and conjunction effects [13, 14].

5 Quantum Scheme of Decision Making

The belief state of a decision maker is represented by a quantum state, typically by a pure state ψ. A question asked to her or task which she has to perform (to select some output) is represented by a quantum observable given by Hermitian operator $\hat{A}$. Its eigenvalues encode the possible answers (outcomes), $a_1, ..., a_k$,

For Alice in the state ψ, the probability of decision a_j is given by the Born rule:

$$P_\psi(A = a_j) = \|E^A_{a_j}\psi\|^2. \tag{7}$$

Then, for a large ensemble of decision makers, the probability is given by this formula. For one fixed question or task, quantumness cannot be detected, only a few incompatible tasks can generate quantum effects.

This scheme was approved for the plenty of cognitive and psychological effects. However, attempts to combine these effects lead to understanding that it is impossible withing the standard quantum measurement theory based on Hermitian operators and the state update of the projection form.

6 Interference in Decision Making

6.1 *Savage Sure Thing Principle as the Rationality Axiom*

In classical theory of decision making, rational behavior of agents is formalized with the *Savage Sure Thing Principle* (STP) [29]:

If you prefer prospect b_+ to prospect b_- if a possible future event A happens ($a = +1$); and you prefer prospect b_+ still if future event A does not happen ($a = -1$);

then you should prefer prospect b_+, despite having no knowledge of whether or not event A will happen.

Savage's illustration refers to a person deciding whether or not to buy a certain property shortly before a presidential election, the outcome of which could radically affect the property market:

"Seeing that he would buy in either event, he decides that he should buy, even though he does not know which event will obtain".

STP is considered as the axiom of rationality of decision makers [29]. It plays the important role in decision making and economics in the framework of Savage's subjective utility theory. In the latter, probability is formalized in the CP-framework [15] and it is endowed with the subjective interpretation.

One of the first violation of the Savage Sure Thing Principle was found by Tversky and Shafir [30]. This study was motivated by behaviour of their own students in San-Francisco who had the custom to take their after-exam vocation in Hawaii. Tversky and Shafir showed that significantly more students report that they would purchase a nonrefundable Hawaii vacation if they knew that they had passed or failed an important exam as opposed to the case when they did not know the outcome of the exam.

We stress that STP is a simple consequence of FTP:

$$\mathbf{P}(b = \beta) = \sum_{\alpha} \mathbf{P}(a = \alpha)\mathbf{P}(b = \beta | a = \alpha). \tag{8}$$

Here the b-variable determine the decision variable, say $b = +1$, "buy house", $b = -1$, "not buy house", and the a-variable represents conditions, say outcomes of elections. Violation of FTP implies violation of STP. Thus, the degree of satisfaction of FTP can be used as a statistical test of classical (STP-type) rationality.

In cognitive psychology, violation of STP is known as the *disjunction effect.* A plenty of statistical data was collected in cognitive psychology in experiments demonstrating disjunction effect. For example, in experiments of the Prisoners' Dilemma type [1–4]. Such data violate FTP. The latter implies irrationality (from classical viewpoint) of agents participating in experiments (mainly students).

We recall that FTP is derived from two assumptions that are firmly incorporated into the Kolmogorov axiomatics:

1. Additive law for probability.
2. Baeys formula for conditional probability.

Therefore, violation of FTP and, hence, of STP (and classical rationality) is generated either by violation of additivity of probability or the Bayes law for conditional probability or by the combination of these factors. Generally, this leads to the impossibility to use in decision making Bayesian inference. Quantum(-like) agents proceed with more general inference machinery based on the quantum state update.

Hence, classical rationality is Bayesian inference rationality and quantum rationality is non-Bayesian inference rationality.[1]

6.2 Is CP-Irrationality Just QP-Rationality?

In the light of above considerations, one can ask:

Are quantum agents irrational?

As was discussed, by using QP it is possible to violate FTP and hence STP. Therefore, generally quantum-like agents are (classically) irrational. However, we can question the CP-approach to mathematical formalization of decision making and, consequently, the corresponding notion of rationality. We define quantum(-like) rationality as respecting the quantum calculus of probabilities and the quantum formula for interference of probabilities, FTP with the interference term.

In the framework of quantum-like modeling, violation of the CP-laws including the Bayes formula for conditional probability or even additivity of probability are not exotic at all. Moreover, the situation in which the probabilistic data satisfies FTP seems to be rather an exception than the norm. We can speculate that CP-processing of information was resulted from evolution of biological systems, not only humans, but even animals and simple bio-organisms. The question whether the "genuine human behavior" should be characterized by classical rationality taken as the normative theory for the rational decision making is very complicated.

Violation of FTP implies violation of STP. Thus satisfaction of FTP can be used as a statistical test of rationality.

A plenty of statistical data was collected in cognitive psychology in experiments demonstrating disjunction effect. For example, in experiments of the Prisoners' Dilemma type. (works of Shafir, Tversky, Kahneman). Such data violate FTP. The latter implies irrationality of agents participating in experiments (mainly students).

6.3 Logic and Rationality

CP is based on classical Boolean logic and QP is based on quantum non-Boolean logic. The latter leads to relaxation of the basic rules of classical logic, in particular, the distributivity law for conjunction and disjunction. Roughly speaking, quantum logic information processing is simpler and it needs less computational resources: one can operate in partial Boolean subalgebra of quantum logic, one need not to construct conjunctions, disjunctions, and negations for all possible events. Such information

[1] Of course, non-Bayesian probability updates are not reduced to quantum, given by state transformations in the complex Hilbert space. One may expect that human decision making violates not only classical, but even quantum rationality (Sect. 6.2).

processing is superior in the situation of *information overload.* It saves a lot of information resources.

However, this kind of information processing can lead to inconsistencies between conclusions which can be derived on the basis of different subalgebras. This inconsistencies are behavioral irrationalities (from the viewpoint of Boolean logic), probability fallacies—the signs that people cannot operate properly with probabilities. However, it seems that people can operate with probabilities, but they use non-classical probability calculus.

6.4 Social Laser

One of the consequences of information overload is that information loses its content. A human has no possibility to analyze deeply the content of communications delivered by mass-media and social networks. People process information without even attempting to construct an extended Boolean algebra of events. They operate with labels such as say covid-19, vaccination, pandemy without trying to go deeper beyond this labels. Contentless information behaves as a bosonic quantum field which is similar to the quantum electromagnetic field. Interaction of humans with such quantum information field can generate a variety of quantum-like behavioral effects. One of them is social lasing, *stimulated amplification of social actions* (SASA) [31–36]. In social laser theory, humans play the role of atoms, social atoms (s-atoms). Interaction of the information field composed of indistinguishable (up to some parameters, as say social energy) excitations with gain medium composed of s-atoms generate the cascade type process of emission of social actions. SASA describes well, e.g., color revolutions and other types of mass protests (see monograph [?] for detailed presentation).

Over the past years, our society has been constantly shaken by high-amplitude information waves. These are waves of enormous social energy. They are often destructive and are a kind of information tsunami. The main distinguishing features of these waves are their high amplitude, coherence (the homogeneous nature of the social actions they generate) and the short time required for their generation and relaxation, huge singular spikes.

We showed [31–36] that such waves can be modeled using the social laser, which describes stimulated amplification of coordinated social actions. "Actions" are interpreted very broadly, from mass protests, in particular, leading to color revolutions such as the Orange or Maidan revolutions in Ukraine or the recent mass protests in USA (anti-Baiden protests for fair votes), Belarus (anti-Lukashenko protests for fair votes), Russia (anti-Putin protests for liberation of Naval'nii), Germany, UK, Australia, Canada, Sweden (protests against the corona-restrictions and violation of the basic human rights with pandemic-justification) as well as generating of "right voting" and other collective decisions as, e.g., acceptance of lockdown and support of the total vaccination against covid-19 by the majority of population [31].

7 Quantum-Like AI

In the quantum-like approach, We proceed with the formal mathematical model in which bio or AI systems' states are given by density operators. Here we do not pre-assume that functioning of an AI system is based on real quantum physical devices such a say quantum computers. (The author is skeptical to the possibility of wide use such devices.) We are interested in quantum-like AI systems based on classical computational devices, but information processing is based on the formalism of quantum theory: information states and their dynamics are described by this formalism. One of the first successful realizations of such an approach was the EU-project QUARTZ on quantum information retrieval [37].

Quantum-like information processing can have advantages not only withing quantum computing perspective, i.e., not only through over performing classical digital computers. Computational resources can be saved not only via such tremendous increase of computational power, but also with using the possibility to process in parallel incompatible quantum observables, without constructing a "big common Boolean algebra". The latter can be done at the software level, without attracting expensive quantum computer hardware.

Notes and Comments. In physics, Bell type inequalities are the important tests separating CP and QP. For humans, such tests were also done and they demonstrated the violation of such inequalities [38–40].

References

1. Kahneman, D., Tversky, A.: Subjective probability: a judgment of representativeness. Cogn. Psych. **3**(3), 430–454 (1972)
2. Kahneman, D., Tversky, A.: Prospect theory: an analysis of decision under risk. Econometrica **47**, 263–291 (1979)
3. Kahneman, D., Tversky, A.: Choices, values and frames. Am. Psych. **39**(4), 341–350 (1984)
4. Kahneman, D.: Maps of bounded rationality: psychology for behavioral economics. Am. Econ. Rev. **93**(5), 1449–1475 (2003)
5. Khrennikov, A.: Classical and quantum mechanics on information spaces with applications to cognitive, psychological, social and anomalous phenomena. Found. Phys. **29**, 1065–1098 (1999)
6. Khrennikov, A.: Information Dynamics in Cognitive, Psychological, Social, and Anomalous Phenomena, Series: Fundamental Theories of Physics. Kluwer, Dordrecht (2004)
7. Khrennikov, A.: Ubiquitous Quantum Structure: From Psychology to Finances. Springer, Berlin (2010)

8. Haven, E., Khrennikov, A.: Quantum Social Science. Cambridge University Press, Cambridge (2013)
9. Asano, M., Khrennikov, A., Ohya, M., Tanaka, Y., Yamato, I.: Quantum Adaptivity in Biology: From Genetics to Cognition. Springer, Heidelberg (2015)
10. Haven, E., Khrennikov, A., Robinson, T.R.: Quantum Methods in Social Science: A First Course. WSP, Singapore (2017)
11. Asano, M., Basieva, I., Khrennikov, A., Ohya, M., Tanaka, Y., Yamato, I.: Quantum information biology: from information interpretation of quantum mechanics to applications in molecular biology and cognitive psychology. Found. Phys. **45**(10), 1362–1378 (2015)
12. Asano, M., Basieva, I., Khrennikov, A., Ohya, M., Tanaka, Y.: A quantum- like model of selection behavior. J. Math. Psychol. **78**, 2–12 (2017)
13. Accardi, L., Khrennikov, A., Ohya, M.: Quantum Markov model for data from Shafir-Tversky experiments in cognitive psychology. Open. Syst. Inf. Dyn. **16**, 371–385 (2009)
14. Haven, E., Khrennikov, A.: Quantum mechanics and violation of the sure-thing principle: the use of probability interference and other concepts. J. Math. Psychol. **53**, 378–388 (2009)
15. Kolmolgorov, A.N.: Foundations of the Probability Theory. Chelsea Publishing Company, New York (1956)
16. von Neuman, J.: Mathematical Foundations of Quantum Mechanics. Princeton University Press, Princenton (1955)
17. Khrennikov, A.Y.: Interpretations of Probability. VSP Int. Sc Publishers, Utrecht/Tokyo (1999)
18. Feynman, R. P.: The concept of probability in quantum mechanics. In: Berkeley Symposium on Mathamatics Statistics and Probability. Proceedings of the Second Berkeley Symposium on Mathamatics Statistics and Probability, pp. 533–541. University of Califonia Press (1951)
19. Feynman, R., Hibbs, A.: Quantum Mechanics and Path Integrals. McGraw-Hill, New York (1965)
20. Khrennikov, A.: On quantum-like probabilistic structure of mental information. Open. Syst. Inf. Dyn. **11**(3), 267–275 (2004)
21. Khrennikov, A.: The principle of supplementarity: a contextual probabilistic viewpoint to complementarity, the interference of probabilities, and the incompatibility of variables in quantum mechanics. Found. Phys. **35**(10), 1655–1693 (2005)
22. Khrennikov, A.: Quantum-like brain: interference of minds. Biosystems **84**, 225–241 (2006)
23. Davies, E.B.: Quantum Theory of Open Systems. Academic, London (1976)
24. Ozawa, M.: Quantum perfect correlations. An. Phys. **321**, 744–769 (2006)
25. Ozawa, M.: Probabilistic interpretation of quantum theory. N. Gener. Comput. **34**, 125–152 (2016)
26. Ozawa, M., Khrennikov, A.: Application of theory of quantum instruments to psychology: combination of question order effect with response replicability effect. Entropy **22**(1), 37–52 (2020)
27. Khrennikov, A.: Quantum-like model of unconscious-conscious dynamics. Front. Psychol. **6**, art. 997 (2015)
28. Khrennikov, A.: Quantum Bayesianism as the basis of general theory of decision-making. Phil. Trans. R. Soc. A **374**, 20150245 (2016)
29. Savage, L.J.: The Foundations of Statistics. Wiley, New York (1954)
30. Tversky, A., Shafir, E.: The disjunction effect in choice under uncertainty. Psycholo. Sci. **3**, 305–309 (1992)
31. Khrennikov, A.: Social Laser. Jenny Stanford Publication, Singapore (2020)
32. Khrennikov, A.: Towards information lasers. Entropy **17**(10), 6969–6994 (2015)
33. Khrennikov, A.: Social laser: action amplification by stimulated emission of social energy. Phil. Trans. R. Soc. **374**(2054), 20150094 (2016)
34. Khrennikov, A.: Social laser model: from color revolutions to Brexit and election of Donald Trump. Kybernetes **47**(2), 273–278 (2018)
35. Khrennikov, A., Alodjants, A., Trofimova, A., Tsarev, D.: On interpretational questions for quantum-Like modeling of social lasing. Entropy **20**(12), 921–942 (2018)

36. Khrennikov, A., Tofano, Z., Dubois, F.: Concept of information laser: from quantum theory to behavioural dynamics. Eur. Phys. J. Spec. Top. **227**(15–16), 2133–2153 (2019)
37. Aerts, D., Khrennikov, A., Melucci, M., Toni, B. (eds.), Quantum-like models for information retrieval and decision-making. STEAM-H: Science, Technology, Engineering, Agriculture, Mathematics and Health. Springer, Berlin (2020)
38. Conte, E., Khrennikov, A., Todarello, O., Federici, A.: A preliminary experimental verification on the possibility of Bell inequality violation in mental states. Neuroquantology **6**(3), 214–221 (2008)
39. Asano, M., Khrennikov, A., Ohya, M., Tanaka, Y., Yamato, I.: Violation of contextual generalization of the Leggett-Garg inequality for recognition of ambiguous figures. Phys. Scr. **T163**, 014006 (2014)
40. Basieva, I., Cervantes, V.H., Dzhafarov, E.N., Khrennikov, A.: True contextuality beats direct influences in human decision making. J. Exp. Psych.: General **148**, 1925–1937 (2019)

Part II
Quantum Information Methods

Note on Complexity of Communication Processes

Noboru Watanabe

Dedicated to the memories of Professor Robin Hudson
Professor Andrzej Kossakowski and Professor W. von Waldenfels

Abstract In order to study the efficiency of information transmission of the quantum communication processes consistently, we discuss the entropy type functional and the mutual entropy type functional with respect to the initial state and the quantum communication channel in general quantum systems. The mutual entropy type measures are formulated by the compound states between the initial and final systems. In this paper, we modify the compound state and investigate the entropy functional and the mutual entropy functional constructed by the modified compound states by means of the initial state and the completely positive channel to examine the efficiency of information transmission of the general quantum communication processes

Keywords Quantum entropy · Quantum compound states · Quantum communication theory

1 Introduction

The quantum entropy was defined by von Neumann [11] in 1932 and the quantum relative entropy was introduced by Umegaki [31], and this relative entropy was extended to general quantum systems by Araki [2, 3], Uhlmann [32] and Donald [4]. In information theory, the efficiency of the information transmission of the communication processes with respect to the initial state p and the channel Λ^* is

N. Watanabe (✉)
Tokyo University of Science, Noda City, Chiba 278-8510, Japan
e-mail: watanabe@is.noda.tus.ac.jp

L. Accardi et al. (eds.), *Infinite Dimensional Analysis, Quantum Probability and Applications*, Springer Proceedings in Mathematics & Statistics 390,
https://doi.org/10.1007/978-3-031-06170-7_5

discussed by using the entropy and the mutual entropy (information) [30], which satisfy the fundamental inequalities such as

$$0 \le I(p; \Lambda^*) \le \min \left\{ S(p), S(\Lambda^* p) \right\}.$$

In quantum information theory, it is necessary to construct the quantum mutual entropy for the quantum systems by introducing joint states (compound states) [33] in quantum compound systems. The Ohya compound state defined by using the von Neumann - Schatten decomposition of the initial state ρ and the quantum channel Λ^*[1, 5]. By applying the Ohya compound state Φ_μ^S, the quantum mutual entropy $I^S(\varphi; \Lambda^*)$ [6, 13, 14, 17, 18, 24, 28] was defined by taking the supremum of the quantum relative entropy between the Ohya compound state Φ_μ^S and the trivial compound state Φ_0 for any maximal measure μ of the initial state φ pseudosupported on $ex\mathcal{S}$, the set of all extreme points of $\mathcal{S}$, which is a weak* compact and convex subset of $\mathfrak{S}(\mathcal{A})$. They holds the fundamental inequalities $0 \le I^S(\varphi; \Lambda^*) \le S^S(\varphi)$ [13, 14]. The entropy and the mutual entropy of C*-systems were introduced by Ohya in [15, 17] and it was extended in [8, 9], and these properties are discussed in [6, 10] and in a book [18].

The study of entangled state is an important topic in quantum information theory. One of the remarkable results to discuss the entanglement is the Jamiołkowski's isomorphism [17]. The study of quantum capacity is investigated in [19–21].

In this paper, we briefly explain the formulation of the Ohya compound state based on the Jamiołkowski isomorphism, and we define the mutual entropy type complexities consisted of the quantum relative entropy between the modified entangled compound states and trivial compound states in general quantum systems. The purpose of this paper is to consider the efficiency of information transmission measured by the quantum mutual entropy type complexities formulated by the modified entangled compound states in general quantum systems.

2 Quantum Channels

Let $\mathcal{A}_1$ (resp. $\mathcal{A}_2$) be a C*-algebra or $\mathbf{B}(\mathcal{H}_1)$ (resp. $\mathbf{B}(\mathcal{H}_1)$) the set of all bounded operators on a separable complex Hilbert space $\mathcal{H}_1$ (resp. $\mathcal{H}_2$) . We denote the input (resp. output) quantum system by $\mathfrak{S}(\mathcal{A}_1)$ (resp. $\mathfrak{S}(\mathcal{A}_2)$) . ($\mathcal{A}_1$, $\mathfrak{S}(\mathcal{A}_1)$) (resp. ($\mathcal{A}_2$, $\mathfrak{S}(\mathcal{A}_2)$) is the input (resp. output) quantum system. Let Λ is a linear mapping from $\mathcal{A}_2$ to $\mathcal{A}_1$ with $\Lambda(I_2) = I_1$, where I_k is the identity operator in $\mathcal{A}_k$ $(k = 1, 2)$. The dual map Λ^* of Λ is a linear quantum channel from $\mathfrak{S}(\mathcal{A}_1)$ to $\mathfrak{S}(\mathcal{A}_1)$ given by $\Lambda^*(\varphi)(B) = \varphi(\Lambda(B))$ for any $\varphi \in \mathfrak{S}(\mathcal{A}_1)$ and any $B \in \mathcal{A}_2$. If Λ holds

$$\sum_{i,j=1}^{n} A_i^* \Lambda(B_i^* B_j) A_j \ge 0$$

for all $n \in \mathbf{N}$, all $B_j \in \mathcal{A}_2$ and all $A_j \in \mathcal{A}_1$ is said to be a completely positive (C.P.) channel [6, 12, 16, 18, 24, 28].

2.1 Quantum Communication Processes

$\mathcal{K}_1$ and $\mathcal{K}_2$ are two Hilbert spaces representing noise and loss systems. Let $\mathfrak{S}(\mathcal{H}_k)$ (resp. $\mathfrak{S}(\mathcal{K}_k)$) be the set of all density operators on $\mathcal{H}_k$ (resp. $\mathcal{K}_k$) $(k = 1, 2)$. Put $\mathcal{A}_1 = \mathbf{B}(\mathcal{H}_1)$, $\mathcal{A}_2 = \mathbf{B}(\mathcal{H}_2)$, $\mathfrak{S}(\mathcal{A}_1) = \mathfrak{S}(\mathcal{H}_1)$, $\mathfrak{S}(\mathcal{A}_2) = \mathfrak{S}(\mathcal{H}_2)$, $\mathcal{B}_2 = \mathbf{B}(\mathcal{K}_1)$, $\mathcal{B}_2 = \mathbf{B}(\mathcal{K}_2)$, $\mathfrak{S}(\mathcal{B}_1) = \mathfrak{S}(\mathcal{K}_1)$ and $\mathfrak{S}(\mathcal{B}_2) = \mathfrak{S}(\mathcal{K}_2)$.

Ohya [13] formulated a model of quantum channel with respect to quantum communication process considering noise and loss.

Let γ^* be a CP channel from $\mathfrak{S}(\mathcal{A}_1)$ to $\mathfrak{S}(\mathcal{A}_1 \otimes \mathcal{B}_1)$ defined by

$$\gamma^* (\varphi) = \varphi \otimes \psi$$

for any input state $\varphi \in \mathfrak{S}(\mathcal{A}_1)$ and any noise state $\psi \in \mathfrak{S}(\mathcal{B}_1)$ and a^* be a CP channel from $\mathfrak{S}(\mathcal{A}_2 \otimes \mathcal{B}_2)$ to $\mathfrak{S}(\mathcal{A}_2)$ given by

$$a^* (\Psi) (A) = \Psi (A \otimes I_2)$$

for any $\Psi \in \mathfrak{S}(\mathcal{A}_2 \otimes \mathcal{B}_2)$ and any $A \in \mathcal{A}_2$, where I_k is the identity operator in $\mathcal{B}_k$ $(k = 1, 2)$. π^* is a CP channel from $\mathfrak{S}(\mathcal{A}_1 \otimes \mathcal{B}_1)$ to $\mathfrak{S}(\mathcal{A}_2 \otimes \mathcal{B}_2)$ depending on the physical properties of the communication device. For all input state $\varphi \in \mathfrak{S}(\mathcal{A}_1)$ and all $A \in \mathcal{A}_2$, he quantum channel Λ^* with respect to the communication process is defined by

$$\begin{aligned} \Lambda^* (\varphi) (A) &\equiv \pi^* (\varphi \otimes \psi) (A \otimes I_2) \\ &= \left(a^* \circ \pi^* \circ \gamma^*\right) (\varphi) (A) . \end{aligned}$$

The example of the quantum communication channels are studied in [25, 27]

3 Entropy and Mutual Entropy for General Quantum Systems

In this section, I briefly explain entropies of general quantum systems [10, 14, 15, 17, 18, 22, 24].

Let $\mathcal{A}$ be a C* algebra, $\mathfrak{S}(\mathcal{A})$ be the set of all normal states on $\mathcal{A}$, and $\mathcal{S}$ be a weak* compact and convex subset of $\mathfrak{S}(\mathcal{A})$. Every state $\varphi \in \mathcal{S}$ has a maximal measure μ pseudosupported on $ex\mathcal{S}$ such that

$$\varphi = \int_{(ex\mathcal{S})} \omega d\mu, \tag{1}$$

where $ex\mathcal{S}$ is the set of all extreme points of $\mathcal{S}$. The measure μ satisfying the above decomposition is not unique unless $\mathcal{S}$ is a Choquet simplex. We describe the set of all such measures by $M_{\varphi}(\mathcal{S})$. We define a subset of $M_{\varphi}(\mathcal{S})$ by $D_{\varphi}(\mathcal{S})$ such as

$$D_{\varphi}(\mathcal{S}) = \Big\{ M_{\varphi}(\mathcal{S}); \quad \exists \mu_k \subset \mathbb{R}^+ \text{ and } \{\varphi_k\} \subset ex\mathcal{S} \\ s.t. \ \sum_k \mu_k = 1, \quad \mu = \sum_k \mu_k \delta(\varphi_k) \Big\}, \tag{2}$$

where $\delta(\varphi)$ is the Dirac measure concentrated on an initial state φ. We put a functional H by

$$H(\mu) = -\sum_k \mu_k \log \mu_k \tag{3}$$

for a measure $\mu \in D_{\varphi}(\mathcal{S})$. The $\mathcal{S}$-mixing entropy of a state $\varphi \in \mathcal{S}$ with respect to $\mathcal{S}$ is defined by

$$S^{\mathcal{S}}(\varphi) = \begin{cases} \inf \{ H(\mu); \quad \mu \in D_{\varphi}(\mathcal{S}) \} \\ +\infty \qquad \text{if } D_{\varphi}(\mathcal{S}) = \varnothing. \end{cases} \tag{4}$$

It describes the amount of information of the state φ measured from the subsystem $\mathcal{S}$. For example, $\mathcal{S}$ is given by $\mathfrak{S}(\mathcal{A})$ which is the set of all states on $\mathcal{A}$, $I(\alpha)$ which is the set of all invariant states for α and $K(\alpha)$ which is the set of all KMS states. If $\mathcal{S}$ is given by $\mathfrak{S}(\mathcal{A})$, we denote $S^{\mathfrak{S}(\mathcal{A})}(\varphi)$ by $S(\varphi)$. It is an extension of von Neumann's entropy [11]

$$S(\rho) = -tr \rho \log \rho, \quad \forall \rho \in \mathfrak{S}(\mathcal{H}). \tag{5}$$

We here briefly explain the mutual entropy in C*−systems defined by Ohya [17]. For any $\varphi \in \mathcal{S} \subset \mathfrak{S}(\mathcal{A}_1)$ and a quantum channel $\Lambda^* : \mathfrak{S}(\mathcal{A}_1) \to \mathfrak{S}(\mathcal{A}_2)$, define the compound states by

$$\Phi_0 = \varphi \otimes \Lambda^* \varphi$$

and

$$\Phi_{\mu}^{\mathcal{S}} = \int_{(ex\mathcal{S})} \omega \otimes \Lambda^* \omega d\mu$$

The compound state $\Phi_{\mu}^{\mathcal{S}}$ generalizes the joint probability in classical systems and it denotes the correlation between the initial state φ and the final state $\Lambda^* \varphi$. The mutual entropy w.r.t. $\mathcal{S}$ is defined by

$$I^{\mathcal{S}}(\varphi; \Lambda^*) = \lim_{\varepsilon \to 0} \sup \left\{ I_{\mu}^{\mathcal{S}}(\varphi; \Lambda^*); \mu \in F_{\varphi}^{\varepsilon}(\mathcal{S}) \right\},$$

where $I_{\mu}^{S}(\varphi; \Lambda)$ is the mutual entropy w.r.t. $\mathcal{S}$ and μ described by

$$I_{\mu}^{S}(\varphi; \Lambda) = S(\Phi_{\mu}^{S}, \Phi_0)$$

and $S\left(\Phi_{\mu}^{S}, \Phi_0\right)$ is the quantum relative entropy by Araki [2, 3], Uhlmann [32]. The quantum relative entropy of two states was introduced by Umegaki in [31] and Lindblad for σ-finite and semifinite von Neumann algebras, and

$$F_{\varphi}^{\varepsilon}(\mathcal{S}) = \begin{cases} \{\mu \in D_{\varphi}(\mathcal{S}); S^{\mathcal{S}}(\varphi) \leq H(\mu) \leq S^{\mathcal{S}}(\varphi) + \varepsilon < +\infty\} \\ M_{\varphi}(\mathcal{S}) \ \text{ if } \ S^{\mathcal{S}}(\varphi) = +\infty \end{cases}$$

Then the following fundamental inequality is satisfied

$$0 \leq I^{\mathcal{S}}(\varphi; \Lambda^*) \leq S^{\mathcal{S}}(\varphi).$$

4 Compound States

Based on [35], we will briefly explain some of the consequences of entangled compound states.

Consider some structures of compound states. Given the initial state φ and the quantum channel Λ^* (the signal is sent from the initial system to the final system), the compound state Φ should hold the following marginal conditions:

$$tr_2\Phi = \varphi \quad \text{(marginal condition 1)},$$
$$tr_1\Phi = \Lambda^*\varphi \quad \text{(marginal condition 2)}.$$

In [36], we introduced the construction of the hybrid entangled compound state $\Phi_E^{(\Delta)}$. However, they were not sufficient as the representation of the compound states. Therefore we here modify the compound states by means of the initial and final systems.

Let $\mathcal{S}_k$ be a weak * convex subset of the state space $\mathfrak{S}(\mathcal{A}_k)$ of $\mathcal{A}_k = \mathbb{B}(\mathcal{H}_k)$ $(k = 1, 2)$. Based on Choquet's theorem, for any normal state $\varphi \in \mathfrak{S}(\mathcal{A}_1)$, there exits regular Borel probability maximal measures representing extremal decomposition of φ

$$\varphi = \int_{(ex\mathcal{S}_1)} \omega d\mu(\omega)$$

Let $M_{\varphi}(\mathcal{S}_1)$ be the set of such measures μ on $\mathfrak{S}(\mathcal{A}_1)$ and $F_{\varphi}^{\varepsilon}(\mathcal{S})$ is defined by

$$F_{\varphi}^{\varepsilon}(\mathcal{S}) = \begin{cases} \{\mu \in D_{\varphi}(\mathcal{S}); S^{\mathcal{S}}(\varphi) \leq H(\mu) \leq S^{\mathcal{S}}(\varphi) + \varepsilon < +\infty\} \\ M_{\varphi}(\mathcal{S}) \ \text{ if } \ S^{\mathcal{S}}(\varphi) = +\infty \end{cases}$$

There exists a density operator $\rho \in \mathfrak{S}(\mathcal{H}_1)$ satisfying

$$\varphi(A) = tr\rho A, \quad (\forall A \in \mathcal{A}_1).$$

For a Schatten decomposition [29] $\sum_{n\in Q} \lambda_n E_n$ of the density operator ρ, one can be denoted by

$$\varphi(A) = tr\sum_{n\in Q} \lambda_n E_n A = \sum_{n\in Q} \lambda_n tr\omega_{n,n}(A), \quad (\forall A \in \mathcal{A}_1)$$

It means that for the state $\omega_{n,n}$, there exists the trace class operator $E_n = |x_n\rangle\langle x_n|$ such that

$$\omega_{n,n}(A) = trE_n A$$

for any $A \in \mathcal{A}_1$. Let $\mathcal{A}$ be a partition of the total index set Q with respect to a Schatten - von Neumann decomposition of the initial state ρ given by

$$\left.\begin{aligned}\mathcal{A} = \{A_j \subset Q;\ \#(A_j) \neq \varnothing\ \ (j \in J),\\ A_i \cap A_j = \varnothing\ (i \neq j),\quad Q = \bigcup_{j\in J} A_j\end{aligned}\right\}$$

$$\begin{aligned}
\mathcal{A}_\Delta &= \{A_j \in \mathcal{A};\ \#(A_j) \geq 2\ \ (j \in J)\},\\
\mathcal{A}_1 &= \{A_j \in \mathcal{A};\ \#(A_j) = 1\ \ (j \in J)\},\\
\mathcal{A} &= \mathcal{A}_1 \cup \mathcal{A}_\Delta,\\
J_\ell &= \{n_i \in J;\ \sharp(A_{n_i}) = \ell\,(\geq 2)\}\\
J_1 &= \{n_i \in J;\ \sharp(A_{n_i}) = 1\}\\
\Delta_\ell &= \bigcup_{n_i\in J_\ell} A_{n_i},\quad \Delta = \bigcup_{\ell=2} \Delta_\ell,\quad Q\diagdown\Delta = \bigcup_{n_i\in J_1} A_{n_i}\\
\mathcal{A}_{\Delta_\ell} &= \{A_j \in \mathcal{A};\ \#(A_j) = \ell\ \ (j \in J)\} \subset \mathcal{A}_\Delta\\
\mathcal{A}_\Delta &= \bigcup_{\ell=2} \mathcal{A}_{\Delta_\ell}
\end{aligned}$$

We denote the set of all partitions of the total index set Q by $\mathcal{P}(Q)$ such as

$$\begin{aligned}\mathcal{P}(Q) = \Big\{\mathcal{A};\ \mathcal{A} = \mathcal{A}_1 \cup \mathcal{A}_\Delta, \forall A_i, A_j\,(\neq \varnothing) \in \mathcal{A}\ (i \neq j),\\
A_i \cap A_j = \varnothing,\ Q = \bigcup_{j\in J} A_j = \bigcup_{\ell=2}\left(\bigcup_{n_i\in J_\ell} A_{n_i}\right) \cup \left(\bigcup_{n_i\in J_1} A_{n_i}\right)\Big\}\end{aligned}$$

By using the trace class operator $\left|x_{n_i}\right\rangle\left\langle x_{n_i'}\right|$ on $\mathcal{H}_1$ and $\left(\sum_{n_i\in A_{n_\ell}}\sqrt{\lambda_{n_i}}\left|x_{n_i}\right\rangle\right)\left(\sum_{n_j\in A_{n_\ell}}\sqrt{\lambda_{n_j}}\left\langle x_{n_j}\right|\right)\in\mathfrak{S}(\mathcal{H}_1)$, we denote a linear functional $\omega_{n_i,n_i'}$ on $\mathcal{A}_1$ and $\omega_{A_{n_\ell},A_{n_\ell}}\in\mathfrak{S}(\mathcal{A}_1)$ by

$$\omega_{n_i,n_i'}(A)=tr\left|x_{n_i}\right\rangle\left\langle x_{n_i'}\right|A$$

$$\sum_{n_i\in A_{n_\ell}}\sum_{n_j\in A_{n_\ell}}\sqrt{\lambda_{n_i}}\sqrt{\lambda_{n_j}}\omega_{n_i,n_j}(A)=//tr\left(\sum_{n_i\in A_{n_\ell}}\sqrt{\lambda_{n_i}}\left|x_{n_i}\right\rangle\right)\left(\sum_{n_j\in A_{n_\ell}}\sqrt{\lambda_{n_j}}\left\langle x_{n_j}\right|\right)A$$

for any $A\in\mathcal{A}_1$.

(1) The separable compound state $\widetilde{\Psi}_{\mu,E,\Lambda^*}^{\mathfrak{S}(1)}$:
For a given extremal decomposition measure $\mu\in F_\varphi^\varepsilon(\mathcal{S})$ of φ, the separable compound state $\widetilde{\Psi}_{\mu,E,\Lambda^*}^{\mathfrak{S}(1)}$ of φ with respect to the Schatten decomposition $\sum_{n\in Q}\lambda_nE_n$ of the density operator ρ and a completely positive channel Λ^* from $\mathfrak{S}(\mathcal{A}_1)$ to $\mathfrak{S}(\mathcal{A}_2)$ is defined by

$$\widetilde{\Psi}_{\mu,E,\Lambda^*}^{\mathfrak{S}(1)}=\sum_{n_k\in Q}\lambda_{n_k}\omega_{n_k,n_k}\otimes\Lambda^*\left(\omega_{n_k,n_k}\right).$$

(2) The full entangled compound state $\widetilde{\Psi}_{\mu,E,\Lambda_t^*}^{\mathfrak{S}(Q)}$:For a given extremal decomposition measure $\mu\in F_\varphi^\varepsilon(\mathcal{S})$ of φ, the full entangled compound state $\widetilde{\Psi}_{\mu,E,\Lambda^*}^{\mathfrak{S}(Q)}$ with respect to the Schatten decomposition of the density operator ρ and the completely positive channel Λ^* is defined by

$$\widetilde{\Psi}_{\mu,E,\Lambda^*}^{\mathfrak{S}(Q)}=\sum_{n_i\in Q}\sum_{n_j\in Q}\sqrt{\lambda_{n_i}}\sqrt{\lambda_{n_j}}\omega_{n_i,n_j}\otimes\Lambda^*\left(\omega_{n_i,n_j}\right).$$

(4) The modified compound state $\widetilde{\Psi}_{\mu,E,\Lambda^*}^{\mathfrak{S}(\mathcal{A})}$:

For a given extremal decomposition measure $\mu\in F_\varphi^\varepsilon(\mathcal{S})$ of φ, the modified compound state $\widetilde{\Psi}_{\mu,E,\Lambda^*}^{\mathfrak{S}(\mathcal{A})}$ concerned with the partitions $\mathcal{A}$ of the total index set Q with respect to the Schatten decomposition of the density operator ρ and the completely positive channel Λ^* is defined by

$$\widetilde{\Psi}_{\mu,E,\Lambda^*}^{\mathfrak{S}(\mathcal{A})}=\sum_{\ell=2}\sum_{n_\ell\in J_\ell}\sum_{n_i\in A_{n_\ell}}\sum_{n_j\in A_{n_\ell}}\sqrt{\lambda_{n_i}}\sqrt{\lambda_{n_j}}\omega_{n_i,n_j}\otimes\Lambda^*\left(\omega_{n_i,n_j}\right)$$
$$+\sum_{n_k\in Q\setminus\Delta}\lambda_{n_k}\omega_{n_k,n_k}\otimes\Lambda^*\left(\omega_{n_k,n_k}\right).$$

For a given extremal decomposition measure $\mu\in F_\varphi^\varepsilon(\mathcal{S})$ of φ and a Schatten decomposition $\sum_{n\in Q}\lambda_nE_n$ of the density operator ρ, let $\widetilde{\Psi}_{\mu,E,\Lambda^*}^{\mathfrak{S}(\mathcal{A})}$ be an entangled compound state with respect to a partition $\mathcal{A}$ of the total index set Q and the CP channel $\Lambda^*(\varphi)(\bullet)=\varphi\circ\Lambda(\bullet)=\varphi\circ\sum_nV_n^*(\bullet)V_n$ for any normal states $\varphi\in\mathfrak{S}(\mathcal{A}_1)$

with $\sum_n V_n^* V_n = I$ and $V_m^* V_n = \delta_{mn} V_n^* V_n$. The mutual entropy type measure $I_E^{(\mathcal{A})}(\varphi; \Lambda^*)$ with respect to a partition $\mathcal{A}$ of the total index set Q and the CP channel Λ^* is defined by taking the supremum of $\mu \in F_\varphi^\varepsilon(\mathcal{S})$ for the relative entropy between $\widetilde{\Psi}_{\mu,E,\Lambda^*}^{\mathfrak{S}(\mathcal{A})}$ and $\varphi \otimes \Lambda^*(\varphi)$

$$I_E^{(\mathcal{A})}\left(\varphi; \Lambda^*\right) = \lim_{\varepsilon \to 0} \sup \left\{ S\left(\widetilde{\Psi}_{\mu,E,\Lambda^*}^{\mathfrak{S}(\mathcal{A})}, \varphi \otimes \Lambda^*(\varphi)\right); \quad \mu \in F_\varphi^\varepsilon(\mathcal{S}) \right\}.$$

The mutual entropy type measure $I^{(\mathcal{A})}(\varphi; \Lambda^*)$ with respect to a partition $\mathcal{A}$ of the total index set Q and the CP channel Λ^* is defined by taking the supremum of for the relative entropy between $\widetilde{\Psi}_{\mu,E,\Lambda^*}^{\mathfrak{S}(\mathcal{A})}$ and $\varphi \otimes \Lambda^*(\varphi)$ for all Schatten decomposition $E = \{E_n\}$ of the density operator ρ

$$I^{(\mathcal{A})}\left(\varphi; \Lambda^*\right) = \sup \left\{ I_E^{(\mathcal{A})}\left(\varphi; \Lambda^*\right); \quad E = \{E_n\} \right\}.$$

Theorem 1 *For a given extremal decomposition measure $\mu \in F_\varphi^\varepsilon(\mathcal{S})$ of φ and a Schatten decomposition $\sum_{n \in Q} \lambda_n E_n$ of the density operator ρ, let $\widetilde{\Psi}_{\mu,E,\Lambda^*}^{\mathfrak{S}(\mathcal{A})}$ be an entangled compound state with respect to a partition $\mathcal{A}$ of the total index set Q and the CP channel $\Lambda^*(\varphi)(\bullet) = \varphi \circ \Lambda(\bullet) = \varphi \circ \sum_n V_n^*(\bullet) V_n$ for any normal states $\varphi \in \mathfrak{S}(\mathcal{A}_1)$ with $\sum_n V_n^* V_n = I$ and $V_m^* V_n = \delta_{mn} V_n^* V_n$ It satisfies two marginal conditions*

$$\widetilde{\Psi}_{\mu,E,\Lambda^*}^{\mathfrak{S}(\mathcal{A})}(A \otimes I) = \varphi(A) \quad \text{and} \quad \widetilde{\Psi}_{\mu,E,\Lambda^*}^{\mathfrak{S}(\mathcal{A})}(I \otimes B) = \Lambda^*(\varphi)(B)$$

for any $A \in \mathcal{A}_1$ and any $B \in \mathcal{A}_2$.

$$I^{(\mathcal{A})}\left(\varphi; \Lambda^*\right) \geq I^{(1)}\left(\varphi; \Lambda^*\right),$$

where $I^{(1)}(\varphi; \Lambda^*)$ in the right hand side is the Ohya mutual entropy.

Proof One has

$$\begin{aligned} &\widetilde{\Psi}_{\mu,E,\Lambda^*}^{\mathfrak{S}(\mathcal{A})} \\ &= \sum_{\ell=2} \sum_{n_\ell \in J_\ell} \sum_{n_i \in A_{n_\ell}} \sum_{n_j \in A_{n_\ell}} \sqrt{\lambda_{n_i}} \sqrt{\lambda_{n_j}} \omega_{n_i,n_j} \otimes \Lambda^*\left(\omega_{n_i,n_j}\right) \\ &+ \sum_{n_k \in Q \setminus \Delta} \lambda_{n_k} \omega_{n_k,n_k} \otimes \Lambda^*\left(\omega_{n_k,n_k}\right) \end{aligned}$$

$$\begin{aligned} &\varphi \otimes \Lambda^*(\varphi) \\ &= \sum_{n_k \in Q} \sum_{m_k \in Q} \lambda_{n_k} \lambda_{m_k} \omega_{n_k,n_k} \otimes \Lambda^*\left(\omega_{m_k,m_k}\right) \end{aligned}$$

Since $\widetilde{\Psi}_{\mu,E,\Lambda^*}^{\mathfrak{S}(\mathcal{A})}$ is written by, for any $A \otimes B \in \mathcal{A}_1 \otimes \mathcal{A}_2$

$$\widetilde{\Psi}_{\mu,E,\Lambda^*}^{\mathfrak{S}(\mathcal{A})}(A \otimes B) = tr\Psi_{E,\Lambda^*}^{(\mathcal{A})}(A \otimes B),$$

where

$$\Psi_{E,\Lambda^*}^{(\mathcal{A})} = \sum_n \sum_{\ell=2} \sum_{n_\ell \in J_\ell} \left\| \left|\phi_n^{(A_{n_\ell})}\right\rangle \right\|^2 \frac{\left|\phi_n^{(A_{n_\ell})}\right\rangle}{\left\| \left|\phi_n^{(A_{n_\ell})}\right\rangle \right\|} \frac{\left\langle\phi_n^{(A_{n_\ell})}\right|}{\left\| \left|\phi_n^{(A_{n_\ell})}\right\rangle \right\|}$$
$$+ \sum_n \sum_{n_k \in Q \backslash \Delta} \lambda_{n_k} \left\| V_n \left|x_{n_k}\right\rangle \right\|^2 E_{n_k} \otimes \frac{V_n \left|x_{n_k}\right\rangle}{\left\| V_n \left|x_{n_k}\right\rangle \right\|} \frac{\left\langle x_{n_k}\right| V_n^*}{\left\| V_n \left|x_{n_k}\right\rangle \right\|},$$

and

$$\left|\phi_n^{(A_{n_\ell})}\right\rangle = \sum_{n_i \in A_{n_\ell}} \sqrt{\lambda_{n_i}} \left|x_{n_i}\right\rangle \otimes V_n \left|x_{n_i}\right\rangle,$$
$$\left\| \left|\phi_n^{(A_{n_\ell})}\right\rangle \right\|^2 = \sum_{n_k \in A_{n_\ell}} \lambda_{n_k} \left\| V_n \left|x_{n_k}\right\rangle \right\|^2$$

then one has

$$\begin{aligned}
&\widetilde{\Psi}_{\mu,E,\Lambda^*}^{\mathfrak{S}(\mathcal{A})}(A \otimes I) \\
&= tr\Psi_{E,\Lambda^*}^{(\mathcal{A})}(A \otimes I) \\
&= tr \sum_n \sum_{\ell=2} \sum_{n_\ell \in J_\ell} \left\| \left|\phi_n^{(A_{n_\ell})}\right\rangle \right\|^2 \frac{\left|\phi_n^{(A_{n_\ell})}\right\rangle}{\left\| \left|\phi_n^{(A_{n_\ell})}\right\rangle \right\|} \frac{\left\langle\phi_n^{(A_{n_\ell})}\right|}{\left\| \left|\phi_n^{(A_{n_\ell})}\right\rangle \right\|}(A \otimes I) \\
&\quad + tr \sum_n \sum_{n_k \in Q \backslash \Delta} \lambda_{n_k} \left\| V_n \left|x_{n_k}\right\rangle \right\|^2 E_{n_k} \otimes \frac{V_n \left|x_{n_k}\right\rangle}{\left\| V_n \left|x_{n_k}\right\rangle \right\|} \frac{\left\langle x_{n_k}\right| V_n^*}{\left\| V_n \left|x_{n_k}\right\rangle \right\|}(A \otimes I) \\
&= \varphi(A)
\end{aligned}$$

and

$$\begin{aligned}
&\widetilde{\Psi}_{\mu,E,\Lambda^*}^{\mathfrak{S}(\mathcal{A})}(I \otimes B) \\
&= tr\Psi_{E,\Lambda^*}^{(\mathcal{A})}(I \otimes B) \\
&= tr \sum_n \sum_{\ell=2} \sum_{n_\ell \in J_\ell} \left\| \left|\phi_n^{(A_{n_\ell})}\right\rangle \right\|^2 \frac{\left|\phi_n^{(A_{n_\ell})}\right\rangle}{\left\| \left|\phi_n^{(A_{n_\ell})}\right\rangle \right\|} \frac{\left\langle\phi_n^{(A_{n_\ell})}\right|}{\left\| \left|\phi_n^{(A_{n_\ell})}\right\rangle \right\|}(I \otimes B)
\end{aligned}$$

$$
\begin{aligned}
&+tr\sum_{n}\sum_{n_k\in Q\backslash\Delta}\lambda_{n_k}\left\|V_n\left|x_{n_k}\right\rangle\right\|^2 E_{n_k}\otimes\frac{V_n\left|x_{n_k}\right\rangle}{\left\|V_n\left|x_{n_k}\right\rangle\right\|}\frac{\left\langle x_{n_k}\right|V_n^*}{\left\|V_n\left|x_{n_k}\right\rangle\right\|}(I\otimes B)\\
&=tr\left(\sum_{n_k\in Q}\lambda_{n_k}E_{n_k}\right)\Lambda(B)\\
&=\varphi\circ\Lambda(B)\\
&=\Lambda^*(\varphi)(B)
\end{aligned}
$$

Since

$$
S\left(\widetilde{\Psi}_{\mu,E,\Lambda^*}^{\mathfrak{S}(\mathcal{A})},\varphi\otimes\Lambda^*(\varphi)\right)=tr\Psi_{E,\Lambda^*}^{(\mathcal{A})}\left(\log\Psi_{E,\Lambda^*}^{(\mathcal{A})}-\log\rho\otimes\Lambda^*\rho\right)
$$

then one can obtain

$$
\begin{aligned}
&tr\Psi_{E,\Lambda^*}^{(\mathcal{A})}\log\Psi_{E,\Lambda^*}^{(\mathcal{A})}\\
&=\sum_{n}\sum_{\ell=2}\sum_{n_\ell\in J_\ell}\left(\sum_{n_i\in A_{n_\ell}}\lambda_{n_i}\left\|V_n\left|x_{n_i}\right\rangle\right\|^2\right)\log\left(\sum_{n_j\in\Delta}\lambda_{n_j}\left\|V_n\left|x_{n_j}\right\rangle\right\|^2\right)\\
&\quad+\sum_{n}\sum_{n_k\in Q\backslash\Delta}\lambda_{n_k}\left\|V_n\left|x_{n_k}\right\rangle\right\|^2\log\left\|V_n\left|x_{n_k}\right\rangle\right\|^2\\
&\quad+\sum_{n_k\in Q\backslash\Delta}\lambda_{n_k}\log\lambda_{n_k},
\end{aligned}
$$

$$
-tr\Psi_{E,\Lambda^*}^{(\mathcal{A})}\log\rho\otimes\Lambda^*\rho=S(\rho)+S\left(\Lambda^*\rho\right)
$$

Therefore we have

$$
\begin{aligned}
&S\left(\widetilde{\Psi}_{\mu,E,\Lambda^*}^{\mathfrak{S}(\mathcal{A})},\varphi\otimes\Lambda^*(\varphi)\right)\\
&=S\left(\sum_{\ell=2}\sum_{n_\ell\in J_\ell}\sum_{n_i\in A_{n_\ell}}\sum_{n_j\in A_{n_\ell}}\sqrt{\lambda_{n_i}}\sqrt{\lambda_{n_j}}\omega_{n_i,n_j}\otimes\Lambda^*\left(\omega_{n_i,n_j}\right)+\right.\\
&\quad\left.+\sum_{n_k\in Q\setminus\Delta}\lambda_{n_k}\omega_{n_k,n_k}\otimes\Lambda^*\left(\omega_{n_k,n_k}\right),\varphi\otimes\Lambda^*(\varphi)\right)\\
&=\sum_{\ell=2}\sum_{n_\ell\in J_\ell}S\left(\sum_{n_i\in A_{n_\ell}}\sum_{n_j\in A_{n_\ell}}\sqrt{\lambda_{n_i}}\sqrt{\lambda_{n_j}}\omega_{n_i,n_j}\otimes\Lambda^*\left(\omega_{n_i,n_j}\right),\varphi\otimes\Lambda^*(\varphi)\right)\\
&\quad+\sum_{n_k\in Q\setminus\Delta}S\left(\lambda_{n_k}\omega_{n_k,n_k}\otimes\Lambda^*\left(\omega_{n_k,n_k}\right),\varphi\otimes\Lambda^*(\varphi)\right)
\end{aligned}
$$

$$
\begin{aligned}
&= \sum_n \sum_{\ell=2} \sum_{n_\ell \in J_\ell} \left(\sum_{n_i \in A_{n_\ell}} \lambda_{n_i} \left\| V_n \left| x_{n_i} \right\rangle \right\|^2 \right) \log \left(\sum_{n_j \in A_{n_\ell}} \lambda_{n_j} \left\| V_n \left| x_{n_j} \right\rangle \right\|^2 \right) \\
&\quad + \sum_n \sum_{n_k \in Q \setminus \Delta} \lambda_{n_k} \left\| V_n \left| x_{n_k} \right\rangle \right\|^2 \log \left\| V_n \left| x_{n_k} \right\rangle \right\|^2 \\
&\quad + \sum_{n_k \in Q \setminus \Delta} \lambda_{n_k} \log \lambda_{n_k} + S(\rho) + S\left(\Lambda^* \rho \right) \\
&= S\left(\Lambda^* \varphi \right) - \sum_n \sum_{n_k \in Q} \lambda_{n_k} S\left(\Lambda^* \omega_{n_k, n_k} \right) \\
&\quad - \sum_n \sum_{\ell=2} \sum_{n_\ell \in J_\ell} \sum_{n_i \in A_{n_\ell}} \lambda_{n_i} \left\| V_n \left| x_{n_i} \right\rangle \right\|^2 \log \frac{\lambda_{n_i} \left\| V_n \left| x_{n_i} \right\rangle \right\|^2}{\left(\sum_{n_j \in A_{n_\ell}} \lambda_{n_j} \left\| V_n \left| x_{n_j} \right\rangle \right\|^2 \right)} \\
&= S\left(\widetilde{\Psi}^{\mathfrak{S}(1)}_{\mu, E, \Lambda^*}, , \varphi \otimes \Lambda^*(\varphi) \right) \\
&\quad - \sum_n \sum_{\ell=2} \sum_{n_\ell \in J_\ell} \sum_{n_i \in A_{n_\ell}} \lambda_{n_i} \left\| V_n \left| x_{n_i} \right\rangle \right\|^2 \log \frac{\lambda_{n_i} \left\| V_n \left| x_{n_i} \right\rangle \right\|^2}{\left(\sum_{n_j \in A_{n_\ell}} \lambda_{n_j} \left\| V_n \left| x_{n_j} \right\rangle \right\|^2 \right)}
\end{aligned}
$$

Taking supremum of $\mu \in F_\varphi^\varepsilon(S)$, $\lim_{\varepsilon \to 0}$ and $E = \{E_n\}$, one has the inequality

$$I^{(\mathcal{A})}\left(\varphi; \Lambda^*\right) \geq I^{(1)}\left(\varphi; \Lambda^*\right)$$

If $\Lambda^* = id$ is held, then we obtain the following inequality:

$$I^{(\mathcal{A})}(\varphi; id) \geq I^{(1)}(\varphi; id) = S(\varphi)$$

It shows that the mutual entropy defined by using the entangled compound state $\widetilde{\Psi}^{\mathfrak{S}(\mathcal{A})}_{\mu, E, \Lambda^*}$ with respect to a partition $\mathcal{A}$ of the total index set Q, the initial state φ and the quantum channel $\Lambda^* = id$ greater than the entropy $S(\varphi)$ of the initial state φ.

If the above completely positive channel Λ^* has orthogonality (i.e., $\omega_{n,n} \perp \omega_{m,m}$ $(n \neq m) \Rightarrow \Lambda^*(\omega_{n,n}) \perp \Lambda^*(\omega_{n,n})$) then we have the following theorem.

Theorem 2 *For a given extremal decomposition measure $\mu \in F_\varphi^\varepsilon(S)$ of φ and a Schatten decomposition $\sum_{n \in Q} \lambda_n E_n$ of the density operator ρ, let $\widetilde{\Psi}^{\mathfrak{S}(\mathcal{A})}_{\mu, E, \Lambda^*}$ be an entangled compound state with respect to a partition $\mathcal{A}$ of the total index set Q and the CP channel $\Lambda^*(\varphi)(\bullet) = \varphi \circ \Lambda(\bullet) = \varphi \circ \sum_n V_n^*(\bullet) V_n$ for any normal states $\varphi \in \mathfrak{S}(\mathcal{A}_1)$ with $\sum_n V_n^* V_n = I$ and $V_m^* V_n = \delta_{mn} V_n^* V_n$ and orthogonality (i.e., $\omega_{n,n} \perp \omega_{m,m}$ $(n \neq m) \Rightarrow$ $\Lambda^*\left(\omega_{n,n}\right) \perp \Lambda^*\left(\omega_{m,m}\right)$). It satisfies two marginal conditions*

$$\widetilde{\Psi}^{\mathfrak{S}(\mathcal{A})}_{\mu, E, \Lambda^*}(A \otimes I) = \varphi(A) \quad \text{and} \quad \widetilde{\Psi}^{\mathfrak{S}(\mathcal{A})}_{\mu, E, \Lambda^*}(I \otimes B) = \Lambda^*(\varphi)(B)$$

for any $A \in \mathcal{A}_1$ *and any* $B \in \mathcal{A}_2$*. The following inequality is held:*

$$I^{(\mathcal{A})}\left(\varphi; \Lambda^*\right) \geq S\left(\widetilde{\Psi}^{\mathfrak{S}(\mathcal{A})}_{\mu,E,\Lambda^*}, \varphi \otimes \Lambda^*\left(\varphi\right)\right) \geq S\left(\varphi\right).$$

Proof

$$\begin{aligned}
&\widetilde{\Psi}^{\mathfrak{S}(\mathcal{A})}_{\mu,E,\Lambda^*} \\
&= \sum_{\ell=2} \sum_{n_\ell \in J_\ell} \sum_{n_i \in A_{n_\ell}} \sum_{n_j \in A_{n_\ell}} \sqrt{\lambda_{n_i}} \sqrt{\lambda_{n_j}} \omega_{n_i,n_j} \otimes \Lambda^*\left(\omega_{n_i,n_j}\right) \\
&\quad + \sum_{n_k \in Q \setminus \Delta} \lambda_{n_k} \omega_{n_k,n_k} \otimes \Lambda^*\left(\omega_{n_k,n_k}\right)
\end{aligned}$$

$$\begin{aligned}
&\varphi \otimes \Lambda^*\left(\varphi\right) \\
&= \sum_{n_k \in Q} \sum_{m_k \in Q} \lambda_{n_k} \lambda_{m_k} \omega_{n_k,n_k} \otimes \Lambda^*\left(\omega_{m_k,m_k}\right)
\end{aligned}$$

Since $\widetilde{\Psi}^{\mathfrak{S}(\mathcal{A})}_{\mu,E,\Lambda^*}$ is written by, for any $A \otimes B \in \mathcal{A}_1 \otimes \mathcal{A}_2$

$$\widetilde{\Psi}^{\mathfrak{S}(\mathcal{A})}_{\mu,E,\Lambda^*}\left(A \otimes B\right) = tr \Psi^{(\mathcal{A})}_{E,\Lambda^*}\left(A \otimes B\right),$$

where

$$\begin{aligned}
\Psi^{(\mathcal{A})}_{E,\Lambda^*} &= \sum_{n} \sum_{\ell=2} \sum_{n_\ell \in J_\ell} \left\| \left|\phi_n^{(A_{n_\ell})}\right\rangle \right\|^2 \frac{\left|\phi_n^{(A_{n_\ell})}\right\rangle}{\left\| \left|\phi_n^{(A_{n_\ell})}\right\rangle \right\|} \frac{\left\langle\phi_n^{(A_{n_\ell})}\right|}{\left\| \left|\phi_n^{(A_{n_\ell})}\right\rangle \right\|} \\
&+ \sum_{n} \sum_{n_k \in Q \setminus \Delta} \lambda_{n_k} \left\| V_n \left|x_{n_k}\right\rangle \right\|^2 E_{n_k} \otimes \frac{V_n \left|x_{n_k}\right\rangle}{\left\| V_n \left|x_{n_k}\right\rangle \right\|} \frac{\left\langle x_{n_k}\right| V_n^*}{\left\| V_n \left|x_{n_k}\right\rangle \right\|},
\end{aligned}$$

and

$$\begin{aligned}
\left|\phi_n^{(A_{n_\ell})}\right\rangle &= \sum_{n_i \in A_{n_\ell}} \sqrt{\lambda_{n_i}} \left|x_{n_i}\right\rangle \otimes V_n \left|x_{n_i}\right\rangle, \\
\left\| \left|\phi_n^{(A_{n_\ell})}\right\rangle \right\|^2 &= \sum_{n_k \in A_{n_\ell}} \lambda_{n_k} \left\| V_n \left|x_{n_k}\right\rangle \right\|^2
\end{aligned}$$

then one has

$$
\begin{aligned}
& \widetilde{\Psi}_{\mu,E,\Lambda^*}^{\mathfrak{S}(\mathcal{A})}(A \otimes I) \\
&= tr\Psi_{E,\Lambda^*}^{(\mathcal{A})}(A \otimes I) \\
&= tr\sum_{n}\sum_{\ell=2}\sum_{n_\ell \in J_\ell} \left\| \left|\phi_n^{(A_{n_\ell})}\right\rangle \right\|^2 \frac{\left|\phi_n^{(A_{n_\ell})}\right\rangle}{\left\| \left|\phi_n^{(A_{n_\ell})}\right\rangle \right\|} \frac{\left\langle\phi_n^{(A_{n_\ell})}\right|}{\left\| \left|\phi_n^{(A_{n_\ell})}\right\rangle \right\|}(A \otimes I) \\
&\quad + tr\sum_{n}\sum_{n_k \in Q\setminus\Delta} \lambda_{n_k} \left\| V_n \left|x_{n_k}\right\rangle \right\|^2 E_{n_k} \otimes \frac{V_n\left|x_{n_k}\right\rangle}{\left\|V_n\left|x_{n_k}\right\rangle\right\|} \frac{\left\langle x_{n_k}\right| V_n^*}{\left\|V_n\left|x_{n_k}\right\rangle\right\|}(A \otimes I) \\
&= tr\left(\sum_{n_k \in Q} \lambda_{n_k} E_{n_k}\right) A \\
&= \varphi(A)
\end{aligned}
$$

and

$$
\begin{aligned}
& \widetilde{\Psi}_{\mu,E,\Lambda^*}^{\mathfrak{S}(\mathcal{A})}(I \otimes B) \\
&= tr\Psi_{E,\Lambda^*}^{(\mathcal{A})}(I \otimes B) \\
&= tr\sum_{n}\sum_{\ell=2}\sum_{n_\ell \in J_\ell} \left\| \left|\phi_n^{(A_{n_\ell})}\right\rangle \right\|^2 \frac{\left|\phi_n^{(A_{n_\ell})}\right\rangle}{\left\| \left|\phi_n^{(A_{n_\ell})}\right\rangle \right\|} \frac{\left\langle\phi_n^{(A_{n_\ell})}\right|}{\left\| \left|\phi_n^{(A_{n_\ell})}\right\rangle \right\|}(I \otimes B) \\
&\quad + tr\sum_{n}\sum_{n_k \in Q\setminus\Delta} \lambda_{n_k} \left\| V_n \left|x_{n_k}\right\rangle \right\|^2 E_{n_k} \otimes \frac{V_n\left|x_{n_k}\right\rangle}{\left\|V_n\left|x_{n_k}\right\rangle\right\|} \frac{\left\langle x_{n_k}\right| V_n^*}{\left\|V_n\left|x_{n_k}\right\rangle\right\|}(I \otimes B) \\
&= tr\left(\sum_{n_k \in Q} \lambda_{n_k} E_{n_k}\right) \Lambda(B) \\
&= \varphi \circ \Lambda(B) \\
&= \Lambda^*(\varphi)(B)
\end{aligned}
$$

Since

$$
S\left(\widetilde{\Psi}_{\mu,E,\Lambda^*}^{\mathfrak{S}(\mathcal{A})}, \varphi \otimes \Lambda^*(\varphi)\right) = tr\Psi_{E,\Lambda^*}^{(\mathcal{A})}\left(\log \Psi_{E,\Lambda^*}^{(\mathcal{A})} - \log \rho \otimes \Lambda^*\rho\right)
$$

then one can obtain

$$
\begin{aligned}
&tr \Psi_{E,\Lambda^*}^{(\mathcal{A})} \log \Psi_{E,\Lambda^*}^{(\mathcal{A})} \\
&= \sum_n \sum_{\ell=2} \sum_{n_\ell \in J_\ell} \left(\sum_{n_i \in A_{n_\ell}} \lambda_{n_i} \left\| V_n \left| x_{n_i} \right\rangle \right\|^2 \right) \log \left(\sum_{n_j \in \Delta} \lambda_{n_j} \left\| V_n \left| x_{n_j} \right\rangle \right\|^2 \right) \\
&\quad + \sum_n \sum_{n_k \in Q \setminus \Delta} \lambda_{n_k} \left\| V_n \left| x_{n_k} \right\rangle \right\|^2 \log \left\| V_n \left| x_{n_k} \right\rangle \right\|^2 \\
&\quad + \sum_{n_k \in Q \setminus \Delta} \lambda_{n_k} \log \lambda_{n_k},
\end{aligned}
$$

$$
-tr \Psi_{E,\Lambda^*}^{(\mathcal{A})} \log \rho \otimes \Lambda^* \rho = S(\rho) + S\left(\Lambda^* \rho\right)
$$

Since

$$
S\left(\widetilde{\Psi}_{\mu,E,\Lambda^*}^{\mathfrak{S}(\mathcal{A})}, \varphi \otimes \Lambda^*(\varphi)\right) = tr \Psi_{E,\Lambda^*}^{(\mathcal{A})} \left(\log \Psi_{E,\Lambda^*}^{(\mathcal{A})} - \log \rho \otimes \Lambda^* \rho\right)
$$

then one can obtain

$$
\begin{aligned}
&tr \Psi_{E,\Lambda^*}^{(\mathcal{A})} \log \Psi_{E,\Lambda^*}^{(\mathcal{A})} \\
&= \sum_n \sum_{\ell=2} \sum_{n_\ell \in J_\ell} \left(\sum_{n_i \in A_{n_\ell}} \lambda_{n_i} \left\| V_n \left| x_{n_i} \right\rangle \right\|^2 \right) \log \left(\sum_{n_j \in \Delta} \lambda_{n_j} \left\| V_n \left| x_{n_j} \right\rangle \right\|^2 \right) \\
&\quad + \sum_n \sum_{n_k \in Q \setminus \Delta} \lambda_{n_k} \left\| V_n \left| x_{n_k} \right\rangle \right\|^2 \log \left\| V_n \left| x_{n_k} \right\rangle \right\|^2 \\
&\quad + \sum_{n_k \in Q \setminus \Delta} \lambda_{n_k} \log \lambda_{n_k},
\end{aligned}
$$

$$
-tr \Psi_{E,\Lambda^*}^{(\mathcal{A})} \log \rho \otimes \Lambda^* \rho = S(\rho) + S\left(\Lambda^* \rho\right)
$$

$$
\begin{aligned}
&S\left(\widetilde{\Psi}_{\mu,E,\Lambda^*}^{\mathfrak{S}(\mathcal{A})}, \varphi \otimes \Lambda^*(\varphi)\right) \\
&= S\left(\sum_{\ell=2} \sum_{n_\ell \in J_\ell} \sum_{n_i \in A_{n_\ell}} \sum_{n_j \in A_{n_\ell}} \sqrt{\lambda_{n_i}} \sqrt{\lambda_{n_j}} \omega_{n_i,n_j} \otimes \Lambda^*\left(\omega_{n_i,n_j}\right) + \right. \\
&\quad \left. + \sum_{n_k \in Q \setminus \Delta} \lambda_{n_k} \omega_{n_k,n_k} \otimes \Lambda^*\left(\omega_{n_k,n_k}\right), \varphi \otimes \Lambda^*(\varphi) \right) \\
&= \sum_{\ell=2} \sum_{n_\ell \in J_\ell} S\left(\sum_{n_i \in A_{n_\ell}} \sum_{n_j \in A_{n_\ell}} \sqrt{\lambda_{n_i}} \sqrt{\lambda_{n_j}} \omega_{n_i,n_j} \otimes \Lambda^*\left(\omega_{n_i,n_j}\right), \varphi \otimes \Lambda^*(\varphi) \right)
\end{aligned}
$$

$$
\begin{aligned}
&+ \sum_{n_k \in Q \setminus \Delta} S\left(\lambda_{n_k} \omega_{n_k,n_k} \otimes \Lambda^*\left(\omega_{n_k,n_k}\right), \varphi \otimes \Lambda^*(\varphi)\right) \\
&= \sum_n \sum_{\ell=2} \sum_{n_\ell \in J_\ell} \left(\sum_{n_i \in A_{n_\ell}} \lambda_{n_i} \left\| V_n \left| x_{n_i} \right\rangle \right\|^2 \right) \log \left(\sum_{n_j \in A_{n_\ell}} \lambda_{n_j} \left\| V_n \left| x_{n_j} \right\rangle \right\|^2 \right) \\
&+ \sum_n \sum_{n_k \in Q \setminus \Delta} \lambda_{n_k} \left\| V_n \left| x_{n_k} \right\rangle \right\|^2 \log \left\| V_n \left| x_{n_k} \right\rangle \right\|^2 \\
&+ \sum_{n_k \in Q \setminus \Delta} \lambda_{n_k} \log \lambda_{n_k} + S(\rho) + S\left(\Lambda^* \rho\right) \\
&= S(\varphi) \\
&- \sum_n \sum_{\ell=2} \sum_{n_\ell \in J_\ell} \sum_{n_i \in A_{n_\ell}} \lambda_{n_i} \left\| V_n \left| x_{n_i} \right\rangle \right\|^2 \log \frac{\lambda_{n_i} \left\| V_n \left| x_{n_i} \right\rangle \right\|^2}{\left(\sum_{n_j \in A_{n_\ell}} \lambda_{n_j} \left\| V_n \left| x_{n_j} \right\rangle \right\|^2 \right)} \\
&\geq S(\varphi)
\end{aligned}
$$

Therefore we obtain the following inequality:

$$
I^{(\mathcal{A})}(\varphi; \Lambda^*) \geq S\left(\widetilde{\Psi}^{\mathfrak{S}(\mathcal{A})}_{\mu,E,\Lambda^*}, \varphi \otimes \Lambda^*(\varphi)\right) \geq S(\varphi).
$$

It is shown that the mutual entropy $I^{(\mathcal{A})}(\varphi; \Lambda^*)$ defined using the entangled compound state $\widetilde{\Psi}^{\mathfrak{S}(\mathcal{A})}_{\mu,E,\Lambda^*}$ with respect to a partition $\mathcal{A}$ of the total index set Q and the CP channel Λ^* for any initial state $\varphi \in \mathfrak{S}(\mathcal{A}_1)$ greater than the entropy $S(\varphi)$ in the initial state φ. It does not satisfy the fundamental inequality.

5 Conclusion

As mentioned above, we discuss quantum mutual entropy-type measure with entangled compound states $\widetilde{\Psi}^{\mathfrak{S}(\mathcal{A})}_{\mu,E,\Lambda^*}$. The mutual entropy type measure $I^{(\mathcal{A})}(\varphi; \Lambda^*)$ defined using an entangled compound state $\widetilde{\Psi}^{\mathfrak{S}(\mathcal{A})}_{\mu,E,\Lambda^*}$ for a partition $\mathcal{A}$ of all index sets Q greater than the Ohya mutual entropy $I^{(1)}(\varphi; \Lambda^*)$. Moreover, under some conditions, the mutual entropy type measure $I^{(\mathcal{A})}(\varphi; \Lambda^*)$ greater than the entropy $S(\varphi)$ of the initial state φ. It does not satisfy the fundamental inequality. It seems reasonable to think that a quantum mutual entropy type measure defined using an entangled compound state is not useful in discussing the efficiency of information transfer from the initial system to the final system.

References

1. Accardi, L., Ohya, M.: Compound channels, transition expectations, and liftings. Appl. Math. Optim. **39**, 33–59 (1999)
2. Araki, H.: Relative Entropy of States of von Neumann Algebras, vol. 11, pp. 809–833. Publication of RIMS, Kyoto University (1976)
3. Araki, H.: Relative Entropy for States of von Neumann Algebras II, vol. 13, pp. 173–192. Publication of RIMS, Kyoto University (1977)
4. Donald, M.J.: On the relative entropy. Commun. Math. Phys. **105**, 13–34 (1985)
5. Fichtner, K.H., Freudenberg, W., Liebscher, V.: Beam splittings and time evolutions of Boson systems. Fakultat fur Mathematik und Informatik, Math/Inf/96/ 39, Jena, 105 (1996)
6. Ingarden, R. S., Kossakowski, A., Ohya, M.: Information Dynamics and Open Systems. Kluwer (1997)
7. Jamiołkowski, A.: Linear transformations which preserve trace and positive semidefiniteness of operators. Rep. Math. Phys. **3**, 275–278 (1972)
8. Mukhamedov, F. Watanabe, N.: On S-mixing entropy of quantum channels. Quantum Inf. Process. **17**, 148, (2018). https://doi.org/10.1007/s11128-018-1916-8
9. Mukhamedov, F., Ohmura, K., Watanabe, N.: A formulation of Rènyi entropy on C*-algebras. Quantum Inf. Process. **18**, 318 (2019). https://doi.org/10.1007/s11128-019-2430-3
10. Muraki, N., Ohya, M., Petz, D.: Entropies of general quantum systems. Open Sys. Infor. Dyn. **1**, 43–56 (1992)
11. von Neumann, J.: Die Mathematischen Grundlagen der Quantenmechanik. Springer, Berlin (1932)
12. Ohya, M.: Quantum ergodic channels in operator algebras. J. Math. Anal. Appl. **84**, 318–327 (1981)
13. Ohya, M.: On compound state and mutual information in quantum information theory. IEEE Trans. Infor. Theor. **29**, 770–777 (1983)
14. Ohya, M.: Note on quantum probability. L. Nuo. Cimen. **38**, 402–404 (1983)
15. Ohya, M.: Entropy transmission in C^*-dynamical systems. J. Math. Anal. Appl. **100**, 222–235 (1984)
16. Ohya, M.: State change and entropies in quantum dynamical systems. Springer Lect. Not. Math. **1136**, 397–408 (1985)
17. Ohya, M.: Some aspects of quantum information theory and their applications to irreversible processes. Rep. Math. Phys. **27**, 19–47 (1989)
18. Ohya, M., Petz, D.: Quantum Entropy and Its Use. Springer (1993)
19. Ohya, M., Petz, D., Watanabe, N.: On capacities of quantum channels. Prob. Math. Stat. **17**, 179–196 (1997)
20. Ohya, M., Petz, D., Watanabe, N.: Numerical computation of quantum capacity. Inter. J. Theor. Phys. **38**, 507–510 (1998)
21. Ohya, M.: Fundamentals of quantum mutual entropy and capacity. Open Sys. Infor. Dyn. **6**, 69–78 (1999)
22. Ohya, M. : Information dynamics and its application to optical communication processes. In: Book: Lecture Notes in Physics, vol. 378, pp. 81–92. Springer (1991)
23. Ohya, M., Volovich, I.V.: On quantum entropy and its bound. Infin. Dimen. Anal. Quant. Prob. Relat. Top. **6**, 301–310 (2003)
24. Ohya, M., Volovich, I.V.: Mathematical Foundations of Quantum Information and Computation and Its Applications to Nano- and Bio-Systems. Springer, Berlin (2011)
25. Ohya, M., Watanabe, N.: Construction and analysis of a mathematical model in quantum communication processes
26. Ohya, M., Watanabe, N.: A new treatment of communication processes with Gaussian channels. Japan J. Appl. Math. **3**, 197–206 (1986)
27. Ohya, M., Watanabe, N.: A mathematical study of information transmission in quantum communication processes. Quant. Commun. Measur. **2**, 371–378 (1995)

28. Ohya, M., Watanabe, N.: Comparison of mutual entropy - type measures. TUS preprint (2003)
29. Schatten, R.: Norm Ideals of Completely Continuous Operators. Springer (1970)
30. Shannon, C. E.: A mathematical theory of communication. Bell Syst. Tech. J. **27**, 379–423 and 623–656 (1948)
31. Umegaki, H.: Conditional expectations in an operator algebra IV (entropy and information). Kodai Math. Sem. Rep. **14**, 59–85 (1962)
32. Uhlmann, A.: Relative entropy and the Wigner-Yanase-Dyson-Lieb concavity in interpolation theory. Commun. Math. Phys. **54**, 21–32 (1977)
33. Urbanik, K.: Joint probability distribution of observables in quantum mechanics. Stud. Math. **21**, 317 (1961)
34. Watanabe, N.: Efficiency of optical modulations with coherent states. Springer Lect. Notes Phys. **378**, 350–360 (1991)
35. Watanabe, N.: Note on entropies for compound systems. Open. Syst. Inf. Dyn. **22**, 1550001–1550021 (2015)
36. Watanabe, N.: A note on complexities by means of quantum compound systems. Entropy **22**(3), 298 (2020). https://doi.org/10.3390/e22030298

Trace Decreasing Quantum Dynamical Maps: Divisibility and Entanglement Dynamics

Sergey N. Filippov

Abstract Trace decreasing quantum operations naturally emerge in experiments involving postselection. However, the experiments usually focus on dynamics of the conditional output states as if the dynamics were trace preserving. Here we show that this approach leads to incorrect conclusions about the dynamics divisibility, namely, one can observe an increase in the trace distance or the system-ancilla entanglement although the trace decreasing dynamics is completely positive divisible. We propose solutions to that problem and introduce proper indicators of the information backflow and the indivisibility. We also review a recently introduced concept of the generalized erasure dynamics that includes more experimental data in the dynamics description. The ideas are illustrated by explicit physical examples of polarization dependent losses.

Keywords Quantum operation · Postselection · Polarization dependent losses · Entanglement dynamics · Divisibility

1 Introduction

Many quantum physics experiments involve a postselection, i.e., an analysis of only those events that are conditioned by some measurement outcomes. Examples include the approximate preparation of a single-photon state via detecting a heralding photon in the parametric down-conversion process [1], the implementation of the conditional

S. N. Filippov (✉)
Steklov Mathematical Institute of Russian Academy of Sciences, Gubkina Street 8, Moscow 119991, Russia

Valiev Institute of Physics and Technology of Russian Academy of Sciences, Nakhimovskii Prospect 34, Moscow 117218, Russia
e-mail: sergey.filippov@phystech.edu

L. Accardi et al. (eds.), *Infinite Dimensional Analysis, Quantum Probability and Applications*, Springer Proceedings in Mathematics & Statistics 390,
https://doi.org/10.1007/978-3-031-06170-7_6

quantum gates [2], the generation of high-dimensional maximally entangled orbital-angular-momentum states [3], the conditional subtraction of a single photon [4] or multiple photons [5] from the electromagnetic field, the conditional addition of a single photon to the electromagnetic field [6], the experimental evaluation of so-called weak values [7], and the simulation of quantum collision models [8]. A proper mathematical description of any postselection procedure is given by a quantum operation, i.e., a trace nonincreasing completely positive map Λ (see, e.g., [9]) that is associated with a specific measurement outcome. If ϱ is an initial density operator before postselection, then $\Lambda[\varrho]$ is a subnormalized density operator such that the trace $\mathrm{tr}\big[\Lambda[\varrho]\big]$ is exactly the probability to get that specific measurement outcome.

In general, a measurement-induced transformation of the system state is fully described in terms of the quantum instrument that assigns a quantum operation to each measurement outcome [10–12]. A quantum operation Λ corresponds to a specific measurement outcome, say, a successful detection of the heralding photon in quantum optics experiments [1, 2, 4, 6]. In mathematical description of experiments with photon subtraction or photon addition [4, 6], however, the conventionally used transformations $\varrho \to t a \varrho a^\dagger$ and $\varrho \to t a^\dagger \varrho a$ are not legitimate quantum operations for any constant $t > 0$ due to the violation of the trace nonincreasing property [13] (a and $a^\dagger$ are the photon annihilation operator and the photon creation operator for a considered mode, respectively). The transformations above approximate the faithful quantum operations of photon subtraction and addition for density operators from some class [13].

To illustrate a simple trace nonincreasing operation, let us consider an experiment with polarization qubits. Suppose ϱ is a density operator for polarization degrees of freedom of single photons and those single photons propagate through a lossy optical fiber with the intensity attenuation factor p, then $\Lambda[\varrho] = p\varrho$ and $\mathrm{tr}\big[\Lambda[\varrho]\big] = p$. The attenuation factor is experimentally estimated as $N_{\mathrm{out}}/N_{\mathrm{in}}$, where N_{out} is the number of photons detected at the output of the fiber and N_{in} is the number of photons that enter the fiber. If the photon detector has a finite quantum efficiency η, then the number of successfully detected photons at the input (the output) of the fiber approximately equals $N_{\mathrm{in}}' = \eta N_{\mathrm{in}}$ ($N_{\mathrm{out}}' = \eta N_{\mathrm{out}}$) so that $N_{\mathrm{out}}'/N_{\mathrm{in}}' = N_{\mathrm{out}}/N_{\mathrm{in}}$, i.e., $\mathrm{tr}\big[\Lambda[\varrho]\big]$ can be correctly estimated even with the use of imperfect devices. Alternatively, the photon pairs can be produced in the parametric down-conversion process, with one photon being sent to a lossy channel (to count N_{out}' at the output) and the other one being used as a herald (to count N_{in}'). These ideas enable one to reconstruct a trace nonincreasing operation based on experimental data. For instance, the experimental quantum process tomography of a trace decreasing map Λ describing the partially transmitting polarizing beam splitter is reported in Ref. [14].

The whole idea of postselection is focused on the conditional output state

$$\varrho_{\mathrm{out}} = \frac{\Lambda[\varrho]}{\mathrm{tr}\big[\Lambda[\varrho]\big]} \equiv \Lambda_{\mathcal{D}}[\varrho]$$

that is a valid density operator provided $\text{tr}\big[\Lambda[\varrho]\big] > 0$. (If $\text{tr}\big[\Lambda[\varrho]\big] = 0$, then there is no sense in postselection because a desired measurement outcome is never observed.) The introduced map $\Lambda_{\mathcal{D}}$ defines a nonlinear transformation of density operators that finds applications in the stroboscopic implementation of effective non-Hermitian Hamiltonians [15, 16]. In this work, we are primarily interested in biased operations Λ that have the property $\text{tr}\big[\Lambda[\varrho_1]\big] \neq \text{tr}\big[\Lambda[\varrho_2]\big]$ for at least two density operators ϱ_1 and ϱ_2 [17]. Examples of biased quantum operations include the partially transmitting polarizing beam splitter [14] and the polarization-dependent losses [18].

Interestingly, the optical simulator of quantum collisions [8] produces a sequence of quantum operations $\Lambda(t_m)$ in the single-photon sector that correspond to transformations from time $t = 0$ to a number of different time moments $t = t_m, m = 1, 2, \ldots$. Similarly, one can consider losses in optical fibers of different length to get a one-parameter family of quantum operations $\{\Lambda(t)\}_{t\geq 0}$ that we refer to as a trace decreasing dynamical map if $\text{tr}\big[\Lambda(t)[\varrho]\big] < 1$ for some density operator ϱ and some $t \geq 0$. Note that the relation $\text{tr}\big[\Lambda(t_1)[\varrho]\big] \geq \text{tr}\big[\Lambda(t_2)[\varrho]\big]$ for $t_2 \geq t_1 \geq 0$ need not hold in general. Suppose $\Lambda(t_2) = \Theta(t_2, t_1)\Lambda(t_1)$ for all $t_2 \geq t_1 \geq 0$, where $\Theta(t_2, t_1)$ is a legitimate quantum operation (i.e., a completely positive and trace nonincreasing map). Then $\Lambda(t)$ is called completely positive divisible (CP-divisible). Physical meaning of CP-divisibility is that the dynamics $\Lambda(t)$ can be effectively viewed as a concatenation of independent subevolutions.

As far as trace preserving dynamical maps $\{\Phi(t)\}_{t\geq 0}$ are concerned, CP-divisibility is studied in a number of papers and many different indicators of CP-indivisibility are proposed too (see, e.g., [19–24]). However, in the following Sections. we show that these indicators become misleading if carelessly applied to postselected states of a trace decreasing dynamical map $\{\Lambda(t)\}_{t\geq 0}$. Note that some authors associate CP-divisibility with a quantum version of the Markovian process [19]; however, a more complicated relation takes place operationally [25].

2 Trace Distance Approach to Non-Markovianity

Let ϱ_1 and ϱ_2 be density operators, then the trace distance $\frac{1}{2}\|\Phi(t)[\varrho_1] - \Phi(t)[\varrho_2]\|_1$ is a nonincreasing function of time t if the trace preserving dynamical map $\{\Phi(t)\}_{t\geq 0}$ is CP-divisible [24]. The physical meaning of the trace distance is related with the maximum success probability to distinguish the initially equiprobable states ϱ_1 and ϱ_2 after time t. The maximum success probability equals $\frac{1}{2} + \frac{1}{4}\|\Phi(t)[\varrho_1] - \Phi(t)[\varrho_2]\|_1$ in this case (see, e.g., [12]). If the trace distance diminishes, then the success probability diminishes too, which is treated as a flow of information from the open system to its environment. On the other hand, the increase of the trace distance is an indication of the backflow of information according to Ref. [24].

If one experimentally reconstructs the conditional output states $\Lambda_{\mathcal{D}}(t)[\varrho_1] = \Lambda(t)[\varrho_1]/\text{tr}\big[\Lambda(t)[\varrho_1]\big]$ and $\Lambda_{\mathcal{D}}(t)[\varrho_2] = \Lambda(t)[\varrho_2]/\text{tr}\big[\Lambda(t)[\varrho_2]\big]$, then it is tempting to use the conventional trace distance

$$\frac{1}{2}\|\Lambda_{\mathcal{D}}(t)[\varrho_1] - \Lambda_{\mathcal{D}}(t)[\varrho_2]\|_1 \tag{1}$$

to analyze the information flow. However, this approach encounters a number of problems. First, the probabilities $p_1(t) := \mathrm{tr}\big[\Lambda(t)[\varrho_1]\big]$ and $p_2(t) := \mathrm{tr}\big[\Lambda(t)[\varrho_2]\big]$ differ in general. This means that by using Eq. (1) an experimentalist disregards some extra information on how often the states $\Lambda_{\mathcal{D}}(t)[\varrho_1]$ and $\Lambda_{\mathcal{D}}(t)[\varrho_2]$ are actually produced. Second, as we show in the example below, the quantity (1) may increase even for a semigroup trace decreasing dynamics $\Lambda(t)$ that is obviously CP-divisible.

Example 1 Consider the polarization qubit dynamics $\Lambda(t) = e^{Lt}$, where

$$L[\varrho] = -\frac{1}{2}\Big\{\gamma_H \,|H\rangle\,\langle H| + \gamma_V \,|V\rangle\,\langle V|\,, \varrho\Big\}, \tag{2}$$

$\{\cdot,\cdot\}$ stands for the anticommutator; $(|H\rangle\,, |V\rangle)$ is a conventional basis composed of the horizontally and vertically polarized states; γ_H and γ_V are the attenuation rates for horizontally and vertically polarized photons, respectively. Suppose $\varrho_1 = |H\rangle\,\langle H|$ and $\varrho_2 = \frac{1}{2}(|H\rangle + |V\rangle)(\langle H| + \langle V|)$, then we explicitly find the subnormalized states in the basis $(|H\rangle\,, |V\rangle)$,

$$\Lambda(t)[\varrho_1] = \begin{pmatrix} e^{-\gamma_H t} & 0 \\ 0 & 0 \end{pmatrix}, \quad \Lambda(t)[\varrho_2] = \frac{1}{2}\begin{pmatrix} e^{-\gamma_H t} & e^{-(\gamma_H+\gamma_V)t/2} \\ e^{-(\gamma_H+\gamma_V)t/2} & e^{-\gamma_V t} \end{pmatrix},$$

and the corresponding conditional output states,

$$\Lambda_{\mathcal{D}}(t)[\varrho_1] = \begin{pmatrix} 1 & 0 \\ 0 & 0 \end{pmatrix}, \quad \Lambda_{\mathcal{D}}(t)[\varrho_2] = \frac{1}{e^{-\gamma_H t} + e^{-\gamma_V t}}\begin{pmatrix} e^{-\gamma_H t} & e^{-(\gamma_H+\gamma_V)t/2} \\ e^{-(\gamma_H+\gamma_V)t/2} & e^{-\gamma_V t} \end{pmatrix}.$$

Therefore, $\frac{1}{2}\|\Lambda_{\mathcal{D}}(t)[\varrho_1] - \Lambda_{\mathcal{D}}(t)[\varrho_2]\|_1 = \big[1 + e^{(\gamma_V - \gamma_H)t}\big]^{-1/2}$, which monotonically increases with time t if $\gamma_H > \gamma_V$. △

Example 1 shows that Eq. (1) cannot be used in quantification of the information flow for trace decreasing dynamical maps. An alternative approach is to utilize the conditional probabilities $\frac{p_1(t)}{p_1(t)+p_2(t)}$ and $\frac{p_2(t)}{p_1(t)+p_2(t)}$ to get the postselected state $\Lambda_{\mathcal{D}}(t)[\varrho_1]$ and $\Lambda_{\mathcal{D}}(t)[\varrho_2]$, respectively, provided the desired quantum operation $\Lambda(t)$ is successfully fulfilled (for either of the input states, $\varrho = \varrho_1$ or $\varrho = \varrho_2$). Then the modified trace distance

$$\left\| \frac{p_1(t)}{p_1(t) + p_2(t)}\Lambda_{\mathcal{D}}(t)[\varrho_1] - \frac{p_2(t)}{p_1(t) + p_2(t)}\Lambda_{\mathcal{D}}(t)[\varrho_2] \right\|_1 \tag{3}$$

is a relevant candidate to track the information flow. However, as we reveal in the example below, the quantity (3) is still unsatisfactory because it does not monotonically decrease under the semigroup dynamics.

Example 2 Let $\Lambda(t) = e^{Lt}$, where L is given by Eq. (2). Suppose $\varrho_1 = |H\rangle\langle H|$ and $\varrho_2 = \frac{1}{2}(|H\rangle + |V\rangle)(\langle H| + \langle V|)$, then

$$\frac{\|p_1(t)\Lambda_{\mathcal{D}}(t)[\varrho_1] - p_2(t)\Lambda_{\mathcal{D}}(t)[\varrho_2]\|_1}{p_1(t) + p_2(t)} = \sqrt{1 - \frac{8e^{-2\gamma_H t}}{(3e^{-\gamma_H t} + e^{-\gamma_V t})^2}},$$

which increases with the increase of time t if $\gamma_H > \gamma_V$. △

The solution of the problem is to multiply the conditional success probability of the discrimination task,

$$\frac{1}{2}\left\{1 + \frac{\|p_1(t)\Lambda_{\mathcal{D}}(t)[\varrho_1] - p_2(t)\Lambda_{\mathcal{D}}(t)[\varrho_2]\|_1}{p_1(t) + p_2(t)}\right\}, \tag{4}$$

by the average probability to implement the operation $\Lambda(t)$, i.e., $\frac{1}{2}[p_1(t) + p_2(t)]$. Then we get $\frac{1}{4}\{p_1(t) + p_2(t) + \|p_1(t)\Lambda_{\mathcal{D}}(t)[\varrho_1] - p_2(t)\Lambda_{\mathcal{D}}(t)[\varrho_2]\|_1\}$. Note that $p_i(t)\Lambda_{\mathcal{D}}(t)[\varrho_i] = \Lambda(t)[\varrho_i]$, $i = 1, 2$, so the obtained expression reduces to

$$p_{\text{succ.dist.}} = \frac{1}{4}\Big(\text{tr}\big[\Lambda(t)[\varrho_1]\big] + \text{tr}\big[\Lambda(t)[\varrho_2]\big] + \|\Lambda(t)[\varrho_1] - \Lambda(t)[\varrho_2]\|_1\Big), \tag{5}$$

which is a nonincreasing function of time t if $\Lambda(t)$ is CP-divisible. Indeed, following the lines of Ref. [26], we readily see that the trace distance $\frac{1}{2}\|\Lambda(t)[\varrho_1] - \Lambda(t)[\varrho_2]\|_1$ remains a nonincreasing function of time t if we extend the result of Theorem 1 in Ref. [26] to the case of trace decreasing completely positive maps. If $\Lambda(t)$ is CP divisible, then the probabilities $\text{tr}\big[\Lambda(t)[\varrho_1]\big]$ and $\text{tr}\big[\Lambda(t)[\varrho_2]\big]$ to successfully implement the operation $\Lambda(t)$ for the input states ϱ_1 and ϱ_2, respectively, are nonincreasing functions of time t too. Therefore, the information flow in a trace decreasing quantum dynamics is to be associated with the change in Eq. (5), whereas the naive expressions (1) and (3) should be avoided.

The operational meaning of Eq. (5) is depicted in Fig. 1a. Let $p(j, k|i)$ be a joint probability to implement an operation $\mathcal{I}_k$ via a quantum instrument $\mathcal{I}$ and to observe the outcome j while measuring the output quantum state, provided the input state is ϱ_i. Suppose $\mathcal{I}_{k_*} = \Lambda(t)$, then the maximum value of $\frac{1}{2}\sum_{i=1,2} p(j = i, k = k_*|i)$ exactly equals $p_{\text{succ.dist.}}$ in Eq. (5).

Fig. 1 **a** Operational meaning of Eq. (5), where $\Lambda(t) = \mathcal{I}_{k_*}$. **b** Discrimination problem in the case of the generalized erasure dynamics

3 System-Ancilla Entanglement Dynamics

Suppose that in addition to the system of interest we also have access to an ancillary system, which is isolated from the decoherence sources. Provided the system dynamics is described by a trace preserving dynamical map $\Phi(t)$, the total system-ancilla aggregate undergoes an evolution given by the dynamical map $\Phi(t) \otimes \mathrm{Id}$, see Fig. 2a.

Suppose that the system and the ancilla are initially entangled, then the system-ancilla entanglement generally changes in time. To quantify the system-ancilla entanglement, we use some entanglement monotone $\mathcal{E}$ that does not increase under trace preserving local operations and classical communication [27]. A seminal observation of Ref. [21] is that the entanglement monotone $\mathcal{E}(t)$ is a nonincreasing function of time t provided the trace preserving map $\Phi(t)$ is CP-divisible. Any increase in $\mathcal{E}(t)$ is an indication of CP-indivisibility (non-Markovianity according to Ref. [19]). Although this approach to CP-indivisibility detection is absolutely legitimate in the case of trace preserving dynamical maps, it turns out to be wrong if one tracks the entanglement dynamics of postselected system-ancilla states [see setup in Fig. 2b with postselection $k = k_*$]. The following example justifies this claim.

Example 3 Consider the polarization dependent losses that are time dependent too, i.e., the master equation of the form

$$\frac{d\varrho(t)}{dt} = -\frac{1}{2}\Big\{\gamma_H(t)\,|H\rangle\,\langle H| + \gamma_V(t)\,|V\rangle\,\langle V|\,,\,\varrho(t)\Big\}. \tag{6}$$

A solution of this master equation is

$$\Lambda(t)[\varrho(0)] \equiv \varrho(t) = \begin{pmatrix} e^{-\Gamma_H(t)}\varrho_{HH}(0) & e^{-\frac{1}{2}[\Gamma_H(t)+\Gamma_V(t)]}\varrho_{HV}(0) \\ e^{-\frac{1}{2}[\Gamma_H(t)+\Gamma_V(t)]}\varrho_{VH}(0) & e^{-\Gamma_V(t)}\varrho_{VV}(0) \end{pmatrix},$$

where $\Gamma_H(t) = \int_0^t \gamma_H(t')dt'$ and $\Gamma_V(t) = \int_0^t \gamma_V(t')dt'$. If $\gamma_H(t)$ and $\gamma_V(t)$ are both nonnegative, then the trace decreasing dynamical map $\Lambda(t)$ is CP-divisible.

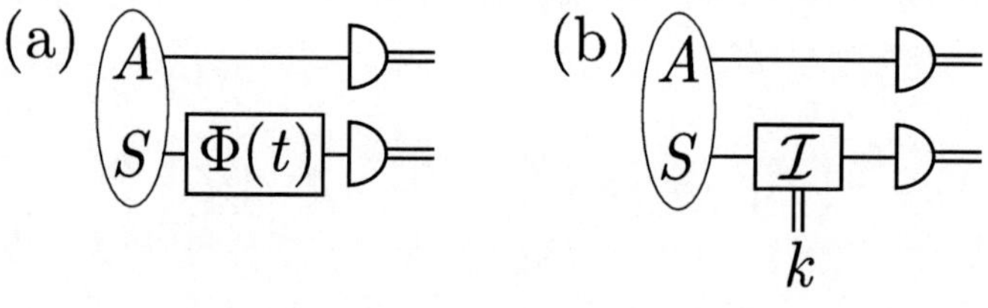

Fig. 2 **a** Joint trace preserving dynamics of the system (S) and the ancilla (A). **b** Schematic of an experiment with trace decreasing dynamics, $\mathcal{I}_{k_*} = \Lambda(t)$

Let us consider the specific rates

$$\gamma_H(t) = \gamma \left(1 - \frac{\omega \cos \omega t}{\sqrt{\gamma^2 + \omega^2} + \gamma \sin \omega t} \right), \tag{7}$$

$$\gamma_V(t) = \gamma \left(1 + \frac{\omega \cos \omega t}{\sqrt{\gamma^2 + \omega^2} - \gamma \sin \omega t} \right), \tag{8}$$

such that $\gamma_H(t) \geq 0$ and $\gamma_V(t) \geq 0$ if $\gamma > 0$, $\omega > 0$. Then

$$p_H(t) := e^{-\Gamma_H(t)} = e^{-\gamma t} \left(1 + \frac{\gamma}{\sqrt{\gamma^2 + \omega^2}} \sin \omega t \right), \tag{9}$$

$$p_V(t) := e^{-\Gamma_V(t)} = e^{-\gamma t} \left(1 - \frac{\gamma}{\sqrt{\gamma^2 + \omega^2}} \sin \omega t \right) \tag{10}$$

are both nonincreasing functions of time t. The physical meaning of $p_H(t)$ is the probability to successfully detect the horizontally polarized photon at time t provided the photon was initially horizontally polarized. Eq. (10) has a similar meaning for vertically polarized photons. CP-divisibility of $\Lambda(t)$ means that the time evolution can be represented as a sequence of independent subevolutions, with each of them being a valid trace decreasing quantum operation. In this example, all the subevolutions are polarization dependent losses with different attenuation factors.

Let $|\psi_+\rangle \langle\psi_+|$ be a maximally entangled initial state of the qubit system and a two-dimensional ancilla. This is exactly the case in the experimental scenario of Ref. [8], where the system-ancilla polarization state of two photons is produced via the spontaneous parametric down-conversion process. To be precise, $|\psi_+\rangle = \frac{1}{\sqrt{2}}(|H\rangle \otimes |H\rangle + |V\rangle \otimes |V\rangle)$. The system-ancilla trace decreasing and CP-divisible dynamics $\Lambda(t) \otimes \mathrm{Id}[|\psi_+\rangle \langle\psi_+|]$ results in the following conditional dynamics:

$$\frac{\Lambda(t) \otimes \mathrm{Id}[|\psi_+\rangle \langle\psi_+|]}{\mathrm{tr}\big[\Lambda(t) \otimes \mathrm{Id}[|\psi_+\rangle \langle\psi_+|]\big]} = |\varphi(t)\rangle \langle\varphi(t)|, \tag{11}$$

$$|\varphi(t)\rangle = \frac{1}{\sqrt{p_H(t) + p_V(t)}} \left(\sqrt{p_H(t)}\, |H\rangle \otimes |H\rangle + \sqrt{p_V(t)}\, |V\rangle \otimes |V\rangle \right). \tag{12}$$

Note that the conditional system-ancilla state remains pure for all times $t \geq 0$, so its entanglement is readily quantified by an entanglement monotone called concurrence [28]. We use that quantifier for our results to be comparable with Ref. [8], where the dynamics of concurrence is analyzed by an experimental reconstruction of the conditional system-ancilla states at different time moments. In our example, we have

$$\mathcal{E}(t) = \frac{2\sqrt{p_H(t) p_V(t)}}{p_H(t) + p_V(t)} = \sqrt{\frac{\gamma^2 \cos^2 \omega t + \omega^2}{\gamma^2 + \omega^2}}, \tag{13}$$

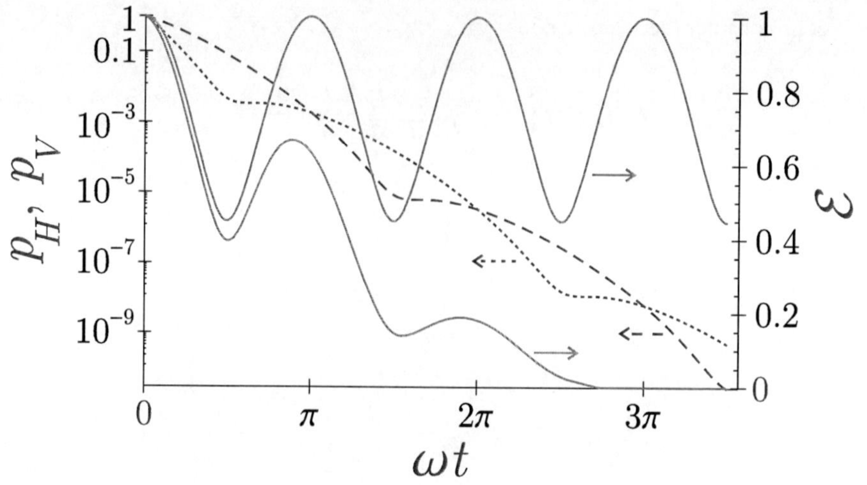

Fig. 3 Monotonic dynamics of the probability p_H (*dashed line*) and the probability p_V (*dotted line*) in the trace decreasing CP-divisible evolution governed by the master equation (6) for polarization dependent losses; the rates $\gamma_H(t)$ and $\gamma_V(t)$ are given by Eqs. (7) and (8), respectively, where $\gamma = 2\omega$. Nonmonotonic dynamics of the concurrence $\mathcal{E}$ for the normalized system-ancilla state in Example 3 with $\gamma = 2\omega$ (*upper solid line*), Eq. (13). Nonmonotonic dynamics of the concurrence $\mathcal{E}$ for the normalized system-ancilla state in Example 4 with $\gamma = 2\omega$ and the depolarization rate $\lambda = 0.05\omega$ (*lower solid line*)

which is clearly a nonmonotonic function of time if $\gamma > 0$, $\omega > 0$. Note that $\max_{t \geq 0} \mathcal{E}(t) = \mathcal{E}(\frac{\pi n}{\omega}) = 1$ and $\min_{t \geq 0} \mathcal{E}(t) = \mathcal{E}(\frac{\pi}{2\omega} + \frac{\pi n}{\omega}) = \frac{\omega}{\sqrt{\gamma^2 + \omega^2}}$, $n \in \mathbb{N}$, so there are infinitely many time periods of the entanglement increase (see the upper solid line in Fig. 3). The minimum value $\min_{t \geq 0} \mathcal{E}(t)$ can be arbitrarily close to 0 if $\omega \ll \gamma$. △

Therefore, the entanglement increase in the experimentally reconstructed post-selected system-ancilla state cannot be a valid indicator of CP-indivisibility (non-Markovianity) for trace decreasing dynamical maps. This fact was overlooked in Ref. [8].

The dynamical map $\Lambda(t)$ considered in Example 3 has Kraus rank 1, which preserves the purity of the postselected state. In the following example, we consider a more sophisticated dynamics, where the conditional system-ancilla state becomes mixed.

Example 4 Suppose the polarization dependent losses in Example 3 are accompanied by a standard depolarization with the rate $\lambda > 0$. Then the master equation takes the form

$$\frac{d\varrho(t)}{dt} = -\frac{1}{2}\Big\{\gamma_H(t)\,|H\rangle\langle H| + \gamma_V(t)\,|V\rangle\langle V|\,,\varrho(t)\Big\} + \frac{\lambda}{4}\sum_{i=1}^{3}\big(\sigma_i\varrho(t)\sigma_i - \varrho(t)\big), \tag{14}$$

where $\gamma_H(t)$ and $\gamma_V(t)$ are given by Eqs. (7) and (8), respectively; $(\sigma_1, \sigma_2, \sigma_3)$ is the conventional set of Pauli operators, i.e., $\sigma_1 = |H\rangle\langle V| + |V\rangle\langle H|$, $\sigma_2 = -i|H\rangle\langle V| + i|V\rangle\langle H|$, $\sigma_3 = |H\rangle\langle H| - |V\rangle\langle V|$. Eq. (14) defines a trace decreasing CP-divisible dynamical map $\Lambda(t)$. The analytical expression for the postselected state $\Lambda(t) \otimes \mathrm{Id}[|\psi_+\rangle\langle\psi_+|]/\mathrm{tr}\big[\Lambda(t) \otimes \mathrm{Id}[|\psi_+\rangle\langle\psi_+|]\big]$ is rather involved though straightforward. We depict the postselected state concurrence $\mathcal{E}(t)$ for the particular choice of parameters γ, ω, and λ in Fig. 3 (see the lower solid line). The concurrence $\mathcal{E}(t)$ is a nonmonotonic function of time t despite the fact that $\Lambda(t)$ is CP-divisible. △

We can make another important observation in Fig. 3: If the depolarization rate is strictly positive in Example 4 (i.e., $\lambda > 0$), then the concurrence $\mathcal{E}(t)$ vanishes at some time moment and remains zero after. In what follows we prove that this is a general phenomenon.

Proposition 1 *Let $\Lambda(t)$ be a trace decreasing CP-divisible dynamical map and $\varrho_{SA}(0)$ be an initial system-ancilla state. If the postselected system-ancilla state $\Lambda(t_*) \otimes \mathrm{Id}[\varrho_{SA}(0)]/\mathrm{tr}\big[\Lambda(t_*) \otimes \mathrm{Id}[\varrho_{SA}(0)]\big]$ becomes separable at time moment t_*, then all future postselected states ($t \geq t_*$) remain separable.*

Proof Since the postselected system-ancilla state is separable at time t_*, we have $\Lambda(t_*) \otimes \mathrm{Id}[\varrho_{SA}(0)]/\mathrm{tr}\big[\Lambda(t_*) \otimes \mathrm{Id}[\varrho_{SA}(0)]\big] = \sum_i p_i \varrho_S^{(i)} \otimes \varrho_A^{(i)}$, where $\{p_i\}_i$ is a probability distribution, $\{\varrho_S^{(i)}\}_i$ and $\{\varrho_A^{(i)}\}_i$ are the sets of density operators for the system and the ancilla, respectively. Let $t \geq t_*$. Since $\Lambda(t)$ is CP-divisible, then there exists a map $\Theta(t, t_*)$ such that $\Lambda(t) = \Theta(t, t_*)\Lambda(t_*)$. Hence,

$$\frac{\Lambda(t) \otimes \mathrm{Id}[\varrho_{SA}(0)]}{\mathrm{tr}\big[\Lambda(t) \otimes \mathrm{Id}[\varrho_{SA}(0)]\big]} = \sum_i q_i \frac{\Theta(t, t_*)[\varrho_S^{(i)}]}{\mathrm{tr}\Big[\Theta(t, t_*)[\varrho_S^{(i)}]\Big]} \otimes \varrho_A^{(i)},$$

$$q_i = \frac{p_i \mathrm{tr}\Big[\Theta(t, t_*)[\varrho_S^{(i)}]\Big]}{\sum_j p_j \mathrm{tr}\Big[\Theta(t, t_*)[\varrho_S^{(j)}]\Big]},$$

i.e., the postselected state at time moment $t \geq t_*$ is separable too. □

Corollary 1 *Suppose the system experiences a trace decreasing dynamics $\Lambda(t)$. The system-ancilla entanglement death followed by the entanglement revival is an indication of CP-indivisibility for $\Lambda(t)$.*

Using the results of Ref. [29], the above claims can be reformulated and generalized in terms of the Schmidt rank as follows: A CP-divisible trace decreasing system dynamics $\Lambda(t)$ does not increase the Schmidt rank of the system-environment normalized states $\Lambda(t) \otimes \mathrm{Id}[\varrho_{SA}(0)]/\mathrm{tr}\big[\Lambda(t) \otimes \mathrm{Id}[\varrho_{SA}(0)]\big]$.

Clearly, the examples and the statements above hold true not only for continuous time evolutions but also in the case of the discrete time evolutions, e.g., collision models that are reviewed in Ref. [30].

4 Generalized Erasure Dynamics

In Ref. [17], a generalized erasure channel has been introduced as a physically motivated extension of a trace decreasing operation. In fact, the successful implementation of a quantum operation Λ for an input state ϱ happens with the probability $\mathrm{tr}\big[\Lambda[\varrho]\big]$, so the failure probability equals

$$p_{\mathrm{fail}} = 1 - \mathrm{tr}\big[\Lambda[\varrho]\big] = \mathrm{tr}\big[\varrho - \Lambda[\varrho]\big] = \mathrm{tr}\left[\left(I - \Lambda^{\dagger}[I]\right)\varrho\right], \tag{15}$$

where $\Lambda^{\dagger}$ is a dual map with respect to Λ such that $\mathrm{tr}\big[\Lambda[X]Y\big] = \mathrm{tr}\big[X\Lambda^{\dagger}[Y]\big]$ for all trace-class operators X and bounded operators Y. The probability p_{fail} is readily accessible in a physical experiment. For instance, in the experimental setup of Ref. [8], this probability can be evaluated as $(N_A - N_S)/N_S$, where N_A is a number of the detector clicks for the ancilla and N_S is a number of the detector clicks for the system, see Fig. 2b. Although this extra information is free, it is usually disregarded in the postselection-oriented experiments. We argue that the unsuccessful implementation of operation Λ can be viewed as an erasure event. In Fig. 2b, this event is associated with the click of the ancilla detector and no click of the system detector. We extend the Hilbert space and include the erasure flag state $|e\rangle$ into it so that the map

$$\Gamma_{\Lambda}[\varrho] = \Lambda[\varrho] \oplus \mathrm{tr}\left[\left(I - \Lambda^{\dagger}[I]\right)\varrho\right] |e\rangle\langle e| \equiv \begin{pmatrix} \Lambda[\varrho] & \mathbf{0} \\ \mathbf{0}^{\top} & \mathrm{tr}\left[\left(I - \Lambda^{\dagger}[I]\right)\varrho\right] \end{pmatrix} \tag{16}$$

is trace preserving and completely positive [17]. Therefore, any trace decreasing dynamics $\Lambda(t)$ can be alternatively described by virtue of the corresponding trace preserving dynamical map $\Gamma_{\Lambda}(t) \equiv \Gamma_{\Lambda(t)}$. We refer to $\Gamma_{\Lambda}(t)$ as the generalized erasure dynamics.

CP-indivisibility identification problems discussed in Sects. 2 and 3 were caused by the postselection. Instead, if we use the generalized erasure dynamics, then those problems are automatically resolved.

Proposition 2 *Suppose $\Lambda(t)$ is a CP-divisible trace decreasing dynamical map, then $\Gamma_{\Lambda}(t)$ is a CP-divisible trace preserving dynamical map.*

Proof Let $t_2 \geq t_1 \geq 0$, then $\Lambda(t_2) = \Theta(t_2, t_1)\Lambda(t_1)$, where $\Theta(t_2, t_1)$ is a quantum operation. Consider the map

$$\Xi(t_2, t_1)\left[\begin{pmatrix} \varrho & \vdots \\ \cdots & c \end{pmatrix}\right] = \begin{pmatrix} \Theta(t_2, t_1)[\varrho] & \mathbf{0} \\ \mathbf{0}^{\top} & c + \mathrm{tr}\left[\rho(I - \Theta(t_2, t_1)^{\dagger}[I])\right] \end{pmatrix}, \tag{17}$$

which is automatically trace preserving and, in addition, completely positive because a quantum operation $\Theta(t_2, t_1)$ is trace nonincreasing, i.e., $\Theta(t_2, t_1)^{\dagger}[I] \leq I$. We have $\Gamma_{\Lambda}(t_2) = \Xi(t_2, t_1)\Gamma_{\Lambda}(t_1)$, i.e., $\Gamma_{\Lambda}(t)$ is CP-divisible. □

Due to Proposition 2, the distinguishability of equiprobable states $\Gamma_{\Lambda}(t)[\varrho_1]$ and $\Gamma_{\Lambda}(t)[\varrho_2]$ monotonically decreases if $\Lambda(t)$ is CP-divisible. Remarkably, the probability to successfully distinguish these states, $\frac{1}{2} + \frac{1}{4}\|\Gamma_{\Lambda}(t)[\varrho_1] - \Gamma_{\Lambda}(t)[\varrho_2]\|_1$, exceeds the probability (5) by $\frac{1}{2}\left(1 - \min_{i=1,2} \mathrm{tr}\big[\Lambda(t)[\varrho_i]\big]\right)$ because the modified scenario takes into account the erasure events too [see Fig. 1b, where ν is a coarse-graining postprocessing with two outcome events: $k = k_*$ and $k \neq k_*$].

Similarly, Proposition 2 implies that the system-ancilla entanglement monotonically decreases during the trace preserving dynamics $\Gamma_{\Lambda}(t) \otimes \mathrm{Id}$ if $\Lambda(t)$ is CP-divisible.

5 Conclusions

In the present work, we have paid attention to some simple yet important questions of how to analyze the divisibility problem in experiments that involve postselection. By a number of examples we have demonstrated the complete inconsistency of some approaches that operate with the postselected states as though they were obtained as a result of a trace preserving dynamics. In particular, none of the Eqs. (1) and (3) is adequate in the analysis of the information backflow because either of these quantities can increase during a CP-divisible dynamics. The original idea behind the information backflow is properly described by Eq. (5) that has a clear physical meaning. Similarly, we show that the increase in the system-ancilla entanglement is not a valid indicator of CP-indivisibility if the postselection takes place. Unfortunately, these facts are sometimes overlooked in research papers, e.g., a nonmonotonic concurrence behavior for the system-ancilla postselected state was mistreated as an indicator of non-Markovianity in the paper [8] (this paper has an extra drawback related with non-unitarity of the inter-environment collision that we do not discuss in this work). We advocate that a correct indicator of non-Markovianity should be related with an increase of the system-ancilla Schmidt rank, e.g., the revival of the entanglement after the entanglement death (Corollary 1). A general analysis of the two-qubit entanglement dynamics under local trace decreasing maps can be accomplished by following the lines of Ref. [31] and using the quantum Sinkhorn theorem [32]. Finally, we have described the framework of generalized erasure dynamics that takes into account how often a desired trace decreasing operation is actually implemented and how often the implementation fails. This extra information is readily available in experiments though it is rarely used. On the other hand, including this information into the description allows us to return to the fold of trace preserving maps. Those trace preserving maps inherit some properties of the original operations, e.g., CP-divisibility (Proposition 2). Moreover, the induced trace preserving maps have an interesting property of superadditivity of coherent information [17].

Our results for quantum dynamics are intimately related with properties of the classical process described by a joint distribution $\mathbb{P}(X_n, t_n; \ldots, X_0, t_0)$ for random variables $\{X_m\}_{m\in\mathbb{Z}_+}$ at times $\{t_m\}_{m\in\mathbb{Z}_+}$. The process is Markovian if

$\mathbb{P}(X_n, t_n | X_{n-1}, t_{n-1}; \ldots; X_0, t_0) = \mathbb{P}(X_n, t_n | X_{n-1}, t_{n-1})$ for all n, which implies the classical divisibility in the form of the Chapman-Kolmogorov equation: $\mathbb{P}(X, t | Z, r) = \sum_Y \mathbb{P}(X, t | Y, s)\mathbb{P}(Y, s | Z, r)$ for any three times $t > s > r$. A rigorous formulation of the Markov condition for quantum dynamics was introduced in Ref. [33], where it was shown that quantum Markovianity implies CP-divisibility $\Theta(t, r) = \Theta(t, s)\Theta(s, r)$ but not the other way round (see also [25, 34]). In this paper, we considered a Markovian and CP-divisible generalized erasure quantum dynamics (trace preserving one), which may be wrongly identified as CP-indivisible in the presence of postselection (when it effectively becomes trace decreasing). This is also true in the classical case. Consider a classical Markovian process, where $X_m = (x_m, k_m)$, x_m is a measurement outcome at time t_m for the system in interest and k_m is a measurement outcome at time t_m for some ancillary system interacting with the studied one. Consider a conditional process in which measurements on the ancillary system always yield the same result k_*. Clearly, the conditional process does not have to be Markovian, i.e., $\mathbb{P}(x_n, t_n | x_{n-1}, t_{n-1}; \ldots; x_0, t_0; k_n = \ldots = k_0 = k_*) \neq \mathbb{P}(x_n, t_n | x_{n-1}, t_{n-1}; k_n = \ldots = k_0 = k_*)$ in general, i.e., the conditional process may violate the Chapman-Kolmogorov equation. In this paper, we focused our attention on the quantum version of this effect.

References

1. U'Ren, A.B., Silberhorn, C., Banaszek, K., Walmsley, I.A.: Efficient conditional preparation of high-fidelity single photon states for fiber-optic quantum networks. Phys. Rev. Lett. **93**, 093601 (2004)
2. Kiesel, N., Schmid, C., Weber, U., Ursin, R., Weinfurter, H.: Linear optics controlled-phase gate made simple. Phys. Rev. Lett. **95**, 210505 (2005)
3. Kovlakov, E.V., Straupe, S.S., Kulik, S.P.: Quantum state engineering with twisted photons via adaptive shaping of the pump beam. Phys. Rev. A **98**, 060301(R) (2018)
4. Wenger, J., Tualle-Brouri, R., Grangier, P.: Non-Gaussian statistics from individual pulses of squeezed light. Phys. Rev. Lett. **92**, 153601 (2004)
5. Bogdanov, Y.I., Katamadze, K.G., Avosopiants, G.V., Belinsky, L.V., Bogdanova, N.A., Kalinkin, A.A., Kulik, S.P.: Multiphoton subtracted thermal states: description, preparation, and reconstruction. Phys. Rev. A **96**, 063803 (2017)
6. Zavatta, A., Viciani, S., Bellini, M.: Quantum-to-classical transition with single-photon-added coherent states of light. Science **306**, 660–662 (2004)
7. Pryde, G.J., O'Brien, J.L., White, A.G., Ralph, T.C., Wiseman, H.M.: Measurement of quantum weak values of photon polarization. Phys. Rev. Lett. **94**, 220405 (2005)
8. Cuevas, Á., Geraldi, A., Liorni, C., Bonavena, L.D., De Pasquale, A., Sciarrino, F., Giovannetti, V., Mataloni, P.: All-optical implementation of collision-based evolutions of open quantum systems. Sci. Rep. **9**, 3205 (2019)
9. Kraus, K.: States, Effects, and Operations. Springer, Berlin (1983)
10. Davies, E.B., Lewis, J.T.: An operational approach to quantum probability. Comm. Math. Phys. **17**, 239–260 (1970)
11. Holevo, A.S.: Quantum Systems, Channels, Information. A Mathematical Introduction. De Gruyter, Berlin (2012)
12. Heinosaari, T., Ziman, M.: The Mathematical Language of Quantum Theory. Cambridge University Press, Cambridge (2012)

13. Filippov, S.N.: On quantum operations of photon subtraction and photon addition. Lobachevskii J. Math. **40**, 1470–1478 (2019)
14. Bongioanni, I., Sansoni, L., Sciarrino, F., Vallone, G., Mataloni, P.: Experimental quantum process tomography of non-trace-preserving maps. Phys. Rev. A **82**, 042307 (2010)
15. Luchnikov, I.A., Filippov, S.N.: Quantum evolution in the stroboscopic limit of repeated measurements. Phys. Rev. A **95**, 022113 (2017)
16. Grimaudo, R., Messina, A., Sergi, A., Vitanov, N.V., Filippov, S.N.: Two-qubit entanglement generation through non-Hermitian Hamiltonians induced by repeated measurements on an ancilla. Entropy **22**, 1184 (2020)
17. Filippov, S.N.: Capacity of trace decreasing quantum operations and superadditivity of coherent information for a generalized erasure channel. J. Phys. A: Math. Theor. **54**, 255301 (2021)
18. Gisin, N., Huttner, B.: Combined effects of polarization mode dispersion and polarization dependent losses in optical fibers. Opt. Commun. **142**, 119 (1997)
19. Rivas, Á., Huelga, S.F., Plenio, M.B.: Quantum non-Markovianity: characterization, quantification and detection. Rep. Prog. Phys. **77**, 094001 (2014)
20. Chruściński, D., Rivas, Á., Størmer, E.: Divisibility and information flow notions of quantum Markovianity for noninvertible dynamical maps. Phys. Rev. Lett. **121**, 080407 (2018)
21. Rivas, Á., Huelga, S.F., Plenio, M.B.: Entanglement and non-Markovianity of quantum evolutions. Phys. Rev. Lett. **105**, 050403 (2010)
22. Chruściński, D., Macchiavello, C., Maniscalco, S.: Detecting non-Markovianity of quantum evolution via spectra of dynamical maps. Phys. Rev. Lett. **118**, 080404 (2017)
23. Filippov, S.N., Chruściński, D.: Time deformations of master equations. Phys. Rev. A **98**, 022123 (2018)
24. Breuer, H.-P., Laine, E.-M., Piilo, J.: Measure for the degree of non-Markovian behavior of quantum processes in open systems. Phys. Rev. Lett. **103**, 210401 (2009)
25. Milz, S., Kim, M.S., Pollock, F.A., Modi, K.: Completely positive divisibility does not mean Markovianity. Phys. Rev. Lett. **123**, 040401 (2019)
26. Ruskai, M.B.: Beyond strong subadditivity? Improved bounds on the contraction of generalized relative entropy. Rev. Math. Phys. **6**, 1147–1161 (1994)
27. Plenio, M.B., Virmani, S.: An introduction to entanglement measures. Quant. Inform. Comput. **7**, 1 (2007)
28. Hill, S., Wootters, W.K.: Entanglement of a pair of quantum bits. Phys. Rev. Lett. **78**, 5022 (1997)
29. Sperling, J., Vogel, W.: The Schmidt number as a universal entanglement measure. Phys. Scr. **83**, 045002 (2011)
30. Campbell, S., Vacchini, B.: Collision models in open system dynamics: a versatile tool for deeper insights? EPL **133**, 60001 (2021)
31. Filippov, S.N., Frizen, V.V., Kolobova, D.V.: Ultimate entanglement robustness of two-qubit states against general local noises. Phys. Rev. A **97**, 012322 (2018)
32. Filippov, S.N.: Entanglement robustness in trace decreasing quantum dynamics. Quanta **10**, 15 (2021)
33. Pollock, F.A., Rodrıguez-Rosario, C., Frauenheim, T., Paternostro, M., Modi, K.: Operational Markov condition for quantum processes. Phys. Rev. Lett. **120**, 040405 (2018)
34. Accardi, L., Frigerio, A., Lewis, J.T.: Quantum stochastic processes. Publ. Rest. Inst. Math. Sci. **18**, 97 (1982)

Compound State, Its Conditionality and Quantum Mutual Information

Takashi Matsuoka and Dariusz Chruściński

Abstract It is shown that quantum states on composite system may be represented by a composition of two types of complete positive (CP) maps, especially one of them is called by quantum conditionality, and such construction can be regarded as the quantum analogue of Bayes formula between a classical joint probability and a conditional probability. Then the mutual information of quantum state (the quantum analogue of Shannon's mutual information) may be related to the channel given by the dual map of quantum conditionality.

Keywords Quantum analogue Bayes formula · Quantum conditionality · Quantum mutual entropy

1 Introduction

In classical information theory [1, 2] the mutual entropy $I(X \otimes Y)$ on a classical composite system $(X \otimes Y,\ r)$ measures the correlation of the joint probability distribution of two random variables X and Y which is regarded as a common information encoded into X and Y. The mutual entropy can be also regarded as the information gain from Y when we observe X. This interpretation is due to the fact that the mutual entropy can be represented by the conditional entropy $S(Y|X)$ which measures how uncertain we are of Y on the average when we know X. Then Shannon applied the mutual entropy to a communication system in which it can be regarded as a proper measure of the amount of information transmitted from an input system $(X,\ p)$ to an output system $(Y,\ q)$ through a (classical) channel Λ_c^* given as a transition probability matrix corresponding to a conditional probability distributions, i.e. Λ_c^* changes

T. Matsuoka (✉)
Suwa University of Science, Toyohira 5001 Chino City, Nagano 391-0292, Japan
e-mail: matsuoka@rs.sus.ac.jp

D. Chruściński
Institute of Physics, Faculty of Physics, Astronomy and Informatics, Nicolaus Copernicus University, Grudziądzka 5/7, 87–100 Toruń, Poland

L. Accardi et al. (eds.), *Infinite Dimensional Analysis, Quantum Probability and Applications*, Springer Proceedings in Mathematics & Statistics 390,
https://doi.org/10.1007/978-3-031-06170-7_7

an input state (or a probability) p into an output state $q = \Lambda_c^* (p)$. In this scheme the mutual entropy $I (X \otimes Y)$ may be denoted by $I \left(p, \Lambda_c^*\right)$ and the validity of this interpretation is caused by the following inequality which is called the Shannon fundamental inequality

$$0 \leq I(p, : \Lambda_c^*) \leq \min \left\{S(p), S(\Lambda_c^* \rho)\right\}, \tag{1}$$

where $S (p)$ is the Shannon entropy of the probability p. Such Shannon's construction is based on the Bayes formula between joint probabilities and conditional probabilities.

For a quantum composite system $\mathcal{G} \otimes \mathcal{H}$ given by Hilbert space $\mathcal{G}$ and $\mathcal{H}$, we can define a quantum analogue of mutual entropy $I (\mathcal{G} \otimes \mathcal{H})$ by using the von Neumann entropy for density matrices of composite system and its marginal systems:

$$I (\mathcal{G} \otimes \mathcal{H}) \equiv S (\rho_{\mathcal{G}}) + S (\rho_{\mathcal{H}}) - S (\rho_{\mathcal{G} \otimes \mathcal{H}}), \tag{2}$$

where $S (\rho_{\mathcal{G}}) = -\mathrm{tr}\rho_{\mathcal{G}} \log \rho_{\mathcal{G}}$ is the von Neumann entropy of a density matrix $\rho_{\mathcal{G}}$ on the quantum system $\mathcal{G}$. The quantum analogue can be also a measure of the correlation encoded into quantum composite system and it is also called information correlation [5]. Note, however, that the notion of conditional probability is not unique defined in quantum theory and hence, unlike the classical one, it is not clear how to relate the quantum analogue (2) to the associated channel. Moreover it is well known that $I (\mathcal{G} \otimes \mathcal{H})$ in general violates the Shannon fundamental inequality (1). Simple example is provided by the pure entangled state $\rho_{\mathcal{G} \otimes \mathcal{H}} = |\Phi\rangle \langle\Phi|, \Phi \in \mathcal{G} \otimes \mathcal{H}$. In this case $I (\mathcal{G} \otimes \mathcal{H}) = 2S (\rho_{\mathcal{G}}) = 2S (\rho_{\mathcal{H}})$. The quantum picture of information theory is much more complicated because of the existence of entangled states [3–6].

There have been mathematical studies of the entanglements in Werner [7, 8], Bennett et al. [9] and Schumacher [10, 11] in which the entangled state is defined by a compound state which can not be written as convex combination of separable states. Horodeckis [12] and Peres [13] gave the another classification of quantum compound states by using the so-called positive partial transposition (PPT) condition.

On the other hand the notion of quantum conditional probability were introduced by some authors. Cref and Adami [14] introduced the conditional amplitude operator by the Trotter decomposition of compound density matrix. Then they defined the quantum analogue of conditional entropy. The another one was proposed by Asory-Kossakowski-Marmo-Sudarshan [15]. They introduced the construction method of operator which can be regarded as the quantum analogue of conditional probability given by applying CP maps to Choi-Jamiołkowski isomorphism [16, 17], and they define the compound state by the product of the marginal density and the quantum analogue of conditional operator. See also the references [18, 19].

Belavkin-Ohya and Belavkin [20–22]: they gave also the method of construction of compound state by using CP maps and co-CP maps. Both of above methods (Asory et. al. and Belavkin-Ohya) are based on the duality between quantum maps and quantum states on composite system, and the both characterize the PPT condition (see [15, 23]). However they did not discuss the relation between both methods.

In this paper we will make clear the relation of both methods and show that quantum states on composite system may be represented by the composition of two types of CP maps, especially one of them may be called by quantum conditionality, and such construction can be regarded as the quantum analogue of Bayes formula in classical probability theory. Applying such construction it is shown that the quantum analogue of mutual entropy (2) can be related to the channel given as the dual map of quantum conditionality. The same result was given in our previous paper [24]. However in that paper we omitted the discussion of general scheme of the quantum analogue of Bayes formula. Here we will show the detail of application to the mutual entropy.

2 Quantum Compound States and CP Maps

2.1 Preliminaries

Consider a quantum system living in a Hilbert space $\mathcal{H}$ (resp. $\mathcal{G}$). Denote a set of all bounded operators and a set of trace class operators in $\mathcal{H}$ by $\mathcal{B}(\mathcal{H})$ and by $\mathcal{T}(\mathcal{H})$ respectively. Normally a quantum state ς on $\mathcal{B}(\mathcal{H})$ is defined with a linear functional

$$\varsigma(b) := \mathrm{tr} b\sigma, \tag{3}$$

giving the expectation of $b \in \mathcal{B}(\mathcal{H})$. Here σ is a trace class operator, so called a density operator, satisfying the positivity and normalized condition which is in one-to-one correspondence $\varsigma \mapsto \sigma$ with the state ς. Let $\mathcal{S}(\mathcal{H})$ be a set of density operators corresponding to normal states

$$\mathcal{S}(\mathcal{H}) = \{\sigma \in \mathcal{T}(\mathcal{H});\ \sigma \geq 0,\ \mathrm{tr}\sigma = 1\}. \tag{4}$$

For simplicity in the subsequent argument we call the linear functional ς and its density σ the same name "state" if without confusion.

Let us recall the definition of completely positive map. A linear map $\chi : \mathcal{B}(\mathcal{H}) \to \mathcal{B}(\mathcal{G})$ is said to be completely positive (CP) if the map

$$\chi_n : M_n(\mathcal{B}(\mathcal{H})) \to M_n(\mathcal{B}(\mathcal{G}));\ \left[a_{ij}\right] \to \left[\chi\left(a_{ij}\right)\right] \tag{5}$$

is positive for all $n \in \mathbb{N}$, where $M_n(\mathcal{B}(\mathcal{H}))$ stands for $n \times n$ matrices with entries from $\mathcal{B}(\mathcal{H})$. A linear map $\chi : \mathcal{B}(\mathcal{H}) \to \mathcal{B}(\mathcal{G})$ is said to be co-completely positive (CCP) if its composition with transposition τ, i.e. $\chi \circ \tau$ is CP.

Here let us give the definition of transpose operation τ on $\mathcal{B}(\mathcal{H})$ or $\mathcal{B}(\mathcal{G})$. For any element $b \in \mathcal{B}(\mathcal{H})$ let $\tau_{\mathcal{H}}$ be a transposition on a some (fixed) complete orthonormal system (CONS) $\{e_i\} \subseteq \mathcal{H}$, for example one can choice the canonical basis, defined by

$$b^{\mathrm{t}} := \tau_{\mathcal{H}}(b) \equiv J_{\mathcal{H}} b^* J_{\mathcal{H}}, \tag{6}$$

where $J_{\mathcal{H}}$ denotes an isometric complex conjugation defined on $\{e_i\}$ as follows:

$$J_{\mathcal{H}} y = J_{\mathcal{H}} \sum_i \langle e_i, y\rangle e_i \equiv \sum_i \overline{\langle e_i, y\rangle} e_i = \sum_i \langle y, e_i\rangle e_i =: \overline{y}. \tag{7}$$

In the matrix elements representation on the basis $\{e_i\}$ we have

$$\begin{aligned} b^{\mathrm{t}} &= \sum_{i,j} |e_i\rangle \langle e_i, J_{\mathcal{H}} b^* J_{\mathcal{H}} e_j\rangle \langle e_j| \\ &= \sum_{i,j} |e_i\rangle \langle b^* e_j, e_i\rangle \langle e_j| = \sum_{i,j} |e_i\rangle \langle e_j, b e_i\rangle \langle e_j| . \end{aligned}$$

A transposition $\tau_{\mathcal{G}}$ on $\mathcal{B}(\mathcal{G})$ with an isometric complex conjugation $J_{\mathcal{G}}$ are defined analogously on a some (fixed) CONS in $\mathcal{G}$. In the following, for simplicity, simple notation "t" would be the transposition in the corresponding space even if the space is given by composite system.

2.2 Duality Between Quantum States and Linear Maps, and Their Classification

In this subsection and the next subsection we recall the construction methods of compound state given in [20–22] in a little bit modified way.

Now let us consider a state $\theta \in \mathcal{S}(\mathcal{G} \otimes \mathcal{H})$ and let $\varphi : \mathcal{B}(\mathcal{H}) \to \mathcal{T}(\mathcal{G})$ (or $\mathcal{B}(\mathcal{G})_*$, i.e. it is the pre-dual space of $\mathcal{B}(\mathcal{G})$) be a linear map defined by

$$\varphi(b) \equiv \mathrm{tr}_{\mathcal{H}}(I_{\mathcal{G}} \otimes b)\theta \tag{8}$$

for any $b \in \mathcal{B}(\mathcal{H})$, and its dual map $\varphi^* : \mathcal{B}(\mathcal{G}) \to \mathcal{T}(\mathcal{H})$ reads

$$\varphi^*(a) \equiv \mathrm{tr}_{\mathcal{G}}(a \otimes I_{\mathcal{H}})\theta \tag{9}$$

for any $a \in \mathcal{B}(\mathcal{G})$, where $I_{\mathcal{G}}$ and $I_{\mathcal{H}}$ are identity operators respectively. Then the compound state ϑ and its expectation $\vartheta(a \otimes b)$ may be rewritten as follows;

$$\vartheta(a \otimes b) = \mathrm{tr}(a \otimes b)\theta = \mathrm{tr}_{\mathcal{G}} a\varphi(b) = \mathrm{tr}_{\mathcal{H}} \varphi^*(a) b. \tag{10}$$

It is clear that the marginal states read

$$\rho = \varphi(I_{\mathcal{H}}) = \mathrm{tr}_{\mathcal{H}}\theta \in \mathcal{T}(\mathcal{G}), \qquad \sigma = \varphi^*(I_{\mathcal{G}}) = \mathrm{tr}_{\mathcal{G}}\theta \in \mathcal{T}(\mathcal{H}). \tag{11}$$

Belavkin and Ohya observed that if $\theta \in S(\mathcal{G} \otimes \mathcal{H})$, then both φ and φ^* are always CCP but not always CP [20, 21]. From the (physical) operational points of view, however, it is suitable to deal with these maps as CP operations. According to this motivation Belavkin [22] introduced the notion of contra-variant densities (see Remark 2) and gave another map by reformulating (8) as follows:

$$\phi(b) \equiv \mathrm{tr}_{\mathcal{H}}(I_{\mathcal{G}} \otimes b^{\mathrm{t}})\theta^{\mathrm{t}}. \tag{12}$$

Using (12) we have

$$\begin{aligned}\vartheta(a \otimes b) &= \mathrm{tr}(a \otimes b)\theta = \mathrm{tr}(a^{\mathrm{t}} \otimes b^{\mathrm{t}})\theta^{\mathrm{t}} \\ &= \mathrm{tr}_{\mathcal{G}} a^{\mathrm{t}}\phi(b) = \mathrm{tr}_{\mathcal{H}}\phi^*(a)b^{\mathrm{t}}.\end{aligned} \tag{13}$$

Actually we can show that ϕ may be a CP map and the complete positivity of ϕ gives the classification of composite system. Let us check this fact in the case of pure compound states, i.e. pure entangled states.

Let ϑ_{P} be a pure state on $\mathcal{B}(\mathcal{G} \otimes \mathcal{H})$ given by

$$\vartheta_{\mathrm{P}}(a \otimes b) = \langle \Phi, (a \otimes b)\, \Phi \rangle = \mathrm{tr}\,(a \otimes b)\, |\Phi\rangle \langle \Phi| \,, \tag{14}$$

where Φ is an entangled vector state, i.e. its Schmidt decomposition is given by $\Phi = \sum \lambda_i^{\frac{1}{2}} g_i \otimes h_i$, $\{g_i\} \subset \mathcal{G}$, $\{h_i\} \subset \mathcal{H}$ are orthonormal systems corresponding to Schatten decompositions, i.e. the spectral decomposition into one-dimensional orthogonal projectors, of marginal state ρ and σ respectively, and $\{\lambda_n\}$ denotes a probability distribution: $\lambda_i \geq 0$ and $\sum \lambda_i = 1$. In this case the explicit formulas of the map $\phi : \mathcal{B}(\mathcal{H}) \to \mathcal{T}(\mathcal{G})$ is given by the following simple computation:

$$\begin{aligned}\phi(b) &= \mathrm{tr}_{\mathcal{H}}(I_{\mathcal{G}} \otimes b^{\mathrm{t}})\, (|\Phi\rangle \langle \Phi|)^{\mathrm{t}} \\ &= \sum_{i,j} |\bar{g}_i\rangle\, \lambda_i^{\frac{1}{2}} \lambda_j^{\frac{1}{2}} \langle \bar{g}_j| \langle \bar{h}_j, b^{\mathrm{t}}\bar{h}_i\rangle \\ &= \sum_{i,j} |\bar{g}_i\rangle\, \lambda_i^{\frac{1}{2}} \langle h_i, bh_j\rangle\, \lambda_j^{\frac{1}{2}} \langle \bar{g}_j| = \kappa^* b \kappa,\end{aligned} \tag{15}$$

where $\kappa \equiv \sum |h_i\rangle\, \lambda_i^{\frac{1}{2}} \langle \bar{g}_i|$ which is the Hilbert-Schmidt operator from $\mathcal{G}$ to $\mathcal{H}$, and $\phi(\cdot) \equiv \kappa^* \cdot \kappa$ is a CP map given in the Steinspring form. However ϕ is not always CCP, because the transposition τ is positive but not CP unless the algebra on which τ is defined is Abelian. In the pure state case, it can be shown that ϕ is not CCP unless θ_{P} is given as a simple product state, i.e., θ_{P} is a separable state (see [23]). Namely the non separability of θ_{P} is characterized by the non co-complete positivity of its associated map ϕ.

In the same way its dual map ϕ^* is given by

$$\phi^*(a) \equiv \mathrm{tr}_{\mathcal{G}}(a^{\mathrm{t}} \otimes I_{\mathcal{H}})\, |\Phi\rangle \langle \Phi|^{\mathrm{t}} = \kappa^{\mathrm{t}*} a \kappa^{\mathrm{t}} \tag{16}$$

where $\kappa^{\mathrm{t}} := J_{\mathcal{G}}\kappa^* J_{\mathcal{H}} = \sum |g_i\rangle \lambda_i^{\frac{1}{2}} \langle \bar{h}_i|$ from $\mathcal{H}$ to $\mathcal{G}$. Furthermore the entangled vector Φ may be represented by κ as follows:

$$\langle \zeta \otimes \eta, \Phi \rangle = \langle \eta, \kappa J_{\mathcal{G}} \zeta \rangle, \tag{17}$$

where $\zeta \in \mathcal{G}$, $\eta \in \mathcal{H}$. Then $\vartheta_{\mathrm{P}}(a \otimes b)$ is represented by

$$\begin{aligned} \vartheta_{\mathrm{P}}(a \otimes b) &= \mathrm{tr}_{\mathcal{G}} a^{\mathrm{t}} \kappa^* b \kappa = \mathrm{tr}_{\mathcal{G}} a^{\mathrm{t}} \phi(b) \\ &= \mathrm{tr}_{\mathcal{H}} \kappa^{\mathrm{t}*} a \kappa^{\mathrm{t}} b^{\mathrm{t}} = \mathrm{tr}_{\mathcal{H}} \phi^*(a)\, b^{\mathrm{t}} \end{aligned} \tag{18}$$

Its marginals are obtained by the operator κ:

$$\rho^{\mathrm{t}} = \phi(I_{\mathcal{H}}) = \kappa^* \kappa = \sum |\bar{g}_i\rangle \lambda_i \langle \bar{g}_i|, \tag{19}$$

$$\sigma^{\mathrm{t}} = \phi^*(I_{\mathcal{G}}) = \kappa^{\mathrm{t}*} \kappa^{\mathrm{t}} = \sum |\bar{h}_i\rangle \lambda_i \langle \bar{h}_i|. \tag{20}$$

Notice that the map ϕ may be recognized as the CP map form $\mathcal{B}(\mathcal{H})$ to the transposed algebra $\widetilde{\mathcal{T}(\mathcal{G})} \equiv \mathcal{T}(\mathcal{G})^{\mathrm{t}} = \{\rho^{\mathrm{t}};\ \rho \in \mathcal{T}(\mathcal{G})\}$. In usual argument we may assume that the transposition leaves invariant the sub-algebra $\mathcal{A} \subseteq \mathcal{B}(\mathcal{G})$ such that $\mathcal{A} = \widetilde{\mathcal{A}}$, however from the notation and operational points of view, it is preferable to distinguish the algebra $\mathcal{A}$ from the transposed algebra $\tilde{\mathcal{A}}$.

We can extend the above construction to mixture states. In general the map ϕ can be given by using a set of stochastic operators $\kappa_i : \mathcal{G} \to \mathcal{H}$ satisfying the constraints

$$\mathrm{tr}\left(\sum \kappa_i^* \kappa_i\right) = 1. \tag{21}$$

Let $\theta \in \mathcal{S}(\mathcal{G} \otimes \mathcal{H})$ be a mixed compound state with the Schatten decomposition $\theta = \sum |\Phi_i\rangle p_i \langle \Phi_i| = \sum |\Psi_i\rangle \langle \Psi_i|$ where $\Psi_i = p_i^{\frac{1}{2}} \Phi_i$. Applying (15) to the density θ the expectation $\vartheta(a \otimes b)$ may be represented by ϕ or ϕ^* as follows:

$$\begin{aligned} \vartheta(a \otimes b) &= \mathrm{tr}\,(a \otimes b)^{\mathrm{t}} \theta^{\mathrm{t}} = \sum \mathrm{tr}\left(a^{\mathrm{t}} \otimes b^{\mathrm{t}}\right)(|\Psi_i\rangle \langle \Psi_i|)^{\mathrm{t}} \\ &= \sum \mathrm{tr}_{\mathcal{G}} a^{\mathrm{t}} \kappa_i^* b \kappa_i = \mathrm{tr}_{\mathcal{G}} a^{\mathrm{t}} \phi(b) \qquad (22) \\ &= \sum \mathrm{tr}_{\mathcal{H}} \kappa_i^{\mathrm{t}*} a \kappa_i^{\mathrm{t}} b^{\mathrm{t}} = \mathrm{tr}_{\mathcal{H}} \phi^*(a)\, b^{\mathrm{t}}, \qquad (23) \end{aligned}$$

where

$$\phi(\cdot) = \sum \kappa_i^* \cdot \kappa_i, \quad \phi^*(\cdot) = \sum \kappa_i^{\mathrm{t}*} \cdot \kappa_i^{\mathrm{t}}, \tag{24}$$

which are CP maps given in the Kraus-Sudarshan form, and the stochastic operator κ_i is derived from the stochastic (unnormalized) vector state Ψ_i. Then ϕ and ϕ^* satisfy the following marginal normalized conditions:

$$\rho^{\mathrm{t}} = \phi\left(I_{\mathcal{H}}\right) = \sum \kappa_i^* \kappa_i, \quad \sigma^{\mathrm{t}} = \phi^*\left(I_{\mathcal{G}}\right) = \sum \kappa_i^{\mathrm{t}*} \kappa_i^{\mathrm{t}}. \tag{25}$$

Notice that the map given by (8) or (9) are both always CCP but not always CP as mentioned above. For example, in the case of pure state with same calculation as (15), it is easy to read

$$\varphi \circ \tau_{\mathcal{H}}(b) = \mathrm{tr}_{\mathcal{H}}(I_{\mathcal{G}} \otimes b^{\mathrm{t}}) \left(|\Phi\rangle \langle\Phi|\right) = \chi^* b \chi$$

where $\chi \equiv \sum |h_i\rangle \lambda_i^{\frac{1}{2}} \langle g_i|$, which means that φ is always CCP.

Now let us consider the partial positive transpose of compound state $\theta \in \mathcal{S}\left(\mathcal{G} \otimes \mathcal{H}\right)$. Recall that θ is a PPT state if $\theta_{\mathrm{PPT}} := \left(1_{\mathcal{G}} \otimes \tau_{\mathcal{H}}\right) \theta \in \mathcal{S}\left(\mathcal{G} \otimes \mathcal{H}\right)$. This means that its associated map ϕ_{PPT} is CP. This condition is equivalent to the co-complete positivity of ϕ [23], because it can be shown that the following equality holds by simple computation:

$$\phi_{\mathrm{PPT}}\left(b\right) = \mathrm{tr}_{\mathcal{H}}(I_{\mathcal{G}} \otimes b^{\mathrm{t}})\theta_{\mathrm{PPT}}^{\mathrm{t}} = \phi \circ \tau_{\mathcal{H}}\left(b\right). \tag{26}$$

Corollary 1 ([23]) *PPT states are completely characterized by ϕ which are both CP and CCP.*

Let us summarize the above observation as the following:

Definition 1 ([22]) For a given CP map (or a CP operation) $\phi : \mathcal{B}(\mathcal{H}) \to \widetilde{\mathcal{T}(\mathcal{G})}$ normalized to a state density $\rho^{\mathrm{t}} = \phi(I_{\mathcal{H}})$ the generalized compound state ϑ in contravariant form is defined as follows:

$$\mathrm{tr}_{\mathcal{G}} a^{\mathrm{t}} \phi\left(b\right) = \vartheta\left(a \otimes b\right) = \mathrm{tr}_{\mathcal{H}} b^{\mathrm{t}} \phi^*\left(a\right). \tag{27}$$

Remark 1 Belavkin and Ohya [21] called the CCP map $\varphi : \mathcal{B}(\mathcal{H}) \to \mathcal{T}(\mathcal{G})$ normal coupling of the state $\varsigma\left(b\right) = \mathrm{tr}_{\mathcal{H}} b\sigma$ to the state $\varrho\left(a\right) = \mathrm{tr}_{\mathcal{G}} a\rho$. On the other hand the CP map $\phi^* : \mathcal{B}(\mathcal{G}) \to \widetilde{\mathcal{T}(\mathcal{H})}$ normalized to the density $\sigma^{\mathrm{t}} = \phi^*\left(I_{\mathcal{G}}\right)$ of state ς was called normal entanglement of the system $\left(\mathcal{B}\left(\mathcal{G}\right), \varrho\right)$ with the density $\rho^{\mathrm{t}} = \phi\left(I_{\mathcal{H}}\right)$ to $\left(\mathcal{B}\left(\mathcal{H}\right), \varsigma\right)$. However, such names may give the misunderstanding or confusion on the usual definition of entangled state. So that we avoid to use such naming here.

Remark 2 On the base of results in [21] Belavkin introduced the notion of contravariant density in order to avoid the consideration of the CCP operation and to deal only with the CP operation [22]. He described the normal state ς not using by the usual density operator σ, called covariant density, but using by the transposed density denoted by $\tilde{\sigma} := \sigma^{\mathrm{t}}$ such that

$$\varsigma\left(b\right) := \langle b,\, \tilde{\sigma}\rangle \equiv \mathrm{tr}_{\mathcal{H}} b^{\mathrm{t}} \tilde{\sigma} =: b_{ij} \tilde{\sigma}^{ij}, \tag{28}$$

where b_{ij} is the matrix element of b, $\tilde{\sigma}$ is called a contravariant density of the state ς with respect to the natural tensor paring (28) (see also Ref. [25]). On this scheme he defined the CP map ϕ in Definition 1 and he called it generalized entanglement. However, here, for the same reason as mentioned in Remark 1, in Definition 1, we call it the generalized compound state ϑ a contravariant form. In the subsequent argument, we call also a CP map $\phi : \mathcal{B}(\mathcal{H}) \to \widetilde{\mathcal{T}(\mathcal{G})}$ normalized to a state density $\rho^{\mathrm{t}} = \phi(I_{\mathcal{H}})$ in (12) a contravariant form, and call a CCP map $\varphi : \mathcal{B}(\mathcal{H}) \to \mathcal{T}(\mathcal{G})$ normalized to a state density $\rho = \varphi(I_{\mathcal{H}})$ in (8) a covariant form.

2.3 Quantum Compound State Versus Classical Joint Probability

As mentioned in the introduction Cerf and Adami [14] introduced the following definition of the conditional amplitude operator

$$\rho_{B|A} = \lim_{n\to\infty} [\rho_{AB}^{1/n} \cdot \rho_A^{-1/n} \otimes I_B]^n, \tag{29}$$

and shown that such definition reproduces the formula for the conditional von Neumann entropy $S(B|A) = S(\rho_{AB}) - S(\rho_B)$, that is,

$$S(B|A) = -\mathrm{Tr}(\rho_{AB} \log \rho_{B|A}). \tag{30}$$

Asorey, Kossakowski, Marmo and Sudarshan, 2005: They introduced another approach of notion of quantum conditional probability operator (QCP for short) [15] (see also Ref. [18, 19].) In this subsection it will be shown that the generalized compound state ϑ has a similar structure as the classical Bayes formula via the QCP in contravariant form. For this purpose, at first, we recall the theorem by Belalvkin and Dai [26] about the decomposition of normalized CP map ϕ in contravariant form.

The normalized contravariant CP map ϕ is called standard for the state ς on $\mathcal{B} := \mathcal{B}(\mathcal{H})$ if it associates σ with the transposed state $\rho = \sigma^{\mathrm{t}}$ on $\mathcal{A} := \tilde{\mathcal{B}} \equiv \{b^{\mathrm{t}} : b \in \mathcal{B}(\mathcal{H})\}$ by

$$\phi(b) = \sigma^{\frac{1}{2}} b \sigma^{\frac{1}{2}} = \rho^{\mathrm{t}\frac{1}{2}} b \rho^{\mathrm{t}\frac{1}{2}} =: \rho(b), \ \forall b \in \mathcal{B} = \tilde{\mathcal{A}}. \tag{31}$$

According to the the argument of the previous section the standard CP map ρ in contravariant form may be derived from the pure entangled state with marginal density ρ in the case of $\mathcal{G} = \mathcal{H}$. Then (see (33))

$$\rho^*(a) = \sigma^{\mathrm{t}\frac{1}{2}} a \sigma^{\mathrm{t}\frac{1}{2}} =: \sigma(a), \ \forall a \in \mathcal{A} = \tilde{\mathcal{B}}. \tag{32}$$

In (31) and (32) we use the same symbols ρ and σ for both the density operators and their CP maps in contravariant form.

Given ς, the standard pure compound state ϑ_s on $\mathcal{A} \otimes \mathcal{B}$ in contravariant form is defined by

$$\mathrm{tr}_{\mathcal{H}} a' \rho(b) = \vartheta_s(a \otimes b) = \mathrm{tr}_{\mathcal{H}} b' \sigma(a). \tag{33}$$

Belalvkin and Dai [26] showed that every normalized contravariant CP map ϕ can be decomposed to composition of a standard CP map ρ with an another type of CP map π. For convenient of readers we show its proof.

Theorem 1 ([26]) *Every CP map ϕ on $\mathcal{B}(\mathcal{H})$ to the state $\rho^{\mathrm{t}} \in \widetilde{\mathcal{T}(\mathcal{G})}$ in cotravariant form has a decomposition,*

$$\phi(b) = \rho^{\mathrm{t}\frac{1}{2}} \pi(b) \rho^{\mathrm{t}\frac{1}{2}}, \tag{34}$$

where π is normal CP contraction $\mathcal{B}(\mathcal{H}) \to \widetilde{\mathcal{B}(\mathcal{G})}$ such that $I_{\mathcal{G}} \geq \pi(I_{\mathcal{H}}) \geq E_{\rho^{\mathrm{t}}}$ on the minimal orthoprojector denoted by $E_{\rho^{\mathrm{t}}} \in \widetilde{\mathcal{B}(\mathcal{G})}$ supporting the density operator ρ^{t}. This decomposition is unique by the normalization condition $\pi(I_{\mathcal{H}}) = E_{\rho^{\mathrm{t}}}$.

Proof: π can be found as the solution to the equation

$$\rho^{\mathrm{t}-\frac{1}{2}} \pi(b) \rho^{\mathrm{t}-\frac{1}{2}} = \phi(b) \ \forall b \in \mathcal{B}(\mathcal{H}) \tag{35}$$

which is unique if $\tilde{\rho}$ is non-degenerate:

$$\pi(b) = \rho^{\mathrm{t}-\frac{1}{2}} \phi(b) \rho^{\mathrm{t}-\frac{1}{2}} \tag{36}$$

If ρ^{t} is degenerate, one should reduce $\mathcal{G}$ to the Hilbert subspace $\mathcal{G}_{\rho^{\mathrm{t}}} = E_{\rho^{\mathrm{t}}}\mathcal{G}$ given by $E_{\rho^{\mathrm{t}}}$, then $\pi(I_{\mathcal{H}}) = E_{\rho^{\mathrm{t}}} = I_{\mathcal{G}_{C^{\mathrm{t}}}}$. □

Let us recall the notion of QCP. Given a compound density operator $\theta \in \mathcal{S}(\mathcal{G} \otimes \mathcal{H})$ with its marginal $\rho = \mathrm{tr}_{\mathcal{H}}\theta$, one can define the operator $\Pi \in \mathcal{T}(\mathcal{G} \otimes \mathcal{H})$ as

$$\Pi \equiv \left(\rho^{-\frac{1}{2}} \otimes I\right) \theta \left(\rho^{-\frac{1}{2}} \otimes I\right) \tag{37}$$

which has the properties

$$\Pi \geq 0, \tag{38}$$

$$\mathrm{tr}_{\mathcal{H}} \Pi = I_{\mathcal{G}}. \tag{39}$$

Notice that if ρ is degenerate, one should reduce $\mathcal{G}$ to the Hilbert subspace $\mathcal{G}_\rho = E_\rho \mathcal{G}$, then $\mathrm{tr}_{\mathcal{H}} \Pi = I_{\mathcal{G}_C}$. It is clear that

$$\theta = \left(\rho^{\frac{1}{2}} \otimes I\right) \Pi \left(\rho^{\frac{1}{2}} \otimes I\right). \tag{40}$$

In the sense of (38), (39) the operator Π can be called by the quantum analogue of conditional probability. In the same sense the decomposition (40) may be corresponding to the classical construction of joint probability by conditional probability.

Asorey et.al. called the operator $\Pi \in \mathcal{T}(\mathcal{G} \otimes \mathcal{H})$ satisfying (38) and (39) the quantum conditional probability (QCP). In [18] it is called conditional quantum state.

Using the transposed density θ^{t} in (40) we can represent the compound state ϑ in contravariant form corresponding to the transposed QCP as follows:

$$\begin{aligned}\vartheta(a \otimes b) &= \mathrm{tr}\left(a^{\mathrm{t}} \otimes b^{\mathrm{t}}\right)\theta^{\mathrm{t}} \\ &= \mathrm{tr}\left(a^{\mathrm{t}} \otimes b^{\mathrm{t}}\right)\left(\rho^{\mathrm{t}\frac{1}{2}} \otimes I_{\mathcal{H}}\right)\Pi^{\mathrm{t}}\left(\rho^{\mathrm{t}\frac{1}{2}} \otimes I_{\mathcal{H}}\right) \qquad (41)\\ &= \mathrm{tr}_{\mathcal{G}} a^{\mathrm{t}} \rho^{\mathrm{t}\frac{1}{2}} \pi_{\Pi}(b) \rho^{\mathrm{t}\frac{1}{2}} = \mathrm{tr}_{\mathcal{G}} a^{\mathrm{t}} \phi_{\Pi}(b), \qquad (42)\end{aligned}$$

where $\pi_{\Pi}(b) \equiv \mathrm{tr}_{\mathcal{H}}\left(I_{\mathcal{G}} \otimes b^{\mathrm{t}}\right)\Pi^{\mathrm{t}}$ and $\phi_{\Pi}(b) = \rho^{\mathrm{t}\frac{1}{2}} \pi_{\Pi}(b) \rho^{\mathrm{t}\frac{1}{2}} = \rho \circ \pi_{\Pi}(b)$. Comparing (42) with the decomposition of ϕ in (34), due to the Theorem 1, it is clear that $\phi = \phi_{\Pi}$, $\pi = \pi_{\Pi}$.

Definition 2 A normal unital CP map π from $\mathcal{B}(\mathcal{H})$ to $\widetilde{\mathcal{B}(\mathcal{G})}$ is called a quantum conditionality in cotravariant form. Then a normalized contravariant CP map ϕ is given by the composition of a standard CP map ρ and a quantum conditionality π:

$$\phi = \rho \circ \pi \qquad (43)$$

In classical probability theory a conditional probability gives a classical channel denoted by Λ_{c}^{*} defined as the transition probability matrix. A quantum conditionality π defines also the quantum channel Λ_{q}^{*} on the $\mathcal{T}(\mathcal{G})$ into $\widetilde{\mathcal{T}(\mathcal{H})}$ as its dual map $\pi^{*} : \mathcal{B}(\mathcal{G}) \rightarrow \widetilde{\mathcal{B}(\mathcal{H})}$ given by the following duality:

$$\varsigma(b) = \mathrm{tr}_{\mathcal{G}} \rho^{\mathrm{t}} \pi(b) = \mathrm{tr}_{\mathcal{H}} b^{\mathrm{t}} \pi^{*}(\rho) = \mathrm{tr}_{\mathcal{H}} b^{\mathrm{t}} \sigma^{\mathrm{t}}, \qquad (44)$$

i.e. $\Lambda_{\mathrm{q}}^{*} \equiv \pi^{*}|_{\mathcal{T}(\mathcal{G})}$. This duality is given by the same dual relation between ϕ and ϕ^{*} in (22) and (23).

We summarize the above observation.

Theorem 2 *There exists the similar structure between normalized CP map ϕ in contravariant form and classical joint probability distributions $r = \left(r_{ij}\right)$ with its marginals $p = (p_i)$, $q = \left(q_j\right)$ as follows:*

$$\text{Joint probability distribution} : \left(r_{ij}\right) = \left(p_i p_{j|i}\right) \longleftrightarrow \phi = \rho \circ \pi \qquad (45)$$

$$\text{Conditional probability distribution} : \left(p_{j|i}\right) = \left(\frac{r_{ij}}{p_i}\right) \longleftrightarrow \pi = \rho^{-1} \circ \phi \qquad (46)$$

$$\text{Channel} : \left(q_j\right) = \Lambda_{\mathrm{c}}^{*}(p) = \left(\sum_i p_i p_{j|i}\right) \longleftrightarrow \sigma^{\mathrm{t}} = \Lambda_{\mathrm{q}}^{*}(\rho) = \pi^{*}(\rho) \qquad (47)$$

Definition 3 We call the above relations (45), (46), (47) the quantum analogue of Bayes formula.

3 Quantum Mutual Information on Quantum Bayes Formula

3.1 Compound State via Choi-Jamiołkowski Isomorphism

From now on we reduce our argument to a finite dimensional case for simplicity.

Let d be a dimensionality of $\mathcal{G}$ and $\mathcal{H}$, i.e, $d = \dim\mathcal{G} = \dim\mathcal{H} < \infty$, and $\{g_i\} \subset \mathcal{G}$ and $\{h_i\} \subset \mathcal{H}$ be a CONS on which the transpositions $\tau_{\mathcal{G}}$ and $\tau_{\mathcal{H}}$ are defined respectively. Given a CP map ϕ (or ϕ^*) normalized to $\rho^{\mathrm{t}} = \phi\left(I_{\mathcal{H}}\right)$ (or $\sigma^{\mathrm{t}} = \phi^*\left(I_{\mathcal{G}}\right)$) in cotravariant form we can construct a compound state θ_ϕ in $\mathcal{B}\left(\mathcal{G} \otimes \mathcal{H}\right)$ via Choi-Jamiolkowski isomorphism [16, 17] as follows:

$$\theta_\phi \equiv (\phi \otimes I_{\mathcal{H}})P_{\mathcal{H}} = \sum_{i,j} \phi(|h_i\rangle\langle h_j|) \otimes |h_i\rangle\langle h_j| \tag{48}$$

$$= (I_{\mathcal{G}} \otimes \phi^*)P_{\mathcal{G}} = \sum_{i,j} |g_i\rangle\langle g_j| \otimes \phi^*\left(|g_i\rangle\langle g_j|\right), \tag{49}$$

where $P_{\mathcal{H}} = \sum_{i,j} |h_i\rangle\langle h_j| \otimes |h_i\rangle\langle h_j|$, $P_{\mathcal{G}} = \sum_{i,j} |g_i\rangle\langle g_j| \otimes |g_i\rangle\langle g_j|$.

Due to the duality given by Choi-Jamiolkowski isomorphism we know that the complete positivity of ϕ leads the positivity of θ_ϕ, and it is clear that $\rho^{\mathrm{t}} = \mathrm{tr}_{\mathcal{G}}\theta_\phi$, $\sigma^{\mathrm{t}} = \mathrm{tr}_{\mathcal{H}}\theta_\phi$ which means the normality condition tr`Œ $= 1$, i.e. $\theta_\phi \in \widetilde{\mathcal{S}}\left(\mathcal{G} \otimes \mathcal{H}\right)$. Then the expectation $\vartheta_\phi\left(a \otimes b\right)$ in contravariant form is given by

$$\vartheta_\phi\left(a \otimes b\right) = \mathrm{tr}\left(a^{\mathrm{t}} \otimes b^{\mathrm{t}}\right)\theta_\phi$$

$$= \mathrm{tr}_{\mathcal{G}} a^{\mathrm{t}} \sum_{i,j} \phi(|h_i\rangle\langle h_j|)\left\langle h_j, b^{\mathrm{t}} h_i\right\rangle = \mathrm{tr}_{\mathcal{G}} a^{\mathrm{t}} \phi\left(b\right) \tag{50}$$

$$= \mathrm{tr}_{\mathcal{H}} \sum_{i,j} \left\langle g_j, a^{\mathrm{t}} g_i\right\rangle \phi^*(|g_i\rangle\langle g_j|) b^{\mathrm{t}} = \mathrm{tr}_{\mathcal{H}} b^{\mathrm{t}} \phi^*\left(a\right) \tag{51}$$

for any $a \in \mathcal{B}\left(\mathcal{G}\right)$ and $b \in \mathcal{B}\left(\mathcal{H}\right)$. If its partial transpose $(1_{\mathcal{G}} \otimes \tau_{\mathcal{H}})\theta_\phi$ is positive, i.e. $\theta_\phi \in \mathcal{S}_{\mathrm{PPT}}$, then it can be represented by

$$(1_{\mathcal{G}} \otimes \tau_{\mathcal{H}})\theta_\phi = (\phi \otimes \tau_{\mathcal{H}})P_{\mathcal{H}}$$

$$= \sum_{i,j} \phi(|h_i\rangle\langle h_j|) \otimes \tau_{\mathcal{H}}\left(|h_i\rangle\langle h_j|\right)$$

$$= \sum_{i,j} \phi(|h_i\rangle \langle h_j|) \otimes |h_j\rangle \langle h_i|$$

$$= \sum_{i,j} \phi \circ \tau_{\mathcal{H}}(|h_j\rangle \langle h_i|) \otimes |h_j\rangle \langle h_i| > 0. \tag{52}$$

Therefore the condition $\theta_\phi \in \mathcal{S}_{\text{PPT}}$ is equivalent to the co-complete positivity of ϕ. In finite dimensional case, via the Choi-Jamiolkowski criterion of CP map, we reach the characterization of $\mathcal{S}_{\text{PPT}}$ corresponding to Corollary 1 in the previous section.

Applying the decomposition $\phi = \rho \circ \pi$, due the duality of quantum conditionality π, the compound state θ_ϕ may be represented by its channel π^* or QCP $\Pi_{\mathcal{H}|\mathcal{G}}$ as follows:

$$\theta_\phi = \sum_{i,j} \rho(|g_i\rangle \langle g_j|) \otimes \pi^* \left(|g_i\rangle \langle g_j|\right) = \left(\rho^{\text{t}\frac{1}{2}} \otimes I_{\mathcal{H}}\right) \Pi_{\mathcal{H}|\mathcal{G}} \left(\rho^{\text{t}\frac{1}{2}} \otimes I_{\mathcal{H}}\right) \tag{53}$$

where $\Pi_{\mathcal{H}|\mathcal{G}} = \sum_{i,j} |g_i\rangle \langle g_j|) \otimes \pi^* \left(|g_i\rangle \langle g_j|\right)$. The channel $\Lambda^*_{\text{Œ}}$ may be corresponding to the partial trace operation:

$$\sigma^{\text{t}} = \text{tr}_{\mathcal{G}} \theta_\phi = \sum_{i,j} \langle g_j, \rho^{\text{t}} g_i \rangle \pi^* \left(|g_i\rangle \langle g_j|\right) = \pi^* (\rho) =: \Lambda^*_{\text{Œ}} (\rho) \tag{54}$$

Notice : Without loss of generality, given a normalized CP map ϕ in contravariant form, we can choice the CONS $\{g_i\}$ and $\{h_j\}$ extending the orthonormal system determined by the Schatten decomposition of $\rho = \tau_{\mathcal{G}} \circ \phi (I_{\mathcal{H}}) = \sum |g_i\rangle \lambda_i \langle g_i|$ and $\sigma = \tau_{\mathcal{H}} \circ \phi^* (I_{\mathcal{G}}) = \sum |h_j\rangle \mu_j \langle h_j|$. Then the marginal densities satisfy the symmetry condition w.r.t. $\tau_{\mathcal{G}}$ and $\tau_{\mathcal{H}}$, i.e. $\rho = \rho^{\text{t}}$ and $\sigma = \sigma^{\text{t}}$. In this condition we have

$$\theta_\phi = \sum_{i,j} \rho^{\frac{1}{2}} |g_i\rangle \langle g_j| \rho^{\frac{1}{2}} \otimes \pi^* \left(|g_i\rangle \langle g_j|\right) \tag{55}$$

$$= \sum_{i,j} |g_i\rangle \sqrt{\lambda_i}\sqrt{\lambda_j} \langle g_j| \otimes \pi^* \left(|g_i\rangle \langle g_j|\right), \tag{56}$$

$$\sigma = \Lambda^*_\phi (\rho) = \pi^* (\rho). \tag{57}$$

This condition gives the simple representation. In the next subsection we assume the symmetry condition of the reduced marginal states.

3.2 *Quantum Mutual Information via Quantum Channel* π^*

Using the decomposition $\phi = \rho \circ \pi$ one can redefine the quantum analogue of mutual entropy $I\left(\mathcal{G} \otimes \mathcal{H}\right)$ in (2) as follows:

$$I\left(\mathcal{G} \otimes \mathcal{H}\right) =: I_\phi\left(\mathcal{G} \otimes \mathcal{H}\right) = S(\rho) + S(\pi^* (\rho)) - S(\theta_\phi). \tag{58}$$

In (58) we can express the mutual information by using the associated channel π^*, however it is not clear its interpretation in communication scheme.

Now let us recall the definition of mutual information in a different scenario where one introduces the notion of entropy exchange [27]. We have a quantum channel $\Gamma : \mathcal{B}(\mathcal{G}) \to \mathcal{B}(\mathcal{H})$ and the input state $\rho \in \mathcal{S}(\mathcal{G})$. Being complete positive and trace-preserving map Γ it allows for the Stienspring representation [28, 29]

$$\Gamma(\rho) = \mathrm{tr}_{\mathcal{E}} V \rho V^*, \tag{59}$$

where $V : \mathcal{G} \to \mathcal{H} \otimes \mathcal{E}$ is a linear isometry, and $\mathcal{E}$ may be interpreted as an environmental Hilbert space. This representation allows to define a complementary channel $\Gamma^{c} : \mathcal{B}(\mathcal{G}) \to \mathcal{B}(\mathcal{E})$:

$$\Gamma^{c}(\rho) = \mathrm{tr}_{\mathcal{G}} V \rho V^*. \tag{60}$$

Complementary channel gives raise to the entropy exchange $S(\rho, \Gamma)$ defined by

$$S(\rho, \Gamma) = S(\rho_{\mathcal{E}}), \tag{61}$$

with $\rho_{\mathcal{E}} = \Gamma^{c}(\rho)$. The entropy exchange $S(\rho, \Gamma)$ characterizes the information exchange between the original (i.e. input) system and the environment during the process described by quantum channel Γ. It is a function of the input state ρ and the channel Γ and it was proved that $S(\rho, \Gamma)$ does not depend on the particular Stinespring representation [27].

Given the input state ρ and the channel Γ one defines the quantum mutual information

$$I(\rho, \Gamma) = S(\rho) + S(\Gamma(\rho)) - S(\rho, \Gamma). \tag{62}$$

Proposition 1 ([5]) *Quantum mutual information is subadditive*

$$I(\rho_{12}, \Gamma_1 \otimes \Gamma_2) \le I(\rho_1, \Gamma_1) + I(\rho_2, \Gamma_2), \tag{63}$$

and satisfies the following data processing inequality

$$I(\rho, \Gamma_1 \circ \Gamma_2) \le \min\{I(\rho, \Gamma_1), I(\sigma, \Gamma_2)\}, \tag{64}$$

where $\Gamma_1 : \mathcal{B}(\mathcal{H}_1) \to \mathcal{B}(\mathcal{H}_2)$, $\Gamma_2 : \mathcal{B}(\mathcal{H}_2) \to \mathcal{B}(\mathcal{H}_2)$, and $\sigma = \Gamma_1(\rho)$.

Let us discuss the entropy exchange on the quantum analogue of Bayes formula. The quantum conditionality $\pi : \mathcal{B}(\mathcal{H}) \to \widetilde{\mathcal{B}(\mathcal{G})}$ as a normal unital CP map can be always written as [22]

$$\pi(b) = \mathrm{tr}_{\mathcal{E}} K b K^*, \tag{65}$$

where $K : \mathcal{H} \to \mathcal{G} \otimes \mathcal{E}$ is a linear co-isometry. Then, taking the symmetry condition of marginal densities into consideration, the channel Λ_{ϕ}^* as the dual map $\pi^* : \mathcal{B}(\mathcal{G}) \to \widetilde{\mathcal{B}(\mathcal{H})}$ is given by the following duality:

$$\varsigma(b) = \mathrm{tr}_{\mathcal{G}}\pi(b)\rho = \mathrm{tr}_{\mathcal{G}\otimes\mathcal{E}}KbK^*(\rho\otimes I_{\mathcal{E}}) \tag{66}$$
$$= \mathrm{tr}_{\mathcal{H}\otimes\mathcal{E}}\left(b^{\mathrm{t}}\otimes I_{\mathcal{E}}\right)V\rho V^* \tag{67}$$
$$= \mathrm{tr}_{\mathcal{H}}b^{\mathrm{t}}\pi^*(\rho) = \mathrm{tr}_{\mathcal{H}}b^{\mathrm{t}}\sigma, \tag{68}$$

where $V : \mathcal{G}\to\mathcal{H}\otimes\mathcal{E}$ is a linear co-isometry as the partially transpose of K given by

$$\langle\eta\otimes\xi,\ VJ_{\mathcal{G}}\zeta\rangle = \langle\zeta\otimes\xi,\ KJ_{\mathcal{H}}\eta\rangle,\ \forall\xi\in\mathcal{E},\ \zeta\in\mathcal{G},\ \eta\in\mathcal{H}. \tag{69}$$

Namely we have

$$\Lambda_\phi^*(\rho) = \pi^*(\rho) = \mathrm{tr}_{\mathcal{E}}V\rho V^* = \sigma. \tag{70}$$

Theorem 3 ([24]) *Given a normalized contravariant CP map $\phi = \rho\circ\pi$ one has*

$$I_\phi(\mathcal{G}\otimes\mathcal{H}) = I\left(\rho,\ \pi^*\right). \tag{71}$$

Proof: let $|\Psi_{\mathcal{R}\otimes\mathcal{G}}\rangle$ be the purification of ρ in $\mathcal{R}\otimes\mathcal{G}$, that is

$$|\Psi_{\mathcal{R}\otimes\mathcal{G}}\rangle = \sum_{i=1}^{d}\sqrt{\lambda_i}\,|g_i\rangle\otimes|g_i\rangle, \tag{72}$$

where $\mathcal{R}$ is a reference Hilbert space. Define $|\Phi_{\mathcal{R}\otimes\mathcal{H}\otimes\mathcal{E}}\rangle=(I_{\mathcal{R}}\otimes V)\,|\Psi_{\mathcal{R}\otimes\mathcal{G}}\rangle\in\mathcal{E}\otimes\mathcal{H}\otimes\mathcal{R}$. Since $|\Phi_{\mathcal{R}\otimes\mathcal{H}\otimes\mathcal{E}}\rangle$ is a pure entangled vector one has $S(\rho_{\mathcal{E}}) = S(\rho_{\mathcal{R}\otimes\mathcal{H}})$ for its marginal densities, and hence the mutual information (62) reads

$$I\left(\rho,\ \pi^*\right) = S(\rho) + S\left(\pi^*(\rho)\right) - S(\rho_{\mathcal{E}}) = S(\rho) + S\left(\pi^*(\rho)\right) - S(\rho_{\mathcal{R}\otimes\mathcal{H}}) \tag{73}$$

One easily finds

$$\rho_{\mathcal{R}\otimes\mathcal{H}} = \mathrm{tr}_{\mathcal{E}}\,|\Phi_{\mathcal{R}\otimes\mathcal{H}\otimes\mathcal{E}}\rangle\langle\Phi_{\mathcal{R}\otimes\mathcal{H}\otimes\mathcal{E}}| \tag{74}$$
$$= \mathrm{tr}_{\mathcal{E}}\,(I_{\mathcal{R}}\otimes V)\,|\Psi_{\mathcal{R}\otimes\mathcal{G}}\rangle\langle\Psi_{\mathcal{R}\otimes\mathcal{G}}|\left(I_{\mathcal{R}}\otimes V^*\right) \tag{75}$$
$$= \sum_{i,j}\sqrt{\lambda_i}\sqrt{\lambda_j}\,|g_i\rangle\langle g_j|\otimes\mathrm{tr}_{\mathcal{E}}V\,|g_i\rangle\langle g_j|\,V^* \tag{76}$$
$$= \sum_{i,j}|g_i\rangle\sqrt{\lambda_i}\sqrt{\lambda_j}\langle g_j|\otimes\pi^*\left(|g_i\rangle\langle g_j|\right) = \theta_\phi, \tag{77}$$

and hence it is clear $S(\rho_{\mathcal{R}\otimes\mathcal{H}}) = S\left(\theta_\phi\right)$ which implies

$$I\left(\rho,\ \pi^*\right) = S(\rho) + S\left(\pi^*(\rho)\right) - S\left(\rho,\ \pi^*\right)$$
$$= S(\rho) + S\left(\pi^*(\rho)\right) - S\left(\theta_\phi\right) = I_\phi(\mathcal{G}\otimes\mathcal{H}). \qquad \square$$

4 Conclusions

In classical information theory Shannon's mutual information has three types of interpretation at the same time: (I) the common information (or correlation), (II) the information gain by observation, (III) the transmitted information via channel. Such interpretations can be done on the base of Bayes formula of probabilities.

In quantum theory, however, due to the lack of general notion of joint probability or conditional probability, there are several types of compound state corresponding to several types of quantum analogue of conditional probability, hence we have several types of mutual information or mutual entropy, and it is not clear the relations between some of them.

In this paper we showed that the quantum state of composite system also have a similar structure as classical Bayes formula in the terminology of CP maps. Applying the quantum channel derived from the quantum conditionality via the quantum analogue of Bayes formula it was shown that the quantum mutual information defined in (2) recognized as a straightforward generalization of Shannon mutual information has not only the interpretation (I), but also the interpretation (III) in the sense of Theorem 3. In our previous paper [24] we also analyzed the relation between another types of mutual information on the base of channel representation of compound state in (53), which is given by the duality of quantum conditionality, in there the interpretation (II) of mutual information was discussed in the context of quantum discord.

Acknowledgements T.M. is grateful to the late Professor V. P. Belavkin for fruitful discussions on normalized contravariant CP maps. D.C. is supported by the National Science Center project 2018/30/A/ST2/00837. We also wish to thank the referee who kindly offered nice suggestion to improve the original manuscript.

References

1. Shannon, C.: Mathematical theory of communication. Bell Syst. Tech. J. **27**(379–423), 623–656 (1948)
2. Cover, T., Thomas, J.A.: Elements of Information Theory, 2nd edn. Wiley-Interscience, New York (2006)
3. Wilde, M.: Quantum Information Theory. CUP, Cambridge (2013)
4. Nielsen, M.A., Chuang, I.L.: Quantum Computation and Quantum Information. Cambridge University Press, Cambridge (2000)
5. Holevo, A.: Quantum Systems, Channels, Information: A Mathematical Introduction. Studies in Mathematical Physics, vol. 16. De Gruyter (2012)
6. Ohya, M., Volovich, I.: Mathematical Foundations of Quantum Information and Computation and Its Applications to Nano- and Bio-systems. Springer, Dordrecht (2011)
7. Werner, R.F.: Lett. Math. Phys. **17**, 359 (1989)
8. Werner, R.F.: Phys. Rev. A **58**, 1827 (1998)
9. Bennett, C.H., Brassard, G., Popescu, S., Schumacher, B., Smolin, J.A., Wootters, W.K.: Phys. Rev. Lett. **76**, 722 (1996)
10. Schumacher, B.: Phys. Rev. A **51**, 2614 (1993)

11. Schumacher, B.: Phys. Rev. A **51**, 2738 (1993)
12. Horodecki, M., Horodecki, P., Horodecki, R.: Phys. Lett, A **223**, 1 (1996)
13. Peres, A.: Phys. Rev. Lett. **77**, 1413 (1996)
14. Cerf, N.J., Adami, C.: Phys. Rev. Lett. **79**, 5194 (1997)
15. Asorey, M., Kossakowski, A., Marmo, G., Sudarshan, E.C.G.: Open. Syst. Inf. Dyn. **12**, 319 (2005)
16. Choi, M.D.: Linear Algebra Appl. **10**, 285 (1975)
17. Jamiołkowski, A.: Rep. Math. Phys. **3**, 275 (1972)
18. Leifer, M.S.: Phys. Rev. A **74**, 042310 (2006)
19. Chruściński, D.: Phys. Lett. A **377**, 606 (2013)
20. Belavkin, V.P., Ohya, M.: Infinite Dimen. Anal., Quantum Probab., Relat. Top. **4**, 137 (2001)
21. Belavkin, V.P., Ohya, M.: Proc. R. Soc. Lond. A. **458**, 209 (2002)
22. Belavkin, V.P.: In: Quantum Bio-Informatics IV, Accardi, L. et al., (eds.). World Scientific, Singapore (2011)
23. Majewski, W.A., Matsuoka, T., Ohya, M.: J. Math. Phys. **50**, 113509 (2009)
24. Chruściński, D., Matsuoka, T.: Rep. Math. Phys. **86**, 115 (2020)
25. Belavkin, V.P.: Rep. Math. Phys. **55**, 61 (2005)
26. Belavkin, V.P., Dai, X.: Int. J. Quantum Inf. **6**, 357 (2008)
27. Schumacher, B.: Phys. Rev. A **54**, 2614 (1996)
28. Paulsen, V.: Completely Bounded Maps and Operator Algebras. CUP, Cambridge (2003)
29. Størmer, E.: Positive Linear Maps of Operator Algebras. Springer, Berlin (2013)

Block Markov Chains on Trees

Abdessatar Souissi

Abstract We introduce a new class of tree-indexed Markov processes, so-called block Markov chains (BMCs). We clarify the structure of BMCs in connection with Markov chains (MCs) and Markov random fields (MRFs). Mainly, we show that the probability measures, which are BMCs for every root, are indeed Markov chains (MCs). Conversely, we characterize a class of MCs which are BMCs. Moreover, we establish that the class of BMCs coincides with MCs only in the one-dimensional lattice.

1 Introduction

Markov random fields (MRFs) on lattices [3, 5, 6] have become standard tools for several branches of science and technology such as statistical physics, computer science, machine learning, graphical models, image recognition and in an increasing number of biological and neurological models. Namely, MRFs are known to provide pertinent models for interacting particles systems in statistical mechanics. The reader is referred to [10, 11, 13] and the references cited therein for further applications.

Markov random fields consist a multi-dimensional extensions of Markov chains [12] but with a deeper Markovian structure. Indeed, even in the one dimensional case MRFs were shown to be distinct from MCs [7].

In the present paper, we introduce the notion of block Markov chains on trees, as a multi-dimensional extension of 1D Markov chains [12]. We stress that this class of Markov processes does not reduces to the notions studied in the literature [13, 14]. Indeed, the difference manifests from the fact that the transition matrices for Markov chains in the sense of [13] are usual square stochastic matrices, but

A. Souissi (✉)
Department of Accounting, College of Business Management, Qassim University, Ar Rass, Saudi Arabia
e-mail: a.souaissi@qu.edu.sa; abdessattar.souissi@ipest.rnu.tn

Preparatory Institute for Scientific and Technical Studies La Marsa, Carthage University, Carthage, Tunisia

L. Accardi et al. (eds.), *Infinite Dimensional Analysis, Quantum Probability and Applications*, Springer Proceedings in Mathematics & Statistics 390,
https://doi.org/10.1007/978-3-031-06170-7_8

the transition matrices for BMCs are rectangular stochastic matrices. Note that the hierarchical structure of rooted trees plays a crucial role in the mere definition of BMCs. Therefore, the results are proper to trees. The main purpose of the present paper is to clarify the structure of BMCs in connection with MCs and MRFs. Mainly, we show that every probability measure which is BMC for every root $o \in V$ is a MC in the sense of [6]. It turns out that a class of Markov chains on trees are block Markov chains. On the other hand, we show that the notions of MCs and BMCs coincide in the one-dimensional case. This coincidence Makes BMCs a strict sub-class of MRFs in the one dimensional case. However, we emphasize that a slight modification of the 1D lattice leads to a counter-examples, that confirms the the distinction between MCs and BMCs over multi-dimensional trees.

In [1, 2] we clarified the structure of quantum Markov states on a tree-indexed quasi-local algebra $\mathcal{A}_V$ in terms of classical Markovian measure and Gibbs measures on the spectrum of a maximal abelian subalgebra. We stress that this classical Markovian measure is indeed a BMC. This will be starting point of a potential new bridge between classical and quantum Markov fields.

Let us mention the outlines of the paper. Section 2 is devoted to some preliminaries on rooted trees. In Sect. 3, we reminder the notions of MCs and MRFs on graphs. Section 4 is devoted to BMCs as far as their correlation functions. Section 5 is dedicated to the main results of the paper. In Sect. 6, we study the 1D BMCs. In Sect. 7, we develop a counter-example for a BMC which is not a MC. The last section is devoted for some concluding remarks.

2 Rooted Trees

Let be given a locally finite, infinite tree $T = (V, E)$. We fix a root $o = x_0 \in V$. For $x, y \in V$, we write $x \sim y$ to say that $< x, y > \in E$. A list of the vertices $x \sim x_1 \sim \cdots \sim x_{d-1} \sim y$ is called *edge-path* from the vertex x to the site y. The distance $d(x, y)$ on the tree is the length of the shortest path from x to y. Due to the absence of cycles in the considered tree, there exists a unique edge-path from x to y whose length equals $d(x, y)$, this path will be denoted by xy.

The direct successors of a vertex x with respect to the root o is defined by

$$S_o(x) := \{y \in V \ : \ x \sim y \text{ and } d(y, o) > d(x, o)\} \tag{1}$$

and its kth successors w.r.t. the root o is defined by induction as follows

$$S^1{}_o(x) := S_o(x); \quad S_o^{k+1}(x) = \bigcup_{y \in S_o^k(x)} S_o(y), \ \forall k \geq 1.$$

The "future" w.r.t. the vertex x is defined by:

$$T_o(x) = \bigcup_{k \geq 1} S_o^k(x); \quad T_o^{'}(x) = T_o(x) \setminus \{x\}. \tag{2}$$

Consider the map r from V into itself characterized by

$$r(o) = o, \quad r(y) = x \quad \text{if} \quad y \in S_o(x)$$

Let $x \in V$. If $n = d(x, o)$ then

$$o = r^n(x) = x_0 \sim r^{n-1}(x) \sim \cdots \sim r(x) \sim r_0(x) = x \tag{3}$$

is the minimal edge-path joining the root o to the vertex x, where $r^k = \underbrace{r \circ \cdots \circ r}_{k\ times}$.

The set

$$R_o(x) := \{r(x), r^2(x), \ldots, r^n(x) = o\} \tag{4}$$

represents the "past" of the vertex x for the root o

The set of *nearest-neighbors vertices* of x is given as follows:

$$N_x = \{y \in V \ : \ x \sim y\}. \tag{5}$$

It is clear that $N_x = \{r(x)\} \cup S_o(x)$. In the sequel, the tree T is assumed to be locally finite, i.e. $|N_x| < \infty$ for each $x \in V$, in this case the integer $d_x := |N_x|$ is called *degree* of x.

The tree can be regarded as growing (upward) away from its fixed root o. Each vertex $x \in V$ then has branches leading to its "children", which are represented here by $S_o(x)$ and $T_o^{'}(x)$. With the possibility of leaves, that is, vertices x without children i.e. $S(x) = \emptyset$.

3 Some Reminders on Markov Fields

Let $\Xi = \{1, \ldots, q\}$. By a stochastic process we mean a family of random variables $(Z_u)_{u \in V}$ defined on a probability space $(\Omega, \mathcal{F}, \mathbb{P})$ and valued in a finite set $\Xi := \{1, 2, \ldots, q\}$. The process $(Z_u)_{u \in V}$ is defined through its joined probability measure μ on the Borel space $(\Xi^V, \mathcal{B}_V)$ where $\mathcal{B}_V$ is the cylindrical σ-algebra, which is generated by the cylinder sets of the following form

$$C(a_x, \ x \in \Lambda) = \left\{\xi \in \Xi^V \ : \ \xi(x) = a_x, \ \forall x \in \Lambda\right\}, \tag{6}$$

where $\Lambda \subset V$ finite and $(a_x)_{x \in \Lambda} \in \Xi^{|\Lambda|}$.

For $\Lambda \subset V$, we denote $\Omega_\Lambda = \Xi^\Lambda$. For the sake of shortness, we denote Ω instead of Ω_V and $\mathcal{B}$ instead of $\mathcal{B}_V$. Recall that

$$\mu\left[\xi(.) \text{ on } \Lambda\right] = \mathbb{P}\left[\bigcap_{u \in \Lambda} (Z_u = \xi(u))\right], \tag{7}$$

where $\xi \in \Omega_\Lambda$.

The conditional probability is defined as follows

$$\mu\left[\xi(.) \text{ on } \Lambda \mid \xi(.) \text{ on } \Lambda^{'}\right] = \frac{\mu\left[\xi(.) \text{ on } \Lambda \cup \Lambda^{'}\right]}{\mu\left[\xi(.) \text{ on } \Lambda^{'}\right]}, \tag{8}$$

where $\Lambda, \Lambda^{'} \subseteq V$ and $\xi \in \Xi^{V}$ such that

$$\mu\left[\xi(.) \text{ on } \Lambda^{'}\right] > 0.$$

Denoting

$$\mathcal{F}_u := \sigma(Z_u) \ ; \ \mathcal{F}_\Lambda = \sigma\left(Z_u \ ; \ u \in \Lambda\right) \tag{9}$$

the σ-algebra generated by Z_u and $(Z_v, \ v \in \Lambda)$, respectively.

Definition 3.1 ([3]) A probability measure μ on $(\Omega, \mathcal{B})$ is said to be Markov random field (MRF) if it takes strictly positive values on finite cylinder sets of the form (6) and such that for every $\xi \in \Omega$

$$\mu\left[\xi(u) \mid \xi(.) \text{ on } V \setminus \{u\}\right] = \mu\left[\xi(u) \mid \xi(.) \text{ on } N_u\right] \tag{10}$$

for any $u \in V$. The set of Markov random fields over T will be denoted by $\mathcal{MF}(T)$.

Definition 3.2 ([4]) A probability measure μ on $(\Omega, \mathcal{B})$ is said to be Markov chain (MC) over the tree $T = (V, E)$ if for each subtree $T' = (V', E')$ the restriction of μ on the measurable space $(\Omega_{V'}, \mathcal{B}_{V'})$ defines a Markov random field. i.e.

$$\mu[\xi(x) \mid \xi(.) \text{ on } V' \setminus \{x\}] = \mu[\xi(x) \mid \xi(.) \text{ on } N_x \cap V^{'}] \tag{11}$$

for all $x \in V'$ and all $\xi \in \Omega_{V'}$. The set of Markov chains over T will be denoted by $\mathcal{MC}(T)$.

Remark 3.3 The class $\mathcal{MC}(T)$ is clearly included in $\mathcal{MF}(T)$. Conversely, in [6] it was proven that if the tail σ-field is trivial then the considered Markov field is indeed a MC.

4 Structure of Block Markov Chains on Trees

In what follows, a root o for the tree $T = (V, E)$ is fixed. For each $n \in \mathbb{N}$, we denote $\Lambda_n := S_o^n(o)$ the set of vertices whose distance to the root o equals n. Let $\Lambda_{n]} = S_o^{[0,n]}(o) = \bigcup_{k=0}^{n} \Lambda_k$. For the sake of shortness, when confusion seems impossible, we will use the notations $S(x)$, $T(x)$, $T^{'}(x)$ and $r(x)$ instead of $S_o(x)$, $T_o(x)$, $T_o^{'}(x)$ and $r_o(x)$, respectively.

Let us set a random enumeration for elements of Λ_n as follows

$$\Lambda_n := \left(x_{\Lambda_n}^{(1)}, x_{\Lambda_n}^{(2)}, \ldots, x_{\Lambda_n}^{(|\Lambda_n|)}\right),$$

where $|\Lambda_n|$ denotes the cardinality of Λ_n.

Definition 4.1 A measure μ on $(\Omega, \mathcal{B})$ is called o-block Markov chain (o-BMC) if it satisfies

$$\mu\left[\xi(.) \text{ on } S(x) \;|\xi(.) \text{ on } V \setminus T'(x)\right] = \mu\left[\xi(.) \text{ on } S(x) \;|\xi(x)\right], \tag{12}$$

for all $x \in V$ and $\xi \in \Omega$. The Eq. (12) will be referred as *block Markov property*. The set of o-block Markov chains over the tree T will be denoted $o - \mathcal{BMC}(T)$.

In [4] a triplet of σ-algebras $(\mathcal{F}_1, \mathcal{F}_2, \mathcal{F}_3)$ such that

$$\mathbb{P}(A \mid \mathcal{F}_1 \vee \mathcal{F}_2) = \mathbb{P}(A \mid \mathcal{F}_2), \quad \forall A \in \mathcal{F}_3 \tag{13}$$

was referred as *Markov triple*. In these notations (12) means that $(\mathcal{F}_{V \setminus T(x)}, \mathcal{F}_x, \mathcal{F}_{S(x)})$ is a Markov triple.

Remark 4.2 The word "block" in Definition 4.1 comes from the conditioning w.r.t. the σ-algebra $\mathcal{F}_{V\setminus T(x)}$ rather then the σ-algebra $\mathcal{F}_{R(x)}$, while this latter represents the past of the vertex x w.r.t. the root o.

The following elementary formula for conditional probabilities will be used frequently in the sequel.

$$\mathbb{P}(A \cap B \mid C) = \mathbb{P}(A \mid B \cap C)\mathbb{P}(B \mid C). \tag{14}$$

Let μ is an o-BMC. According to (14), we have

$$\begin{aligned}\mu[\xi(.) \text{ on } \Lambda_{n]}] &= \mu[\xi(.) \text{ on } \Lambda_n \mid \xi(.) \text{ on } \Lambda_{n-1]}] \times \mu[\xi(.) \text{ on } \Lambda_{n-1]}] \\ &= \mu[\xi(.) \text{ on } \Lambda_0] \prod_{k=0}^{n-1} \mu[\xi(.) \text{ on } \Lambda_{k+1} \mid \xi(.) \text{ on } \Lambda_{k]}].\end{aligned}$$

For $k = 1, \ldots, n-1$, the same reason as above implies that

$$\mu[\xi(.) \text{ on } \Lambda_{k+1} \mid \xi(.) \text{ on } \Lambda_{k]}] = \prod_{j=1}^{|\Lambda_k|} \mu[\xi(.) \text{ on } S(x_{\Lambda_k}^{(j)}) \mid \xi(.) \text{ on } \Lambda_{n-1]} \cup \bigcup_{i=j+1}^{|\Lambda_k|} S(x_{\Lambda_k}^{(i)})].$$

Since $x_{\Lambda_k}^{(i)} \in \Lambda_{n-1]} \cup \bigcup_{i=j+1}^{|\Lambda_k|} S(x_{\Lambda_k}^{(i)}) \subset V \setminus T'(x_{\Lambda_k}^{(i)})$ then the block Markov property (12) leads to

$$\mu[\xi(.)\text{ on }S(x_{\Lambda_k}^{(j)})\mid\xi(.)\text{ on }\Lambda_{n-1}]\cup\bigcup_{i=j+1}^{|\Lambda_k|}S(x_{\Lambda_k}^{(i)})]=\mu[\xi(.)\text{ on }S(x_{\Lambda_k}^{(j)})\mid\xi(x_{\Lambda_k}^{(j)})].$$

Therefore,

$$\mu[\xi(.)\text{ on }\Lambda_{n]}]=\mu[\xi(o)]\prod_{k=0}^{n-1}\prod_{x\in\Lambda_k}\mu\big[\xi(.)\text{ on }S(x)\mid\xi(x)\big].\tag{15}$$

Remark 4.3 The BMC μ is characterized by the initial distribution μ_o on $\Omega_{\{o\}}$ together with the family of transition probabilities $\mu\big[\xi(.)\text{ on }S(x)\mid\xi(x)\big]$. The $d\times(d^{|S(x)|})$ "stochastic" matrices $\Pi_{x,S(x)}=\big(\mu[\xi'(.)\text{ on }S(x)\mid\xi(x)]\big)_{\xi'\in\Xi^{S(x)},\xi\in\Xi^{\{x\}}}$ are clearly in-homogeneous. This lets the measure μ a multi-dimensional markovian process which is inhomogeneous both in space and time contrarily to the notion of Markov chains studied in [13], where the transition matrices are usual square stochastic matrices.

The following Theorem extends the local Markov property (12) to a global one, which concerns the conditional independence of the σ-algebras $\mathcal{F}_{T(x)}$ and $\mathcal{F}_{V\setminus T'(x)}$ given $\mathcal{F}_x$.

Theorem 4.4 *Let μ be a block Markov chain on $(\Omega,\mathcal{B})$. Then*

$$\mu\left[\xi(.)\,on\ T'(x)\,|\,\xi(.)\,on\ V\setminus T'(x)\right]=\mu\left[\xi(.)\,on\ T'(x)\,|\,\xi(x)\right]\tag{16}$$

for all $\xi\in\Omega$ and all $x\in V$.

Proof If $T'(x)=\emptyset$ then (16) holds true.

We will proceed by induction on $S_{[1,n]}(x):=\bigcup_{k=1}^{n}S_k(x)$. One has

$$\mu\left[\xi(.)\text{ on }S_{[1,n+1]}(x)\mid\xi(.)\text{ on }V\setminus T'(x)\right]$$

$$=\mu\left[\xi(.)\text{ on }S_{n+1}(x)\mid\xi(.)\text{ on }S_{[1,n]}(x)\cup V\setminus T'(x)\right]$$

$$\times\mu\left[\xi(.)\text{ on }S_{[1,n]}(x)\mid\xi(.)\text{ on }V\setminus T'(x)\right].$$

Denoting $S_n(x)=\{x_1^{(n)},\dots,x_{|S_n(x)|}^{(n)}\}$, one has

$$\mu\left[\xi(.)\text{ on }S_{n+1}(x)\mid\xi(.)\text{ on }S_{[1,n]}(x)\cup V\setminus T'(x)\right]$$

$$= \prod_{i=1}^{|S_n(x)|} \mu \left[\xi(.) \text{ on } S(x_i^{(n)}) \mid \xi(.) \text{ on } (\bigcup_{k=i+1}^{n} S(x_k^{(n)})) \cup S_{[1,n]}(x) \cup V \setminus T^{'}(x) \right].$$

From (12), one gets

$$\mu \left[\xi(.) \text{ on } S(x_i^{(n)}) \mid \xi(.) \text{ on } (\bigcup_{k=i+1}^{n} S(x_k^{(n)})) \cup S_{[1,n]}(x) \cup V \setminus T^{'}(x) \right]$$

$$= \mu \left[\xi(.) \text{ on } S(x_i^{(n)}) \mid \xi(x_i^{(n)}) \right].$$

Thus

$$\mu \left[\xi(.) \text{ on } S_{n+1}(x) \mid \xi(.) \text{ on } S_{[1,n]}(x) \cup V \setminus T^{'}(x) \right]$$

$$= \prod_{i=1}^{|S^{(n)}(x)|} \mu \left[\xi(.) \text{ on } S(x_i^{(n)}) \mid \xi(x_i^{(n)}) \right].$$

Using the same argument as above, we obtain

$$\mu \left[\xi(.) \text{ on } S_{n+1}(x) \mid \xi(.) \text{ on } S_{[1,n]}(x) \right] \tag{17}$$

$$= \prod_{i=1}^{|S_n(x)|} \mu \left[\xi(.) \text{ on } S(x_i^{(n)}) \mid \xi(x_i^{(n)}) \right].$$

On the other hand, the induction's hypothesis leads to

$$\mu \left[\xi(.) \text{ on } S_{[1,n]}(x) \mid \xi(.) \text{ on } V \setminus T^{'}(x) \right] = \mu \left[\xi(.) \text{ on } S_{[1,n]}(x) \mid \xi(x) \right].$$

Therefore

$$\mu \left[\xi(.) \text{ on } S_{[1,n+1]}(x) \mid \xi(.) \text{ on } V \setminus T^{'}(x) \right]$$

$$= \mu \left[\xi(.) \text{ on } S_{n+1}(x) \mid \xi(.) \text{ on } S_{[1,n]}(x) \right] \times \mu \left[\xi(.) \text{ on } S_{[1,n]}(x) \mid \xi(x) \right]$$

$$= \mu \left[\xi(.) \text{ on } S_{[1,n+1]}(x) \mid \xi(x) \right].$$

Finally, one finds

$$
\begin{aligned}
&\mu\left[\xi(.) \text{ on } T'(x) \mid \xi(.) \text{ on } V \setminus T'(x)\right] \\
&= \lim_{n\to[1,\infty)} \mu\left[\xi(.) \text{ on } S_{[1,n+1]}(x) \mid \xi(.) \text{ on } V \setminus T'(x)\right] \\
&= \lim_{n\to[1,\infty)} \mu\left[\xi(.) \text{ on } S_{[1,n+1]}(x) \mid \xi(x)\right] \\
&= \mu\left[\xi(.) \text{ on } T'(x) \mid \xi(x)\right].
\end{aligned}
$$

Corollary 4.5 *In the notations of Theorem 4.4, if* $\Lambda \subseteq T'(x)$ *then*

$$
\mu\left[\xi(.)\, on\ \Lambda \mid \xi(.)\, on\ V \setminus T'(x)\right] = \mu\left[\xi(.)\, on\ \Lambda \mid \xi(x)\right] \tag{18}
$$

for all $\xi \in \Omega$.

Proof From Theorem 4.4, for each $\xi' \in \Omega_{T'(x)\setminus\Lambda}$

$$
\mu\left[\xi(.) \text{ on } \Lambda,\ \xi'(.) \text{ on } T'(x) \setminus \Lambda \mid \xi(.) \text{ on } V \setminus T'(x)\right]
$$

$$
= \mu\left[\xi(.) \text{ on } \Lambda,\ \xi'(.) \text{ on } T'(x) \setminus \Lambda \mid \xi(x)\right].
$$

Summing up on $\xi' \in \Omega_{T'(x)\setminus\Lambda}$, one finds (18). This finishes the proof.

The following result proposes a multi-dimensional analogue of the Chapmann-Kolmogorov equation.

Theorem 4.6 *Let* μ *be a BMC on* $(\Omega, \mathcal{B})$. *Then for* $x \in V$ *and* $m, n \in \mathbb{N}$ *one has*

$$
\mu\left[\xi(.)\ on\ S_{n+m}(x) \mid \xi(x)\right] \tag{19}
$$

$$
= \sum_{\xi' \in \Xi^{S_n(x)}} \mu\left[\xi(.)\ on\ S_{n+m}(x)\ |\xi'(.)\ on\ S_n(x)\right] \times \mu\left[\xi'(.)\ on\ S_n(x)\ |\xi(x)\right]
$$

for all $\xi \in \Omega$.

Proof For each $\xi' \in \Omega_{S_n(x)}$, using the same reason as in (17), we get

$$
\mu\left[\xi(.) \text{ on } S_{n+m}(x)\ |\xi'(.) \text{ on } S_n(x)\right]
$$

$$
= \mu\left[\xi(.) \text{ on } S_{n+m}(x)\ |\xi'(.) \text{ on } S_n(x),\ \xi(x)\right].
$$

Then

$$
\mu\left[\xi(.) \text{ on } S_{n+m}(x)\ |\xi'(.) \text{ on } S_n(x)\right] \times \mu\left[\xi'(.) \text{ on } S_n(x) \mid \xi(x)\right]
$$

$$= \left[\xi(.) \text{ on } S_{n+m}(x) \,,\, \xi^{'}(.) \text{ on } S_n(x) \mid \xi(x) \right].$$

Summing up, one gets (19). □

5 Connection with MCs and MRFs

Lemma 5.1 *Let $x \in V$. If Λ is a subset of $S(x)$ then the subgraph of the tree $T = (V, E)$, whose set of vertices is $\Lambda \cup (V \setminus T^{'}(x))$ is itself a tree.*

Proof First, we see that if $y \in T^{'}(x)$ then $T^{'}(y) \subseteq T^{'}(x)$. This implies that for each $y \in V \setminus T^{'}(x)$ is connected, the set of its roots $\{r^k(y),\ k = 0, \ldots\}$ (defined in (3)) is disjoint of the set $T^{'}(x)$. Then the path $y \sim r(y) \sim \cdots \sim o$ is in $V \setminus T^{'}(x)$. Therefore, the subgraph whose vertex set $V \setminus T^{'}(x)$ is connected. Since every element of $S(x)$ is joined to x, we conclude that the subgraph $(\Lambda \cup (V \setminus T^{'}(x)), \sim)$ is connected. Taking into account that the fact that every connected subgraph of a tree is a subtree, the proof is complete. □

Theorem 5.2 *Let μ be a Markov chain on Ω. Then for each $x \in V$ the following property holds true.*

$$\mu \left[\xi(.) \text{ on } S(x) \mid \xi(.) \text{ on } V \setminus T^{'}(x) \right] = \prod_{y \in S(x)} \mu \left[\xi(y) \mid \xi(x) \right]. \qquad (20)$$

If in addition, the σ-algebras $(\mathcal{F}_y)_{y \in S(x)}$ are conditionally independent given $\mathcal{F}_x$ then μ is an o-BMC.

Proof First let us write $S(x) := \{y_1, y_2, \ldots, y_{|S(x)|}\}$. According to (14), we have

$$\mu \left[\xi(.) \text{ on } S(x) \mid \xi(.) \text{ on } V \setminus T^{'}(x) \right]$$

$$= \prod_{k=1}^{|S(x)|} \mu \left[\xi(y_k) \mid \xi(.) \text{ on } (V \setminus T^{'}(x)) \cup \{y_{k+1}, \ldots, y_n\} \right].$$

By Lemma 5.1 the subgraph of T whose set of vertices is $V^{'} := (V \setminus T^{'}(x)) \cup \{y_{k+1}, \ldots, y_n\}$ is a tree. Then

$$\prod_{k=1}^{|S(x)|} \mu\left[\xi(y_k) \mid \xi(.) \text{ on } (V \setminus T'(x)) \cup \{y_{k+1}, \ldots, y_n\}\right]$$
$$= \prod_{k=1}^{|S(x)|} \mu\left[\xi(y_k) \mid \xi(.) \text{ on } (V \setminus \{y_k\}) \cap V'\right]$$
$$= \prod_{k=1}^{|S(x)|} \mu\left[\xi(y_k) \mid \xi(.) \text{ on } N_{y_k} \cap V'\right].$$

where the last equality derives from the fact that μ is a Markov chain in the sense of Definition 3.2. Since $N_{y_k} \cap V' = \{x\}$, we get (20). For the second part of the proof, the conditional independence of $\mathcal{F}_y := \sigma(Z_y)$, $y \in S(x)$ leads to

$$\prod_{k=1}^{|S(x)|} \mu\left[\xi(y) \mid \xi(x)\right] = \mu\left[\xi(.) \text{ on } S(x) \mid \xi(x)\right].$$

Hence, (20) leads to (16). Therefore μ is a o-block Markov chain, for any root $o \in V$. This achieves the proof. □

Lemma 5.3 *If μ is an o-BMC on $(\Omega, \mathcal{B})$ and $x \in V$ then*

$$\mu[\xi(x) \mid \xi \text{ on } \Lambda] = \mu[\xi(x) \mid \xi(.) \text{ on } \{r(x)\} \cup (T'(x) \cap \Lambda)] \tag{21}$$

for all $\Lambda \subseteq V \setminus \{x\}$ containing $r(x)$.

Proof Since $x \notin \Lambda$, then according to (16) one gets

$$\mu\left[\xi(x) \mid \xi(.) \text{ on } \Lambda\right] = \frac{\mu\left[\xi(x);\ \xi(.) \text{ on } \Lambda \cap T'(x);\ \xi(.) \text{ on } \Lambda \setminus T'(x)\right]}{\mu\left[\xi(.) \text{ on } \Lambda \cap T'(x);\ \xi(.) \text{ on } \Lambda \setminus T'(x)\right]}$$
$$= \frac{\mu\left[\xi(.) \text{ on } \Lambda \cap T'(x) \mid \xi(x)\right] \mu\left[\xi(x) \mid \xi(r(x))\right]}{\mu\left[\xi(.) \text{ on } \Lambda \cap T'(x) \mid \xi(r(x))\right]}.$$

Again from (16), we have

$$\mu\left[\xi(.) \text{ on } \Lambda \cap T'(x) \mid \xi(x)\right] = \mu\left[\xi(.) \text{ on } \Lambda \cap T'(x) \mid \xi(x)\ ;\ \xi(r(x))\right].$$

This leads to

$$\mu\left[\xi(x) \mid \xi(.) \text{ on } \Lambda\right] = \mu\left[\xi(x) \mid \xi(.) \text{ on } \{r(x)\} \cup \Lambda \cap T'(x)\right]. \square$$

□

Remark 5.4 Notice that Definition 3.2 extends the notion of Markov chain introduced in [4, 5] into inhomogeneous trees and for inhomogeneous transition probabilities. It was shown [5] that the class of homogenous Markov chain is strictly included in the class of Markov random fields. In the inhomogeneous we have the following.

Theorem 5.5 *Let μ be a probability measure on $(\Omega, \mathcal{B})$. If μ is an o-BMC for each $o \in V$ then it is a Markov chain.*

Proof Consider a subtree $T^{'} = (V^{'}, E^{'})$ of T. Let $x \in V^{'}$. If $V^{'} \cap N_x = \emptyset$ then $V^{'} = \{x\}$ and (11) is trivial. Otherwise, let us denote $N_x \cap V' = \{y_1, y_2, \ldots, y_d\}$ with $d = |N_x \cap V^{'}|$. Remark that if $o = y \in N_x \cap V^{'}$ then $r_o(x) = y$ and $N_x \cap V^{'} \setminus \{y\} \subseteq S(x) \cap V^{'} \subseteq T_o^{'}(x) \cap V^{'}$. As μ is an y_1-BMC, by Lemma 5.3 we have

$$\mu[\xi(x) \mid \xi(.) \text{ on } V^{'} \setminus \{x\}] = \mu[\xi(x) \mid \xi(.) \text{ on } \{y_1\} \cup (T_{y_1}^{'}(x) \cap V^{'})].$$

Since μ is an y_2-BMC by Lemma 5.3 , we have

$$\begin{aligned} &\mu[\xi(x) \mid \xi(.) \text{ on } \{y_1\} \cup (T_{y_1}^{'}(x) \cap V^{'})] \\ &= \mu[\xi(x) \mid \xi(.) \text{ on } \{y_2\} \cup \big(\{y_1\} \cup T_{x_1}^{'}(x) \cap T_{y_2}^{'}(x) \cap V^{'}\big)] \\ &= \mu[\xi(x) \mid \xi(.) \text{ on } \{y_1, y_2\} \cup \big(T_{y_1}^{'}(x) \cap T_{y_2}^{'}(x) \cap V^{'}\big)]. \end{aligned}$$

Iterating this procedure, we get

$$\begin{aligned} \mu[\xi(x) \mid \xi(.) \text{ on } V^{'} \setminus \{x\}] &= \mu[\xi(x) \mid \xi(.) \text{ on } \{y_1, y_2, \ldots, y_d\} \cup \big(\bigcap_{i=1}^{d} T_{y_i}^{'}(x) \cap V^{'}\big)] \\ &= \mu[\xi(x) \mid \xi(.) \text{ on } N_x \cap V^{'}] \end{aligned}$$

because

$$\bigcap_{i=1}^{d} T_{y_i}^{'}(x) \cap V^{'} = \bigcap_{y \in N_x} T_y^{'}(x) \cap V^{'} = \emptyset. \tag{22}$$

Therefore, the measure μ satisfies (11). This finishes the proof, the verification of (22) being left to the reader. □

Corollary 5.6

$$\bigcap_{o \in V} o - \mathcal{BMC}(T) \subseteq \mathcal{MC}(T) \subseteq \mathcal{MF}(T).$$

6 One-Dimensional BMC

In this section we consider the one-dimensional lattice $V = \mathbb{Z}$ occupied with its natural structure of tree, where the edge set is $E = \{< k, k+1 >, \quad k \in \mathbb{Z}\}$. Here $\Omega = \Xi^{\mathbb{Z}}$.

Proposition 6.1 *Let μ be a probability measure on $(\Omega, \mathcal{B})$. The following assertions are equivalent:*

(i) *μ is a o-BMC for each $o \in V$;*
(ii) *μ is an o'-BMC, for some root $o' \in \mathbb{Z}$;*
(iii) *μ is a MC.*

In particular, a probability measure on $(\Omega, \mathcal{B})$ is markovian for the backward direction if and only if it is for the forward direction.

Proof $(i) \Rightarrow (ii)$ straightforward.
$(ii) \Rightarrow (i)$ Let $o' \in \mathbb{Z}$, without loss of generality we can assume that $o < o'$. Observe that if $x \geq max(o, o')$ or $x \leq min(o, o')$ then $T_0^{'}(x) = T_{o'}^{'}(x)$. Then (12) is also true if we replace o by o'.

Let us now examine the case $o < x < o'$ then $S^{o'}(x) = \{x-1\}$ and $T_{o'}^{'}(x) = (-\infty, x-1]$. Let $m \in \mathbb{N}$ and $\xi \in \Omega$.

Applying (12) to $y \geq x$, we get

$$\mu[\xi(y) \mid \xi(y-1), \ldots, \xi(x)] = \mu[\xi(y) \mid \xi(y-1)]$$

because $\{x, \ldots, y-1\} \subseteq (-\infty, y-1] = \mathbb{Z} \setminus T_o^{'}(y)$.

According to (14), it follows that

$$\begin{aligned}
\mu\,[\xi(x-1) \mid \xi(.) \text{ on } [x, x+m]] &= \frac{\mu[\xi(.) \text{ on } [x-1, x+m]]}{\mu[\xi(.) \text{ on } [x, x+m]]} \\
&= \frac{\mu[\xi(x-1)] \prod_{k=x}^{x+m} \mu[\xi(k) \mid \xi(k-1)]}{\mu[\xi(x)] \prod_{k=x+1}^{x+m} \mu[\xi(k) \mid \xi(k-1)]} \\
&= \frac{\mu[\xi(x) \mid \xi(x-1)] \times \mu[\xi(x-1)]}{\mu[\xi(x)]} \\
&= \mu[\xi(x-1) \mid \xi(x)].
\end{aligned}$$

Thus

$$\mu[\xi(x) \mid \xi(.) \text{ on } \mathbb{Z} \setminus T_{o'}^{'}(x)] = \mu[\xi(x-1) \mid \xi(x)]$$

for all $x \in \mathbb{Z}$. Hence μ is a $o^{'} - BMC$. $(ii) \Rightarrow (iii)$ Let $x \in \mathbb{Z}$ and $m \in \mathbb{N}$. Since μ is BMC then it is a $o - BMC$ for $o = x - 1$ and $T^{'}(x) = [x+1, \infty)$. By (12), it follows that

$$\mu[\,\xi(x) \mid \xi(x-1), \ldots, \xi(x-m)] = \mu[\xi(x) \mid \xi(x-1)\,].$$

Therefore, μ is a Markov chain.

$(iii) \Rightarrow (i)$ If μ is a Markov chain then

$$\mu[\xi(x) \mid \xi(.) \text{ on } (-\infty, x-1)].$$

Fig. 1 T-graph

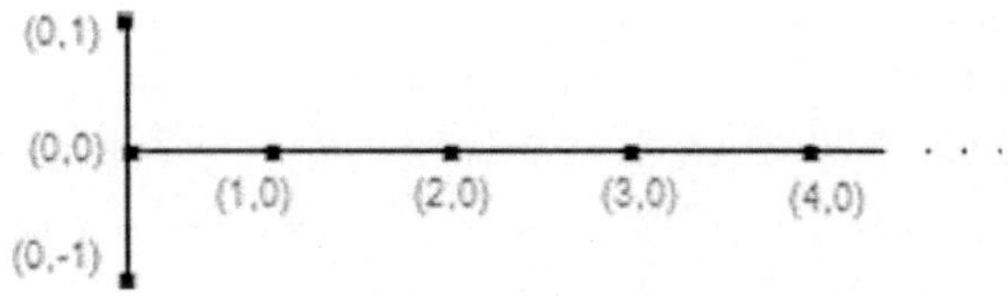

By taking $x > 0$, this implies that μ is 0-block Markov chain, which completes the proof.

Remark 6.2 Proposition 6.1 may be summarized by saying that for each $x \in \mathbb{Z}$ the triple $(\mathcal{F}_{[x+1,\infty)}, \mathcal{F}_x, \mathcal{F}_{(-\infty,x-1]})$ is a Markov triple in the sense of (13). Namely, this result is still true by taking $\mathbb{N}$ instead of $\mathbb{Z}$. However, a slight modification on the one dimensional lattice can provide a counter-example in the multi-dimensional case, in fact we have the following section.

7 Counter-Example

Consider the sets $V = \mathbb{N} \times \{0\} \cup \{(0, 1), (0, -1)\} \subset \mathbb{Z}^2$ and $E = \{\{x, y\} \in V\,;\, |x - y| = 1\}$ where $|(a, b)| = |a| + |b|$. We get then The tree $T = (V, E)$ (see Fig.1).

Consider a $\{0, 1\}$-valued Markov chain $(X_n)_{n\geq 0}$ with initial measure $\mu_0 = \frac{1}{2}(\delta_0 + \delta_1)$ and transition matrix $P = \begin{bmatrix} 1/2 & 1/2 \\ 1 & 0 \end{bmatrix}$. Define the $\{0, 1\}$-valued stochastic process $(Z_u)_{u\in V}$ by

$$Z_u = \begin{cases} X_0, & \text{if } u = (0, -1); \\ X_1, & \text{if } u = (0, 0); \\ X_2, & \text{if } u = (0, 1); \\ X_{n+2}, & \text{if } u = (n, 0),\ n \geq 1. \end{cases}$$

Put $\Xi = \{0, 1\}$ and μ be the probability measure on Ξ^V associated with $(Z_u)_{u\in V}$. Let $o = (0, -1)$ and $o' = (0, 1)$, it easy to check that μ is an o-BMC. However, μ is not a o'-BMC. In fact, if $x = (0, 0)$ we have $S_{o'}(w) = \{(0, -1), (1, 0)\}$, $r(x) = (0, 1)$ and $T'_{o'}(x) = V \setminus \{x, r(x)\}$. Let $\xi \equiv 0 \in \Omega$

$$\mu[\xi(.) \text{ on } S_{o'}(x) \mid \xi(.) \text{ on } V \setminus T'_{o'}(x)] = \frac{1}{6}$$

On the other hand

$$\mu[\xi(.) \text{ on } S_{o'}(x) \mid \xi(x)] = \frac{1}{4}.$$

Therefore

$$\mu[\xi(.) \text{ on } S_{o'}(x) \mid \xi(.) \text{ on } V - T'_{o'}(x)] \neq \mu[\xi(.) \text{ on } S_{o'}(x) \mid \xi(x)].$$

Hence μ is not an o'-BMC.

Furthermore, the probability measure μ is not a MC. In fact, by considering the subtree with vertex set $V_0 = \{(0, 1), (0, 0), (1, 0)\}$. We get

$$\mu[\xi((1,0)) \mid \xi((0,0)), \xi((0,1))] = \frac{1}{2} \neq \frac{3}{4} = \mu[\xi((1,o)) \mid \xi((0,0))].$$

8 Conclusion

Block Markov chains on trees are natural multidimensional extensions of the one-dimensional Markov chains. We stress that, this notion will open new horizons to the study of quantum Markov chains on trees both for the classical and the quantum cases. Namely, in [1, 2] classical probability measures on the spectrum of a diagonal subalgebra play a crucial role. We forecast that theses measure are BMCs. An other possible application of BMCs concerns the Markovian structure of a class of Gibbs measures on trees for an important class of potentials [8, 9]. The study of phenomena of Phase transitions associated with BMCs on trees is another tractable problem in analogy with [5].

References

1. Mukhamedov, F., Souissi, A.: Quantum Markov States on Cayley trees. J. Math. Anal. Appl. **473**, 313–333 (2019). https://doi.org/10.1016/j.jmaa.2018.12.050
2. Mukhamedov F., Souissi A., Diagonalizability of quantum Markov States on trees. J. Stat. Phys. **182**, 9 (2020). https://doi.org/10.1007/s10955-020-02674-1
3. Dobrushin, R.L.: The description of a random field by means of conditional probabilities and conditions of its regularity. Probab. Theory Appl. **13**, 201–22 (1968). https://doi.org/10.1137/1113026
4. Spataru, A.: Construction of a Markov field on an infinite tree. Adv. Math. **81**, 105–116 (1990). https://doi.org/10.1016/0001-8708(90)90007-A
5. Spitzer, F.: Markov random fields on an infinite tree. Ann. Probab. **3**, 387–398 (1987). https://doi.org/10.1214/aop/1176996347
6. Zachary, S.: Countable state space Markov random fields and Markov chains on trees. Ann. Probab. **4**, 894–903 (1983). https://doi.org/10.1214/aop/1176993439
7. Chandgotia, N., Han, G., Marcus, B., Meyerovitch, T., Pavlov, R.: One-dimensional Markov random fields, Markov chains and topological Markov fields. Am. Math. S **142**(1), 227–242 (2014)
8. Rozikov, U.A.: Gibbs Mesures on Cayley trees. World Sci. (2013). https://doi.org/10.1142/8841
9. Mukhamedov, F., Akın, H.: Phase transitions for p-adic Potts model on the Cayley tree of order three. J. Stat. Mech: Theory Exp. **07**, 30 (2013). https://doi.org/10.1063/1.2193118
10. Cao, J., Worsley, K.J.: Applications of Random Fields in Human Brain Mapping, Spatial Statistics: Methodological Aspects and Applications. Lecture Notes in Statistics, vol. 159. Springer, New York (2001)
11. Glymour C.: The Mind's Arrows: Bayes Nets and Graphical Causal Models in Psychology. The MIT Press (2001)

12. Norris, J.R.: Markov Chains, Cambridge Series in Statistical and Probabilistic Mathematics, vol. 2, Cambridge University Press, Cambridge (1998). Reprint of 1997 original. MR1600720 (99c:60144). https://doi.org/10.1017/CBO9780511810633
13. Georgii, H.O.: Gibbs Measures and Phase Transitions. de Gruyter Studies in Mathematics, vol. 9, 2nd edn. Walter de Gruyter, Berlin (2011)
14. Preston, C.: Gibbs States on Countable Sets. Cambridge University Press, London (1974)
15. Ramanan, K., Sengupta, A., Ziedins, I., Mitra, P.: Markov random field models of multicasting in tree networks. Adv. Appl. Probab. **34**(1), 58–84 (2002). https://doi.org/10.1239/aap/1019160950

Part III
Quantum Dynamical Systems

Hilbert von Neumann Modules *Versus* Concrete von Neumann Modules

Michael Skeide

Abstract Apart from presenting some new insights and results, one of our main purposes is to put some records in the development of von Neumann modules straight. The von Neumann or W^*–objects among the Hilbert (C^*–)modules are around since the first papers by Paschke [12] and Rieffel [14, 15] that lift Kaplansky's setting [8] to modules over noncommutative C^*–algebras. While the formal definition of *W^*–modules* is due to Baillet, Denizeau, and Havet [2], the one of *von Neumann modules* as strongly closed operator spaces started with Skeide [19]. It has been paired with the definition of *concrete von Neumann modules* in Skeide [23]. It is well-known that (pre-)Hilbert modules can be viewed as *ternary rings of operators* and in their definition of *Hilbert-von Neumann modules*, Bikram, Mukherjee, Srinivasan, and Sunder [4] take that point of view.

We illustrate that a (suitably nondegenerate) *Hilbert-von Neumann module* [4] is the same thing as a *strongly full concrete von Neumann module* [23]. We also use this occasion to put some things in the papers [4, 19, 23] right. We show that the tensor product of (concrete, or not) von Neumann correspondences **is**, indeed, (a generalization of) the tensor product of Connes correspondences (claimed in Skeide [24]), viewed in a way quite different from [4]. We also present some new arguments that are useful even for (pre-)Hilbert modules.

Keywords Von Neumann modules

1 Von Neumann Modules: Comparison

Let us right start with comparing the notion of *Hilbert-von Neumann modules* in the sense of Bikram, Mukherjee, Srinivasan, and Sunder [4] with that of *concrete von Neumann modules* from Skeide [23]. On the relation with the, still earlier, notion

M. Skeide (✉)
Università degli Studi del Molise, Via de Sanctis, 86100 Campobasso, Italy
e-mail: skeide@unimol.it
URL: http://web.unimol.it/skeide

L. Accardi et al. (eds.), *Infinite Dimensional Analysis, Quantum Probability and Applications*, Springer Proceedings in Mathematics & Statistics 390,
https://doi.org/10.1007/978-3-031-06170-7_9

of *von Neumann modules* in Skeide [19], we comment in Remark 1.2, especially, in Footnotes 2 and 3.

Let G and H be Hilbert spaces. For $E \subset \mathcal{B}(G, H)$ denote by $[E] := \overline{\operatorname{span}}^s E$ the strongly closed subspace of $\mathcal{B}(G, H)$ generated by E.

Let $\mathcal{B}$ be a von Neumann algebra acting nondegenerately on the Hilbert space G. Skeide [23, Definition 2] says, E is a ***concrete von Neumann $\mathcal{B}$–module*** if E is strongly closed, if it is a (right) $\mathcal{B}$–submodule[1] of $\mathcal{B}(G, H)$ (that is, $E + E \subset E$ and $E\mathcal{B} \subset E$), if $E^*E \subset \mathcal{B}$, and if $\overline{\operatorname{span}}\, EG = H$. We say E is ***strongly full*** if $\overline{\operatorname{span}}^s E^*E = \mathcal{B}$.

Bikram, Mukherjee, Srinivasan, and Sunder [4] say on Page 50, E is a ***von Neumann corner*** if $E = [E] \supset EE^*E$. It is ***nondegenerate*** if $\overline{\operatorname{span}}\, EG = H$, $\overline{\operatorname{span}}\, E^*H = G$. They say in [4, Definition 1.2(1)], a ***Hilbert-von Neumann module*** over a von Neumann algebra $\mathcal{A}$ is a von Neumann corner E with a normal isomorphism π from $\mathcal{A}$ onto $[E^*E]$.

Modulo nondegeneracy conditions, these definitions capture the same class of subspaces of $\mathcal{B}(G, H)$:

Theorem 1.1 *Let G and H be Hilbert spaces. For the subset E of $\mathcal{B}(G, H)$ denote by $\mathcal{B}$ the strong closure in $\mathcal{B}(G)$ of the algebra generated by E^*E. Then the following are equivalent:*

1. *E is a nondegenerate von Neumann corner (so that, in particular, $\mathcal{B} = [E^*E]$).*
2. *E is a Hilbert-von Neumann module over $\mathcal{B}$ for $\pi = \mathrm{id}_{\mathcal{B}}$ satisfying $\overline{\operatorname{span}}\, EG = H$ and $\mathrm{id}_G \in \mathcal{B}$.*
3. *E is a(n, automatically, strongly full) concrete von Neumann $\mathcal{B}$–module.*

Moreover, for an isomorphism π from the von Neumann algebra $\mathcal{A}$ onto the von Neumann algebra $\mathrm{id}_G \in \pi(\mathcal{A}) \subset \mathcal{B}(G)$, the subset $E \subset \mathcal{B}(G, H)$ is a Hilbert-von Neumann module over $\mathcal{A}$ satisfying $\overline{\operatorname{span}}\, EG = H$ if and only if E is a strongly full concrete von Neumann $\pi(\mathcal{A})$–module.

Proof It is enough to observe that $\operatorname{span} EE^*E = \operatorname{span}\Big(E \operatorname{span}(E^*E)\Big)$. □

Remark 1.2

1. The statement in the proof is essentially the necessary observation to recover the well-known fact that Hilbert modules and *ternary rings of operators* (*TRO*) are the same thing; see the book [6] by Blecher and Le Merdy (in particular, [6, Sect. 8.7]) and the references therein. (The references go back at least to the to the 1990s.) A ***TRO*** is a (usually, norm closed) linear subspace E of a C^*–algebra (for instance, of $\mathcal{B}(G, H) \subset \mathcal{B}(G \oplus H)$) satisfying $EE^*E \subset E$. Of course, nothing changes under strong closure.
2. Von Neumann modules have been defined in Skeide [19]. A pre-Hilbert module E over a von Neumann algebra $\mathcal{B}$ is a ***von Neumann module*** if its *extended linking algebra* $\begin{pmatrix} \mathcal{B} & E^* \\ E & \mathcal{B}^a(E) \end{pmatrix} = \mathcal{B}^a\begin{pmatrix} \mathcal{B} \\ E \end{pmatrix}$(denoting by $\mathcal{B}^a(E)$ the bounded adjointable

[1] In [23] we omitted to repeat that a module is closed under addition.

operators on E) is a von Neumann algebra.[2] Directly after the definition, in [19, Proposition 4.5] it is stated that this property is equivalent to E being strongly closed, that is, E is a concrete von Neumann module.

The definition of concrete von Neumann $\mathcal{B}$–module frees the definition of von Neumann $\mathcal{B}$–module from the obligation to first construct the (anyway, uniquely determined by $\mathcal{B} \subset \mathscr{B}(G)$ and E) embedding $E \subset \mathscr{B}(G, H)$; see Footnote 2. If E is given as $E \subset \mathscr{B}(G, H)$, then Theorem 1.1 tells that the definition in [4] does not present a mentionable difference to the definition in [23].[3]

2 Self-duality

Von Neumann or W^*–modules can be characterized as those objects in the category of (pre-)Hilbert modules over a von Neumann algebra that are *self-dual*; a (pre-)Hilbert $\mathcal{B}$–module E is ***self-dual*** if every bounded right linear map $\Phi : E \to \mathcal{B}$ has the form $y^* : x \mapsto \Phi(x) = \langle y, x \rangle$ for a suitable (obviously, unique) $y \in E$. Baillet, Denizeau, and Havet [2] take the property to be self-dual to define W^*–modules.[4] For other definitions, of course, self-duality has to be a theorem.[5] Actually, to show

[2] It is to be noted that a von Neumann algebra $\mathcal{B}$ is, as always(!), acting nondegenerately as $\mathcal{B} \subset \mathscr{B}(G)$ on a Hilbert space G. Having said this, we recall the well-known (at least, since Rieffel [15]) fact that every pre-Hilbert module over a concrete operator pre-C^*–algebra $\mathscr{B} \subset \mathscr{B}(G)$ embeds canonically (and uniquely, up to obvious unitary equivalence) into $\mathscr{B}(G, H)$, where $H := E \odot G$ is the Hausdorff completion of the algebraic tensor product $E \underline{\otimes} G$ with the semiinner product determined by $\langle x \otimes g, x' \otimes g' \rangle := \langle g, \langle x, x' \rangle g' \rangle$. Now $\mathscr{B}^a\binom{\mathcal{B}}{E}$ sits naturally in $\mathscr{B}\binom{G}{H}$, and it is clear what it means for $\mathscr{B}^a\binom{\mathcal{B}}{E}$ to be a von Neumann algebra in $\mathscr{B}\binom{G}{H}$.

[3] Calling this an "apparently much simpler manner" avoiding a "two stages completion" in the abstract of [4] seems out of place. The "two stages completion" they are referring to is that of norm closure (to get a C^*–module) and then strong closure. It is not necessary to go through C^*–modules in the construction, and while [19, Definition 4.4] is really (for convenience, in order not to frighten people who are afraid of pre-Hilbert modules) formulated only for Hilbert modules, [20, Definition 3.1.1] (which for some reason is the only work quoted in [4]) is formulated for pre-Hilbert modules. As for the (Hausdorff) completion to obtain H, in [4, Sects. 2+3] quite similar constructions are performed. It would appear strange to "allow" such constructions for Stinespring type theorems and the tensor product of correspondences, but not to obtain the tensor product $E \odot G$. (Anyway, it is well-known (at least, from or Skeide [19], but probably earlier; see also Murphy [11]) how to obtain the Stinespring construction of a CP-map into $\mathcal{B} \subset \mathscr{B}(G)$ from the tensor product of Pascke's GNS-correspondence [12] E and the representation space G of $\mathcal{B}$. In this context, while it is not clear to us why [4, Sect. 2] is restricted to *standard* Hilbert-von Neumann modules, a comparison with [19, Sect. 7] seems interesting.)

[4] Attention, a few authors by *W^*–module* mean just a Hilbert module over a W^*–algebra. But this seems not quite enough to merit to be considered an object in a W^*–category.

[5] One should note that the various definitions may be compared not only as to what extent they allow to check easily if a given pre-Hilbert module E over a von Neumann algebra is a von Neumann module or not; checking strong closure seems to have considerable advantages over checking self-

that a definition of von Neumann module is equivalent to another one, may be done by showing that it is equivalent to self-duality.[6]

Fortunately, we have:

Theorem 2.1 ([19, Theorem 4.16]) *Any von Neumann module is self-dual.*

We also have (more or less [19, Theorem 5.5]) the following (more or less obvious, when taking into account Footnote 2) consequence:

Corollary 2.2 *A (pre-)Hilbert module over a von Neumann algebra is a von Neumann module (or isomorphic to a concrete von Neumann module or isomorphic to a Hilbert-von Neumann module) if and only if it is self-dual.*

The proof of Theorem 2.1 presented in [19] relies on three basic pieces of knowledge:

- Polar decomposition ([19, Proposition 2.10][7]): Every x in a von Neumann $\mathcal{B}$–module E can be written (uniquely) as $v\,|x|$ where v is a partial isometry in E (satisfying $\ker\langle v, v\rangle = \ker\langle x, x\rangle \subset \mathcal{B} \subset \mathcal{B}(G)$).
- Existence of complete quasi-orthonormal systems ([19, Theorem 4.11][8]): In a von Neumann $\mathcal{B}$–module E there exists a family $\left(e_s\right)_{s\in S}$ of partial isometries such that $\sum_{s\in S} e_s e_s^* = \mathsf{id}_E$ (strong convergence in $\mathcal{B}^a(E) \subset \mathcal{B}(H)$).
- Boundedness of amplifications of bounded right linear maps between (pre-) Hilbert $\mathcal{B}$–modules under tensoring from the right (Rieffel [13, Theorem 1.4] or Skeide [22, Remark 3.3]).[9] In our concrete case: If a von Neumann module E

---

duality. But they may also be compared as to what extent it is easy to obtain a von Neumann module from E in case it is not yet. Paschke [12] has constructed a self-dual completion. As a space it is simply $\mathcal{B}^r(E, \mathcal{B})$, the space of all bounded right linear $\Phi\colon E \to \mathcal{B}$. But while it is actually not a priori easy to find that space, it is even much harder to compute the inner products of its elements. The definitions in [4, 19, 20, 23] suggest: Take the strong closure of E – and it is obvious what the inner product of elements x, y in this closure is, namely, $\langle x, y\rangle = x^* y$, and that it takes values in the von Neumann algebra $\mathcal{B}$.

6 In [4, Proposition 1.9] it is shown that Hilbert-von Neumann modules are self-dual, and it is claimed earlier that this shows that this is equivalent to being a von Neumann module in the sense of [19]. But, actually the backwards direction of this statement is missing; it would not be easy to obtain it without the procedure of embedding a (pre-)Hilbert module into $\mathcal{B}(G, H)$ (see Footnote 2), which the authors of [4] are so keen to avoid.

Anyway, it appears a remarkable idea to show equivalence of Hilbert-von Neumann modules and concrete von Neumann modules appealing to the (not so easy) self-duality, while we have the (fairly obvious) Theorem 1.1 (and the easy observation that the procedure explained in Footnote 2 transforms a von Neumann module into a concrete von Neumann module and the observation that a concrete von Neumann module is, obviously, a von Neumann module).

7 See [4, Lemma 1.6].

8 See also the proof of [4, Proposition 1.7].

9 The result [13, Theorem 1.4] is for tensor products of Banach modules over C^*–algebras. [19, Lemma 3.9] is for (pre-)Hilbert modules, but (as sketched in [22, Remark 3.3]; see also the proof of [4, Proposition 1.9]) the proof in [19] works correctly only for pre-Hilbert modules over a von Neumann algebra; in the case of C^*–algebras, there is a gap in the proof. See the appendix for a correct and new proof of this important result.

over $\mathcal{B} \subset \mathcal{B}(G)$ is represented as $E \subset \mathcal{B}(G, H)$, then a bounded right linear map $\Phi\colon E \to \mathcal{B}$ is represented as $\Phi \in \mathcal{B}(H, G)$ via $\Phi(x) = \Phi x$ (operator multiplication); see [22, Remark 3.3].[10] Namely: If $E \subset \mathcal{B}(G, H)$, then (see Footnote 2) $H = E \odot G$; and the general result asserts that the map Φ "=" $\Phi \odot \mathsf{id}_G\colon H = E \odot G \to \mathcal{B} \odot G = G$ is bounded.

With these ingredients, it is more or less plain to see that $y := \sum_{s \in S} e_s \Phi(e_s)^*$ (strong convergence!) is an element of E such that $\langle y, x \rangle = y^* x = \Phi(x)$.

Remark 2.3 As for these three ingredients and their interplay, polar decomposition follows (as explained in [19]) directly from polar decomposition in a von Neumann algebra; it is not necessary to redo its proof for Hilbert-von Neumann modules. It is applied to make sure that the elements e_s that occur in the quasi-orthonormal system (that is, $\langle e_{s'}, e_s \rangle = \langle e_s, e_s \rangle \delta_{s',s}$, but not necessarily complete) can be chosen to be partial isometries. Existence of a complete such system follows by a standard application of *Zorn's lemma* from the observation that $\sum_{s \in S} e_s e_s^*$ still converges strongly to a projection [19, Proposition 4.9][11]).

We have already explained before this remark how boundedness of $\Phi \odot \mathsf{id}_G$, together with the other two ingredients, allows to conclude the proof of self-duality. We wish, however, to emphasize that while the first two ingredients are specific to the proof presented in [19],[12] it seems that establishing the third, boundedness of $\Phi \odot \mathsf{id}_G$, plays a decisive role in **any** proof of self-duality. The proof of [15, Proposition 6.10] is based on identifying E as an intertwiner space, which is easily shown to be self-dual.[13] (We come back to this in Sect. 3.) Since the statement for (pre-)Hilbert modules in [19, Lemma 3.9] is as well-known as it is difficult to actually find a (simple and independent) proof of it,[14] we present a (new, we think) easy proof of this (general and interesting in its own) result in the appendix.

Observation 2.4 Here are some simple standard consequences – *standard* because they are (when choosing the right proof, of course) proved exactly as for Hilbert spaces. Here, $\mathcal{B}$ is a C^*–algebra.

---

[10] See also the proof of [4, Proposition 1.9].

[11] See also the proof of [4, Proposition 1.7].

[12] See also the proof of [4, Proposition 1.9].

[13] The proof of self-duality in [22] has, in fact, a huge overlap with that of [15, Proposition 6.10]. The discussion of intertwiner spaces adds only a little to what was known already to [15]. Only the proof of boundedness of $\Phi \odot \mathsf{id}_G$ (see also the proof of [4, Proposition 1.9]) seems to be really new—and quite a bit simpler than that of the statement about Banach modules in [13] alluded to in [15].

[14] It appears that many books or papers with introductions to the topic of Hilbert modules, either look only at amplifications of adjointable operators (where boundedness is an easy consequences of automatic contractivity of homomorphisms between C^*–algebras) or look at not necessarily adjointable operators but without considering their amplifications. It is likely that Rieffel's original result in [13] (going beyond our scope) has been reproved in the (more general!) context of operator modules over (not necessarily self-adjoint) operator algebras.

1. Of course, a self-dual pre-Hilbert $\mathcal{B}$–module E is necessarily Hilbert. (Corollary 2.2 tells that if $\mathcal{B} \subset \mathcal{B}(G)$ is a von Neumann algebra, then E is necessarily strongly closed in the associated $\mathcal{B}(G, H)$.)
2. If E is self-dual, then every bounded right linear map $E \to F$ has a (bounded right linear) adjoint.
3. Let F be a (pre-)Hilbert submodule of the pre-Hilbert $\mathcal{B}$–module E. If the submodule F is self-dual then there is projection $p \in \mathcal{B}^a(E)$ such that $pE = F$.[15]
4. And, less directly (rather a consequence of the preceding one and the fact that the strong closure of a pre-Hilbert submodule of a von Neumann module is a von Neumann module, hence, self-dual), if A is a subset of a von Neumann module E, then the double-complement $A^{\perp\perp}$ is the von Neumann submodule of E generated by A (that is, $\overline{\operatorname{span}}^s A\mathcal{B} = A^{\perp\perp}$). In particular, if F is a pre-Hilbert submodule, then $F^{\perp\perp} = \overline{F}^s \subset E \subset \mathcal{B}(G, H)$.

3 Von Neumann Correspondences, Connes Correspondences, and Their Tensor Products

We do not know when exactly a full definition of the internal tensor product of two (C^*–)correspondences[16] occurred for the first time; but, in special cases it is present already in Rieffel's paper [14]. The ***(internal) tensor product*** (or ***tensor product over*** $\mathcal{B}$) of an $\mathcal{A}$–$\mathcal{B}$–correspondence E and a $\mathcal{B}$–$\mathcal{C}$–correspondence F it the unique $\mathcal{A}$–$\mathcal{C}$–correspondence $E \odot F$ generated by elements $x \odot y$ ($x \in E$, $y \in F$) subject to the relations

$$\langle x \odot y, x' \odot y' \rangle = \langle y, \langle x, x' \rangle y' \rangle, \qquad a(x \odot y) = (ax) \odot y.$$

A special case is the tensor product $E \odot G$ of a Hilbert $\mathcal{B}$–module (that is, a $\mathbb{C}$–$\mathcal{B}$–correspondence) E and a representation space G of $\mathcal{B}$ (that is, a $\mathcal{B}$–$\mathbb{C}$–correspondence), we have already met (for faithful representations of $\mathcal{B}$ only, but faithfulness is, of course, not necessary).[17]

---

[15] See also [4, Proposition 1.7]. Of course, for strongly closed submodules of von Neumann modules, this can be proved by constructing the projection with the help of a quasi-orthonormal basis for the submodule (which is a von Neumann module in its own right). But for the simple and elegant (standard) proof from self-duality of the submodule, $\mathcal{B}$ may even be only a pre-C^*–algebra.

[16] Recall that a ***correspondence*** from $\mathcal{A}$ to $\mathcal{B}$ (or $\mathcal{A}$–$\mathcal{B}$–***correspondence***) is a Hilbert $\mathcal{B}$–module E with a nondegenerate(!) left action of $\mathcal{A}$ that induces a homomorphism $\mathcal{A} \to \mathcal{B}^a(E)$.

[17] By simple C^*–algebraic arguments it follows that the sesquilinear map on the algebraic tensor product $E \otimes F$ defined as in the first relation above, is positive, hence, defines a semiinner product on $E \otimes F$. (See the appendix.) Doing Hausdorff completion we obtain the Hilbert $\mathcal{C}$–module $E \odot F$, and since the left action of $\mathcal{A}$ induced by the second relation is by adjointable elements, it survives the quotient.

However, already in [20, Sect. 4.2] the following alternative has been observed: It is easy, using cyclic decomposition of the representation of $\mathcal{B}$ on G (see also the proofs [19, Lemma 3.9] and [4,

A ***(concrete) von Neumann correspondence*** ([23, Definition 4.2]) [20, Definition 3.3.1] is a correspondence between von Neumann algebras which is a (concrete) von Neumann module such that the left action is *normal*. (There are various equivalent ways, to say what it means that the left action is normal; see, for instance, [20, Lemma 3.3.2]. Just recall that a (necessarily positive!) map is ***normal*** if it is order-continuous and that, for positive maps, this is equivalent to σ–weak continuity.)

Remark 3.1 There is an—obvious—analogue of Theorem 1.1 for Hilbert-von Neumann bimodules [4] and concrete von Neumann correspondences [23], which we omit stating. And, obviously, every concrete von Neumann correspondence is a von Neumann correspondence, while every von Neumann correspondence can be turned into a concrete one be the procedure mentioned in Footnote 2.

Let $\mathcal{A} \subset \mathcal{B}(K), \mathcal{B} \subset \mathcal{B}(G), \mathcal{C} \subset \mathcal{B}(L)$ be von Neumann algebras and let E and F be a von Neumann $\mathcal{A}$–$\mathcal{B}$–correspondence and a von Neumann $\mathcal{B}$–$\mathcal{C}$–correspondence, respectively. Then:

- Either construct the internal tensor product $E \odot F$, embed it into $\mathcal{B}(L, (E \odot F) \odot L)$ (see Footnote 2) to obtain the strong closure $E \bar{\odot}^s F := \overline{\operatorname{span}}^s E \odot F \subset \mathcal{B}(L, E \odot F \odot L)$.
- Or, construct for $x \in E$ and $y \in F$ the operators $L_y = y \odot \mathsf{id}_L \in \mathcal{B}(L, F \odot L)$ and $L_x = x \odot \mathsf{id}_{F \odot L} \in \mathcal{B}(F \odot L, E \odot (F \odot L))$ (see Footnote 17) to obtain the strong closure $E \bar{\odot}^s F := \overline{\operatorname{span}}^s L_E L_F \subset \mathcal{B}(L, E \odot F \odot L)$.

Since in the tensor category of correspondences the spaces $(E \odot F) \odot L$ and $E \odot (F \odot L)$ are identified and denoted as $E \odot F \odot L$, the two possibilities coincide. So the two ways lead to the same von Neumann $\mathcal{C}$–module $E \bar{\odot}^s F$. It is a remarkable feature of von Neumann modules obtained from strong closures of (pre-)Hilbert modules over a von Neumann algebra, that the bounded right linear operators on the original (pre-)Hilbert module extend as (adjointable) operators on the strong closure. (Taking into account the statement proved in the appendix, the bounded right linear operators are represented as bounded operators between the associated Hilbert spaces, and the composition by operator multiplication extends, of course, to the strong closure.) It is routine to show that the left action of $\mathcal{A}$ on $E \odot F$ extended to $E \bar{\odot}^s F$ is normal. So, $E \bar{\odot}^s F$ is a von Neumann $\mathcal{A}$–$\mathcal{C}$–correspondence, the ***(internal) tensor product*** in the category of von Neumann correspondences.

Remark 3.2 In [4, Sect. 3], the authors call the (internal) tensor product of Hilbert-von Neumann bimodules (that is, of concrete von Neumann correspondences) the '*Connes fusion*'. Reasoning: There are certain bimodules over von Neumann algebras (also known under the name *Connes correspondences*) and a sort of tensor product

Proposition 1.9]), to see positivity of the inner product of $E \odot G$. Using this, (assuming $\mathcal{C} \subset \mathcal{B}(L)$) one gets that $x \odot y$ may be represented as operator $L_x L_y \in \mathcal{B}(L, E \odot (F \odot L))$ where $L_y = y \odot \mathsf{id}_L \in \mathcal{B}(L, F \odot L)$ and $L_x = x \odot \mathsf{id}_{F \odot L} \in \mathcal{B}(F \odot L, E \odot (F \odot L))$; see, in particular, [20, Eq. (4.2.2)]. (See also the discussion leading to [4, Proposition 3.4].)

for them (sometimes going under the name of '*Connes fusion*')[18], so it would be natural to use the same terminology also for other bimodules *of von Neumann type*.

- We find this a rather feeble motivation for changing the name of something, the tensor product over an algebra, that existed quite a bit before Connes correspondences.
- The motivation does not right hit the point. In fact, as mentioned in [23, Remark 4.3], Connes correspondences may be viewed as (a special case of) concrete von Neumann correspondences and, in this translation, the tensor product of the latter **is** Connes' product of the former. These facts are entirely missing in [4].[19]
  We shall explain and prove this in the sequel.

A concrete von Neumann correspondence $E \subset \mathcal{B}(G, H)$ from a von Neumann algebra $\mathcal{A} \subset \mathcal{B}(K)$ to a von Neumann algebra $\mathcal{B} \subset \mathcal{B}(G)$ comes along with a normal and unital representation

$$\rho: \mathcal{A} \longrightarrow \mathcal{B}^a(E) \subset \mathcal{B}(H)$$

(where, with $H = E \odot G$, we identify $\mathcal{B}^a(E) = \mathcal{B}^a(E) \odot \mathrm{id}_G \subset \mathcal{B}(H)$; see Footnote 2), which we call the ***Stinespring representation***[20] of $\mathcal{A}$ on H associated with E. The Stinespring representation comes from amplification of operators in $\mathcal{B}^a(E)$ acting on the left tensor factor E in the tensor product $H = E \odot G$. However, there is also the possibility to amplify operators acting on the right factor G. But, in order that an operator b' can by amplified to an operator $\mathrm{id}_E \odot b'$ on H, it has to be left $\mathcal{B}$–linear, that is exactly, $b' \in \mathcal{B}' \subset \mathcal{B}(G)$. We refer to this normal and unital representation $\rho': \mathcal{B}' \to \mathrm{id}_E \odot \mathcal{B}' \subset \mathcal{B}(H)$ as the ***commutant lifting***[21] of $\mathcal{B}'$ on H associated with E.

We collect some basic facts:

1. $E = C_{\mathcal{B}'}(\mathcal{B}(G, H)) := \{x \in \mathcal{B}(G, H) : \rho'(b')x = xb' \ (b' \in \mathcal{B}')\}$.
2. Conversely, if ρ' is any normal unital representation of $\mathcal{B}'$ on a Hilbert space H, then $E \subset \mathcal{B}(G, H)$ defined as $C_{\mathcal{B}'}(\mathcal{B}(G, H))$ is a concrete von Neumann

[18] We have heard rumors that the creator of that construction is not too fond of that terminology, so we shall not use it.

[19] We did not quite understand, if no relation between Hilbert-von Neumann bimodules and Connes correspondences is made in [4], or if the authors think the relation would be the occurrence of two representations in the definition of Hilbert-von Neumann bimodule. The latter would not be applicable, because for a Hilbert-von Neumann bimodule $E \subset \mathcal{B}(G, H)$ there is the representation π as for Hilbert-von Neumann modules (see the beginning of Sect. 1) acting on G plus an analogue representation acting on H, while a Connes correspondence (see the basic fact 5 below) is a single Hilbert space H with a representation and an anti-representation both acting on H.

[20] Already in [19, Sect. 7] (see also [3, Example 2.16 and Observation 2.17]) it has been explained that when E is the *GNS-correspondence* (Paschke [12]) of a CP-map into a concrete C^*–algebra $\mathcal{B} \subset \mathcal{B}(G)$, then ρ is, indeed, the Stinespring representation. See also [4, Sect. 2].

[21] Arveson's representation discussed in the section "*Lifting Commutants*" in [1], is a special case of our *commutant liftings*.

$\mathcal{B}$–module and its commutant lifting is the ρ' we started with [23, Theorem 3.10]: The category of concrete von Neumann $\mathcal{B}$–modules and the category of normal unital representations of $\mathcal{B}'$ are the same (not just equivalent, but really "the same", in the sense that objects and morphisms correspond one-to-one).

3. $\rho'(\mathcal{B}')' = \mathcal{B}^a(E)$. Since $\rho(\mathcal{A}) \subset \mathcal{B}^a(E)$, we see, in particular, $[\rho(\mathcal{A}), \rho(\mathcal{B}')] = \{0\}$.
4. Conversely, if ρ and ρ' are any commuting pair of normal unital representations of $\mathcal{A}$ and $\mathcal{B}'$, respectively, then $E := C_{\mathcal{B}'}(\mathcal{B}(G, H)) \subset \mathcal{B}(G, H)$ is a concrete von Neumann $\mathcal{B}$–module inheriting a left action of $\mathcal{A}$ via ρ. Also here, we have a true one-to-one correspondence between the categories of such commuting pairs and concrete von Neumann $\mathcal{A}$–$\mathcal{B}$–correspondences.
   But, we have even more. Exchanging the roles of ρ and ρ', we get a concrete von Neumann $\mathcal{B}'$–$\mathcal{A}'$–correspondence as $E' = C_{\mathcal{A}}(\mathcal{B}(K, H)) := \{x' \in \mathcal{B}(K, H) \colon \rho(a)x' = x'a \ (a \in \mathcal{A})\}$ with left action of $\mathcal{B}'$ via ρ'. Also here we have [23, Theorem 4.4]: This ***commutant*** is a bijective functor between the category of concrete von Neumann $\mathcal{A}$–$\mathcal{B}$–correspondences and the category of concrete von Neumann $\mathcal{B}'$–$\mathcal{A}'$–correspondences.
5. In particular, if $\mathcal{B}$ is in *standard representation*, then $\mathcal{B}' \cong \mathcal{B}^\circ$, the opposite von Neumann algebra of $\mathcal{B}$, via *Tomita conjugation*. That is, on H we have a commuting pair of representations ρ of $\mathcal{A}$ and ρ° of $\mathcal{B}^\circ$. In other words, H is an $\mathcal{A}$–$\mathcal{B}$–bimodule, and as such a ***Connes correspondence*** from $\mathcal{A}$ to $\mathcal{B}$; see Connes [7]. Conversely, interpreting the representation ρ° of $\mathcal{B}^\circ$ as representation ρ' of $\mathcal{B}'$, every Connes correspondence gives rise to a concrete von Neumann correspondence.
   But, in general, for us $\mathcal{B}$ need not be in standard representation: Every Connes correspondence is a concrete von Neumann correspondence, but not every concrete von Neumann correspondence is a Connes correspondence.

Remark 3.3 The proofs of (1) and (3) rely on the simple computation

$$\begin{pmatrix} \mathcal{B} & E^* \\ E & \mathcal{B}^a(E) \end{pmatrix}'' = \left\{ \begin{pmatrix} b' & \\ & \rho'(b') \end{pmatrix} : b' \in \mathcal{B}' \right\}' = \begin{pmatrix} \mathcal{B} & C_{\mathcal{B}'}(\mathcal{B}(E \odot G, G)) \\ C_{\mathcal{B}'}(\mathcal{B}(G, E \odot G)) & \rho'(\mathcal{B}')' \end{pmatrix}$$

(which holds for every pre-Hilbert module over $\mathcal{B} \subset \mathcal{B}(G)$, when embedded into $\mathcal{B}(G, H)$ with $H = E \odot G$); see [22, Sect. 2]. The proof of (2), hence, (4) (only nondegeneracy $\overline{\operatorname{span}}\, EG = H$ is an issue), relies, furthermore, on Muhly and Solel [10, Lemma 2.10]. Both have been first observed in [21, Sect. 2]. The statement in (5) is not more than a simple observation that has been made in [23, Remark 4.3].

Theorem 3.4 *The internal tensor product of two Connes correspondences (viewed as von Neumann correspondences) is (when viewed as Connes correspondence) Connes' product of the two Connes correspondences.*

This has been mentioned after Skeide [24, Remark 6.3] without indicating a proof. The argument given in Shalit and Skeide [18, Footnote y] is very sketchy. Therefore:

Proof Let $\mathcal{A}$, $\mathcal{B}$, $\mathcal{C}$ be W^*–algebras, let H be a Connes $\mathcal{A}$–$\mathcal{B}$–correspondence, and let K be a Connes $\mathcal{B}$–$\mathcal{C}$–correspondence. As far as their Connes product $H \otimes_\psi K$ (relative to a faithful normal semi-finite weight ψ on $\mathcal{B}$) is concerned, we use the presentation in Takesaki [27, Sect. IX.3].

For the time being, we do not fix a representation of $\mathcal{A}$ nor of $\mathcal{C}$ turning them into von Neumann algebras. (This is only necessary in the very end, when we wish to consider H and K as von Neumann correspondences which are defined only over von Neumann algebras.) But we assume that $\mathcal{B} \subset \mathcal{B}(G)$ where $G = L^2(\mathcal{B}, \psi)$, so that $\mathcal{B}$ is in standard representation and $\mathcal{B}' = \mathcal{B}^\circ$ via *Tomita conjugation*.

It follows that H carries a normal unital representation of $\mathcal{B}'$ (commuting with the action of $\mathcal{A}$) so that we may define the concrete von Neumann $\mathcal{B}$–module $E := C_{\mathcal{B}'}(\mathcal{B}(G, H))$ (a von Neumann $\mathcal{A}$–$\mathcal{B}$–correspondence as soon as we represent $\mathcal{A}$ as a von Neumann algebra). Note that this E is exactly what has been denoted as $\mathcal{L}(L^2(\mathcal{N})_\mathcal{N}, \mathfrak{H}_\mathcal{N})$ preceding [27, Lemma IX.3.3] for $\mathcal{N} = \mathcal{B}$, $L^2(\mathcal{N}) = G$, and $\mathfrak{H} = H$.

There is a dense ([27, Lemma IX.3.3(iii)]) subspace $\mathcal{D}$ (denoted $\mathfrak{D}(\mathfrak{H}, \psi)$ and defined in [27, Eq. IX.3.(10)]) and an injective ([27, Lemma IX.3.3(iv)]) map $\mathfrak{L}\colon \mathcal{D} \to E$ (denoted L_ψ and defined in [27, Eq. IX.3.(11)]). From the definition [27, Eq. IX.3.(11)], it follows that $\mathfrak{L}(H)G$ is dense in $\mathcal{D}$, hence, in H. Therefore, $\mathfrak{L}(\mathcal{D})$ is a strongly dense subset of E.

Putting in [27, Proposition IX.3.15] $\mathfrak{K} = K$, just looking at [27, Eq. IX.3.(23)], tells us that the (semi)inner product of elements $d_i \otimes k_i$ in $\mathcal{D} \underline{\otimes} K$ defined there is

$$B(d_1 \otimes k_1, d_2 \otimes k_2) \;=\; \langle \mathfrak{L}(d_1) \odot k_1, \mathfrak{L}(d_2) \odot k_2 \rangle,$$

the same as the inner product of the elements $\mathfrak{L}(d_i) \odot k_i$ in $E \odot K$. Now, if also $\mathcal{C} \subset \mathcal{B}(L)$ is in standard representation (so that $\mathcal{C}' = \mathcal{C}^\circ$), then $F := C_{\mathcal{C}'}(\mathcal{B}(L, K))$ is a von Neumann $\mathcal{B}$–$\mathcal{C}$–correspondence such that $K = F \odot L$, including the representations of $\mathcal{B}$ and and $\mathcal{C}'$. It is routine to show that $\mathfrak{H} \otimes_\psi \mathfrak{K}$ in [27, Definition IX.3.16] is $E \odot F \odot L$ including the representations of $\mathcal{A}$ and $\mathcal{C}^\circ = \mathcal{C}'$ that identify $E \bar{\odot}^s F$ as the von Neumann $\mathcal{A}$–$\mathcal{C}$–correspondence coming from the Connes $\mathcal{A}$–$\mathcal{C}$–correspondence $\mathfrak{H} \otimes_\psi \mathfrak{K}$. □

Remark 3.5 The commuting pair picture of von Neumann correspondences is, therefore, a generalization of Connes correspondences and the tensor product of the former includes the Connes product of the latter. It coincides if we only look at von Neumann algebras in standard representation. But, unlike the construction of Connes' product, the construction of the tensor product of von Neumann correspondences does not depend on more advanced tools like *Tomita-Takesaki theory* and *standard representation* and the formulae do not depend manifestly on the choice of a faithful semi-finite normal weight.

This has been mentioned after [24, Remark 6.3]; it has nothing to do with the term '*Connes fusion*' in [4]. See also Thom [28], where the discussion is, however, limited to *bi-finite* Connes correspondences (that is, the module dimensions of H as left and right module are finite). We do not have any limitation.

We also mention that there is a close connection with Sauvageot's *tensor product relative to an algebra* [16, 17].

Acknowledgements Discussions with several people helped a lot to improve this paper. In particular, I would like to thank Malte Gerhold and Orr Shalit.

Appendix

The scope of this appendix is to present a new and 'painless' proof of the following result.

Theorem A.1 *Let E be a (pre-)Hilbert module over a (pre-)C^*–algebra $\mathcal{B}$ and let F be a (pre-) C^*–correspondence from $\mathcal{B}$ to another (pre-)C^*–algebra $\mathcal{C}$. Then for every $a \in \mathcal{B}^r(E)$ we have*

$$\|a \odot \mathsf{id}_F\| \ \le \ \|a\|,$$

so $a \odot \mathsf{id}_F \in \mathcal{B}^r(E \odot F)$.

Observation A.2

- We have not yet specified what we mean by $E \odot F$ in case of 'pre-'. It is simply $\mathsf{span}\,\{x \odot y : x \in E, y \in F\} \subset \overline{E} \odot \overline{F}$.[22] So, since a is bounded we simply may complete the spaces $E, F, \mathcal{B}, \mathcal{C}$, proving the theorem only without 'pre-'.
- Note that we need not have equality. Effectively, $E \odot F$, hence $a \odot \mathsf{id}_F$, may very well be zero independently of $\|a\|$.
- Of course, the theorem remains true for $a \in \mathcal{B}^r(E_1, E_2)$; simply embed $\mathcal{B}^r(E_1, E_2)$ into $\mathcal{B}^r\left(\begin{smallmatrix}E_1\\E_2\end{smallmatrix}\right)$.

Remark A.3 For Banach modules over C^*–algebras, the theorem is [13, Theorem 1.4]. There should be versions for *operator modules* over (non-selfadjoint) *operator algebras* (including similar statements for CB-norms). The theorem is easy, if a is required adjointable. ($a \mapsto a \odot \mathsf{id}_F$ is a homomorphism between C^*–algebras and, therefore, contractive.)

The proof of the theorem is a typical example for the following strategy to prove statements about (internal) tensor products.

---

22 Note that the algebraic tensor product of a right $\mathcal{B}$–module E and a left $\mathcal{B}$–module F ***over*** $\mathcal{B}$ is $E \odot_{\mathcal{B}} F := (E \underline{\otimes} F)/\,\mathsf{span}\{xb \otimes y - x \otimes by\}$. In the setting of the theorem, the semiinner product defined on $E \underline{\otimes} F$ as usual always survives the quotient but still need not be inner on $E \odot_{\mathcal{B}} F$. (It is inner if $\mathcal{B}$ and E are complete; see [20, Proposition 4.2.22].) Therefore, it is pointless to divide first out the relations $xb \otimes y - x \otimes by$ and then a still possible kernel of the semiinner product on $E \odot_{\mathcal{B}} F$, but we divide out immediately the kernel of the semiinner product on $E \underline{\otimes} F$. We emphasize, however, that the relation $xb \odot y = x \odot by$ is satisfied at any stage.

Observation A.4

1. Frequently, a certain statement is easy (or easier) to prove on simple tensors $x \odot y$ instead of general elements $\sum_{i=1}^n x_i \odot y_i$ of the algebraic tensor product.
2. Use the equality of $E \odot F = E_n \odot F^n$ via $\sum_{i=1}^n x_i \odot y_i = X_n \odot Y^n$ (with $X_n = (x_1, \ldots, x_n) \in E_n$ and $Y^n = \begin{pmatrix} y_1 \\ \vdots \\ y_n \end{pmatrix} \in F^n$, where E_n is the row-module with inner product $\langle X_n, X'_n \rangle = \big(\langle x_i, x'_j \rangle\big) \in M_n(\mathcal{B})$ and the obvious action of $M_n(\mathcal{B})$ on F^n) to reduce the problem to simple tensors.

Among the various applications, there is that the inner product on $E \odot F$ is actually positive ($\Big\langle \sum_{i=1}^n x_i \odot y_i, \sum_{i=1}^n x_i \odot y_i \Big\rangle = \langle X_n \odot Y^n, X_n \odot Y^n \rangle = \langle Y^n, \langle X_n, X_n \rangle Y^n \rangle \geq 0$, having established before that the inner product of E_n is positive; see Skeide [25, Sect. 5]). Another application is the proof in Skeide and Sumesh [26, Corollary] of Blecher's result [5, Theorem 4.3] that $E \odot F$ is (completely isometrically) isomorphic to the Haagerup tensor product.

Proof of the theorem. Recall that we may restrict to the case without 'pre-'. The statement of the theorem actually means that we show the estimate for elements of the form $\sum_{i=1}^n x_i \odot y_i$ and, then, extend extend $a \odot \mathsf{id}_F$, well defined on that dense subspace, continuously to all of $E \odot F$. (Recall from Footnote 22 that this subspace is $E \underline{\odot}_{\mathcal{B}} F$.)

The proof decomposes into four parts.

(i) For all $a \in \mathcal{B}^r(E)$ we have $\langle ax, ax \rangle \leq \|a\|^2 \langle x, x \rangle$ in $\mathcal{B}$. (This is the easier direction of [12, Theorem 2.8], which asserts that the inequality actually characterizes among all linear maps on E those that are in $\mathcal{B}^r(E)$. See there for the quick proof of the part we need here; the other direction is much harder.)

(ii) For each $a \in \mathcal{B}^r(E)$ we have $\mathcal{B}^a(E) \ni a \geq 0$ if (and only if) $\langle x, ax \rangle \geq 0$ for all $x \in E$. (In this form, this is Lance [9, Lemma 4.1], while [12, Proposition 6.1] is the special case we would actually need in the next part. Paschke actually shows in his proof Lance' more general statement; but here we prefer Lance' proof.)

(iii) The inequality in Part (i) survives 'amplification' to E_n. More precisely, for each $X_n = (x_1, \ldots, x_n) \in E_n$, we have $\Big(\langle ax_i, ax_j \rangle\Big) \leq \|a\|^2 \Big(\langle x_i, x_j \rangle\Big)$ in $M_n(\mathcal{B})$ or, equivalently,

$$M_n(\mathcal{B}) \ni \Big(\|a\|^2 \langle x_i, x_j \rangle - \langle ax_i, ax_j \rangle \Big) \geq 0.$$

Since $M_n(\mathcal{B})$ acts in the obvious way on the Hilbert $\mathcal{B}$–module $\mathcal{B}^n$, we may apply Part (ii) to see that this element of $M_n(\mathcal{B})$ is positive. Indeed, for $B^n = \begin{pmatrix} b_1 \\ \vdots \\ b_n \end{pmatrix} \in \mathcal{B}^n$ we find

$$\langle B^n, \Big(\|a\|^2 \langle x_i, x_j\rangle - \langle ax_i, ax_j\rangle\Big) B^n\rangle = \sum_{i,j} b_i\big(\|a\|^2 \langle x_i, x_j\rangle - \langle ax_i, ax_j\rangle\big) b_j$$
$$= \|a\|^2 \Big\langle \sum_i x_i b_i, \sum_i x_i b_i \Big\rangle - \Big\langle a \sum_i x_i b_i, a \sum_i x_i b_i \Big\rangle.$$

By Part (i), this is positive.

(iv) By Part (iii) (applied in the step where the $\leq$ occurs), for $Y^n = \begin{pmatrix} y_1 \\ \vdots \\ y_n \end{pmatrix} \in F^n$ we find

$$\Big\| (a \odot \mathrm{id}_F)\Big(\sum_i x_i \odot y_i\Big)\Big\|^2 = \Big\|\Big(\sum_i ax_i \odot y_i\Big)\Big\|^2$$
$$= \|aX_n \odot Y_n\|^2 = \big\|\langle Y^n, \langle aX_n, aX_n\rangle Y^n\rangle\big\|$$
$$\leq \|a\|^2 \big\|\langle Y^n, \langle X_n, X_n\rangle Y^n\rangle\big\| = \|a\|^2 \Big\|\sum_i x_i \odot y_i\Big\|^2,$$

what was to be shown. □

References

1. Arveson, W.: Subalgebras of C^*-algebras. Acta Math. **123**, 141–224 (1969)
2. Baillet, M., Denizeau, Y., Havet, J.F.: Indice d'une esperance conditionnelle. Compositio Math. **66**, 199–236 (1988)
3. Bhat, B., Skeide, M.: Tensor product systems of Hilbert modules and dilations of completely positive semigroups. Infin. Dimens. Anal. Quantum Probab. Relat. Top. **3**, 519–575 (2000) (Rome, Volterra-Preprint 1999/0370)
4. Bikram, P., Mukherjee, K., Srinivasan, R., Sunder, V.: Hilbert von Neumann modules. Commun. Stoch. Anal. **6**, 49–64 (2012). (arXiv: 1102.4663)
5. Blecher, D.: A new approach to Hilbert C^*-modules. Math. Ann. **307**, 253–290 (1997)
6. Blecher, D., Le Merdy, C.: Operator Algebras and Their Modules—an Operator Space Approach. Number 30 in London Mathematical Society Monographs. Oxford University Press, Oxford (2004)
7. Connes, A.: Correspondences. His hand-written notes, unpublished (1980)
8. Kaplansky, I.: Modules over operator algebras. Am. J. Math. **75**, 839–853 (1953)
9. Lance, E.: Hilbert C^*–Modules. Cambridge University Press, Cambridge (1995)
10. Muhly, P., Solel, B.: Quantum Markov processes (correspondences and dilations). Int. J. Math. **51**, 863–906 (2002) (arXiv: math.OA/0203193)
11. Murphy, G.: Positive definite kernels and Hilbert C^*-modules. Proc. Edinburgh Math. Soc. **40**, 367–374 (1997)
12. Paschke, W.: Inner product modules over B^*-algebras. Trans. Am. Math. Soc. **182**, 443–468 (1973)
13. Rieffel, M.: Multipliers and tensor products of L^p-spaces of locally compact groups. Studia Math. **33**, 71–82 (1969)
14. Rieffel, M.: Induced representations of C^*-algebras. Adv. Math. **13**, 176–257 (1974)
15. Rieffel, M.: Morita equivalence for C^*-algebras and W^*-algebras. J. Pure Appl. Algebra **5**, 51–96 (1974)

16. Sauvageot, J.L.: Produites tensoriels de Z–Modules. Preprint 23, Publishing Pierre and Marie Curie University (1980)
17. Sauvageot, J.L.: Sur le produit tensoriel relatif d'espace de Hilbert. J. Oper. Theory **9**, 237–252 (1983)
18. Shalit, O.M., Skeide, M.: CP-Semigroups and dilations, subproduct systems and superproduct systems: The multi-parameter case and beyond. Preprint, arXiv: 2003.05166v3, to appear in Dissertationes Math. (2020)
19. Skeide, M.: Generalized Matrix C^*–Algebras and Representations of Hilbert Modules. In: Mathematical Proceedings of the Royal Irish Academy, vol. 100A, pp. 11–38 (2000) (Cottbus, Reihe Mathematik 1997/M-13)
20. Skeide, M.: Hilbert modules and applications in quantum probability. Habilitationsschrift, Cottbus (2001). http://web.unimol.it/skeide/
21. Skeide, M.: Commutants of von Neumann modules, representations of $\mathcal{B}^a(E)$ and other topics related to product systems of Hilbert modules. In: Price, G., Baker, B., Jorgensen, P., Muhly, P., (eds.), Advances in quantum dynamics. Number 335 in Contemporary Mathematics, American Mathematical Society, pp. 253–262 (2003) (Preprint, Cottbus 2002, arXiv: math.OA/0308231)
22. Skeide, M.: Von Neumann modules, interviwers and self-duality. J. Oper. Theory **54**, 119–124 (2005). (arXiv: math.OA/0308230)
23. Skeide, M.: Commutants of von Neumann Correspondences and Duality of Eilenberg-Watts Theorems by Rieffel and by Blecher, vol. 73, pp. 391–408. Banach Center Publications (2006). (arXiv: math.OA/0502241)
24. Skeide, M.: Product systems; a survey with commutants in view. In Belavkin, V., Guta, M., (eds.), Quantum Stochastics and Information, pp. 47–86. World Scientific (2008). (Preferable version: arXiv: 0709.0915v1!)
25. Skeide, M.: Hilbert modules—square roots of positive maps. In: Rebolledo, R., Orszag, M., (eds.), Quantum Probability and Related Topics — Proceedings of the 30th Conference. Number XXVII in Quantum Probability and White Noise Analysis, pp. 296–322. World Scientific (2011). (arXiv: 1007.0113v1)
26. Skeide, M., Sumesh, K.: CP-H-Extendable maps between Hilbert modules and CPH-semigroups. J. Math. Anal. Appl. **414**, 886–913 (2014) electronically Jan 2014. Preprint, arXiv: 1210.7491v2
27. Takesaki, M.: Theory of Operator Algebras II. Number 125 (Number VI in the Subseries Operator Algebras and Non-Commutative Geometry) in Encyclopedia of Mathematical Sciences. Springer (2003)
28. Thom, A.: A remark about the Connes fusion tensor product. Theory Appl. Categ. **25**, 38–50 (2011) (arXiv: math0601045)

Absorption and Fixed Points for Semigroups of Quantum Channels

Federico Girotti

Abstract In the present work we review and refine some results about fixed points of semigroups of quantum channels. Noncommutative potential theory enables us to show that the set of fixed points of a recurrent semigroup is a W^*-algebra; aside from the intrinsic interest of this result, it brings an improvement in the study of fixed points by means of absorption operators (a noncommutative generalization of absorption probabilities): under the assumption of absorbing recurrent space (hence allowing non-trivial transient space) we can provide a description of the fixed points set and a probabilistic characterization of when it is a W^*-algebra in terms of absorption operators. Moreover we are able to exhibit an example of a recurrent semigroup which does not admit a decomposition of the Hilbert space into orthogonal minimal invariant domains (contrarily to the case of classical Markov chains and positive recurrent semigroups of quantum channels).

Keywords Quantum channels · Quantum Markov semigroups · Fixed points · Absorption operators · Quantum recurrence · Null recurrence · Reducibility

1 Introduction

Quantum channels are mathematical objects used to describe the most general evolution an open quantum system can undergo and semigroups of quantum channels (which in continuous time are known as quantum Markov semigroups) have been used for decades to model time evolution of open quantum systems under reasonable assumptions and approximations. From a mathematical point of view they represent an interesting noncommutative generalization of transition kernels and classical Markov semigroups. Many fundamental concepts and tools from the theory of clas-

---

F. Girotti (✉)
School of Mathematical Sciences, Nottingham University, University Park, Nottingham NG7 2RD, United Kingdom
e-mail: federico.girotti@nottingham.ac.uk

L. Accardi et al. (eds.), *Infinite Dimensional Analysis, Quantum Probability and Applications*, Springer Proceedings in Mathematics & Statistics 390,
https://doi.org/10.1007/978-3-031-06170-7_10

sical Markov processes have been carried to the setting of semigroups of quantum channels and have provided valuable contributions to their study.

The notion of positive recurrence for semigroups of quantum channels dates back to the 70s and it is a fundamental tool for instance for studying the long-time behaviour of quantum systems; taking inspiration from the theory of classical Markov semigroups, transience and the distinction between positive and null recurrence were further analyzed in [8, 17] (in finite dimensional quantum systems, as in the case of Markov chains on a finite state space, null recurrence does not show up and the situation is far less complicated). However, there are still some open issues in the theory of recurrence for semigroups of quantum channels; following on from [3], in this work we present some recent results that improve the understanding of null recurrence in the noncommutative setting.

Another topic which has drown attention in the last years is the study of the fixed points of the evolution: they are relevant for the asymptotics of the evolution of quantum systems [10] and whether or not they are a W^*-algebra has implications, for instance, on the relationship between conserved quantities and symmetries of the semigroup (see [2] about the general problem of when fixed points are an algebra and [12] for a discussion about Noether-type results in the context of quantum channels). Fixed points are well understood in the case of positive recurrent semigroups [4, 9, 13], while less is known for general semigroups [1, 3, 11]. In the present work we review and improve some results of [3] which, under mild assumptions, characterize the fixed points sets in terms of absorption operators, which are a noncommutative generalization of absorption probabilities introduced always in [3].

Let us briefly present the setting. Let $\mathfrak{h}$ be a separable Hilbert space; we denote by $L^1(\mathfrak{h})$ trace class operators and by $B(\mathfrak{h})$ bounded linear operators; positive trace class operators with unit trace are called states and play the role of noncommutative probability densities. We recall that the topological dual of $L^1(\mathfrak{h})$ is isometric to $B(\mathfrak{h})$ via the following correspondence:

$$B(\mathfrak{h}) \ni x \mapsto \operatorname{tr}(x\cdot) \in L^1(\mathfrak{h})^*. \tag{1}$$

A quantum channel $\Phi : B(\mathfrak{h}) \to B(\mathfrak{h})$ is a completely positive unital w^*-w^*-continuous linear map, while the predual map Φ_* (defined via Eq. (1)) is a completely positive trace preserving map acting on $L^1(\mathfrak{h})$. A semigroup of quantum channels is a collection of quantum channels $(\mathcal{P}_t)_{t\in\mathfrak{T}}$ indexed by a semigroup $\mathfrak{T}$ such that

$$\mathcal{P}_0 = \mathrm{Id}, \qquad \mathcal{P}_s\mathcal{P}_t = \mathcal{P}_{t+s} \text{ for any } s, t \in \mathfrak{T}.$$

The collection of predual maps $(\mathcal{P}_{*t})$ forms a semigroup too. We will mainly consider two cases: $\mathfrak{T} = \mathbb{N}$ and the semigroup consists of the powers of a single quantum channel $\mathcal{P} := (\Phi^n)_{n\in\mathbb{N}}$, and $\mathfrak{T} = [0, +\infty)$ and we ask for the map $t \mapsto \mathcal{P}_t$ to be w^*-pointwise continuous (these semigroups are known in the literature as quantum Markov semigroups).

In Sect. 2 we briefly recall some fundamental concepts about reducibility of quantum channels. After that, we recap the basic notions of recurrence and transience for semigroups of quantum channels.

In Sect. 3 we show that the fixed points set of a recurrent semigroup of quantum channels is a W^*-algebra. Moreover, assuming that the recurrent space is absorbing, we provide a description of the fixed points set of the semigroup in terms of absorption operators and we present some relevant consequences (these results have already been proved in [3] under stronger assumptions). We recall that the set of fixed points (sometimes called harmonic operators) of a semigroup of quantum channels is the following set:

$$\mathcal{F}(\mathcal{P}) := \{x \in B(\mathfrak{h}) : \mathcal{P}_t(x) = x, \forall t \in \mathfrak{T}\}.$$

It is well known that in the case of classical Markov chains, recurrent states can be partitioned in communication classes; however, in Sect. 4 we show that recurrent semigroups in general do not admit a decomposition of $\mathfrak{h}$ into orthogonal minimal invariant spaces (hence the result [6, Proposition 7.1] for positive recurrent semigroups does not extend to recurrent ones).

2 Preliminaries on Semigroups of Quantum Channels

Reducibility. The concept of reducibility for quantum channels was introduced in [7] and since then have been intensively studied and exploited. We recall that for every positive operator x, $\operatorname{supp}(x) := \ker(x)^\perp$.

Definition 1 (Enclosure) A closed subspace $\mathcal{V}$ of $\mathfrak{h}$ is an enclosure (sometimes also called invariant domain) for a quantum channel Φ if, for any state ρ,

$$\operatorname{supp}(\rho) \subseteq \mathcal{V} \quad \text{implies} \quad \operatorname{supp}(\Phi_*(\rho)) \subseteq \mathcal{V}.$$

$\mathcal{V}$ is an enclosure for a semigroup $\mathcal{P}$ when it is an enclosure for any channel of the semigroup $\mathcal{P}$.

An enclosure $\mathcal{V}$ is said to be minimal if the only enclosures contained in $\mathcal{V}$ are the trivial ones, i.e. $\{0\}$ and $\mathcal{V}$.

The following are equivalent (see [5, Sect. 3]):

- $\mathcal{V}$ is an enclosure for Φ,
- $p_\mathcal{V}$ is a reducing or subharmonic projection for Φ, i.e. $\Phi(p_\mathcal{V}) \geq p_\mathcal{V}$
- $p_\mathcal{V} L^1(\mathfrak{h}) p_\mathcal{V}$ is hereditary for Φ_*, i.e. $\Phi_*(p_\mathcal{V} L^1(\mathfrak{h}) p_\mathcal{V}) \subseteq p_\mathcal{V} L^1(\mathfrak{h}) p_\mathcal{V}$.

When $\mathcal{V}$ is an enclosure, we can define the restricted quantum channel $\Phi^\mathcal{V}$ and the corresponding predual map:

$$\begin{aligned} \Phi^\mathcal{V}(p_\mathcal{V} x p_\mathcal{V}) &:= p_\mathcal{V} \Phi(p_\mathcal{V} x p_\mathcal{V}) p_\mathcal{V} = p_\mathcal{V} \Phi(x) p_\mathcal{V} \quad && \forall x \in B(\mathfrak{h}), \\ \Phi_*^\mathcal{V}(p_\mathcal{V} x p_\mathcal{V}) &:= p_\mathcal{V} \Phi_*(p_\mathcal{V} x p_\mathcal{V}) p_\mathcal{V} = \Phi_*(p_\mathcal{V} x p_\mathcal{V}) \quad && \forall x \in L^1(\mathfrak{h}). \end{aligned} \tag{2}$$

Thank to Eq. (2), the restrictions of a semigroup of quantum channels $\mathcal{P}^{\mathcal{V}}$ still maintain the semigroup property. Where it does not create confusion, we will often identify $B(\mathcal{V})$ with $p_{\mathcal{V}}B(\mathfrak{h})p_{\mathcal{V}}$.

Recurrence and transience. Recurrence and transience are strictly connected to reducibility properties and, of course, the ergodic behaviour of the semigroup; for the convenience of the reader, we briefly recall some related definitions and relevant properties following [8, 17] (see [11] for the discrete time case). In analogy with the theory of classical Markov chains, the following space is called the positive recurrent space:

$$\mathcal{R}_+ := \sup\{\text{supp}(\rho) : \rho \text{ is an invariant state, i.e.} \forall t \geq 0, \mathcal{P}_{t*}(\rho) = \rho\}. \quad (3)$$

It is well known that $\mathcal{R}_+$ is an enclosure ([17, Proposition 3]). For every positive $x \in B(\mathfrak{h})$, we call *form-potential* of x the following positive symmetric closed quadratic form: $\forall\, v \in \mathfrak{h}$

$$\mathfrak{U}(x)[v] := \int_0^\infty \text{tr}(\mathcal{P}_{*t}(|v\rangle \langle v|)x)dm(t), \quad D(\mathfrak{U}(x)) = \{v \in \mathfrak{h} \,:\, \mathfrak{U}(x)[v] < +\infty\}\,.$$

m stays for the counting measure in the discrete time case and for the Lebesgue measure in the continuous time setting. Loosely speaking, we can say that for any orthogonal projection p, $\mathfrak{U}(p)[v]$ represents the average time spent in $\text{supp}(p)$ by a quantum systems that starts in the state $|v\rangle \langle v|$ and evolves according to $\mathcal{P}$; if $\mathfrak{U}(p)$ is bounded, no matter what is the initial state $|v\rangle \langle v|$, the system spends on average a finite amount of time in $\text{supp}(p)$. We define $\ker(\mathfrak{U}(x)) = \{v \in \mathfrak{h} : \mathfrak{U}(x)[v] = 0\}$ and $\text{supp}(\mathfrak{U}(x)) = \ker(\mathfrak{U}(x))^\perp$; the transient space is given by

$$\mathcal{T} = \sup\{\text{supp}(\mathfrak{U}(x)) : x \in B(\mathfrak{h}), x \geq 0, \mathfrak{U}(x) \text{ bounded}\}.$$

$\mathcal{R} := \mathcal{T}^\perp$ is an enclosure and $\mathcal{T} \perp \mathcal{R}_+$ (see [17, Proposition 7 and Corollary 2]), hence it is meaningful to call $\mathcal{R}$ the recurrent space and to define $\mathcal{R}_0 := \mathcal{R} \cap \mathcal{R}_+^\perp$ the null recurrent space; as in the classical case, it turns out that the null recurrent space is an enclosure too ([3, Theorem 9]). We say that the semigroup $\mathcal{P}$ is positive recurrent (resp. null recurrent/recurrent/transient) if $\mathfrak{h} = \mathcal{R}_+$ (resp. $\mathcal{R}_0$, $\mathcal{R}$, $\mathcal{T}$); in general, a semigroup $\mathcal{P}$ does not need to be of one of the above types, but [17, Theorem 9] shows that it can always be decomposed in its transient and recurrent restrictions, $\mathcal{P}^{\mathcal{T}}$ and $\mathcal{P}^{\mathcal{R}}$, and the recurrent restriction splits again into the positive and null recurrent restrictions, $\mathcal{P}^{\mathcal{R}_+}$ and $\mathcal{P}^{\mathcal{R}_0}$ (since $\mathcal{R}$, $\mathcal{R}_+$ and $\mathcal{R}_0$ are enclosures, the corresponding restrictions are defined as in Eq. (2)).

3 Absorption Operators to Describe Fixed Points

Recurrent semigroups. In the first part of this section, unless differently specified, we assume that the semigroup $\mathcal{P}$ is recurrent, i.e. $\mathfrak{h} = \mathcal{R}$; notice that, in the case of a generic semigroup $\mathcal{P}$, everything that we prove holds true for the recurrent restriction $\mathcal{P}^{\mathcal{R}}$. In the case of recurrent semigroups, $\mathcal{F}(\mathcal{P})$ is tightly related to communication properties of the semigroup: every projection in $\mathcal{F}(\mathcal{P})$ is by definition the projection onto an enclosure and the vice versa is also true.

Theorem 1 (Theorem 9, [3]) *If $\mathcal{V}$, $\mathcal{Z}$ are increasing enclosures included in $\mathcal{R}$, i.e. such that $\mathcal{V} \subseteq \mathcal{Z} \subseteq \mathcal{R}$, then $\mathcal{Z} \cap \mathcal{V}^{\perp}$ is an enclosure.*

Hence projections corresponding to enclosures are harmonic projections. In general $\mathcal{F}(\mathcal{P})$ is only a selfadjoint w^*-closed linear space, however the following result proves that the fixed points set of a recurrent semigroup is a W^*-algebra, hence it is completely determined by its projections.

Proposition 1 *If $\mathcal{P}$ is recurrent, then $\mathcal{F}(\mathcal{P})$ is a W^*-algebra.*

We remark that this fact was already known in case of positive recurrent semigroups, i.e. when $\mathcal{R} = \mathcal{R}_+$ (see [2, Theorem 2.3]).

Proof If $\mathcal{F}(\mathcal{P})$ is not an algebra, [2, Lemma 2.2] shows that there exists $x \in \mathcal{F}(\mathcal{P})$ such that $x^*x \notin \mathcal{F}(\mathcal{P})$; moreover $\mathcal{P}_t(x^*x) \geq \mathcal{P}_t(x^*)\mathcal{P}_t(x) = x^*x$ for all $t \geq 0$ (Kadison-Schwarz inequality). There must exists a time $\bar{t} > 0$ such that the inequality is strict, otherwise x^*x would be a fixed point. x^*x is subharmonic, hence $(\mathcal{P}_t(x^*x))$ is a bounded positive monotone increasing net and the least upper bound y is a fixed point, since $\forall s \in \mathfrak{T}$

$$\mathcal{P}_s(y) = \lim_{t \to +\infty} \mathcal{P}_s(\mathcal{P}_t(x^*x)) = y.$$

Notice that

- $y - x^*x \geq \mathcal{P}_{\bar{t}}(x^*x) - x^*x \neq 0$;
- $\forall t \in \mathfrak{T}, \mathcal{P}_t(y - x^*x) = y - \mathcal{P}_t(x^*x) \leq y - x^*x$;
- $\mathcal{P}_t(y - x^*x) \searrow 0$.

By [8, Theorem 4], there exists a non-null positive operator z such that $\mathfrak{U}(z)$ is bounded, hence $\{0\} \neq \mathrm{supp}(\mathfrak{U}(z)) \subset \mathcal{T}$ and the transient space is non-trivial.

Corollary 1 *For every enclosure $\mathcal{V}$, $p_{\mathcal{V}} \in \mathcal{F}(\mathcal{P})$ and*

$$\mathcal{F}(\mathcal{P}) = \overline{\mathrm{span}\{p_{\mathcal{V}} : \mathcal{V} \textit{ enclosure}\}}^{\| \|_\infty}.$$

Proof As we already pointed out, Theorem 1 shows that projections corresponding to enclosures are harmonic projections and, since $\mathcal{F}(\mathcal{P})$ is a W^*-algebra, it is the norm-closure of the linear span of its projections.

Another immediate consequence is a nice diagonal structure of the elements in $\mathcal{F}(\mathcal{P})$.

Corollary 2 *Let* $x \in \mathcal{F}(\mathcal{P})$, *then* $x = p_{\mathcal{R}_+} x p_{\mathcal{R}_+} + p_{\mathcal{R}_0} x p_{\mathcal{R}_0}$.

Proof By Corollary 2, it is enough to prove the statement for the projections corresponding to the enclosures, hence the result follows from [3, Corollary 17].

Absorbing recurrent space. We now turn our attention to the fixed points set of semigroups with non-trivial transient part. For every enclosure $\mathcal{V}$, one can consider the corresponding absorption operator, defined as

$$A(\mathcal{V}) := w^* - \lim_{t \to +\infty} \mathcal{P}_t(p_{\mathcal{V}}).$$

Absorption operators are a noncommutative generalization of absorption probabilities: for every $v \in \mathfrak{h}$

$$\langle v, A(\mathcal{V})v \rangle := \lim_{t \to +\infty} \mathrm{tr}(\mathcal{P}_{*t}(|v\rangle\langle v|) p_{\mathcal{V}})$$

can be interpreted as the asymptotic probability of finding in $\mathcal{V}$ a quantum systems that starts in the state $|v\rangle\langle v|$ and evolves according to $\mathcal{P}$. Absorption operators have a convenient block structure which is related to recurrence: Theorem 14 in [3] shows that

$$A(\mathcal{V}) = p_{\mathcal{V}} + p_{\mathcal{V}^\perp \cap \mathcal{T}} A(\mathcal{V}) p_{\mathcal{V}^\perp \cap \mathcal{T}}. \tag{4}$$

Moreover absorption operators are fixed points of $\mathcal{P}$ ([3, Proposition 4]) and, since the set of fixed points is norm closed, we know that

$$\overline{\mathrm{span}\{A(\mathcal{V}) : \mathcal{V} \text{ enclosure}\}}^{\|\,\|_\infty} \subset \mathcal{F}(\mathcal{P}).$$

A natural question is to find under which conditions the reverse inclusion is true and so fixed points are completely described by absorption operators. We think it is an interesting problem because it creates a bridge between the fixed points set and the absorption dynamic and communication properties of the semigroup.

Although we still are not able to characterize the situation in full generality, we can improve [3, Theorem 22] and show that fixed points are completely characterized by absorption operators when the recurrent space is absorbing, that is $A(\mathcal{R}) = \mathbf{1}$. It means that asymptotically the evolution gets absorbed in the recurrent space, which, therefore, contains all relevant information regarding asymptotic quantities; it is not a very restrictive hypothesis: it holds true if $\mathfrak{h}$ is finite dimensional and stronger restrictions where often considered in ergodic theory [10, 11]. Furthermore, as pointed out already in [10], there is a wide class of quantum Markov semigroups for which checking its validity reduces to an analogous problem for a classical Markov chains.

Theorem 2 *If $A(\mathcal{R}) = \mathbf{1}$, then*

$$\mathcal{F}(\mathcal{P}) = \overline{\operatorname{span}\{A(\mathcal{V}) : \mathcal{V} \subset \mathcal{R} \text{ enclosure}\}}^{\|\ \|_\infty} = \mathcal{F}(\mathcal{P}^{\mathcal{R}_+}) \oplus \mathcal{F}(\mathcal{P}^{\mathcal{R}_0}) \oplus p_{\mathcal{T}}\mathcal{F}(\mathcal{P})p_{\mathcal{T}}.$$

Proof *(sketch)* Together with Proposition 1, Theorem 2.4 in [2] shows that there exists a conditional expectation $\mathcal{E} : B(\mathfrak{h}) \to \mathcal{F}(\mathcal{P})$ (it cannot be w^*-w^*-continuous unless $\mathcal{R} = \mathcal{R}_+$ by [10, Theorem 2.1]) which can be approximated in the w^*-pointwise topology by the net $\left(\frac{1}{t_\alpha}\int_0^{t_\alpha} \mathcal{P}_s(\cdot)dm(s)\right)_{t_\alpha}$ for a directed set of times (t_α). One can show that $\mathcal{E}_{|\mathcal{F}(\mathcal{P}^{\mathcal{R}})}$ is an isomorphism of Banach spaces as in the proof of [3, Proposition 21], hence, by virtue of Corollary 2,

$$\mathcal{F}(\mathcal{P}) = \overline{\operatorname{span}\{\mathcal{E}(p_{\mathcal{V}}) : \mathcal{V} \subset \mathcal{R} \text{ enclosure}\}}^{\|\ \|_\infty}.$$

Notice that, by definition of absorption operator, for every enclosure $\mathcal{V}$

$$\mathcal{E}(p_{\mathcal{V}}) = w^* - \lim_{t_\alpha \to +\infty} \frac{1}{t_\alpha} \int_0^{t_\alpha} \mathcal{P}_s(p_{\mathcal{V}})dm(s) = A(\mathcal{V}).$$

In order to prove the diagonal structure of $\mathcal{F}(\mathcal{P})$ with respect to $\mathcal{R}_+$ and $\mathcal{R}_0$, it suffices to show that it holds for absorption operators corresponding to recurrent enclosures and this follows from the fact that every enclosure $\mathcal{V} \subset \mathcal{R}$ is of the form $\mathcal{V} = (\mathcal{V} \cap \mathcal{R}_+) \oplus (\mathcal{V} \cap \mathcal{R}_0)$ ([3, Corollary 17]) and from Eq. (4). Equation (2) implies that for every $x \in \mathcal{F}(\mathcal{P})$, $p_{\mathcal{R}_+}xp_{\mathcal{R}_+}$ and $p_{\mathcal{R}_0}xp_{\mathcal{R}_0}$ are fixed points of the corresponding restrictions.

Notice that in the case of recurrent semigroups, by Eq. (4), absorption operators coincide with harmonic projections and Theorem 2 becomes Corollaries 1 and 2. There are some interesting consequences of Theorem 2 (hence we always assume $A(\mathcal{R}) = \mathbf{1}$), whose proofs we skip because they are very similar to the ones of [3, Propositions 23 and 29], with suitable trivial adaptations.

1. Let $\mathcal{V} \subset \mathcal{R}$ be an enclosure; $A(\mathcal{V}) = p_{\mathcal{V}} + p_{\mathcal{T}}A(\mathcal{V})p_{\mathcal{T}}$ (see Eq. (4)) and $p_{\mathcal{T}}A(\mathcal{V})p_{\mathcal{T}}$ is the unique $y \in B(\mathcal{T})$ that solves

$$\mathcal{L}(y) = -p_{\mathcal{T}}\mathcal{L}(p_{\mathcal{V}})p_{\mathcal{T}} \tag{5}$$

where $\mathcal{L}$ is the infinitesimal generator of $\mathcal{P}$ in continuous time or is equal to $\Phi - \mathrm{Id}$ in discrete time.
Equation (5) is the analogous of the characterization of absorption probabilities as solution of a linear system: let $(X_n)_{n \geq 0}$ be a Markov chain on a countable state space E with transition matrix $(p_{xy})_{x,y \in E}$. In the classical setting, enclosures correspond to closed sets: a subset $C \subset E$ is said to be closed if for every $x \in C$, $n \in \mathbb{N}$, $\mathbb{P}(X_n \notin C | X_0 = x) = 0$. If recurrent states are absorbing, for every close set $C \subset \mathcal{R}$, the corresponding absorption probability is the unique solution of

$$\begin{cases} A(C)_x = 1 \text{ if } x \in C, \\ \sum_{y\in\mathcal{T}}(p_{xy} - \delta_{xy})A(C)_y = -\sum_{y\in C} p_{xy} \text{ if } x \in \mathcal{T}. \end{cases} \quad (6)$$

2. Every enclosure $\mathcal{V}$ is of the form

$$\mathcal{V} = (\mathcal{V} \cap \mathcal{R}_+) \oplus (\mathcal{V} \cap \mathcal{R}_0) \oplus (\mathcal{V} \cap \mathcal{T}) \quad (7)$$

   where $p_{\mathcal{V}\cap\mathcal{T}} \leq A(\mathcal{V} \cap \mathcal{R}) - p_{\mathcal{V}\cap\mathcal{R}}$ and $\mathcal{V} \cap \mathcal{R} \neq \{0\}$.
   As we mentioned in the previous section, enclosures play a fundamental role in the study of semigroups of quantum channels and in applications it is useful to find them; Eq. (7) ensures that, at least when $A(\mathcal{R}) = 1$, we can reduce our search to projections which commutes with $p_{\mathcal{R}_+}$, $p_{\mathcal{R}_0}$ and $p_{\mathcal{T}}$. It would be extremely interesting to know whether or not a subharmonic projection always commutes with $p_{\mathcal{R}_+}$, $p_{\mathcal{R}_0}$ and $p_{\mathcal{T}}$.
   The last part of the statement tells us that every enclosure is composed by a non-null recurrent enclosure $\mathcal{V} \cap \mathcal{R}$, plus a linear space which gets completely absorbed into $\mathcal{V} \cap \mathcal{R}$.
3. $\mathcal{F}(\mathcal{P})$ is a W^*-algebra if and only if for every pair of orthogonal enclosures $\mathcal{V}, \mathcal{W} \subset \mathcal{R}$, $A(\mathcal{V})A(\mathcal{W}) = A(\mathcal{W})A(\mathcal{V}) = 0$.
   This last point provides a nice probabilistic characterization of the fact that $\mathcal{F}(\mathcal{P})$ is an algebra: losely speaking, it happens if and only if every state ρ cannot reach two orthogonal recurrent enclosures.

4 Reducibility of Recurrent Semigroups

Whether or not a recurrent semigroup admits a decomposition of the Hilbert space $\mathfrak{h}$ into orthogonal minimal enclosures (DOME) is a natural question because it is true for classical Markov chains and it was shown in many papers (for instance, see [6, Proposition 7.1]) for positive recurrent semigroups. In this section we will exhibit an example of a recurrent semigroup which does not admit a DOME, but before, we need to reformulate the problem thanks to Corollary 1.

Lemma 1 *A recurrent semigroup $\mathcal{P}$ admits a DOME if and only if $\mathcal{F}(\mathcal{P})$ is an atomic W^*-algebra.*

We recall that a W^*-algebra is atomic if for every projection p, there exists a minimal projection $q \leq p$ and this implies that there exists a denumerable family of orthogonal minimal projections $(q_\alpha)_{\alpha\in A}$ such that $\sum_{\alpha\in A} q_\alpha = \mathbf{1}$.

We remark that any DOME of $\mathcal{R}$ must be compatible with $\mathcal{R}_+$ and $\mathcal{R}_0$.

Lemma 2 *Let $(\mathcal{V}_\alpha)$ be a DOME of $\mathcal{R}$, then $(\mathcal{V}_\alpha \cap \mathcal{R}_+)$ and $(\mathcal{V}_\alpha \cap \mathcal{R}_0)$ are DOMEs for $\mathcal{R}_+$ and $\mathcal{R}_0$, respectively.*

Proof Corollary 2 implies that every (sub)harmonic projection $p_\mathcal{V}$ for $\mathcal{P}^\mathcal{R}$ commutes with $p_{\mathcal{R}_+}$, $p_{\mathcal{R}_0}$ and, since $\mathcal{R}_+$ and $\mathcal{R}_0$ are enclosures and the intersection of two enclosures is again an enclosure, $\mathcal{V} \cap \mathcal{R}_+$ and $\mathcal{V} \cap \mathcal{R}_0$ are enclosures. Therefore a DOME of $\mathcal{R}$ $(\mathcal{V}_\alpha)$ induces DOMEs $(\mathcal{V}_\alpha \cap \mathcal{R}_+)$ and $(\mathcal{V}_\alpha \cap \mathcal{R}_0)$ for $\mathcal{R}_+$ and $\mathcal{R}_0$, respectively (since $\mathcal{V}_\alpha$ is minimal, $\mathcal{V}_\alpha \cap \mathcal{R}_+$ and $\mathcal{V}_\alpha \cap \mathcal{R}_0$ are either $\{0\}$ or $\mathcal{V}_\alpha$).

We already know that there always exists a DOME of $\mathcal{R}_+$, hence the existence of a DOME of $\mathcal{R}$ and $\mathcal{R}_0$ are equivalent problems.

Example 1 (Noncommutative symmetric random walk on $\mathbb{Z}$) Let us consider the group G with two generators a, b satisfying $a^2 = b^2 = e$ (e is the identity element) and the corresponding left and right representations defined on $\mathfrak{h} = \ell^2(G)$. For every $g \in G$, $\lambda(g)$ and $\rho(g)$ are the unitary operators acting in the following way on the canonical basis $\mathcal{C} = \{\delta_g : g \in G\}$:

$$\lambda(g)\delta_h = \delta_{gh}, \quad \rho(g)\delta_h = \delta_{hg^{-1}} \quad \forall g, h \in G.$$

Notice that $\lambda(g)^* = \lambda(g^{-1})$ and $\rho(g)^* = \rho(g^{-1})$ for every $g \in G$. We introduce the following notation:

$$L(G) = \{\lambda(g) : g \in G\}'' = \{\lambda(a), \lambda(b)\}'', \quad R(G) = \{\rho(g) : g \in G\}'' = \{\rho(a), \rho(b)\}''.$$

We recall that $R(G) = L(G)'$ ([15, Sect. V.7]). We define the quantum channel

$$\Phi(x) = \frac{1}{2}\lambda(a)x\lambda(a) + \frac{1}{2}\lambda(b)x\lambda(b), \quad x \in B(\mathfrak{h})$$

and we consider the semigroup $\mathcal{P} := (\Phi^n)_{n\in\mathbb{N}}$.

Invariant commutative subalgebra and null recurrence. Consider the commutative W*-subalgebra of operators which are diagonal in the canonical basis $\mathcal{C}$ and its predual:

$$\Delta = \Big\{x \in B(\mathfrak{h}) : x = \sum_{g\in G} x_g \, |\delta_g\rangle\langle\delta_g| \Big\} \simeq \ell^\infty(G),$$

$$\Delta_* = \Big\{x \in L^1(\mathfrak{h}) : x = \sum_{g\in G} x_g \, |\delta_g\rangle\langle\delta_g| \Big\} \simeq \ell^1(G).$$

Since G is countable, we can relabel its elements with integer numbers:

$$\ldots aba \mapsto -3 \quad ba \mapsto -2 \quad a \mapsto -1 \quad e \mapsto 0 \quad b \mapsto 1 \quad ab \mapsto 2 \quad bab \mapsto 3 \ldots$$

and this provides isomorphisms between $\ell^\alpha(G)$ and $\ell^\alpha(\mathbb{Z})$ for $\alpha \in \{1, 2, \infty\}$. $\lambda(a)$, $\lambda(b)$ act on $\mathcal{C}$ in the following way (for any $g \in G$, the label g stays for δ_g):

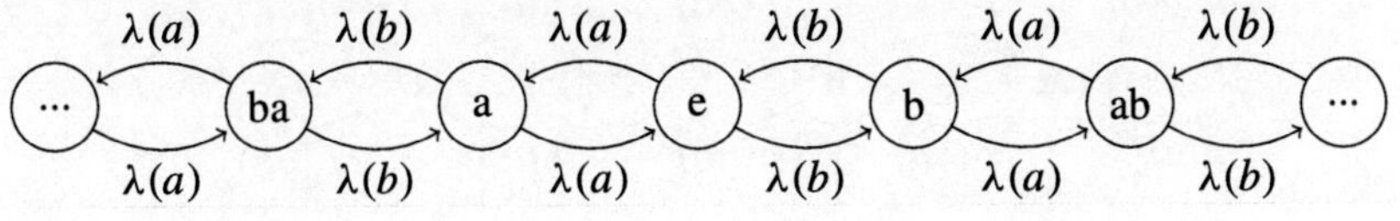

It is easy to see that Φ preserves Δ and Δ_* and its restriction corresponds via the isomorphisms above to the transition matrix of a symmetric random walk on $\mathbb{Z}$: $\Phi(|\delta_g\rangle\langle\delta_g|) = \frac{1}{2}(|\delta_{ag}\rangle\langle\delta_{ag}| + |\delta_{bg}\rangle\langle\delta_{bg}|)$. The symmetric random walk on $\mathbb{Z}$ is null recurrent and this implies that also $\mathcal{P}$ is null recurrent.

Proposition 2 *$\mathcal{P}$ is null recurrent.*

Proof 1. Suppose that there exists a non-null positive operator x such that $\mathfrak{U}(x)$ is bounded; then there must exists some $g \in G$ such that $\langle\delta_g, x\delta_g\rangle \geq c > 0$ for some positive constant c. By the symmetry of the semigroup, we can assume $g = e$.

$$+\infty > \mathfrak{U}(x)[\delta_e] = \sum_{k=0}^{+\infty} \mathrm{tr}(\Phi_*^k(|\delta_e\rangle\langle\delta_e|)x) \geq \sum_{k=0}^{+\infty} p_{0,0}^k\langle\delta_e, x\delta_e\rangle \geq c\sum_{k=0}^{+\infty} p_{0,0}^k = +\infty,$$

where $p_{0,0}^k$ is the probability that a symmetric random walk on $\mathbb{Z}$ that starts in 0 comes back to 0 in k steps. We obtained a contradiction, hence $\mathcal{T} = \{0\}$.
2. Suppose there exists an invariant state ρ; the action on ρ on Δ is represented by a state $\tilde{\rho} \in \Delta_*$, which must be invariant for the symmetric random walk on $\mathbb{Z}$, but this is again a contradiction, hence $\mathcal{R}_+ = \{0\}$.

$\mathcal{F}(\mathcal{P})$ has no minimal projections. When $\mathcal{F}(\mathcal{P})$ is an algebra (this is the case by Proposition 1) and the semigroup is generated by a single quantum channel Φ, Proposition 1 in [4] provides us with a characterization of $\mathcal{F}(\mathcal{P})$ in terms of Kraus operators of Φ:

$$\mathcal{F}(\mathcal{P}) = \{\lambda(a), \lambda(b)\}' = L(G)' = R(G).$$

We will show something stronger than the fact that $R(G)$ is not atomic: namely, we will prove that it has no minimal projections. We denote by $Z(G) := R(G) \cap L(G)$ the center of $R(G)$; notice that $(\rho(ab) + \rho(ba))/2 \in Z(G)$: it clearly belongs to $R(G)$ and it commutes with $\rho(a)$ (by symmetry it commutes with $\rho(b)$ too):

$$\rho(a)(\rho(ab) + \rho(ba)) = \rho(b) + \rho(aba) = (\rho(ab) + \rho(ba))\rho(a).$$

Let us focus on the action of $\rho(ab)$ and $\rho(ba) = \rho(ab)^*$ on $\mathfrak{C}$:

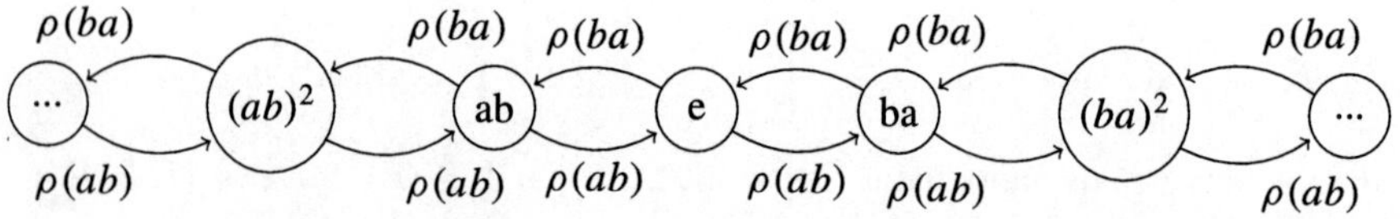

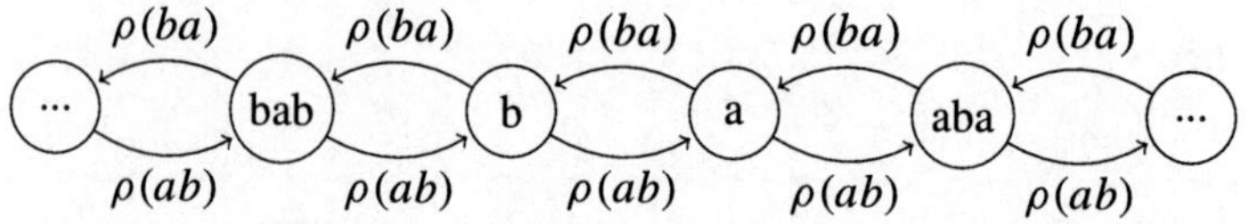

Hence there exists a unitary operator $U : \ell^2(\mathbb{Z}) \otimes \mathbb{C}^2 \to \ell^2(G)$ such that $U^*\rho(ab)U$ is $S \otimes \mathbf{1}_{\mathbb{C}^2}$, where S is the right shift operator. Consider the Fourier transform between the one dimensional torus $\mathbb{T}$ and $\mathbb{Z}$:

$$\begin{aligned}\mathcal{F} &: L^2(\mathbb{T}) \to \ell^2(\mathbb{Z}) \\ f(x) &\mapsto \hat{f}(k) := \frac{1}{2\pi}\int_{\mathbb{T}} f(x)e^{-ikx}dx.\end{aligned}$$

By the Fourier transform properties, it is easy to see that $\mathcal{F}^{-1}S\mathcal{F}$ is the multiplication operator $M_{e^{ix}}$ corresponding to the function e^{ix}. We have the following chain of equalities

$$\begin{aligned}M_{\cos(x)} \otimes \mathbf{1}_{\mathbb{C}^2} &= \left(\frac{M_{e^{ix}} + M_{e^{-ix}}}{2}\right) \otimes \mathbf{1}_{\mathbb{C}^2} = \mathcal{F}^{-1}(S + S^*)\mathcal{F} \otimes \mathbf{1}_{\mathbb{C}^2} = \\ &= (\mathcal{F}^{-1} \otimes \mathbf{1}_{\mathbb{C}^2})U^* \left(\frac{\rho(ab) + \rho(ba)}{2}\right) U(\mathcal{F} \otimes \mathbf{1}_{\mathbb{C}^2}).\end{aligned}$$

Therefore the W*-algebras generated by $M_{\cos(x)}$ and $(\rho(ab) + \rho(ba))/2$ are isomorphic.

Proposition 3 *$R(G)$ has no minimal projection.*

Proof Let us consider the following sequence of projections in the W^*-algebra generated by $M_{\cos(x)}$: for any $n \in \mathbb{N}$, $j = 0, \ldots, 2^{n+1} - 1$ we define $q_{j,n}$ as the characteristic function of

$$\cos^{-1}\left([-1 + j2^{-n}, -1 + (j+1)2^{-n}]\right);$$

we call $p_{j,n}$ the corresponding spectral projection of $(\rho(ab) + \rho(ba))/2$, which is in $Z(G)$. Notice that by construction

1. $\sum_{j=0}^{2^{n+1}-1} p_{j,n} = \mathbf{1}$,
2. $p_{j,n}p_{k,n} = \delta_{jk}p_{j,n}$ for every $n \in \mathbb{N}$, $j = 0, \ldots, 2^{n+1} - 1$,
3. for every $p_{j,n}$ with $n \geq 1$, there exists a unique k such that $p_{j,n}p_{k,n-1} = p_{j,n}$.

Let $f \in R(G)$ be a minimal projection, then for every $n \in \mathbb{N}$, f must be dominated by $p_{j,n}$ for some $j \in \{0, \ldots, n-1\}$, which we call $j(n)$. Hence $f \leq p_{j(n),n}$ for every $n \in \mathbb{N}$ and $p_{j(n+1),n+1} \leq p_{j(n),n}$ for every $n \in \mathbb{N}$ and $p_{j(n),n} \downarrow 0$ (in the w^*-topology), hence $f = 0$.

Remark 1 The following is the GKLS generator of a continuous time counterpart of the same example:

$$\mathcal{L}(x) = \frac{1}{2}(\lambda(a)x\lambda(a) - x) + \frac{1}{2}(\lambda(b)x\lambda(b) - x).$$

The analysis above can be carried out also in this case.

Remark 2 The structure of the decoherence-free subalgebra $\mathcal{N}(\mathcal{P})$ of a semigroup and wether it is atomic or not has been investigated in [4, 9, 14], especially in relation to environmental decoherence. In the present example, even $\mathcal{N}(\mathcal{P})$ has no minimal projection. Indeed, by [4, Proposition 3], we have that

$$\mathcal{N}(\mathcal{P}) = \{\lambda(ab), \lambda(ba)\}'.$$

Consider the decomposition $\ell^2(G) = \bigoplus_{h \in S} \ell^2(Hs) \simeq \ell^2(H) \otimes \ell^2(S)$ where H is the subgroup of G generated by ab, ba and S is a set of representatives of G/H. Notice that $\mathcal{N}(\mathcal{P})$ is $(L(H) \otimes \mathbf{1}_{\ell^2(S)})' = R(H) \otimes B(\ell^2(S))$, hence its center is $L(H) \cap R(H) \otimes \mathbf{1}_{\ell^2(S)} = \{\lambda(ab), \lambda(ba)\}'' \cap \{\rho(ab), \rho(ba)\}''$. $(\rho(ab) + \rho(ba))/2$ is in the center of $\mathcal{N}(\mathcal{P})$ too, hence we can repeat the same proof as for $\mathcal{F}(\mathcal{P})$.

Although Example 1 shows that in general $\mathcal{F}(\mathcal{P}^{\mathcal{R}})$ is not atomic, [2, Theorem 2.4] together with Proposition 1 shows that it is injective. A natural question is if it possible to find *interesting* features of the semigroup ensuring the atomicity of $\mathcal{F}(\mathcal{P}^{\mathcal{R}})$ and, hence, the decomposition of $\mathcal{R}$ into orthogonal minimal enclosures.

Acknowledgements The author acknowledges the support of the INDAM GNAMPA project 2020 "Evoluzioni markoviane quantistiche" and of the Italian Ministry of Education, University and Research (MIUR) for the Dipartimenti di Eccellenza Program (2018- 2022)-Dept. of Mathematics "F. Casorati", University of Pavia. Special acknowledgments go to the organizers of the 41st *International Conference on Quantum Probability and Related Topics* and to Raffaella Carbone for introducing the author to the problem and for many useful suggestions.

References

1. Albert, V.: Asymptotics of quantum channels: conserved quantities, an adiabatic limit, and matrix product states. Quantum **3**, 151 (2019). https://doi.org/10.22331/q-2019-06-06-151
2. Arias, A., Gheondea, A., Gudder, S.: Fixed points of quantum operations. J. Math. Phys. **43**, 5872–5881 (2002). https://doi.org/10.1063/1.1519669
3. Carbone, R., Girotti, F.: Absorption in invariant domains for semigroups of quantum channels. Ann. Henri Poincaré **22**, 2497–2530 (2021). https://doi.org/10.1007/s00023-021-01016-5
4. Carbone, R., Jenčová, A.: Cycles and fixed points of a quantum channel. Ann. Henri Poincaré **21**, 155–188 (2020). https://doi.org/10.1007/s00023-019-00861-9
5. Carbone, R., Pautrat, Y.: Open quantum random walks: reducibility, period, ergodic properties. Ann. Henri Poincaré **17**, 99–135 (2016). https://doi.org/10.1007/s00023-015-0396-y
6. Carbone, R., Pautrat, Y.: Irreducible decompositions and stationary states of quantum channels. Rep. Math. Phys. **77**, 293–313 (2016). https://doi.org/10.1016/S0034-4877(16)30032-5
7. Davies, E.B.: Quantum stochastic processes. II Comm. Math. Phys. **19**, 83–105 (1970)
8. Fagnola, F., Rebolledo, R.: Transience and recurrence of quantum Markov semigroups. Probab. Theory Relat. Fields **126**, 289–306 (2003). https://doi.org/10.1007/s00440-003-0268-0
9. Fagnola, F., Sasso, E., Umanità, V.: The role of the atomic decoherence-free subalgebra in the study of quantum Markov semigroups. J. Math. Phys. **60**, 072703 (2019). https://doi.org/10.1063/1.5030954
10. Frigerio, A., Verri, M.: Long-time asymptotic properties of dynamical semigroups on W^*-algebras. Math. Z. **180**, 275–286 (1982). https://doi.org/10.1007/BF01318911

11. Gärtner, A., Kümmerer, B.: A coherent approach to recurrence and transience for quantum Markov operators. arXiv: 1211.6876 (2012) http://adsabs.harvard.edu/abs/2012arXiv1211.6876G
12. Gheondea, A.: Symmetries versus conservation laws in dynamical quantum systems: a unifying approach through propagation of fixed points. Ann. Henri Poincaré **19**, 1787–1816 (2018). https://doi.org/10.1007/s00023-018-0666-6
13. Jenčová, A., Petz, D.: Sufficiency in quantum statistical inference. Commun. Math. Phys. **263**, 259–276 (2006). https://doi.org/10.1007/s00220-005-1510-7
14. Emanuela Sasso, E., Umanita, V.: The general structure of the Decoherence-free subalgebra for uniformly continuous Quantum Markov semigroups. arXiv: 2101.05121 (2021)
15. Takesaki, M.: Theory of Operator Algebras I. Springer, Berlin (2002)
16. Tomiyama, J.: On the projection of norm one in W^*-algebras. III. Tohoku Math. J. **11**, 125–129 (1959). https://doi.org/10.2748/tmj/1178244633
17. Umanità, V.: Classification and decomposition of quantum Markov semigroups. Probab. Theory Relat. Fields **134**, 603–623 (2006). https://doi.org/10.1007/s00440-005-0450-7

Characterization of Gaussian Quantum Markov Semigroups

Damiano Poletti

Abstract We give a characterization of QMSs on the Bosonic Fock Space $\Gamma(\mathbb{C}^d)$ whose predual preserves the set of gaussian states. We show they can be obtained via certain generalized GKLS generators and they satisfy an explicit formula for their action on Weyl operators.

Keywords Quantum Markov semigroup · Quasi-free semigroup · Gaussian · QMS

1 Introduction

Gaussian Quantum Markov Semigroups (QMS) have appeared in the literature since they arise in relevant models and are advantageous for some explicit formulas they satisfy (see for example [3–5, 15, 17]). They are also called quasi-free semigroups and are usually introduced in two ways: the first one uses a Gorini-Kossakowski-Lindblad-Sudarshan (GKLS) generator which is quadratic in creation and annihilation operators (cf. (20), (21), (22)), while the second one gives their explicit action on Weyl operators (cf. (15)). In [1, 2] it is shown that these two approaches are the same and that gaussian QMSs defined in either of these ways also preserve the set of the so called gaussian states (also known as quasi-free states [16]). The main result of this paper is to prove that the preservation of the set of gaussian state is a defining property of a gaussian QMS. Indeed, we show that a QMS that preserves gaussian states is necessarily a gaussian QMS with its usual definition. Namely, it can be introduced via a generalized GKLS generator quadratic in creation and annihilation operators and it satisfies the explicit action on Weyl operators.

This kind of problems are of interest also in the realm of classical probability, where are studied on Gauss-Markov processes (see [10, 14]). In quantum prob-

D. Poletti (✉)
Dipartimento di Matematica, Politecnico di Milano, Piazza Leonardo da Vinci 32, Milano, Italy
e-mail: damiano.poletti@polimi.it

L. Accardi et al. (eds.), *Infinite Dimensional Analysis, Quantum Probability and Applications*, Springer Proceedings in Mathematics & Statistics 390,
https://doi.org/10.1007/978-3-031-06170-7_11

ability some steps in this direction were already present in the literature. In [12] automorphisms on the CCR Algebra that preserve gaussian states are fully characterized. While in [9, 21] complete positivity of this kind of maps on the CCR algebra and their semigroups are studied along with their extensions to QMSs. Here however we focus on the preservation of gaussian states as the main property of the transformation. Therefore we extend the result of [12] to general transformations of states on the CCR algebra and not just those induced by automorphisms of the algebra itself. Moreover we recover results similar to the ones in [21] but starting from QMSs whose predual preserves gaussian states, effectively characterizing them.

The paper is organized as follows: in Sect. 2 we introduce and recall some properties of gaussian states on the CCR algebra; in Sect. 3 we introduce and characterize gaussian maps, i.e. maps that act on states of the CCR algebra that preserve gaussian states, and prove a characterization theorem for their action on the parameters of such states; in Sect. 4 we introduce Gaussian QMS as semigroups whose predual preserves gaussian states and characterize them in terms of both their action on Weyl operators and their generalized GKLS generator.

2 Gaussian States

Let us consider $\mathbb{C}^d$ as a complex Hilbert space h with the usual inner product given by $\langle\cdot,\cdot\rangle$, antilinear in the first component. There is a natural real Hilbert space associated with h whose inner product is $\mathrm{Re}\,\langle\cdot,\cdot\rangle$. We use $\mathsf{h}_\mathbb{R}$ when referring to h as as the real Hilbert space. We will also denote with $\mathcal{B}(\mathsf{h})$ the complex linear operators on h and with $\mathcal{B}_\mathbb{R}(\mathsf{h}_\mathbb{R})$ the (real) linear operators on $\mathsf{h}_\mathbb{R}$.

On $\mathsf{h}_\mathbb{R}$ we can also define the form $\sigma(z_1, z_2) := \mathrm{Im}\,\langle z_1, z_2\rangle$ for all $z_1, z_2 \in \mathsf{h}_\mathbb{R}$ which is a symplectic form (i.e. bilinear, anti-symmetric, non degenerate). The pair $(\mathsf{h}_\mathbb{R}, \sigma)$ is then a symplectic space and we can construct the C^*-algebra of the canonical commutation relations (CCR-algebra) (see [6, Theorem 5.2.8]) by completion of $\mathrm{span}\{W(z) | z \in \mathsf{h}_\mathbb{R}\}$ where $W(z)$ are the Weyl operators which satisfy the Weyl form of the CCR

$$\begin{aligned} W(z_1)W(z_2) &= \mathrm{e}^{-\mathrm{i}\sigma(z_1,z_2)}W(z_1+z_2)\,, \quad z_1, z_2 \in \mathsf{h}_\mathbb{R} \\ W(z)^* &= W(-z)\,, \quad z \in \mathsf{h}_\mathbb{R}. \end{aligned} \tag{1}$$

We denote the CCR-algebra on $(\mathsf{h}_\mathbb{R}, \sigma)$ as $CCR\,(\mathsf{h}_\mathbb{R}, \sigma)$.

We now introduce gaussian states as those that satisfy a particular formula when acting on Weyl operators.

Definition 1 A state ω on $CCR\,(\mathsf{h}_\mathbb{R}, \sigma)$ is called a gaussian state if there exist $\zeta \in \mathsf{h}_\mathbb{R}$ and $S \in \mathcal{B}_\mathbb{R}(\mathsf{h}_\mathbb{R})$ invertible such that

$$\omega(W(z)) = \exp\left(-\mathrm{i}\,\mathrm{Re}\,\langle \zeta, z\rangle - \frac{1}{2}\mathrm{Re}\,\langle z, Sz\rangle\right), \quad \forall z \in \mathsf{h}_{\mathbb{R}}.$$

In that case we write $\omega = \omega_{(\zeta,S)}$ and we call ζ the mean vector and S the covariance operator of ω.

Let us denote with $\mathcal{Q}$ the set of covariance operators for gaussian states. It is known that not all invertible operators $S \in \mathcal{B}_{\mathbb{R}}(\mathsf{h}_{\mathbb{R}})$ are in $\mathcal{Q}$ (see for example [19]). However there exists an explicit condition that characterizes elements in $\mathcal{Q}$ that we will show in the following. In order to precisely state it we need a short digression to recall some relevant facts about operators in $\mathcal{B}_{\mathbb{R}}(\mathsf{h}_{\mathbb{R}})$. The action of every operator $A \in \mathcal{B}_{\mathbb{R}}(\mathsf{h}_{\mathbb{R}})$ can be uniquely written as the sum of a linear and antilinear part, namely

$$Az = \frac{1}{2}(A - \mathrm{i}A\mathrm{i})\,z + \frac{1}{2}(A + \mathrm{i}A\mathrm{i})\,z = A_1 z + A_2\overline{z}\,, \tag{2}$$

for every $z \in \mathsf{h}_{\mathbb{R}}$, where $2A_1 z := (A - \mathrm{i}A\mathrm{i})\,z$, $2A_2 z := (A + \mathrm{i}A\mathrm{i})\,\overline{z}$ are now complex linear operators. We will denote via $A^{\sharp}$ the adjoint of A with respect to the real scalar product $\mathrm{Re}\,\langle\cdot,\cdot\rangle$, whose action on $z \in \mathsf{h}_{\mathbb{R}}$ is explicitly given by

$$A^{\sharp}z = A_1^{*}z + A_2^{T}\overline{z}\,,$$

where now the adjoint operation is with respect to the usual complex scalar product on h and $A_2^T = \overline{A_2^*}$. Usually one identifies the real Hilbert space $\mathsf{h}_{\mathbb{R}}$ with $\mathbb{R}^{2d}$ by exploiting the decomposition $z = x + \mathrm{i}y$, with $x, y \in \mathbb{R}^d$. The real linear operator A, after this identification acts on $\mathbb{R}^{2d}$ via the matrix action

$$A_{\mathbb{R}^{2d}}\begin{pmatrix} x \\ y \end{pmatrix} = \begin{pmatrix} \mathrm{Re}(A_1 + A_2) & \mathrm{Im}(A_2 - A_1) \\ \mathrm{Im}(A_1 + A_2) & \mathrm{Re}(A_1 - A_2) \end{pmatrix}\begin{pmatrix} x \\ y \end{pmatrix}.$$

This is the usual approach and the one also taken in [19]. Here however we consider a different identification of $\mathsf{h}_{\mathbb{R}}$ with the real subspace $\mathfrak{h}$ of $\mathbb{C}^{2d}$ given by

$$\mathfrak{h} = \left\{\begin{pmatrix} z \\ \overline{z} \end{pmatrix}, z \in \mathsf{h}_{\mathbb{R}}\right\},$$

with scalar product given by $\mathrm{Re}\,\langle\cdot,\cdot\rangle_{2d}/2$, where $\langle\cdot,\cdot\rangle_{2d}$ is the usual complex scalar product on $\mathbb{C}^{2d}$. Now the identification of A acts on $\mathfrak{h}$ as

$$A_{\mathfrak{h}}\begin{pmatrix} z \\ \overline{z} \end{pmatrix} = \begin{pmatrix} A_1 & A_2 \\ \overline{A_2} & \overline{A_1} \end{pmatrix}\begin{pmatrix} z \\ \overline{z} \end{pmatrix}.$$

which is easier to compute, starting from the decomposition (2) for A, than $A_{\mathbb{R}^{2d}}$. We will also need to consider complexification of the space $\mathfrak{h}$, denoted by $\mathfrak{h}_{\mathbb{C}}$, which turns out to be the whole $\mathbb{C}^{2d}$ with complex scalar product $\langle\cdot,\cdot\rangle_{2d}/2$. The complexification of $A_{\mathfrak{h}}$ will be denoted $\mathbf{A}$ in boldface character and has the same matrix expression of $A_{\mathfrak{h}}$. Whenever in the following we speak of the complexification of an operator $A \in \mathcal{B}_{\mathbb{R}}(\mathsf{h}_{\mathbb{R}})$ we always refer to its identification with $\mathbf{A}$. It is also true, via this identification, that the complexification of the adjoint operator $A^\sharp$ is the usual adjoint operation of the complexified operator $\mathbf{A}$, in formulas

$$\mathbf{A}^\sharp = \mathbf{A}^*.$$

A useful example will be the operator $Jz = -\mathrm{i}z$, for every $z \in \mathsf{h}_{\mathbb{R}}$. Here $J_1 = -\mathrm{i}\mathbb{1}$, $J_2 = 0$ therefore

$$\mathbf{J} = \begin{pmatrix} -\mathrm{i}\mathbb{1} & 0 \\ 0 & \mathrm{i}\mathbb{1} \end{pmatrix}. \tag{3}$$

Note that $J^\sharp = J^{-1} = -J$ and we can use J to rewrite the symplectic form σ

$$\sigma(z_1, z_2) = \mathrm{Im}\,\langle z_1, z_2\rangle = \mathrm{Re}\,\langle z_1, Jz_2\rangle\,, \tag{4}$$

for every $z_1, z_2 \in \mathsf{h}_{\mathbb{R}}$. We can now state the characterization theorem for operators in Q whose proof can be found in [19].

Theorem 1 *Let S be a real linear, self adjoint, invertible operator on $\mathsf{h}_{\mathbb{R}}$ then $S \in Q$ if and only if*

$$\mathbf{S} - \mathrm{i}\mathbf{J} \geq 0,$$

where positivity has to be intended on $\mathfrak{h}_{\mathbb{C}}$.

In the final part of this section we define a partial order relation on the set of gaussian states which was introduced by Fannes in [12].

Definition 2 Let $S_1, S_2 \in Q$ and $\zeta_1, \zeta_2 \in \mathsf{h}_{\mathbb{R}}$, we say

$$\omega_{(\zeta_1, S_1)} \preceq \omega_{(\zeta_2, S_2)}\,,$$

if there exists $\gamma \in \mathbb{R}$ such that

$$\omega_{(\zeta_1, S_1)} \leq \gamma \omega_{(\zeta_2, S_2)}\,.$$

We refer to [12] for the proof that this is really a partial order. In general it is not easy to translate $\preceq$ in terms of the parameters ζ and S of the gaussian states. However, this can be done if we restrict ourselves to comparing covariance operators that belong to the same gauge, in the following sense. Consider $S \in Q$ and define the scalar product $\langle\cdot,\cdot\rangle_S$ via

$$\langle z_1, z_2\rangle_S = \mathrm{Re}\,\langle z_1, Sz_2\rangle\,, \quad z_1, z_2 \in \mathsf{h}_{\mathbb{R}}\,.$$

One can consider the polar decomposition of J^*S with respect to $\langle\cdot,\cdot\rangle_S$, given by $J^*S = K_S|S|_S$.

Definition 3 We say $S_1 \in Q$ belongs to the same gauge of S if and only if the polar decomposition of $J^* S_1$ with respect to $\langle \cdot, \cdot \rangle_{S_1}$ is $K_S |S_1|_{S_1}$.

If we restrict to covariance operators belonging to the same gauge we can completely characterize the partial ordering $\preceq$. We refer to [12] for the proof of the following proposition that states this result.

Proposition 1 *Let $S_1, S_2 \in Q$ belong to the same gauge K and let $\zeta_1, \zeta_2 \in \mathsf{h}_{\mathbb{R}}$. Then $\omega_{(\zeta_1, S_1)} \preceq \omega_{(\zeta_2, S_2)}$ if and only if $S_1 \leq S_2$ and $\zeta_1 - \zeta_2 \in (\ker(S_2 - S_1))^{\perp}$, where orthogonality has to be intended with respect to the scalar product on $\mathsf{h}_{\mathbb{R}}$.*

3 Gaussian Maps

In this section we will introduce the main object of this article by considering maps that preserves the set of gaussian states. Most of the results in this section follow again the ideas in [12], with a notable difference. We do not restrict ourselves to considering just automorphism of the CCR algebra that preserve gaussian states, instead we just focus on this last property to be defining for a gaussian map. We consider therefore gaussian maps as acting on functionals and only later recover their action on the CCR algebra. We proceed now to recall the definition of normal states which will be the actual domain of gaussian maps. Let us consider the symmetric Fock Space (see [7])

$$\mathcal{H} = \Gamma(\mathbb{C}^d) = \Gamma(\mathbb{C}) \otimes \cdots \otimes \Gamma(\mathbb{C}) ,$$

where we have the canonical orthonormal basis of $(e(n_1, \ldots, n_d))_{n_1, \ldots, n_d \geq 0}$ given by

$$e(n_1, \ldots, n_d) = e_{n_1} \otimes \cdots \otimes e_{n_d} .$$

Recall that the exponential vectors are defined for $f \in \mathbb{C}^d$ as

$$e(f) = \sum_{n_1, \ldots, n_d} \frac{f_1^{n_1} \ldots f_d^{n_d}}{\sqrt{n_1! \ldots n_d!}} e_{n_1} \otimes \cdots \otimes e_{n_d}$$

and are a total set on $\Gamma(\mathbb{C}^d)$. We can then introduce the operators $W(z)$ with $z \in \mathbb{C}^d$ that act on the set of exponential vectors as

$$W(z)e(f) = \exp\left(-\frac{|z|^2}{2} - \langle z, f \rangle\right) e(z + f) .$$

It is well known that $W(z)$ satisfy (1) and that the linear map $\pi : CCR\,(\mathsf{h}_{\mathbb{R}}, \sigma) \to \mathcal{B}(\mathsf{h})$ such that $\pi(W(z)) = W(z)$ extends to a regular, projective, unitary, irreducible representation [7, 18].

Definition 4 We say a functional ω on $CCR\,(\mathsf{h}_{\mathbb{R}}, \sigma)$ is normal if there exists a trace class operator, positive with $\mathrm{tr}\rho = 1$, such that

$$\mathrm{tr}(\rho\pi(x)) = \omega(x) \, ,$$

for every $x \in CCR\,(\mathsf{h}_{\mathbb{R}}, \sigma)$. We will denote with $CCR\,(\mathsf{h}_{\mathbb{R}}, \sigma)_*$ the set of normal functionals and with $\mathcal{S}(CCR\,(\mathsf{h}_{\mathbb{R}}, \sigma))$ the set of normal states.

An important property of normal states is that they can be approximated in the norm topology via gaussian states. We precisely state this result whose proof can be found in [13, Theorem II.4].

Proposition 2 *The set of linear combinations of gaussian states is norm dense in the set of all normal states* $\mathcal{S}(CCR\,(\mathsf{h}_{\mathbb{R}}, \sigma))$.

We can now give the definition of a gaussian map.

Definition 5 Let α be a norm-continuous, linear transformation on the set of normal functionals $CCR\,(\mathsf{h}_{\mathbb{R}}, \sigma)_*$. We say α is a Gaussian map if

1. $\alpha \circ \omega_{(\zeta,S)} = \omega_{(\alpha_1(\zeta,S),\alpha_2(\zeta,S))}$, for every $\zeta \in \mathsf{h}_{\mathbb{R}}$, $S \in Q$.
2. the map $(\zeta, S) \mapsto (\alpha_1(\zeta, S), \alpha_2(\zeta, S))$ is continuous with respect to the product topology.

We will show in Theorem 2 that the maps α_1, α_2 in definition 5 cannot be arbitrary but instead should have a specific expression. In order to prove this result we need some preliminary Lemmas.

Lemma 1 *Let* α *be a gaussian map, then* α *is also positive. Suppose now* α' *is another gaussian map such that*

$$\alpha_1(\zeta, S) = \alpha'_1(\zeta, S) \, , \quad \alpha_2(\zeta, S) = \alpha'_2(\zeta, S) \, ,$$

for all $\zeta \in \mathsf{h}_{\mathbb{R}}$, $S \in Q$. *Then* $\alpha = \alpha'$.

Proof In order to prove positivity of α let $\eta \in CCR\,(\mathsf{h}_{\mathbb{R}}, \sigma)_*$ be positive. Then η is also bounded and therefore $\eta_0 := \eta / \|\eta\| \in \mathcal{S}(CCR\,(\mathsf{h}_{\mathbb{R}}, \sigma))$. Via Proposition 2 we can then find a sequence $(\omega_n)_n$ of linear combinations of gaussian states that converge to η_0 in norm. Eventually, via norm continuity of α we have

$$\alpha(\eta) = \|\eta\| \, \alpha(\eta_0) = \|\eta\| \lim_{n\to\infty} \alpha(\omega_n) \geq 0 \, .$$

This proves positivity of α. Now let ω be a gaussian state. By hypothesis we have

$$\alpha(\omega) = \omega_{(\alpha_1(\zeta,S),\alpha_2(\zeta,S))} = \omega_{(\alpha'_1(\zeta,S),\alpha'_2(\zeta,S))} = \alpha'(\omega) \, .$$

In particular we have

$$\left(\alpha - \alpha'\right)(\omega) = 0 \, ,$$

for every ω gaussian state. Using density of gaussian states, given by Proposition 2 we obtain $\alpha = \alpha'$.

The next two Lemmas come from [12], we show just a sketch of the proof of the first one.

Lemma 2 *Let α be a Gaussian map. Then for every $S \in Q$ and every $\zeta \in \mathsf{h}_{\mathbb{R}}$*

$$\alpha_2(\zeta, S) = \alpha_2(0, S) = \alpha_2(S) .$$

Proof Let K be the gauge corresponding to S and let $S_0 = p_\zeta + p_{K^\sharp\zeta}$ where p_x is the projection on $\mathbb{R}x$. For every $\lambda \geq 0$ we have $S + \lambda S_0 \in Q$ and $S + \lambda C$ belongs to the gauge K. Therefore we can use Proposition 1 and obtain, for $0 < \lambda_1 < \lambda_2$

$$\omega_{(\zeta,S)} \preceq \omega_{(0,S+\lambda_1 C)} \preceq \omega_{(\zeta,S+\lambda_2 C)} .$$

Composition with α preserves the relation $\preceq$, since it is a positive transformation by definition. Therefore by composition and application of Proposition 1 we get

$$\alpha_2(\zeta, S) \leq \alpha_2(0, S + \lambda_1 C) \leq \alpha_2(\zeta, S + \lambda_2 C) .$$

By the continuity property in Definition 5, taking the limit as $\lambda_1, \lambda_2 \to 0$ we obtain

$$\alpha_2(\zeta, S) \leq \alpha_2(0, \zeta) \leq \alpha_2(\zeta, S) ,$$

which concludes the proof.

We omit the proof of the third Lemma that can be found in [12].

Lemma 3 *Suppose $s \in \mathbb{R} \mapsto f(s) \in \mathbb{R}$ is continuous and*

$$\pi^{-\frac{1}{2}} \int_{\mathbb{R}} \mathrm{e}^{-s^2} \mathrm{e}^{itf(s)} \mathrm{d}s = \mathrm{e}^{-\frac{1}{4}t^2}, \quad \forall t \in \mathbb{R}.$$

Then either $f(s) = s$ or $f(s) = -s$ for every $s \in \mathbb{R}$.

We can now state the characterization theorem for Gaussian maps. It extends the result of [12, Theorem 4.5] to maps that are not necessarily automorphisms of the CCR algebra. The proof is exactly the one of [12] except in its final part. We report the entire proof for clarity's sake, since this will be an important result in the following.

Theorem 2 *Let α be a Gaussian map. Then there exist $T_0, C_0 \in \mathcal{B}_{\mathbb{R}}(\mathsf{h}_{\mathbb{R}})$ and $\zeta_0 \in \mathsf{h}_{\mathbb{R}}$ such that*

$$\alpha_1(z, S) = \alpha_1(z) = T_0^* z + \zeta_0 , \quad \alpha_2(S) = T_0^* S T_0 + C_0 . \tag{5}$$

for every $z \in \mathsf{h}_{\mathbb{R}}$ and

$$\mathbf{C_0} - \mathrm{i}(\mathbf{J} - \mathbf{T_0}^*\mathbf{J}\mathbf{T_0}) \geq 0 , \tag{6}$$

where positivity is intended on $\mathfrak{h}_{\mathbb{C}}$.

Proof We use the notation $|x\rangle\langle y|$, $x, y \in \mathsf{h}_{\mathbb{R}}$ to denote the operator acting on $z \in \mathsf{h}_{\mathbb{R}}$ as $|x\rangle\langle y|\, z = \operatorname{Re}\langle y, z\rangle\, x$. For every $r > 0$, $z, z_0 \in \mathsf{h}_{\mathbb{R}}$ and $S \in Q$ one has

$$(\pi r)^{-\frac{1}{2}} \int_{\mathbb{R}} \mathrm{e}^{-\frac{s^2}{r}} \omega_{(sz+z_0,S)} ds = \omega_{(z_0,S+r|z\rangle\langle z|)} \, .$$

Let $x \in \mathsf{h}_{\mathbb{R}}$, $t \in \mathbb{R}$, by composition with α and evaluation on $W(tx)$ we obtain

$$\begin{aligned}&(\pi r)^{-\frac{1}{2}} \int_{\mathbb{R}} \mathrm{e}^{-\frac{s^2}{r}} \mathrm{e}^{itf(s)} \mathrm{d}s \\ &= \mathrm{e}^{it \operatorname{Re}\langle x, \alpha_1(S+r|z\rangle\langle z|, z_0)\rangle} \exp\left(-\frac{1}{4} t^2 \operatorname{Re}\langle x, (\alpha_2(S+r\,|z\rangle\langle z|) - \alpha_2(S))\, x\rangle\right),\end{aligned}$$

where $f(s) := \operatorname{Re}\langle \alpha_1(s\zeta + \zeta_0, S), x\rangle$ and we used Lemma 2. By Lemma 3 we have then that f satisfies $f(s) = f(0) + s(f(1) - f(0))$. Equivalently we get

$$\alpha_1(s\zeta + \zeta_0, S) = \alpha_1(\zeta_0, S) + s\,(\alpha_1(\zeta + \zeta_0, S) - \alpha_1(\zeta_0, S))\, .$$

Substituting it into the integral equation and performing integration we obtain the following conditions

$$\alpha_1(z_0, S) = \alpha_1(z_0, S + r\,|z\rangle\langle z|)\, , \tag{7}$$

$$\begin{aligned}\alpha_2(S) + r\,|\alpha_1(z+z_0, S) - \alpha_1(z_0, S)\rangle\langle \alpha_1(z+z_0, S) - &\alpha_1(z_0, S)| \\ &= \alpha_2(S + r\,|z\rangle\langle z|)\, .\end{aligned} \tag{8}$$

Since the right hand side of (8) does not depend on z_0 it must hold

$$\alpha_1(z+z_0, S) - \alpha_1(z_0, S) = \alpha_1(z, S) - \alpha_1(0, S)\, . \tag{9}$$

subtracting $\alpha_1(0, S)$ on both sides we have the map

$$T_S : \mathsf{h}_{\mathbb{R}} \ni z \mapsto \alpha_1(z, S) - \alpha_1(0, S) \in \mathsf{h}_{\mathbb{R}}$$

is linear. Moreover T_S is also bounded since by definition of α we have α_1 is continuous. Now recall that we can write any $S \in Q$ via its spectral decomposition and, using (7) and continuity of α_1, we get

$$\alpha_1(0, S_1) = \alpha_1(0, S_1 + S_2) = \alpha_1(0, S_2) =: \zeta_0 \in \mathsf{h}_{\mathbb{R}}.$$

In particular we have $T_{S_1} = T_{S_2} =: T_0^*$, since equality between $T_{S_1} = T_{S_2}$ holds for any $S_1, S_2 \in Q$. We can then rewrite (8) using T_0 and (9), obtaining

$$\alpha_2(S) + rT_0\,|z\rangle\langle z|\, T_0^* = \alpha_2(S + r\,|z\rangle\langle z|)\, .$$

Let us define the operator C as

$$C(S) := \alpha_2(S) - T_0^* S T_0 ,$$

we then have $C(S) = C(S + r\,|z\rangle\langle z|)$. By a similar argument than before we obtain $C(S_1) = C(S_2) =: C_0$, for every $S_1, S_2 \in Q$. This completes the proof of (5). In order to prove (6) suppose there exists $z \in \mathsf{h}_{\mathbb{R}}$ such that

$$\left\langle z, \left(\mathbf{C_0} - \mathrm{i}\left(\mathbf{J} - \mathbf{T_0^* J T_0}\right)\right) z\right\rangle < 0 .$$

Then we can choose $S \in Q$ such that

$$\langle \mathbf{T_0} z, (\mathbf{S} - \mathrm{i}\mathbf{J})\, \mathbf{T_0} z\rangle < -\left\langle z, \left(\mathbf{C_0} - \mathrm{i}\left(\mathbf{J} - \mathbf{T_0^* J T_0}\right)\right) z\right\rangle .$$

But this is a contradiction since it would imply $\alpha_2(S) \notin Q$.

4 Gaussian Quantum Markov Semigroups

In this section we will extend the definitions and the results obtained in the previous sections to QMSs on $\mathcal{B}(\mathcal{H})$, i.e. a weak* continuous family of weak* continuous, completely positive, identity preserving maps. We start by recalling the usual definition for a gaussian state on $L_1(\mathcal{H})$ (see for example [1, 2, 19, 20]).

Definition 6 A trace class operator ρ on $\mathcal{H}$, positive and such that $\mathrm{tr}\rho = 1$ is called a gaussian state if there exist $\zeta \in \mathsf{h}_{\mathbb{R}}$ and $S \in \mathcal{B}_{\mathbb{R}}(\mathsf{h}_{\mathbb{R}})$, an invertible operator, such that

$$\mathrm{tr}(\rho W(z)) = \exp\left(-\mathrm{i}\,\mathrm{Re}\,\langle \zeta, z\rangle - \frac{1}{2}\,\mathrm{Re}\,\langle z, Sz\rangle\right), \quad \forall z \in \mathsf{h}_{\mathbb{R}} .$$

In this case we write $\rho = \rho_{(\zeta, S)}$.

The set of possible covariance operators in this case is always Q (see [19, 20]). Therefore every gaussian state ρ on $\mathcal{H}$ defines a normal gaussian state $\omega \in \mathcal{S}(CCR\,(\mathsf{h}_{\mathbb{R}}, \sigma))$. Viceversa every gaussian state on $CCR\,(\mathsf{h}_{\mathbb{R}}, \sigma)$ is normal.

Recall now that every QMS $\mathcal{T} = (\mathcal{T}_t)_{t\geq 0}$ uniquely defines a semigroup $\mathcal{T}_* = (\mathcal{T}_{*t})_{t\geq 0*}$ on trace class operators on $\mathcal{H}$, called the predual semigroup, via the pairing

$$\mathrm{tr}(\mathcal{T}_{*t}(\rho)x) = \mathrm{tr}(\rho \mathcal{T}_t(x)) ,$$

for every $t \geq 0$, trace class operator ρ and $x \in \mathcal{B}(\mathcal{H})$. We give now the definition of a gaussian QMS.

Definition 7 Let $\mathcal{T} = (\mathcal{T}_t)_{t\geq 0}$ be a QMS on $\mathcal{B}(\mathcal{H})$. We say it is gaussian if its predual $\mathcal{T}_*$ leaves the set of gaussian states invariant.

By virtue of the results we have showed in the previous sections we can prove the following.

Proposition 3 *Let $\mathcal{T} = (\mathcal{T}_t)_{t\geq 0}$ be a gaussian QMS. Then for every $t \geq 0$ there exist $T_t, C_t \in \mathcal{B}_{\mathbb{R}}(\mathsf{h}_{\mathbb{R}})$ and $\zeta_t \in \mathsf{h}_{\mathbb{R}}$ satisfying*

$$\mathbf{C_t} - \mathrm{i}\left(\mathbf{J} - \mathbf{T_t^*JT_t}\right) \geq 0\,, \tag{10}$$

as operators on $\mathfrak{h}_{\mathbb{C}}$, such that

$$\mathcal{T}_{*t}(\omega_{(\zeta,S)}) = \omega_{(T_t^*\zeta+\zeta_t, T_t^*ST_t+C_t)}\,, \tag{11}$$

for every $\zeta \in \mathsf{h}_{\mathbb{R}}$ and $S \in \mathcal{Q}$. Moreover we have

$$\mathcal{T}_t(W(z)) = \exp\left(-\frac{1}{2}\operatorname{Re}\langle z, C_t z\rangle - \mathrm{i}\operatorname{Re}\langle \zeta_t, z\rangle\right) W(T_t z)\,, \tag{12}$$

for every $z \in \mathsf{h}_{\mathbb{R}}$.

Proof We start by noticing that each $\mathcal{T}_{*t}$ can be extended to all normal functionals on $CCR\,(\mathsf{h}_{\mathbb{R}}, \sigma)$. Indeed, as noted immediately after Definition 6, gaussian states on $\mathcal{H}$ are identified with normal gaussian states on $CCR\,(\mathsf{h}_{\mathbb{R}}, \sigma)$ and each $\mathcal{T}_{*t}$ is norm continuous. Therefore, by Proposition 2, $\mathcal{T}_{*t}$ can be extended to $\mathcal{S}(CCR\,(\mathsf{h}_{\mathbb{R}}, \sigma))$ and from it to the whole set of normal functionals on $CCR\,(\mathsf{h}_{\mathbb{R}}, \sigma)$. Let us denote with $\bar{\mathcal{T}}_{*t}$ this extension and note that it is still norm continuous and preserves the set of gaussian states . Therefore each $\bar{\mathcal{T}}_{*t}$ is a gaussian map on $CCR\,(\mathsf{h}_{\mathbb{R}}, \sigma)_*$. We can now apply Theorem 2 to show existence of $T_t, C_t \in \mathcal{B}_{\mathbb{R}}(\mathsf{h}_{\mathbb{R}})$ and $\zeta_t \in \mathsf{h}_{\mathbb{R}}$ satisfying (11) and such that

$$\bar{\mathcal{T}}_{*t}(\omega_{(\zeta,S)}) = \omega_{(T_t^*\zeta+\zeta_t, T_t^*ST_t+C_t)}\,,$$

for every $t \geq 0$, $\zeta \in \mathsf{h}_{\mathbb{R}}$ and $S \in \mathcal{Q}$. But $\bar{\mathcal{T}}_*$ coincides with $\mathcal{T}_*$ on gaussian states and (11) follows. Eventually one can see that

$$\operatorname{tr}\left(\omega_{(\zeta,S)}\mathcal{T}_t(W(z))\right) = \operatorname{tr}\left(\omega_{(\zeta,S)}\exp\left(-\frac{1}{2}\operatorname{Re}\langle z, C_t z\rangle - \mathrm{i}\operatorname{Re}\langle \zeta_t, z\rangle\right) W(T_t z)\right),$$

for every $z, \zeta \in \mathsf{h}_{\mathbb{R}}$ and $S \in \mathcal{Q}$. By Proposition 2 the equality holds for every normal state and thus for every normal functional. Therefore equation (12) holds.

We have therefore restricted the possible transformations that a gaussian QMS can induce on the parameters of a gaussian state to the one of the kind (11). Although this is not the final result of this paper it is an essential step for its proof. It provides a link between the qualitative definition of a gaussian semigroup introduced in this article and the usual quantitative one that uses the GKLS generator or the explicit action on Weyl operators. Proposition 3 also yields an even more explicit expression for the action of the QMS on Weyl operators. By assuming some regularity conditions and exploiting the fact that $(\mathcal{T}_t)_{t\geq 0}$ is a semigroup we can further specify the action (12) and obtain what is usually seen in the literature (see [1, 2, 21]).

Theorem 3 *Let $\mathcal{T} = (\mathcal{T}_t)_{t\geq 0}$ be a gaussian QMS and let $Z_t, C_t \in \mathcal{B}_{\mathbb{R}}(\mathsf{h}_{\mathbb{R}})$ and $\zeta_t \in \mathsf{h}_{\mathbb{R}}$ be as given by Proposition 3. Suppose $t \mapsto T_t$ is continuous and $t \mapsto C_t$, $t \mapsto \zeta_t$ are differentiable. Then there exists $Z \in \mathcal{B}_{\mathbb{R}}(\mathsf{h}_{\mathbb{R}})$ such that*

$$T_t = \mathrm{e}^{tZ} , \quad \zeta_t = \int_0^t \mathrm{e}^{sZ^*}\zeta \mathrm{d}s , \quad C_t = \int_0^t \mathrm{e}^{sZ^*}C\mathrm{e}^{sZ}\mathrm{d}s , \tag{13}$$

where

$$\zeta := \left.\frac{\mathrm{d}}{\mathrm{d}t}\zeta_t\right|_{t=0} , \quad C := \left.\frac{\mathrm{d}}{\mathrm{d}t}C_t\right|_{t=0} .$$

Moreover, in this case,

$$\mathbf{C} + \mathrm{i}\left(\mathbf{Z}^*\mathbf{J} + \mathbf{J}\mathbf{Z}\right) \geq 0 , \tag{14}$$

as operator on $\mathfrak{h}_{\mathbb{C}}$, and, for every $z \in \mathsf{h}_{\mathbb{R}}$, we can write $\mathcal{T}_t(W(z)) = f_t(z)W(\mathrm{e}^{tZ}z)$ where

$$f_t(z) = \exp\left(-\frac{1}{2}\int_0^t \mathrm{Re}\left\langle z, \mathrm{e}^{sZ^*}C\mathrm{e}^{sZ}z\right\rangle \mathrm{d}s - \mathrm{i}\int_0^t \mathrm{Re}\left\langle \mathrm{e}^{sZ^*}\zeta, z\right\rangle \mathrm{d}s\right) . \tag{15}$$

Proof Proposition 3 allows us to write $\mathcal{T}_t(W(z)) = f_t(z)W(T_t z)$ where

$$f_t(z) = \exp\left(-\frac{1}{2}\mathrm{Re}\left\langle z, C_t z\right\rangle - \mathrm{i}\,\mathrm{Re}\left(\left\langle \zeta_t, z\right\rangle\right)\right).$$

Differentiability of $t \mapsto C_t$ and $t \mapsto \zeta_t$ implies differentiability of $f_t(z)$ for every $z \in \mathsf{h}_{\mathbb{R}}$. Since $\mathcal{T}$ is a semigroup it must hold

$$W(T_{t_1+t_2}z)f_{t_1+t_2}(z) = f_{t_2}(z)f_{t_1}(T_{t_2}z)W(T_{t_1}T_{t_2}z) , \tag{16}$$

for every $t_1, t_2 \geq 0$ and $z \in \mathsf{h}_{\mathbb{R}}$. This is equivalent to the requirements

$$T_{t_1+t_2}z = T_{t_1}T_{t_2}z, \quad f_{t_1+t_2}(z) = f_{t_2}(z)f_{t_1}(T_{t_2}z) , \tag{17}$$

for every $z \in \mathsf{h}_{\mathbb{R}}, t_1, t_2 \geq 0$. We also have $T_0 = \mathbb{1}$, therefore $(T_t)_{t\geq 0}$ is a uniformly continuous semigroup and $T_t = \exp(tZ)$ where $Z \in \mathcal{B}_{\mathbb{R}}(\mathsf{h}_{\mathbb{R}})$ is its generator. Using the second equation of (17) we can find the derivative of $f_t(z)$

$$\frac{\mathrm{d}}{\mathrm{d}t}f_t(z) = f_t(z)\left.\frac{\mathrm{d}}{\mathrm{d}s}f_s\left(\mathrm{e}^{tZ}z\right)\right|_{s=0} . \tag{18}$$

We can explicitly evaluate the derivative at $s = 0$ obtaining

$$\frac{\mathrm{d}}{\mathrm{d}s} f_s\left(\mathrm{e}^{tZ}z\right)\Bigg|_{s=0} = -\mathrm{i}\,\mathrm{Re}\left\langle \zeta, \mathrm{e}^{tZ}z\right\rangle - \frac{1}{2}\,\mathrm{Re}\left\langle \mathrm{e}^{tZ}z, C\mathrm{e}^{tZ}z\right\rangle,$$

since $f_0(\exp(tZ)\,z) = 1$. Solving the differential equation (18) for $f_t(z)$ with initial condition $f_0(z) = 1$ yelds

$$f_t(z) = \exp\left(-\mathrm{i}\,\mathrm{Re}\left\langle \int_0^t \mathrm{e}^{sZ^*}\zeta \mathrm{d}s, z\right\rangle - \frac{1}{2}\,\mathrm{Re}\left\langle z, \int_0^t \mathrm{e}^{sZ^*}C\mathrm{e}^{sZ}z\mathrm{d}s\right\rangle\right). \tag{19}$$

This proves (13) and (15). In order to show (14) we use the condition (10) for the operators. The quantity involved in (10) with the new expressions for C_t, T_t becomes

$$\int_0^t \mathrm{e}^{s\mathbf{Z}^*}\mathbf{C}\mathrm{e}^{s\mathbf{Z}}\mathrm{d}s + \mathrm{i}\left(\mathrm{e}^{t\mathbf{Z}^*}\mathbf{J}\mathrm{e}^{t\mathbf{Z}} - \mathbf{J}\right) = \int_0^t \mathrm{e}^{s\mathbf{Z}^*}\left(\mathbf{C} + \mathrm{i}\left(\mathbf{Z}^*\mathbf{J} + \mathbf{J}\mathbf{Z}\right)\right)\mathrm{e}^{s\mathbf{Z}}\mathrm{d}s\,.$$

Its positivity for every $t \geq 0$ is equivalent to (14).

This result is key to the study of many related problems since it gives an explicit action for the semigroup on Weyl operators and differentiability assumptions are very natural. These problems include invariant states for the semigroups, which turn out to be gaussian states (see [1]), and the study of the decoherence free subalgebra (see [2, 8]). Here however we focus on the characterization result by showing the converse implication to the one of Theorem 3. This means showing that, whenever we fix $Z, C \in \mathcal{B}_{\mathbb{R}}(\mathsf{h}_{\mathbb{R}})$ and $\zeta \in \mathsf{h}_{\mathbb{R}}$ satisfying (14), we can define a QMS via the explicit formula (15) which is gaussian. In [9, 21] it is shown that such a semigroup exists proving at first the maps $\mathcal{T}_t$, defined on the CCR algebra, are completely positive and then extending them to $\mathcal{B}(\mathcal{H})$. Here we follow a different approach using a generalized GKLS generator for the QMS and showing that it generates a QMS that satisfies (15). Recall at first that we can define creation and annihilation operators $a_j^\dagger, a_j$ on $\mathcal{H}$ via

$$\begin{aligned} a_j^\dagger e(n_1,\ldots,n_d) &= \sqrt{n_j+1}\,e(n_1,\ldots n_j+1,\ldots,n_d)\,,\\ a_j e(n_1,\ldots,n_d) &= \sqrt{n_j}\,e(n_1,\ldots n_j-1,\ldots,n_d)\,. \end{aligned}$$

Now let $\Omega, \kappa \in \mathcal{B}(\mathbb{C}^d)$, $\zeta \in \mathbb{C}^d$ and $U, V \in M_{m\times d}(\mathbb{C})$, with $m \leq 2d$, $\Omega = \Omega^*$ and $\kappa = \kappa^T$. Let us define

$$H = \sum_{j,k=1}^d \left(\Omega_{jk}a_j^\dagger a_k + \frac{\kappa_{jk}}{2}a_j^\dagger a_k^\dagger + \frac{\overline{\kappa_{jk}}}{2}a_j a_k\right) + \sum_{j=1}^d \left(\frac{\zeta}{2}a_j^\dagger + \frac{\overline{\zeta}}{2}a_j\right), \tag{20}$$

$$L_\ell = \sum_{j=1}^d \left(u_{\ell j}a_j^\dagger + \overline{v_{\ell j}}a_j\right), \quad \ell = 1,\ldots,m\,. \tag{21}$$

Following the approach of [11] we can construct a QMS via the generalized GKLS generator

$$\mathcal{L}(x) = \mathrm{i}\,[H, x] - \frac{1}{2}\sum_{\ell=1}^{m}\left(L_\ell^* L_\ell x - 2L_\ell^* x L_\ell + x L_\ell^* L_\ell\right). \tag{22}$$

Namely we have the following, whose proof can be found in [1] for $d = 1$ or in [2] for the generic case.

Proposition 4 *Let $\mathcal{L}$ be the unbounded operator defined as in* (22) *with H, L_ℓ given by* (20), (21). *Then there exists a QMS $\mathcal{T} = (\mathcal{T}_t)_{t\geq 0}$ such that $\mathcal{T}_t$ satisfies* (15) *for every $t \geq 0$ with*

$$Zz = \left[\left(\overline{U^*U - V^*V}\right)/2 + \mathrm{i}\Omega\right]z + \left[\left(U^T V - V^T U\right)/2 + \mathrm{i}\kappa\right]\overline{z}\,, \tag{23}$$

$$Cz = \left(\overline{U^*U + V^*V}\right)z + \left(U^T V + V^T U\right)\overline{z}\,, \tag{24}$$

for every $z \in \mathsf{h}_{\mathbb{R}}$. Moreover

$$\mathbf{C} + \mathrm{i}(\mathbf{Z}^*\mathbf{J} + \mathbf{J}\mathbf{Z}) \geq 0 \tag{25}$$

on $\mathfrak{h}_{\mathbb{C}}$ and $\mathcal{T}$ is a Gaussian QMS.

Inversion of Theorem 3 is now just a matter of showing that we can obtain any pair of operators Z, C satisfying (25) via a suitable choice of U, V, Ω, κ.

Theorem 4 *Let $Z, C \in \mathcal{B}_{\mathbb{R}}(\mathsf{h}_{\mathbb{R}})$ satisfy* (25). *Then there exist $m \leq 2d$, $U, V \in M_{m\times d}(\mathbb{C})$, $\Omega, \kappa \in M_d(\mathbb{C})$ such that Z, C coincide with the operators given by* (23) *and* (24).
In particular, for every $Z, C \in \mathcal{B}_{\mathbb{R}}(\mathsf{h}_{\mathbb{R}})$ and $\zeta \in \mathsf{h}_{\mathbb{R}}$, there exists a gaussian QMS $\mathcal{T} = (\mathcal{T}_t)_{t\geq 0}$ such that $\mathcal{T}_t$ satisfies (15) *for every $t \geq 0$.*

Proof This result will follow for the most part from Proposition 4. The only missing step is to show that there is a suitable choice of U, V, Ω, κ such that Z, C are given by formulae (23), (24). Every real linear operator, as a linear operator on the Hilbert space $\mathsf{h}_{\mathbb{R}}$, can be uniquely decomposed into the sum of a selfadjoint and anti-selfadjoint part. In particular there exist $A, B, C \in \mathcal{B}_{\mathbb{R}}(\mathsf{h}_{\mathbb{R}})$ such that, for every $z \in \mathsf{h}_{\mathbb{R}}$,

$$Zz = Az + Bz = (A_1 z + A_2\overline{z}) + (B_1 z + B_2\overline{z})\,, \quad C = C_1 z + C_2\overline{z}\,, \tag{26}$$

where $A = A^\sharp$, $B = -B^\sharp$ and therefore $A_1 = A_1^*$, $A_2 = A_2^T$, $B_1 = -B_1^*$, $B_2 = -B_2^T$, while $C_1 = C_1^*$, $C_2 = C_2^T$. Comparing (26) with (23), (24), by the uniqueness of the decomposition we get $\Omega = -\mathrm{i}B_1$, $\kappa = -\mathrm{i}A_2$ and

$$U^*U = \overline{A_1} + \frac{\overline{C_1}}{2}\,, \quad V^*V = \frac{\overline{C_1}}{2} - \overline{A_1}\,, \quad U^T V = \frac{C_2}{2} + B_2\,. \tag{27}$$

To conclude the proof we will now show there is a suitable choice of U, V that satisfies (27). Since Z, C satisfy (25) there is $\mathbf{x_0} \in \mathcal{B}(\mathfrak{h}_\mathbb{C})$ such that

$$\mathbf{C} + \mathrm{i}\left(\mathbf{Z}^*\mathbf{J} + \mathbf{J}\mathbf{Z}\right) = \mathbf{x_0}^*\mathbf{x_0} \geq 0\,. \tag{28}$$

Let $m := \dim \mathbf{x_0}(\mathfrak{h}_\mathbb{C})$, by restricting the codomain of $\mathbf{x_0}$ to its range we can assume $\mathbf{x_0} \in M_{m\times 2d}(\mathbb{C})$ and it still satisfies (28). Via direct computation, using (26), we get

$$\mathbf{x_0}^*\mathbf{x_0} = \mathbf{C} + \mathrm{i}\left(\mathbf{Z}^*\mathbf{J} + \mathbf{J}\mathbf{Z}\right) = \begin{pmatrix} C_1 + 2A_1 & C_2 + 2B_2 \\ \overline{C_2 - 2B_2} & \overline{C_1 - 2A_1} \end{pmatrix}.$$

While using (23) and (24) we get

$$\mathbf{C} + \mathrm{i}\left(\mathbf{Z}^*\mathbf{J} + \mathbf{J}\mathbf{Z}\right) = \begin{pmatrix} \overline{U^*U} & U^T V \\ V^*\overline{U} & V^*V \end{pmatrix} = \left(\overline{U},\, V\right)^* \left(\overline{U},\, V\right),$$

where $(\overline{U}, V) \in M_{m\times 2d}(\mathbb{C})$. Therefore it is sufficient to choose U, V such that $(\overline{U}, V) = \mathbf{x_0}/\sqrt{2}$ in order to satisfy (27) and conclude the proof of the theorem.

Acknowledgements I would like to thank Prof. Franco Fagnola for suggesting the problem and assisting me during the writing of this paper.

References

1. Agredo, J., Fagnola, F., Poletti, D.: Gaussian quantum markov semigroups on a one-mode Fock space: irreducibility and normal invariant states. Open Syst. Inf. Dyn. **28**(1) (2021)
2. Agredo, J., Fagnola, F., Poletti, D.: The decoherence-free subalgebra of Gaussian QMSs. In preparation
3. Alicki, R., Lendi, K.: Quantum Dynamical Semigroups and Applications. Springer, Berlin (2007)
4. Badveli, R., Jagadish, V., Akshaya, S., Srikanth, R., Petruccione, F.: Dynamics of quantum correlations in a Qubit-Oscillator system interacting via a dissipative bath. Open Syst. Inf. Dyn. **27**(1), 2050004 (2020). https://doi.org/10.1142/s1230161220500043
5. Barchielli, A., Vacchini, B.: Quantum Langevin equations for optomechanical systems. New J. Phys. **17**(8), 083004 (2015). https://doi.org/10.1088/1367-2630/17/8/083004
6. Bratteli, O., Robinson, D.W.: Operator Algebras and Quantum Statistical Mechanics I. Springer, Berlin (1979)
7. Bratteli, O., Robinson, D.W.: Operator Algebras and Quantum Statistical Mechanics II. Springer, Berlin (1997)
8. Carbone, R., Sasso, E., Umanitá, U.: Environment induced decoherence for Markovian evolutions. J. Math. Phys. **56**(9), 092704 (2015). https://doi.org/10.1063/1.4931479
9. Demoen, B., Vanheuverzwijn, P., Verbeure, A.: Completely positive maps on the CCR-algebra. Lett. Math. Phys. **2**(2), 161–166 (1977). https://doi.org/10.1007/bf00398582
10. Doob, J.L.: Heuristic approach to the Kolmogorov-Smirnov theorems. Ann. Math. Statist. **20**(3), 393–403 (1949). https://doi.org/10.1214/aoms/1177729991

11. Fagnola, F.: Quantum Markov semigroups and quantum Markov flows. Proyecciones **18**(3), 1–144 (1999)
12. Fannes, M.: Quasi-free states and automorphisms of the CCR-algebra. Commun. Math. Phys. **51**(3), 55–66 (1976). https://doi.org/10.1007/bf01609051
13. Fannes, M., Verbeure, A.: Gauge transformations and normal states of the CCR. J. Math. Phys. **16**(10), 2086–2088 (1975). https://doi.org/10.1063/1.522423
14. Goldys, B., Peszat, S., Zabczyk, J.: Gauss-Markov processes on Hilbert spaces. Trans. Am. Math. Soc. **368**(1), 89–108 (2016). https://doi.org/10.1090/tran/6329
15. Isar, A., Sandulescu, A., Scheid, W.: Density matrix for the damped harmonic oscillator within the Lindblad theory. J. Math. Phys. **34**(9), 3887–3900 (1993). https://doi.org/10.1063/1.530013
16. Manuceau, J., Verbeure, A.: Quasi-free states of the C.C.R.-algebra and Bogoliubov transformations. Commun. Math. Phys. **9**(4), 293–302 (1968). https://doi.org/10.1007/bf01654283
17. Mihaescu, T., Isar, A.: Evolution of quantum steering in a Gaussian noisy channel, Eur. Phys. J. D. **72**(6), 104 (2018). https://doi.org/10.1140/epjd/e2018-90068-0
18. Parthasarathy, K.R.: An introduction to Quantum Stochastic Calculus. Birkhauser, Basel (1992)
19. Parthasarathy, K.R.: What is a Gaussian state? Commun. Stoch. Anal. **4**(2), 143–160 (2010). https://doi.org/10.31390/cosa.4.2.02
20. Tiju, J.C.: Infinite Mode Quantum Gaussian States. Ph.D Thesis (2018)
21. Vanheuverzwijn, P.: Generators for quasi-free completely positive semi-groups. Annales de l'I.H.P. Physique théorique **29**(1), 123–138 (1978)

A Mean-Field Laser Quantum Master Equation

Franco Fagnola and Carlos M. Mora

Abstract We discuss the properties of a master equation on density matrices with nonlinear mean-field Hamiltonian that models a simple laser. We show the existence of a unique regular family $(\rho_t)_{t\geq 0}$ of density matrices which is a stationary solution. In case a relevant parameter C_b is less than 1, we prove that any regular solution converges exponentially fast to the equilibrium. A locally exponential stable limit cycle arises at the regular stationary state as C_b crosses the critical value 1 so that the nonlinear GKLS equation has a Poincaré-Andronov-Hopf bifurcation at $C_b = 1$ of supercritical-like type. Moreover, the limit cycle is locally exponentially stable if another function of relevant parameters is between the first and second laser thresholds appearing in the semiclassical laser theory. Furthermore, applying our main results we find the long-time behaviour of the von Neumann entropy, the photon-number statistics, and the quantum variance of the quadratures.

Keywords Open quantum system · Mean-field quantum master equation · Laser dynamics · Hopf bifurcation · Exponential convergence

1 Introduction

In this note, based on our recent papers [12, 13], we consider a GKLS (Gorini, Kossakowski, Lindblad, Sudharshan) master equation on density matrices with nonlinear mean-field Hamiltonian that models a simple laser and illustrate the occurrence of

---

WWW home page: https://www.mate.polimi.it/qp/

F. Fagnola (✉)
Dipartimento di Matematica, Politecnico di Milano, 20133 Milano, Italy
e-mail: franco.fagnola@polimi.it

C. M. Mora
Departamento de Ingeniería Matemática, Universidad de Concepción, 4089100, 160-C
Concepción, Casilla, Chile
e-mail: cmora@ing-mat.udec.cl

L. Accardi et al. (eds.), *Infinite Dimensional Analysis, Quantum Probability and Applications*, Springer Proceedings in Mathematics & Statistics 390,
https://doi.org/10.1007/978-3-031-06170-7_12

a supercritical Poincaré-Andronov-Hopf bifurcation (Hopf bifurcation, for short) at the level of density matrices.

The laser is composed of many identical two-level atoms interacting with an electromagnetic field with resonance frequency ω (see [5] Sect. 3.7.3). The atoms undergo downward and upward transitions at rates κ_- and κ_+, respectively. The photons leave the radiation field at rate 2κ due to the light output. Under the mean field approximation, as the number of non-interacting two-level atoms goes to ∞, the density matrix of the state of the system is described by the following effective nonlinear GKLS equation

$$\frac{d\rho_t}{dt} = -\mathrm{i}\left[\omega\left(a^\dagger a + \sigma^3/2\right) + g\mathcal{H}(\rho_t), \rho_t\right] + \kappa\left(2a\,\rho_t a^\dagger - a^\dagger a\rho_t - \rho_t a^\dagger a\right) \tag{1}$$
$$+\frac{\kappa_-}{2}\left(2\sigma^-\rho_t\,\sigma^+ - \sigma^+\sigma^-\rho_t - \rho_t\,\sigma^+\sigma^-\right) + \frac{\kappa_+}{2}\left(2\sigma^+\rho_t\,\sigma^- - \sigma^-\sigma^+\rho_t - \rho_t\,\sigma^-\sigma^+\right)$$

with mean-field Hamiltonian $\mathcal{H}$, depending on the density matrix ϱ, given by

$$\mathcal{H}(\varrho) = \mathrm{i}\left(\mathrm{tr}\left(\sigma^-\varrho\right)a^\dagger - \mathrm{tr}\left(\sigma^+\varrho\right)a\right) + \mathrm{i}\left(\mathrm{tr}\left(a^\dagger\varrho\right)\sigma^- - \mathrm{tr}\left(a\,\varrho\right)\sigma^+\right) \tag{2}$$

(see, e.g., [5] Sect. 3.7.3 and also [28]), where the real and non-zero constant g corresponds to strength of the coupling between atoms and the field mode. The unknown ρ_t is a non-negative trace-class operator with unit trace on the tensor product $\Gamma(\mathbb{C}) \otimes \mathbb{C}^2$ of the one-mode Fock space $\Gamma(\mathbb{C})$ and two-level atom space $\mathbb{C}^2$, ω is a real number, $\kappa, \kappa_+, \kappa_- > 0$, σ^+, σ^- are the two-level atom raising and lowering operators and σ^3 the Pauli matrix σ_z are defined as $\sigma^+ = \begin{pmatrix} 0 & 1 \\ 0 & 0 \end{pmatrix}$, $\sigma^- = \begin{pmatrix} 0 & 0 \\ 1 & 0 \end{pmatrix}$, $\sigma^3 = \begin{pmatrix} 1 & 0 \\ 0 & -1 \end{pmatrix}$. The closed operators $a^\dagger, a$ on $\Gamma(\mathbb{C})$ are the usual creation and annihilation operators. They are defined by means of the standard orthonormal basis $(e_n)_{n\geq 0}$ of $\Gamma(\mathbb{C})$ on the domain of vectors $\xi = \sum_{n\geq 0} \xi_n e_n$ such that $\sum_{n\geq 0} |\xi_n|^2 < \infty$. They are determined by their action on the orthonormal basis $(e_n)_{n\geq 0}$ by $a^\dagger e_n = \sqrt{n+1}\, e_{n+1}$ for all $n \in \mathbb{Z}_+$ and $ae_0 = 0$, $ae_n = \sqrt{n}\, e_{n-1}$ for $n \geq 1$. By means of the orthonormal basis $(e_n)_{n\geq 0}$ one can also define the unitary isomorphism of $\Gamma(\mathbb{C})$ with the Hilbert space $\ell^2(\mathbb{N}; \mathbb{C})$ of complex-valued, square summable sequences. Each vector e_n corresponds to the sequence with all zeros but the n-th term which is equal to 1.

The non-linear quantum master equation (1) reproduces the Dicke-Haken-Lax model of the laser, and can be formally obtained from the dissipative Tavis-Cummings model with n_{atom} atoms by taking the limit as $n_{atom} \to \infty$ of the partial trace with respect to $n_{atom} - 1$ atoms of the full density operator (Sect. 3.7.3 of [5] and Sect. V.E of [28]).

We will discuss the long-term behaviour of the solution to (1). However, in order to get an idea of the physical phenomena and our results we begin by summarizing some simple facts. Lasers can have stable or unstable behaviours according to some tuning parameters. Qualitative properties of the laser dynamics have been investigated through the application of linear stability analysis of systems of ordinary differential equations describing expected values of some observables in semiclassical laser models (see, e.g., [2, 14, 15] and the references therein). In the present framework, a regular enough solution to (1) defines scalar means, $A(t) = \operatorname{tr}(\rho_t\, a)$, $S(t) = \operatorname{tr}\left(\rho_t\, \sigma^-\right)$ and $D(t) = \operatorname{tr}\left(\rho_t\, \sigma^3\right)$ satisfying

$$\begin{cases} \dfrac{d}{dt} A(t) = -(\kappa + \mathrm{i}\,\omega)\, A(t) + g\, S(t) \\ \dfrac{d}{dt} S(t) = -(\gamma + \mathrm{i}\,\omega)\, S(t) + g\, A(t)\, D(t) \\ \dfrac{d}{dt} D(t) = -4g\, \Re\left(\overline{A(t)} S(t)\right) - 2\gamma\,(D(t) - d) \end{cases}, \qquad (3)$$

where $d = (\kappa_+ - \kappa_-)\,/\,(\kappa_- + \kappa_+)$ and $\gamma = (\kappa_- + \kappa_+)\,/2$ (see, e.g., [12]). We refer to [2, 4, 15], for another derivation of (3) by taking limit in many body linear quantum master equations.

The Maxwell-Bloch equations (3) describe the dynamics of the field, polarization and population inversion (i.e., $\operatorname{tr}(\rho_t\, a)$, $\operatorname{tr}\left(\rho_t\, \sigma^-\right)$ and $\operatorname{tr}\left(\rho_t\, \sigma^3\right)$, respectively) of certain lasers in semiclassical theory (see, e.g., [14, 27]). The system (3) develops a stable set of periodic solutions from the stable fixed point $(0, 0, d)$ as the cavity cooperative parameter

$$C_b := \frac{g^2\, d}{\kappa\, \gamma} = \frac{2g^2\,(\kappa_+ - \kappa_-)}{\kappa\,(\kappa_- + \kappa_+)^2}$$

crosses 1 (see, e.g., [2, 5, 15, 16]), and so (3) undergoes a supercritical Hopf bifurcation at $C_b = 1$.

Quantum master equations in GKLS form, and their mean-field approximations, however, not only describe mean values but also quantum mechanical properties of both the atoms and the light fields, therefore they capture quantum effects such as coherence, correlation, spontaneous emission and photon-number statistics (see, e.g., [14] and the references therein). This motivated numerical studies of the dynamical properties of the evolution of density operators representing laser states (see, e.g., [3, 30]).

In our recent papers [12, 13], we rigorously established qualitative properties of the dynamics of the solution to (1), at the level of density matrices, as the parameter C_b crosses the critical value 1 giving the first (to the best of our knowledge) mathematical proof of a supercritical Hopf bifurcation in an infinite dimensional nonlinear GKLS equation. We also showed the transition from a global attractor state, where the light is not emitted, to a locally stable set of coherent states producing coherent light.

In addition, we proved (Theorem 2.6 [13], Theorem 6 here) that the difference between ρ_t and a certain periodic function of coherent states converges exponentially fast to 0 as $t \to +\infty$ whenever $\kappa^2 + 5\kappa\gamma > \gamma\,(\kappa - 3\gamma)\, C_b$ and ρ_0 is in a neighborhood of certain coherent states. As a consequence, coherent laser light emission persists over time if C_b is between the first (with value 1) and second laser thresholds. Similar properties for mean values have been established also in the semiclassical laser theory by the Maxwell-Bloch equations (3), but a fully quantum foundation was not yet available.

In the present note we expose these results referring to our papers [12, 13] for proofs. In Sect. 2 we present the existence and uniqueness Theorem 1 for the nonlinear quantum master equation, then we consider the stationary state (4) in which no light is emitted, we prove its uniqueness as a stationary state (Theorem 2) and discuss its stability (Theorem 3). In Sect. 3, for $C_b > 1$, we explicitly find periodic solutions corresponding to states in which light is emitted 7 and show their local stability (Theorem 6). Moreover, we study the evolution of von Neumann entropy of solutions. Section 4 contains some final remarks and some comments on open problems.

2 The Quantum Master Equation

The starting point of our analysis was the existence result proved in [12]. Let C be a reference operator that we fix to control unboundedness of creation and annihilation operators in (1), typically $C = aa^\dagger = N + 1$ where N is the number operator, or some power of $aa^\dagger$. A C-weak solution to (1) is a curve $(\rho_t)_{t\geq 0}$ of positive operators with unit trace, such that $t \mapsto \mathrm{tr}(a\rho_t)$ (whence also $t \mapsto \mathrm{tr}(a^\dagger \rho_t)$) is continuous and

$$\frac{\mathrm{d}}{\mathrm{d}t}\mathrm{tr}(X\rho_t) = \mathrm{tr}\left(X\left(\mathcal{L}_\star\, \rho_t - \mathrm{i}g\, [\,\mathcal{H}(\rho_t), \rho_t\,]\right)\right)$$

for all $t \geq 0$ and X bounded operator on $\Gamma(\mathbb{C}) \otimes \mathbb{C}^2$, where

$$\begin{aligned}\mathcal{L}_\star\, \varrho &= -\mathrm{i}\omega\left[\left(a^\dagger a + \sigma^3/2\right), \varrho\right] + \kappa\left(2\, a\, \varrho a^\dagger - a^\dagger a \varrho - \varrho a^\dagger a\right)\\ &+ \frac{\gamma\,(1-d)}{2}\left(2\sigma^- \varrho\, \sigma^+ - \sigma^+\sigma^- \varrho - \varrho\,\sigma^+\sigma^-\right)\\ &+ \frac{\gamma\,(1+d)}{2}\left(2\sigma^+ \varrho\, \sigma^- - \sigma^-\sigma^+ \varrho - \varrho\,\sigma^-\sigma^+\right)\end{aligned}$$

is the linear part of equation (1).

Theorem 1 *If ϱ is a density matrix with* $\mathrm{tr}(\varrho\, C^k) < +\infty$ *for* $k \geq 1$, *there exists a unique* C^k*-weak solution* $(\varrho_t)_{t\geq 0}$ *to (1) with initial condition* $\varrho_0 = \varrho$.

We refer to [12] Sect. A.7 for the proof. Here we just note that the solution is obtained as an mean $\varrho_t = \mathbb{E}\left[\,|Z_t\rangle\,\langle Z_t|\,\right]$ where $(Z_t)_{t\geq 0}$ solves a nonlinear stochastic Schrödinger equation (see [11]) driven by three independent Brownian motions, one for each Kraus operator in the linear part of $\mathcal{L}_\star$, with initial datum $Z_0 = \xi$ for some random variable ξ such that $\varrho = \mathbb{E}\left[|\xi\rangle\,\langle \xi|\right]$. This also allows one to check immediately complete positivity of maps $t \mapsto X_t$ defined by duality $\mathrm{tr}(\varrho X_t) = \mathrm{tr}(\varrho_t X)$.

Next step is to note that

$$\varrho_\infty := |e_0\rangle\langle e_0| \otimes \left(\frac{d+1}{2}|e_+\rangle\langle e_+| + \frac{1-d}{2}|e_-\rangle\langle e_-|\right) \tag{4}$$

is a C^k-regular ($k \geq 1$) stationary state (or, in other words, a time-independent fixed point) for (1) with $\omega \neq 0$, the physical situation we are interested in. One can prove the following

Theorem 2 *The operator ϱ_∞ defined by (4) is a stationary state for (1). If $\omega \neq 0$, then ϱ_∞ is the unique C-regular density operator which is a stationary state for (1).*

We refer to [13] Sect. 4.3 for the proof. We emphasize that (4) is the unique density operator ϱ such that the constant solution $\rho_t \equiv \varrho$ satisfies (1), with $\omega \neq 0$, and, roughly speaking, the trace of $a^\dagger a\, \varrho$ is well defined. This stationary solution also yields the unique stationary solution of (3) via $A = \text{tr}\,(\varrho_\infty\, a)$, $S = \text{tr}\,(\varrho_\infty\, \sigma^-)$ and $D = \text{tr}\,(\varrho_\infty\, \sigma^3)$.

In the case where $C_b < 1$ and $\omega \neq 0$, the equilibrium solution $(0, 0, d)$ of the the Maxwell-Bloch equations (3) is asymptotically stable (see, e.g. [12] Theorem 8). Hence, $\lim_{t\to+\infty} \text{tr}\,(\rho_t\, a) = \lim_{t\to+\infty} \text{tr}\,(\rho_t\, \sigma^-) = 0$, and $\lim_{t\to+\infty} \text{tr}\,(\rho_t\, \sigma^3) = d$. In the state ϱ_∞ the light is not emitted, and hence we quickly perceive a faint light output when the normalized pump parameter d is below $\kappa\,\gamma/g^2$.

At the level of density matrices we can show ([13] Theorem 2.4) that ρ_t converges in the trace norm exponentially fast to ϱ_∞ as $t \to +\infty$, and hence ϱ_∞ is the global attractor for (1) finding, at the same time, the limiting behaviour of the mean values of N-bounded operators like $N \otimes (\sigma^+ + \sigma^-)$.

Theorem 3 *Let $(\rho_t)_{t\geq 0}$ be the N-weak solution to (1) with $\omega \neq 0$ and initial datum ρ_0 with $\text{tr}(\rho_0 N) < +\infty$. Suppose that $C_b < 1$. Then*

$$\text{tr}\,(|\rho_t - \varrho_\infty|) \leq K_{sys}\,(|g|\,, \text{tr}\,(\rho_0\, N)) \exp\left(-\delta_{sys}\, t\right) \qquad \forall t \geq 0, \tag{5}$$

where ϱ_∞ is given by (4),

$$\delta_{sys} = \begin{cases} \min\{\kappa, \gamma\}/2 & \text{if } d < 0 \\ (1 - C_b)\min\{\kappa, \gamma\}/3 & \text{if } d \geq 0 \end{cases}, \tag{6}$$

and $K_{sys}(\cdot, \cdot)$ is a non-decreasing non-negative function of two variables that depends on the parameters d, κ and γ. Fix $\widetilde{K} > 0$. Then, for all $t \geq 0$ we have

$$\left|\text{tr}\,(\rho_t\, X) - \frac{d+1}{2}\langle e_0 \otimes e_+, X\, e_0 \otimes e_+\rangle - \frac{1-d}{2}\langle e_0 \otimes e_-, X\, e_0 \otimes e_-\rangle\right|$$
$$\leq \widetilde{K}\,\hat{K}_{sys}\,(|g|\,, \text{tr}\,(\rho_0\, N)) \exp\left(-\delta_{sys}\, t\right),$$

for any operator X on $\Gamma(\mathbb{C}) \otimes \mathbb{C}^2$ satisfying

$$\max\left\{\|X\,u\|^2, \left\|X^\star\,u\right\|^2\right\} \leq \widetilde{K}^2\left(\|u\|^2 + \|Nu\|^2\right) \qquad \forall u \in \mathcal{D}(N),$$

where $\hat{K}_{sys}(\cdot, \cdot)$ is a non-decreasing non-negative function depending on d, κ, γ.

See [13] Sect. 4.6 for the proof. Our methods are partially inspired by those of [1] where exponentially fast convergence towards the ground state $|e_0\rangle\langle e_0|$ for a field interacting with a zero temperature reservoir is proved. We do not consider the spectral gap in the L^2 space of some invariant state as in [6, 9] which is more useful when the invariant state is faithful.

Theorem 3 shows that, if $C_b < 1$, the radiation field converges very fast to its ground state $|e_0\rangle\langle e_0|$. Furthermore, applying this result one obtains at once the long time behaviour of, for instance, the photon-number statistics, the quantum variance of the quadratures, and the quantum linear entropy, which are relevant physical quantities that are not given by the Maxwell-Bloch equations (3).

Some of these results are summarized in the following

Corollary 1 *Suppose that $C_b < 1$ and $\omega \neq 0$. Let δ_{sys} be defined by (6). Consider the N-weak solution $(\rho_t)_{t\geq 0}$ to (1). Then there exists $K > 0$ such that:*

1. *For all $t \geq 0$ and $n > 0$,*

$$|\mathrm{tr}(\rho_t |e_0\rangle\langle e_0|) - 1| \leq K \exp\left(-\delta_{sys} t\right)$$
$$|\mathrm{tr}(\rho_t |e_n\rangle\langle e_n|)| \leq K \exp\left(-\delta_{sys} t\right).$$

2. *For all $t \geq 0$, if X denotes either the position $\left(a^\dagger + a\right)/\sqrt{2}$ or momentum $\mathrm{i}\left(a^\dagger - a\right)/\sqrt{2}$ operator,*

$$\left|\mathrm{tr}\left(\rho_t X^2\right) - \mathrm{tr}\left(\rho_t X\right)^2 - 1/2\right| \leq K \exp\left(-\delta_{sys} t\right).$$

3. *For all $t \geq 0$,*

$$\left|\left(1 - \mathrm{tr}\left(\rho_t^2\right)\right) - \left(1 - d^2\right)/2\right| \leq K \exp\left(-\delta_{sys} t\right).$$

3 Quantum Hopf Bifurcation

In this section we present our results on bifurcation for the quantum master equation (1). As a first step we consider the free interaction solutions to (1) with $\omega \neq 0$, namely, solutions of (1) that also satisfy the equation

$$\frac{\mathrm{d}\rho_t}{\mathrm{d}t} = -\mathrm{i}\,\omega\left[a^\dagger a + \sigma^3/2, \rho_t\right],$$

which describes the evolution of the physical system in absence of interactions between the laser mode, atoms and the bath. The explicit solution of the above equation is immediately found.

Definition 1 Assume that $(\rho_t)_{t\geq 0}$ is a N-weak solution to (1). We call $(\rho_t)_{t\geq 0}$ free interaction solution to (1) if and only if

$$\rho_t = \exp\left(-\mathrm{i}\omega\left(N + \sigma^3/2\right)t\right)\varrho_0 \exp\left(\mathrm{i}\omega\left(N + \sigma^3/2\right)t\right) \qquad \forall t \geq 0.$$

If $C_b \leq 1$, then ϱ_∞ is the unique N-regular free interaction solution to (1) by Theorem 3 and it is constant in time. But when the cavity cooperative parameter C_b is greater than 1 and $\omega \neq 0$, we find a family of non-constant N-regular free interaction solutions of the form $t \mapsto \varrho_c\left(\omega t - \theta\right)$ for any $\theta \in [0, 2\pi[$, given by

$$\varrho_c\left(\vartheta\right) = \left|\mathcal{E}(\zeta)\right\rangle\left\langle\mathcal{E}(\zeta)\right| \otimes \begin{pmatrix} \frac{1}{2}\left(1 + \frac{d}{C_b}\right) & \frac{\zeta\kappa}{g} \\ \frac{\bar{\zeta}\kappa}{g} & \frac{1}{2}\left(1 - \frac{d}{C_b}\right) \end{pmatrix}, \tag{7}$$

($\vartheta \in \mathbb{R}$) where $\zeta = \mathrm{e}^{-\mathrm{i}\vartheta}\gamma\sqrt{C_b - 1}/\sqrt{2}\,|g|$ and the coherent vector $\mathcal{E}\left(\zeta\right)$ associated with $\zeta \in \mathbb{C}$ is defined by

$$\mathcal{E}\left(\zeta\right) = \exp\left(-\left|\zeta\right|^2/2\right)\sum_{n=0}^{+\infty}\zeta^n e_n/\sqrt{n!}. \tag{8}$$

Thus, the laser emits coherent light when the normalized pump parameter d exceeds the barrier $\kappa\,\gamma/g^2$, and the periodic solutions $t \mapsto \varrho_c\left(\omega t - \theta\right)$ yield periodic solutions of (3). In the dynamical system language, the phase path of all non-constant N-regular free interaction solutions of (1) gives the closed orbit $\{\varrho_c\left(\vartheta\right) : \vartheta \in [0, 2\pi]\}$.

The following result summarizes these properties of (1).

Theorem 4 *Let $\omega \neq 0$, $C_b = dg^2/\left(\gamma\kappa\right)$, and consider ϱ_c is given by (7). If $C_b \leq 1$, then (1) does not have any non-constant N-regular free interaction solution. In case $C_b > 1$, the family $\{t \mapsto \varrho_c\left(\omega t - \theta\right) : \theta \in [0, 2\pi[\}$is composed by all non-constant N-regular free interaction solutions to (1).*

We refer to [13], Sect. 4.4 for the proof.

In particular, as C_b crosses the bifurcation value 1, namely $g^2 d > \kappa\gamma$, tracing the free interaction solution we find the periodic solutions of (3),

$$\operatorname{tr}\left(\rho_t\,\sigma^3\right) = \frac{\gamma\kappa}{g^2}, \quad \operatorname{tr}\left(\rho_t\,a\right) = \frac{z\,\mathrm{e}^{-\mathrm{i}\omega t}\gamma\sqrt{C_b - 1}}{\sqrt{2}\,|g|}, \quad \operatorname{tr}\left(\rho_t\,\sigma^-\right) = \frac{z\,\mathrm{e}^{-\mathrm{i}\omega t}\gamma\kappa\sqrt{C_b - 1}}{\sqrt{2}\,|g|\,g}$$

for any $z \in \mathbb{C}$ with $|z| = 1$.

3.1 Long-Time Behaviour

As a first step we consider the long-time behaviour of solutions to (1) with special initial datum ϱ satisfying $\mathrm{tr}\,(\varrho\, a) = \mathrm{tr}\left(\varrho\, \sigma^-\right) = 0$.

Theorem 5 *Let ϱ be a N-regular density operator in $\Gamma(\mathbb{C}) \otimes \mathbb{C}^2$ such that $\mathrm{tr}\,(\varrho\, a) = \mathrm{tr}\left(\varrho\, \sigma^-\right) = 0$. Suppose that $(\rho_t)_{t\geq 0}$ is the N-weak solution to (1) with initial state ϱ. Then $\mathrm{tr}\,(\rho_t\, a) = \mathrm{tr}\left(\rho_t\, \sigma^-\right) = 0$ for all $t \geq 0$, and*

$$\mathrm{tr}\,(|\rho_t - \varrho_\infty|) \leq 12\, \exp(-\gamma t)\,(1 + |d|) + 4\, \exp(-\kappa t)\, \sqrt{\mathrm{tr}\,(\varrho\, N)}, \tag{9}$$

for all $t \geq 0$, with ϱ_∞ defined by (4).

We refer to [13], Sect. 4.5 for the proof. In particular, if the density ρ_0 with $\mathrm{tr}(\rho_0 N) < +\infty$ also satisfies $\mathrm{tr}\,(\rho_0\, a) = \mathrm{tr}\left(\rho_0\, \sigma^-\right) = 0$, it follows from (3) that $\mathrm{tr}\,(\rho_t\, a) = \mathrm{tr}\left(\rho_t\, \sigma^-\right) = 0$ for all $t \geq 0$, and so $\frac{d}{dt}\left(\mathrm{tr}\left(\rho_t\, \sigma^3\right) - d\right) = -2\gamma\left(\mathrm{tr}\left(\rho_t\, \sigma^3\right) - d\right)$. Therefore, $\mathrm{tr}\left(\rho_t\, \sigma^3\right)$ converges exponentially fast to d as $t \to +\infty$. Theorem 5 provides a fully quantum explanation for this long time behaviour.

When condition $\mathrm{tr}\,(\varrho\, a) = \mathrm{tr}\left(\varrho\, \sigma^-\right) = 0$ is not satisfied for all times t by density matrices of the families $\{t \mapsto \varrho_c\,(\omega t - \theta) : \theta \in [0, 2\pi[\}$, we also showed in [13] that these families are locally exponential stable under some conditions on κ, γ and $C_b > 1$. More precisely

Theorem 6 *Suppose $C_b > 1$, $\omega \neq 0$ and assume that one of the following two conditions holds:*

1. $\kappa \leq 3\gamma$,
2. $\kappa > 3\gamma$ *and* $\kappa^2 + 5\kappa\gamma > \gamma\,(\kappa - 3\gamma)\, C_b$.

Then, there exist constants $\epsilon, \lambda > 0$ such that for any N-weak solution $(\rho_t)_{t\geq 0}$ to (1) with $\omega \neq 0$ we have

$$\mathrm{tr}\,(|\rho_t - \varrho_c\,(\omega t - \theta_\infty)|) \leq K\,(\mathrm{tr}\,(\rho_0\, N))\, \exp(-\lambda t) \qquad \forall t \geq 0 \tag{10}$$

provided that the initial datum ρ_0 with $\mathrm{tr}(\rho_0 N) < +\infty$ also satisfies

$$\left|\,|\mathrm{tr}\,(\rho_0\, a)| - \frac{\gamma\sqrt{C_b - 1}}{\sqrt{2}\,|g|}\right| < \epsilon, \quad \left|\mathrm{tr}\left(\rho_0\, \sigma^-\right) - \frac{\kappa}{g}\frac{\gamma\sqrt{C_b - 1}}{\sqrt{2}\,|g|}\frac{\mathrm{tr}\,(\rho_0\, a)}{|\mathrm{tr}\,(\rho_0\, a)|}\right| < \epsilon,$$
$$\left|\mathrm{tr}\left(\rho_0\, \sigma^3\right) - \frac{d}{C_b}\right| < \epsilon. \tag{11}$$

Here, ϱ_c is given by (7), and $\theta_\infty \in [0, 2\pi[$ is the argument of the unit complex number

$$\frac{\mathrm{tr}\,(\rho_0\, a)}{|\mathrm{tr}\,(\rho_0\, a)|}\exp\left(ig\int_0^{+\infty}\Im\left(\frac{S\,(s)}{A\,(s)}\right)ds\right),$$

where $(A(t), S(t), D(t))$ *is the solution of* (3) *with* $\omega = 0$, $A(0) = \operatorname{tr}(\rho_0 a)$, $S(0) = \operatorname{tr}(\rho_0 \sigma^-)$ *and* $D(0) = \operatorname{tr}(\rho_0 \sigma^3)$. *Moreover, under the condition (11) we have*

$$|\operatorname{tr}(\rho_t X) - \operatorname{tr}(\varrho_c(\omega t - \theta_\infty) X)| \leq \widetilde{K} \cdot K(\operatorname{tr}(\rho_0 N)) \exp(-\lambda t) \qquad \forall t \geq 0 \quad (12)$$

for some $\widetilde{K} > 0$, *for any linear operator* X *on* $\Gamma(\mathbb{C}) \otimes \mathbb{C}^2$ *satisfying*

$$\max\left\{\|X u\|^2, \left\|X^\star u\right\|^2\right\} \leq \widetilde{K}^2\left(\|u\|^2 + \|Nu\|^2\right) \qquad \forall u \in \mathcal{D}(N),$$

and the non-decreasing non-negative function $K(\cdot)$ *does not depend on* X.

We refer to [13], Sect. 4.6 for the proof.

Under the hypotheses of Theorem 6, the laser operates stably because the families $\{t \mapsto \varrho_c(\omega t - \theta) : \theta \in [0, 2\pi[\}$are attractive limit cycles in the phase space of density matrices if $C_b > 1$ is close to 1. Hence, the nonlinear quantum master equation (1) has a Hopf bifurcation at $C_b = 1$ of supercritical-like type. As far as we know, this is the first situation in which the Hopf bifurcation is rigorously established at the level of (infinite dimensional) density matrices in the study of nonlinear evolutions of open quantum systems. The bad-cavity condition $\kappa > 3\gamma$, which means that the relaxation time of the atoms is greater than three times the relaxation time of the field, takes place in the so-called lasers of type C (see, e.g., [27]). In this case, it turns out that the laser beam is stable when the normalized pump parameter is in the interval

$$\frac{\kappa \gamma}{g^2} < d < \frac{\kappa^3 + 5\kappa^2\gamma}{g^2(\kappa - 3\gamma)}.$$

If C_b is bigger than the second threshold $\left(\kappa^2 + 5\kappa\gamma\right) / (\gamma(\kappa - 3\gamma))$, then the set of explicit periodic solutions of (3) loses its stability.

Stability of linear quantum Markov semigroup has been studied (see e.g. [19, 20] and the references therein) but, as far as we know, our results are the first in the nonlinear case.

As a byproduct, applying Theorem 6, we deduce that if the initial density operator is in a small enough neighborhood of the orbit $\{\varrho_c(\vartheta) : \vartheta \in [0, 2\pi]\}$, then, for instance, the probability distribution of finding n photons are close to those of a Poissonian statistics, the product of the standard deviations of the position and momentum operators converges to the lower bound of the Heisenberg's uncertainty principle, and the quantum linear entropy converges exponentially fast to $1/2 + d^2/\left(2C_b^2\right) - d^2/C_b$.

We summarize these results, that extend at the level of density matrices local stability of the periodic solutions of the Maxwell-Bloch equations (3) (see, e.g., [26]), in the following

Corollary 2 *Let* $C_b > 1$ *and* $\omega \neq 0$. *Suppose that* $\kappa \leq 3\gamma$ *or that* $\kappa > 3\gamma$ *and* $\kappa^2 + 5\kappa\gamma > \gamma(\kappa - 3\gamma) C_b$. *Then, there exist constants* $\epsilon, \lambda > 0$ *such that the fulfillment of* (11) *implies that:*

(a) There exists $K > 0$ such that all $t \geq 0$ and $n \in \mathbb{N}$,

$$\left| \operatorname{tr}\left(\rho_t \left|e_n\right\rangle \left\langle e_n\right|\right) - \mathrm{e}^{-\frac{\gamma^2 (C_b - 1)}{2g^2}} \left(\frac{\gamma^2 (C_b - 1)}{2g^2} \right)^n \Big/ n! \right| \leq K \exp\left(-\lambda t\right).$$

(b) For all $t \geq 0$, if X denotes either the position $\left(a^\dagger + a\right)/\sqrt{2}$ or momentum $\mathrm{i}\left(a^\dagger - a\right)/\sqrt{2}$ operator,

$$\left| \operatorname{tr}\left(\rho_t X^2\right) - \operatorname{tr}\left(\rho_t X\right)^2 - 1/2 \right| \leq K \exp\left(-\lambda t\right).$$

(c) For all $t \geq 0$,

$$\left| \left(1 - \operatorname{tr}\left(\rho_t^2\right)\right) - \left(\frac{1}{2} + \frac{1}{2} \frac{d^2}{C_b^2} - \frac{d^2}{C_b} \right) \right| \leq K \exp\left(-\lambda t\right).$$

From the proof of Theorems 3 and 6 we have $\sup_{t \geq 0} \operatorname{tr}\left(\rho_t N\right) < +\infty$, and so combining Lemma 18 of [29] with Theorems 3 and 6 we obtain the long-time limit of the von Neumann entropy.

Corollary 3 *Consider the N-weak solution $\left(\rho_t\right)_{t \geq 0}$ to (1) with $\omega \neq 0$.*

1. *If $C_b < 1$, then*

$$\lim_{t \to +\infty} -\operatorname{tr}\left(\rho_t \log\left(\rho_t\right)\right) = -\frac{1}{2} \log\left(\frac{1}{4} - \frac{d^2}{4} \right) - \frac{d}{2} \log\left(\frac{1 + d}{1 - d} \right).$$

2. *If $C_b > 1$ and $\gamma\left(\kappa - 3\gamma\right) C_b < \kappa^2 + 5\kappa\gamma$, then there exist $\epsilon > 0$ such that $-\operatorname{tr}\left(\rho_t \log\left(\rho_t\right)\right)$ converges to*

$$-\frac{1}{2} \log\left(\frac{1}{4} - \frac{d^2}{4C_b^2}\left(2C_b - 1\right) \right) - \frac{d\sqrt{2C_b - 1}}{2C_b} \log\left(\frac{C_b + d\sqrt{2C_b - 1}}{C_b - d\sqrt{2C_b - 1}} \right)$$

as $t \to +\infty$ in case (11) *holds.*

Remark 1 If we change the phase of the electromagnetic field of the laser modeled by (1), then the evolution of the density operator $\hat{\rho}_t$ describing the laser, under the mean field approximation, is governed by the GKLS equation (1) with the mean-field Hamiltonian $\mathcal{H}$ replaced by

$$\begin{aligned} \mathcal{H}\left(\varrho\right) &= \omega\left(a^\dagger a + \sigma^3/2\right) \\ &\quad + g\left(\mathrm{e}^{\mathrm{i}\phi} \operatorname{tr}\left(\sigma^- \varrho\right) a^\dagger + \mathrm{e}^{-\mathrm{i}\phi} \operatorname{tr}\left(\sigma^+ \varrho\right) a + \mathrm{e}^{\mathrm{i}\phi} \operatorname{tr}\left(a^\dagger \varrho\right) \sigma^- + \mathrm{e}^{-\mathrm{i}\phi} \operatorname{tr}\left(a \varrho\right) \sigma^+\right). \end{aligned}$$

Since $\rho_t = \exp\left(\mathrm{i}\left(\pi/2 - \phi\right) N\right) \hat{\rho}_t \exp\left(-\mathrm{i}\left(\pi/2 - \phi\right) N\right)$ satisfies (1), the long time behaviour of $\hat{\rho}_t$ is characterized by that of ρ_t.

4 Conclusions and Outlook

In this article we presented recent results on nonlinear quantum master equations with exponential convergence to stationary states. This allowed us to deduce the asymptotic behaviour of ρ_t as $t \to +\infty$. In the spirit of perturbation theory, our methods are based on a new variation of constant formula proved in [13] (Theorem 3.2) which is proved by combining the stochastic Schrödinger equations ([11, 22, 23]) with classical arguments from the semigroup theory.

From the dynamical systems viewpoint, (1) is a model problem for understanding the behaviour of the mean-field GKLS master equations, which generate non-linear quantum dynamical semigroups (see, e.g., [18, 21, 24] and also [7, 8, 10] for the linear case). Indeed, (1) plays a role in open quantum systems similar to the one played by the Lorenz equations in finite-dimensional dynamical systems. Deduction of the nonlinear quantum master equation (1) is essentially phenomenological as well as its dilation either through stochastic Schrödinger equations or quantum stochastic differential equations driven by quantum noises. It would be interesting to obtain it from a rigorous mathematical argument in some limit of an Hamiltonian system because on one hand this may lead to a useful formula for physical parameters $\kappa_{\pm}$, κ in terms of other quantities, on the other hand important corrections might arise and reveal new interesting phenomena (see e.g. [17] for quantum transport models). We would like to investigate also dependence on some geometric structure (such as the one of [25], for instance) on the set of two-level atoms.

Finally, it would also be interesting to find further bifurcations when the cavity cooperative parameter grows beyond the second threshold. However, this is a difficult problem also for semiclassical analysis based on systems of ordinary differential equations.

Acknowledgements The authors would like to thank Prof. Farrukh Mukhamedov, Prof. Ahmed Al-Rawashdeh and their colleagues of the Department of Mathematical Sciences, United Arab Emirates University Al Ain for the perfect organization of the 41st International Conference on Quantum Probability and Related Topics. Having visited the beautiful new campus, the Al Ain oasis and enjoyed their warm hospitality in February 2018, F. Fagnola really regrets that the conference took place online.

Financial support by projects VRID-Enlace 218.013.043-1.0 of Universidad de Concepcińn and GNAMPA INdAM 2020 "Processi stocastici quantistici e applicazioni" are gratefully acknowledged.

References

1. Accardi, L., Fagnola, F., Hachicha S.: Generic q-Markov semigroups and speed of convergence of q-algorithms. Infin. Dimens. Anal. Quantum Probab. Relat. Top. **9**, 567–594 (2006). https://doi.org/10.1142/S0219025706002548
2. Alli, G., Sewell, G.L.: New methods and structures in the theory of the multi-mode Dicke laser model. J. Math. Phys. **36**, 5598–5626 (1995). https://doi.org/10.1063/1.531279

3. Armen, M.A., Mabuchi, H.: Low-lying bifurcations in cavity quantum electrodynamics. Phys. Rev. A **73**, 063801 (2006). https://doi.org/10.1103/PhysRevA.73.063801
4. Bagarello, F.: Relations between the Hepp-Lieb and the Alli-Sewell laser models. Ann. Henri Poincaré **3**, 983–1002 (2002). https://doi.org/10.1007/s00023-002-8645-2
5. Breuer, H.P., Petruccione, F.: The Theory of Open Quantum Systems. Oxford University Press, Oxford (2002)
6. Carbone, R., Fagnola, F.: Exponential L^2-convergence of quantum Markov semigroups on $B(h)$. Math. Notes **68**, 452–463 (2000). https://doi.org/10.1007/BF02676724
7. Carbone, R., Sasso, E., Umanità, V.: Environment induced decoherence for Markovian evolutions. J. Math. Phys. **56**(9) 092704 (2015). https://doi.org/10.1063/1.4931479
8. Carbone, R., Sasso, E., Umanità, V.: Structure of generic quantum Markov semigroup Infinite Dimens. Anal. Quantum Probab. Relat. Top. **20**(2), 1750012 (2017). https://doi.org/10.1142/S0219025717500126
9. Cipriani, F., Fagnola, F., Lindsay, J.: Spectral analysis and feller property for quantum Ornstein-Uhlenbeck semigroups. Commun. Math. Phys. **210**, 85–105 (2000). https://doi.org/10.1007/s002200050773
10. Deschamps, J., Fagnola, F., Sasso, E., Umanità, V.: Structure of uniformly continuous quantum Markov semigroups. Rev. Math. Phys. **28**(1), 1650003 (2016). https://doi.org/10.1142/S0129055X16500033
11. Fagnola, F., Mora, C.M.: Stochastic Schrödinger equations and applications to Ehrenfest-type theorems. ALEA, Lat. Am. J. Probab. Math. Stat. **10**, 191–223 (2013). http://alea.impa.br/articles/v10/10-09.pdf
12. Fagnola, F., Mora, C.M.: Basic properties of a mean field laser equation. Open Syst. Inf. Dyn. **26**(3), 1950015 (2019). https://doi.org/10.1142/S123016121950015X
13. Fagnola, F., Mora, C.M.: Supercritical Poincaré-Andronov-Hopf bifurcation in a mean-field quantum laser equation. Ann. Henri Poincaré **22**, 171–217 (2021). https://doi.org/10.1007/s00023-020-00966-6
14. Haken, H.: Light, Vol. Laser Light Dynamics, II. North Holland, Amsterdam (1985)
15. Hepp, K., Lieb, E.H.: On the superradiant phase transition for molecules in a quantized radiation field: the Dicke maser model. Ann. Phys. **76**, 360–404 (1973). https://doi.org/10.1016/0003-4916(73)90039-0
16. Hepp, K., Lieb, E.H.: phase transitions in reservoir-driven open systems with applications to lasers and superconductors. Helv. Phys. Acta **46** 573–602 (1974). https://link.springer.com/chapter/10.1007/978-3-662-06390-3_13
17. Hernández-Cervantes, A., Quezada, R.: Stationary states of weak coupling limit-type Markov generators and quantum transport models. Infin. Dimens. Anal. Quantum Probab. Relat. Top. **23**, 2050003 (2020). https://doi.org/10.1142/S0219025720500034
18. Kolokoltsov, V.N.: Nonlinear Markov Processes and Kinetic Equations. Cambridge Tracts in Mathematics, vol. 182. Cambridge University Press, Cambridge (2010)
19. Kumar, D., Sinha, K.B., Srivastava, S.: Stability of the markov (conservativity) property under perturbations. Infin. Dimens. Anal. Quantum Probab. Relat. Top. **23**, 2050009 (2020). https://doi.org/10.1142/S0219025720500095
20. Li, A.C.Y., Petruccione, F., Koch, J.: Perturbative approach to Markovian open quantum systems. Sci. Rep. **4**, 4887 (2014)
21. Merkli, M., Berman, G.P.: Mean-field evolution of open quantum systems: an exactly solvable model. Proc. R. Soc. A **468**, 3398–3412 (2012). https://doi.org/10.1098/rspa.2012.0327
22. Mora, C.M.: Regularity of solutions to quantum master equations: a stochastic approach. Ann. Probab. **41**, 1978–2012 (2013). https://doi.org/10.1214/11-AOP692
23. Mora, C.M., Rebolledo, R.: Basic properties of non-linear stochastic Schrödinger equations driven by Brownian motions. Ann. Appl. Probab. **18**, 591–619 (2008). https://doi.org/10.1214/105051607000000311
24. Mukhamedov, F., Ganikhodjaev, N.: Quantum Quadratic Operators and Processes. Springer, Berlin (2015)

25. Mukhamedov, F., Souissi, A.: Types of factors generated by quantum Markov states of Ising model with competing interactions on the Cayley tree. Infin. Dimens. Anal. Quantum Probab. Relat. Top. **23**, 2050019 (2020). https://doi.org/10.1142/S0219025720500198
26. Ning, C.Z., Haken, H.: Detuned lasers and the complex Lorenz equations: subcritical and supercritical Hopf bifurcations. Phys. Rev. A **41**, 3826–3837 (1990). https://doi.org/10.1103/PhysRevA.41.3826
27. Ohtsubo, J.: Semiconductor Lasers. Springer Series in Optical Sciences, 3rd edn. Springer, Berlin (2013)
28. Spohn, H.: Kinetic equations from Hamiltonian dynamics: Markovian limits. Rev. Modern Phys. **52**, 569–615 (1980). https://doi.org/10.1103/RevModPhys.52.569
29. Winter, A.: Tight uniform continuity bounds for quantum entropies: conditional entropy, relative entropy distance and energy constraints. Commun. Math. Phys. **347**, 291–313 (2016). https://doi.org/10.1007/s00220-016-2609-8
30. Yusipov, I., Ivanchenko, M.: Quantum Neimark-Sacker bifurcation. Sci. Rep. **9**, 17932 (2019). https://doi.org/10.1038/s41598-019-53526-2

Unique Ergodicity and Weakly Monotone Fock Space

Vitonofrio Crismale

Abstract We show that the C^*-dynamical system given by the so-called weakly monotone C^*-algebra $\mathcal{WM}$, and the shift automorphism on it, is not uniquely ergodic. In fact, the fixed-point subalgebra with respect to such an action is trivial, whereas there are plenty of shift-invariant states. The last assertion is proved using a Hamel basis for the linear structure of a dense $*$-algebra of $\mathcal{WM}$ here exhibited.

Keywords Non-commutative dynamical systems · Unique ergodicity · Weakly monotone Fock spaces

1 Introduction

A dynamical system (Ω, T), where Ω is a compact Hausdorff space and T a homeomorphism on Ω, is said to be uniquely ergodic if there exists a unique T-invariant Borel probability measure on Ω. By the Riesz-Markov Theorem, any measure of this kind is uniquely determined by a positive unit preserving linear functional (usually called state) on $C(\Omega)$, the complex-valued algebra of continuous functions on Ω. In addition, the map $\alpha : f \in C(\Omega) \mapsto f \circ T \in C(\Omega)$ uniquely defines a $*$-automorphism on $C(\Omega)$ (the $*$-map being the complex conjugation). This allows to rephrase the aforementioned unique ergodicity by the existence of a unique α-invariant state on $C(\Omega)$.

Passing to a more general setting given by a non-commutative unital C^*-algebra $\mathcal{A}$, and an abstract $*$-automorphism α on $\mathcal{A}$, one naturally arrives at a non-commutative version of the ergodic property presented above. Furthermore, this condition can be generalised again by requiring the existence of a unique α-invariant conditional expectation from $\mathcal{A}$ to the so-called fixed-point subalgebra. The latter is nothing else than the subalgebra collecting all the α-invariant elements of $\mathcal{A}$. When it just consists of multiples of the unit, this notion reduces to the usual unique ergodicity.

V. Crismale (✉)
Dipartimento di Matematica, Università degli studi di Bari Via E. Orabona 4, 70125 Bari, Italy
e-mail: vitonofrio.crismale@uniba.it

L. Accardi et al. (eds.), *Infinite Dimensional Analysis, Quantum Probability and Applications*, Springer Proceedings in Mathematics & Statistics 390,
https://doi.org/10.1007/978-3-031-06170-7_13

Here, we stress that some ergodic properties with respect to the action of a particular α, namely the shift automorphism, were studied in [6, 9, 11], for some prominent examples coming from Quantum Probability, like q-deformed (where $|q| < 1$), monotone, and Boolean cases. One can also consider the action of the permutation group, or the monoid of strictly increasing maps on $\mathbb{Z}$, obtaining de Finetti or Ryll-Nardzewski Theorems (see [1–5] and references cited therein).

As for other examples coming from Quantum Probability, one can deal with weakly monotone quantisation. Weakly monotone Fock spaces were first introduced by Wysoczanski in [13], where we refer the reader for a general presentation. In the last years the investigation of the vacuum distribution for sums of position operators in such a space has been realised in [7], where it was also shown that the basic operators of annihilation and creation are monotone independent in the vacuum. For a similar study in the monotone case, see [8].

The above considerations naturally lead to the investigation of ergodic states with respect to the action of the shift for the concrete C^*-algebra generated by weakly monotone annihilation operators. This is the main content of these notes. Indeed, after recalling some useful tools from the theory of C^*-dynamical systems based on a group in Sect. 2, in Sect. 3 we present the definitions of weakly monotone Fock space, and the basic operators on it. Furthermore, in analogy to what happens in the monotone case [4], we also exhibit a Hamel basis for the dense unital $*$-algebra generated by annihilators. Without surprise, almost all the elements of the basis are Wick-ordered words, i.e. they are a sequence of creators on the left and annihilators on the right. In Sect. 4 we find a decomposition of the weakly monotone C^*-algebra: it is indeed the sum of an irreducible C^*-subalgebra added to the multiples of the identity. Taking into account that the shift automorphism acts on the C^*-algebra generated by the weakly monotone annihilators in a natural way, the decomposition allows us to find, in Proposition 4, that the fixed-point subalgebra with respect to the shift is trivial. Namely, it contains essentially the identity. Thus, one naturally wonders if the related C^*-dynamical system is uniquely ergodic. This is false: In Proposition 5 we show that the weak$*$-compact subset of shift-invariant states is indeed the convex combination of the vacuum state and the so-called state at infinity.

2 Preliminaries

This section is mainly devoted to recall some definitions on discrete C^*-dynamical systems.

First, for a C^*-algebra $\mathcal{A}$, and a positive functional φ on $\mathcal{A}$, we denote by $(\mathcal{H}_\varphi, \pi_\varphi, \xi_\varphi)$ the GNS representation.

A C^*-dynamical system based on a group G is given by the triplet $(\mathcal{A}, G, \alpha)$, where $\mathcal{A}$ is a C^*-algebra which will be ever supposed unital, with unit $1_{\mathcal{A}}$, and α is a group homomorphism

$$\alpha : g \in G \mapsto \alpha_g \in Aut(\mathcal{A}),$$

$Aut(\mathcal{A})$ being the group of $*$-automorphisms on $\mathcal{A}$. As α_g is multiplicative for any $g \in G$, the closed space

$$\mathcal{A}_G := \{a \in \mathcal{A} \mid \alpha_g(a) = a\,,\ g \in G\}$$

is indeed a $*$-subalgebra, called the fixed-point subalgebra.

A state $\varphi \in \mathcal{S}(\mathcal{A})$ is called G-invariant if $\varphi \circ \alpha_g = \varphi$ for any $g \in G$. Denote by $\mathcal{S}_G(\mathcal{A})$ the set of G-invariant states, which is convex and weak*-compact. Its extremal points are said to be ergodic.

A discrete C^*-dynamical system is given by the pair $(\mathcal{A}, \alpha)$, where α is a single automorphism on $\mathcal{A}$. In this case, if $\mathcal{S}_\alpha(\mathcal{A}) = \{\omega\}$, then the C^*-dynamical system $(\mathcal{A}, \alpha)$ is called *uniquely ergodic*. This condition is equivalent to the following convergence in norm

$$\lim_{n\to\infty} \frac{1}{n}\sum_{k=1}^{n} \alpha^k(a) = \omega(a)1_{\mathcal{A}}\,, \quad a \in \mathcal{A}$$

or to the following weak limit of averages

$$\lim_{n\to\infty} \frac{1}{n}\sum_{k=1}^{n} f(\alpha^k(a)) = \omega(a) f(1_{\mathcal{A}})\,, \quad a \in \mathcal{A}\,,\ f \in \mathcal{A}^*\,.$$

In this case, one naturally finds that the fixed-point subalgebra is trivial, namely $\mathcal{A}_\alpha = \mathbb{C}1_{\mathcal{A}}$. When the structure of $\mathcal{A}_\alpha$ is richer, one naturally wonders if, for any state in $\mathcal{A}_\alpha$, there exists a unique α-invariant extension to $\mathcal{A}$. This condition, which reduces to the usual notion when $\mathcal{A}_\alpha = \mathbb{C}1_{\mathcal{A}}$, by Theorem 2.1 in [10] is seen to be equivalent to the existence of a unique invariant conditional expectation[1] $E_\alpha : \mathcal{A} \to \mathcal{A}_\alpha$ given by

$$E_\alpha(a) = \lim_{n\to\infty} \frac{1}{n}\sum_{k=1}^{n} \alpha^k(a)\,, \quad a \in \mathcal{A}\,.$$

In such a case the C^*-dynamical system $(\mathcal{A}, \alpha)$ is called *uniquely ergodic with respect to its fixed-point subalgebra*.

In these notes we deal with invariant states on a suitable C^*-algebra with respect to the group generated by the one-step shift $\tau(i) := i + 1$ on $\mathbb{Z}$, which is canonically isomorphic to $\mathbb{Z}$ itself.

---

[1] By a theorem of Tomiyama [12], any conditional expectation from a C^*-algebra $\mathcal{A}$ to its C^*-subalgebra $\mathcal{B}$ is a norm 1 projection.

3 Weakly Monotone Fock Space

For any $k \geq 1$, let $I_k := \{(i_1, i_2, \dots, i_k) \mid i_1 \leq i_2 \leq \dots \leq i_k,\ i_j \in \mathbb{Z}\}$. When $k = 0$, we take $I_0 := \{\emptyset\}$, where $\emptyset$ is the empty sequence. If $\mathcal{H}_k := \ell^2(I_k)$, where $\mathcal{H}_0$ is identified with the scalar field $\mathbb{C}$, the weakly monotone Fock space is the Hilbert space $\mathcal{F}_{wm} := \bigoplus_{k=0}^{\infty} \mathcal{H}_k$. The sequence $\Omega := (1, 0, 0, \dots)$ in $\mathcal{F}_{wm}$ is the Fock vacuum vector.

If $\beta = (i_1, i_2, \dots, i_k) \in I_k$, we denote by $e_{(\beta)}$ the generic element of the canonical basis of $\mathcal{F}_{wm}$, namely $e_{(\beta)} := e_{i_1} \otimes e_{i_2} \otimes \dots \otimes e_{i_k}$.

The weakly monotone annihilation and creation operators are respectively firstly defined, for $i \in \mathbb{Z}$ and β as above with $k \neq 0$, by

$$a_i \Omega = 0\,, \quad a_i e_{(\beta)} = \delta_{i,i_1} e_{i_2} \otimes \dots \otimes e_{i_k}\,,$$

where $\delta_{i,j}$ is the Kronecker symbol, and

$$a_i^\dagger \Omega = e_i\,, \quad a_i^\dagger e_{(\beta)} = \alpha_{i,i_1} e_i \otimes e_{i_1} \otimes e_{i_2} \otimes \dots \otimes e_{i_k}\,,$$

where $\alpha_{j,k} = 1$ if $j \leq k$, and 0 otherwise.

They can be extended by linearity and continuity to the whole $\mathcal{F}_{wm}$, are adjoint to one another, and of norm one. Furthermore, they satisfy the following relations

$$a_i^\dagger a_j^\dagger = a_j a_i = 0 \quad if\ i > j\,, \tag{1}$$

$$a_i a_j^\dagger = 0 \quad if\ i \neq j\,. \tag{2}$$

$$a_i a_i^\dagger = I - \sum_{k<i} a_k^\dagger a_k\,, \tag{3}$$

where I is the identity operator on $\mathcal{F}_{wm}$, and (3) is understood in the strong operator topology, see [6], Proposition 3.2.

As it will be useful in the sequel, we report here Lemma 2.1 from [7].

Lemma 1 *For any $k, j \geq 1$, one has*

$$a_k a_j a_j^\dagger = \alpha_{j,k} a_k\,, \quad a_j a_j^\dagger a_k^\dagger = \alpha_{j,k} a_k^\dagger\,. \tag{4}$$

Moreover, for $j \leq k$

$$a_j a_j^\dagger a_k = a_k\,, \quad a_k^\dagger a_j a_j^\dagger = a_k^\dagger\,. \tag{5}$$

Suppose now that $m, n \geq 1$, and $i \in \mathbb{Z}$. Then, (2) and (3) give

$$a_i^m a_i^{\dagger n} = \begin{cases} a_i^{\dagger n-m} & if \ n > m \\ a_i^{m-n} & if \ n < m \\ a_i a_i^\dagger & if \ n = m \end{cases} \tag{6}$$

In addition,

$$a_i^\dagger a_i = a_i a_i^\dagger - a_{i+1} a_{i+1}^\dagger . \tag{7}$$

We denote by $\mathcal{WM}$ and $\mathcal{WM}_o$ the concrete unital C^*-algebra, and with its dense unital $*$-algebra generated by the annihilators $\{a_i \mid i \in \mathbb{Z}\}$ in the weakly monotone Fock space. They are both contained in $\mathcal{B}(\mathcal{F}_{wm})$.

In analogy to the monotone case [4, 6], we say that $X \in \mathcal{WM}_o$ has a $w\lambda$-form if

$$X = a_{i_1}^{\dagger m_1} \cdots a_{i_h}^{\dagger m_h} a_{j_1}^{n_1} \cdots a_{j_k}^{n_k} ,$$

for all integers $h, k \geq 0, m_1, \ldots, m_h, n_1, \ldots, n_k \in \mathbb{N}$, and finally $i_1 < \cdots < i_h, j_1 > \cdots > j_k$. The length of X is $h + k$. When $h = k = 0$, then $X = I$. It is called the empty word, and its length is zero.

Note that every $w\lambda$-form is reduced. Indeed, if $X_1 := a_{i_1}^{\dagger m_1} \cdots a_{i_h}^{\dagger m_h} a_{j_1}^{n_1} \cdots a_{j_k}^{n_k}$ and $X_2 := a_{p_1}^{\dagger l_1} \cdots a_{p_r}^{\dagger l_r} a_{t_1}^{s_1} \cdots a_{t_v}^{s_v}$ are equal, where $h, k, r, v \geq 0$, $i_1 < \cdots < l_h$, $j_1 > \cdots > j_k$, $p_1 < \cdots < p_r$, $t_1 > \cdots > t_v$, $m_1, \ldots, m_h, n_1, \ldots, n_k, l_1, \ldots, l_r, s_1, \ldots, s_v \in \mathbb{N}$, the identity

$$X_1(e_{j_k}^{n_k} \otimes \ldots \otimes e_{j_1}^{n_1}) = X_2(e_{j_k}^{n_k} \otimes \ldots \otimes e_{j_1}^{n_1})$$

gives that $k = v$, $j_u = t_u$, $n_u = s_u$ for each $u = 1, \ldots, k$, and moreover $h = r$, $i_u = p_u$, and finally $m_u = l_u$ for each $u = 1, \ldots, h$.

Define the index-set Λ such that $\{X_\lambda\}_{\lambda \in \Lambda}$ gives the set of all the $w\lambda$-forms. As any word of type $a_i^\dagger a_i$ is identified with $i \in \mathbb{Z}$, in a natural way one has $\mathbb{Z} \subset \Lambda$, and we denote $\Gamma := \Lambda \setminus \mathbb{Z}$.

Since by (3) for each $i \in \mathbb{Z}$, $a_i a_i^\dagger$ is the orthogonal projection onto

$$\mathcal{F}_{wm}^{\leq i} := \overline{\{span\{e_{l_1} \otimes \cdots e_{l_m} \mid i \leq l_1 \, , \; l_1 \leq \cdots \leq l_m, m \geq 1\}} \, ,$$

the word $a_i a_i^\dagger$ is reduced. The $w\lambda$-forms in Γ, as well as these projections, are prominent in the linear structure of $\mathcal{WM}_o$, as the following proposition shows.

Proposition 1 *The family* $Ł := \{X_\lambda\}_{\lambda \in \Gamma} \bigcup \{a_i a_i^\dagger\}_{i \in \mathbb{Z}}$ *is a basis for* $\mathcal{WM}_o$.

Proof First, we show that any non-null operator in $\mathcal{WM}_o$ is a finite linear combination of words in $\{X_\lambda\}_{\lambda \in \Gamma} \bigcup \{a_i a_i^\dagger\}_{i \in \mathbb{Z}}$. To this goal, take $X := a_{i_1}^{\natural h_1} \cdots a_{i_n}^{\natural h_n}$, $n \in \mathbb{N}$, $i_1, \ldots, i_n \in \mathbb{Z}$, $h_1, \ldots, h_n \in \mathbb{N}$, $\natural \in \{1, \dagger\}$.

When the length of X is at most 2, the identities (1), (6) and (7) give that X can be reduced to a sum of elements in Ł.

Denote now $X = X_1 a_{i_n}^{\natural h_n}$, X_1 having length at least 2, and suppose by induction that X_1 belongs to the family above. If $X_1 = a_i a_i^\dagger$, $i \in \mathbb{Z}$, and $\natural = \dagger$, then $X = 0$ when $i > i_n$, or by (4) $X = a_{i_n}^{\dagger h_n}$ if $i \leq i_n$. If $\natural = 1$, then by (1) and (3), $X = a_{i_n}^{h_n} - \sum_{i_n \leq k < i} a_k^\dagger a_k a_{i_n}^{h_n}$ when $i > i_n$, or by (5) $X = a_{i_n}^{h_n}$ when $i \leq i_n$. Suppose now that X_1 is a $w\lambda$-form, namely $X_1 = a_{i_1}^{\dagger m_1} \cdots a_{i_r}^{\dagger m_h} a_{j_1}^{n_1} \cdots a_{j_k}^{n_k}$. If $k > 0$, then X is null when $j_k < i_n$. When $j_k \geq i_n$, (4), (5) and (6) give that X is a $w\lambda$-form in the case it does not vanish. Finally, for $k = 0$, one sees that X is a $w\lambda$-form, except the vanishing case given by $\natural = \dagger$, and $i_n < i_r$.

Thus, we have shown that the word in Ł, which are reduced, generate the whole $\mathcal{WM}_o$. Therefore, to complete the proof, it remains to prove that the set $\{X_\lambda\}_{\lambda \in \Gamma} \bigcup \{a_i a_i^\dagger\}_{i \in \mathbb{Z}}$ is linearly independent in $\mathcal{WM}_o$. This is essentially obtained by rephrasing the arguments presented in the proof of [4], Lemma 3.2. □

4 Shift-Invariant States on Weakly Monotone C^*-Algebra

In this section we present the convex set of shift-invariant states on the $\mathcal{WM}$, and show that there is no unique ergodicity for the dynamical system based on the shift automorphism.

Denote $\mathcal{W}_o := span\{X \in Ł \mid l(X) > 0\}$, and take $\mathcal{W} := \overline{\mathcal{W}_o}$. We first show that $I \notin \mathcal{W}$.

Proposition 2 *The identity $I \in \mathcal{B}(\mathcal{F}_{wm})$ does not belong to $\mathcal{W}$.*

Proof Let X be an element of $\mathcal{W}_o$. Thus, $X = \sum_{j \in F} \beta_j X_j$, where F is a finite set, any word X_j belongs to Ł, and $l(X_j) > 0$. After writing $X_j = Y_j a_{l_j}^\natural$, $\natural \in \{1, \dagger\}$, we take an arbitrary unit vector $e_n \in \mathcal{F}_{wm}$ such that $n < \min l_j$. This gives that, for any $\alpha \in \mathbb{C}$

$$\|X + \alpha I\| \geq \|(X + \alpha I)e_n\| = |\alpha|\,.$$

The thesis then follows by a standard approximation argument. □

Proposition 3 *The C^*-algebra $\mathcal{W}$ is irreducible.*

Proof First note that $\sum_{k \in \mathbb{Z}} a_k^\dagger a_k$ vanishes when applied to Ω, and leaves invariant any vector $\xi := e_{(\beta)}$, $\beta \in I_k$ for some $k > 0$. Since the annihilators $(a_k)_{k \in \mathbb{Z}}$ have orthogonal ranges, it follows that $\sum_{k \in \mathbb{Z}} a_k^\dagger a_k$ converges to $I - P_\Omega$ in the strong topology, P_Ω being the projection onto the space generated by the vacuum vector Ω. Thus, if $\omega := \langle \cdot \Omega, \Omega \rangle$, P_Ω belongs to $\mathcal{W}'' = \pi_\omega(\mathcal{W})''$, and then $B\Omega \in \mathbb{C}\Omega$ for any $B \in \mathcal{W}'$. Since Ω is separating for $\mathcal{W}'$, the thesis follows. □

As a consequence of the previous proposition, one has the decomposition

$$\mathcal{WM} = \mathcal{W} \oplus \mathbb{C}I\,. \tag{8}$$

In what follows we investigate ergodic properties of a discrete dynamical system. To this goal, consider the unitary U_τ on $\mathcal{F}_{wm}$ given by

$$U_\tau e_{(\beta)} := e_{i_1+1} \otimes e_{i_2+1} \cdots \otimes e_{i_n+1}\,, \quad U_\tau \Omega = \Omega$$

where $\beta = (i_1, \ldots, i_n) \in I_n$ for some $n > 0$. The map α on $\mathcal{WM}$ unitarily implemented by U_τ, namely the automorphism α defined by $\alpha(X) := U_\tau X U_\tau^*$, $X \in \mathcal{WM}$, gives rise to the discrete C^*-dynamical system $(\mathcal{WM}, \alpha)$, and from now on we refer to α as the shift automorphism.

Proposition 4 *The fixed-point subalgebra of $\mathcal{WM}$ for the action of the shift automorphism is trivial, namely $\mathcal{WM}_\alpha = \mathbb{C}I$.*

Proof As $\alpha(\mathcal{W}) \subseteq \mathcal{W}$, we fix $X \in \mathcal{W}$ and show that $\alpha(X) \neq X$. Indeed, for any arbitrary non-null $X_o \in \mathcal{W}_o$, there exists a non decreasing sequence $\beta = (i_1, i_2, \ldots, i_j) \in I_j$ for some $j > 0$, for which $X_o e_{(\beta)} \neq 0$. As $\alpha^k(X_o)$ vanishes on $e_{(\beta)}$ for sufficiently large $k \in \mathbb{N}$, it follows that $\alpha(X_o) \neq X_o$.

Take now a sequence $(X_n)_{n\geq 1}$ in $\mathcal{W}_o$ converging to X in the uniform topology. The argument above shows that, for any n, there exists $\varepsilon_n > 0$ such that $\|\alpha(X_n) - X_n\| = \varepsilon_n$. Now denote $\varepsilon := \inf_n \varepsilon_n$, and suppose that $\alpha(X) = X$. As α is contractive, for m large enough the inequalities $\|\alpha(X_m - X)\| < \frac{\varepsilon}{2}$, and $\|X_m - X\| < \frac{\varepsilon}{2}$ hold. Hence one finds the following contradiction

$$\varepsilon \leq \|\alpha(X_m) - X_m\| \leq \|\alpha(X_m - X)\| + \|\alpha(X) - X\| + \|X - X_m\| < \frac{\varepsilon}{2} + \frac{\varepsilon}{2} = \varepsilon$$

□

By Proposition 3, the C^*-algebra $\mathcal{W}$ does not have an identity, and by (8) it follows that the weak*-compact set of all states $\mathcal{S}(\mathcal{WM})$ on the weakly monotone C^*-algebra is the one-point compactification of positive functionals on $\mathcal{W}$ with norm not greater than one. The shift-invariant convex subset of $\mathcal{S}(\mathcal{WM})$ contains at least two states. One is the vacuum state ω, which is as usual given by

$$\omega(X) := \langle X\Omega, \Omega\rangle\,, \quad X \in \mathcal{WM}\,.$$

The other one is the so-called "state at infinity" ω_∞, namely

$$\omega_\infty(A + cI) := c \quad A \in \mathcal{W}\,,\ c \in \mathbb{C}\,.$$

Gluing these considerations with Proposition 4, one figures out that $(\mathcal{WM}, \alpha)$ is not uniquely ergodic. The next result shows that two the states above are exactly the only ergodic shift-invariant.

Proposition 5 *The weak*-compact convex of shift-invariant states $\mathcal{S}_{\mathbb{Z}}(\mathcal{WM})$ of $\mathcal{WM}$ is*

$$\mathcal{S}_{\mathbb{Z}}(\mathcal{WM}) = \{\gamma\omega + (1-\gamma)\omega_\infty \mid \gamma \in [0,1]\}.$$

Proof We first take an arbitrary element X in the dense subalgebra $\mathcal{W}_o \oplus \mathbb{C}I$. By Proposition 1, one has

$$X = cI + Y + \sum_{j\in F} \beta_j a_j a_j^\dagger .$$

where $c \in \mathbb{C}$, Y is a finite linear combination of $w\lambda$-forms in Ł with non-null length, F is a finite set, and finally β_j is a complex scalar. One easily finds that

$$\omega(X) = c + \sum_{j\in F} \beta_j , \quad \omega_\infty(X) = c . \tag{9}$$

We further observe that for any family of integers $i_1, \dots, i_n$, and vectors $(\eta_j)_{j=1}^n$ in $\mathcal{H}_k$, by (2) it follows that

$$\Big\| \sum_{j=1}^n a_{i_j}^\dagger \eta_j \Big\|^2 = \sum_{j,l=1}^n \big\langle a_{i_l} a_{i_j}^\dagger \eta_j, \eta_l \big\rangle = \sum_{j=1}^n \big\| a_{i_j}^\dagger \eta_j \big\|^2 \leq n \max_{j=1,\dots,n} \|\eta_j\|^2 , \tag{10}$$

as the creators have unit norm.

Fix now the operator $a_{i_1}^\dagger a_{i_2}^\natural \cdots a_{i_m}^\natural$, where $i_1, \dots, i_m \in \mathbb{Z}$, $m \in \mathbb{N}$ and $\natural \in \{1, \dagger\}$. If l is the number of annihilators in $a_{i_1}^\dagger a_{i_2}^\natural \cdots a_{i_m}^\natural$, choose a vector $\eta \in \mathcal{H}_k$, where $\|\eta\| \leq 1$ and $k \geq l$. As $\langle e_{i+j}, e_{i+l}\rangle = \delta_{j,l}$, one has

$$\Big\| \sum_{h=1}^n \alpha^{k_h}\big(a_{i_1}^\dagger a_{i_2}^\natural \cdots a_{i_m}^\natural\big)\eta \Big\| = \Big\| \sum_{h=1}^n a_{i_1+k_h}^\dagger a_{i_2+k_h}^\natural \cdots a_{i_m+k_h}^\natural \eta \Big\| = \Big\| \sum_{h=1}^n a_{i_1+k_h}^\dagger \eta_h \Big\| ,$$

where $\eta_h := a_{i_2+k_h}^\natural \cdots a_{i_m+k_h}^\natural \eta$. As $\|\eta_h\| \leq 1$ for any h, by (10) the last term above is less than or equal to $\sqrt{n}$. Likewise, by taking the adjoint, one achieves that $\| \sum_{h=1}^n \alpha^{k_h}\big(a_{i_1}^\natural \cdots a_{i_{m-1}}^\natural a_{i_m}\big)\eta \| \leq \sqrt{n}$. Consequently,

$$\lim_{n\to\infty} \frac{1}{n} \sum_{k=0}^{n-1} \alpha^k(Y) = 0$$

in norm. If φ is a shift-invariant state on $\mathcal{WM}$, then

$$\varphi(Y) = \varphi\Big(\frac{1}{n}\sum_{k=0}^{n-1}\alpha^k(Y)\Big) = \lim_{n\to\infty} \varphi\Big(\frac{1}{n}\sum_{k=0}^{n-1}\alpha^k(Y)\Big) = \varphi\Big(\lim_{n\to\infty}\frac{1}{n}\sum_{k=0}^{n-1}\alpha^k(Y)\Big) = 0 .$$

In addition, by invariance, $\varphi(a_i a_i^\dagger)$ is constant for any i, and we denote by γ this common value. Hence, by (9)

$$\varphi(X) = c + \gamma \sum_{j\in F} \beta_j = \gamma\omega(X) + (1-\gamma)\omega_\infty(X)\,.$$

The thesis follows by a standard density argument. □

References

1. Crismale, V., Fidaleo, F., De: Finetti theorem on the CAR algebra. Commun. Math. Phys. **315**, 135–152 (2012)
2. Crismale, V., Fidaleo, F.: Exchangeable stochastic processes and symmetric states in quantum probability. Ann. Mat. Pura Appl. **194**, 969–993 (2015)
3. Crismale, V., Fidaleo, F.: Symmetries and ergodic properties in quantum probability. Colloq. Math. **149**, 1–20 (2017)
4. Crismale, V., Fidaleo, F., Griseta, M.E.: Wick order, spreadability and exchangeability for monotone comutation relations. Ann. Henri Poincaré **19**, 3179–3196 (2018)
5. Crismale, V., Fidaleo, F., Griseta, M.E.: Spreadability for quantum stochastic processes, with an application to Boolean commutation relations. Entropy **22**, 532 (2020)
6. Crismale V., Fidaleo F., Lu Y. G.: Ergodic theorems in quantum probability: an application to monotone stochastic processes. Ann. Sc. Norm. Super. Pisa Cl. Sci. **XVII**(5), 113–141 (2017)
7. Crismale, V., Griseta, M.E., Wysoczański, J.: Weakly Monotone fock space and monotone convolution of the Wigner Law. J. Theor. Probab. **33**, 268–294 (2020)
8. Crismale, V., Lu, Y.G.: Vacuum distribution, norm and spectral properties for sums of monotone position operators. J. Oper. Theory **83**, 495–515 (2020)
9. Dykema, K., Fidaleo, F.: Unique mixing of the shift on the C^*-algebras generated by the q-canonical commutation relations. Houston J. Math. **36**, 275–281 (2010)
10. Fidaleo, F., Mukhamedov, F.: Strict weak mixing of some C^*-dynamical systems based on free shifts. J. Math. Anal. Appl. **336**, 180–187 (2007)
11. Fidaleo, F., Mukhamedov, F.: Ergodic properties of Bogoliubov automorphisms in free probability. Inf. Dim. Anal. Quantum Probab. Relat. Top. **13**, 393–411 (2010)
12. Tomiyama, J.: On the projection of norm one in W^*-algebras. Proc. Japan Acad. **33**, 608–612 (1957)
13. Wysoczanski J.: Monotonic independence on the weakly monotone Fock space and related Poisson type theorem. Infin. Dimens. Anal. Quantum Probab. Rel. Top. **8**, 259–275 (2005)

Part IV
Infinite Dimensional Analysis

Solutions of Infinite Dimensional Partial Differential Equations

Olfa Draouil and Habib Ouerdiane

Abstract In this work we study some evolution equations associated with the power of the adjoint of the Gross Laplacian Δ_G^p and a potential function V on an infinite dimensional distributions space. The technique we use is the representation of this infinite dimensional Laplacian as a convolution operator. This representation enables us to apply the convolution calculus on a suitable distribution space to obtain explicit solution of some perturbed evolution equation.

Keywords Gross Laplacian · Infinite dimensional distribution space · Convolution calculus · Perturbed evolution equation

1 Introduction and Background

In the paper [7] Hochberg studied the parabolic partial differential equation

$$\frac{\partial u}{\partial t} = (-1)^{n+1} \frac{\partial^{2n} u}{\partial x^{2n}}, \quad n \geq 2.$$

He showed that the fundamental solution of this equation is the density of a finitely additive signed measure of unbounded variation. In this work we will consider the infinite dimensional generalization of the above equation with an additional potential function

$$\frac{\partial U_t}{\partial t} = (-1)^{p+1} \frac{1}{2} (\Delta_G^*)^p U_t + V_t, \quad p \in \mathbb{N}, \tag{1}$$

where Δ_G^* is the adjoint of the Gross Laplacian [5, 10]. We will show that the solutions are generalized functions (rather than ordinary function), which reflect the above fact

O. Draouil · H. Ouerdiane (✉)
University of Tunis El Manar, Tunis, Tunisia
e-mail: habib.ouerdiane@fst.utm.tn

O. Draouil
e-mail: olfa.draouil@fst.utm.tn

L. Accardi et al. (eds.), *Infinite Dimensional Analysis, Quantum Probability and Applications*, Springer Proceedings in Mathematics & Statistics 390,
https://doi.org/10.1007/978-3-031-06170-7_14

that the corresponding "measure" are finitely additive signed measure (rather than countably additive measure) of unbounded variation.

In Sect. 3 we will study this equation with an initial condition and show that the unique solution is a generalized function. The main tool is the interpretation of the Gross Laplacian as a convolution operator. In the next we will briefly described the infinite dimensional framework in order to setup Eq. (1). First we review from the papers [4, 8, 9] basic concepts, notation, and some results which will be needed in the present paper. Let X be a real nuclear Fréchet space with topology given by an increasing family $\{|\cdot|_p;\ p \in \mathbb{N}_0\}$ of Hilbertian norms, $\mathbb{N}_0$ being the set of nonnegative integers. Then X is represented as $X = \bigcap_{p\in\mathbb{N}_0} X_p$, where X_p is the completion of X with respect to the norm $|\cdot|_p$. We use X_{-p} to denote the dual space of X_p. Then the dual space X' of X can be represented as $X' = \bigcup_{p\in\mathbb{N}_0} X_{-p}$ and is equipped with the inductive limit topology. Let $N = X + iX$ and $N_p = X_p + iX_p,\ p \in \mathbb{Z}$, be the complexifications of X and X_p, respectively.

For $n \in \mathbb{N}_0$ we denote by $N^{\widehat{\otimes}n}$ the n-fold symmetric tensor product of N equipped with the π-topology and by $N_p^{\widehat{\otimes}n}$ the n-fold symmetric Hilbertian tensor product of N_p. We will preserve the notation $|\cdot|_p$ and $|\cdot|_{-p}$ for the norms on $N_p^{\widehat{\otimes}n}$ and $N_{-p}^{\widehat{\otimes}n}$, respectively.

1.1 Young Functions

Let θ be a Young function, i.e., it is a continuous, convex, and increasing function defined on $\mathbb{R}_+$ such that $\theta(0) = 0$ and $\lim_{x\to\infty} \theta(x)/x = \infty$

We define the conjugate function (or the Legendre Transform) θ^* of θ by

$$\theta^*(x) = \sup_{t\geq 0} \big(tx - \theta(t)\big), \quad x \geq 0.$$

1.2 Functional Spaces

For a Young function θ, we denote by $\mathcal{F}_\theta(N')$ the space of holomorphic functions on N' with exponential growth of order θ and of minimal type. Similarly, let $\mathcal{G}_\theta(N)$ denote the space of holomorphic functions on N with exponential growth of order θ and of arbitrary type. Moreover, for each $p \in \mathbb{Z}$ and $m > 0$, define $\mathrm{Exp}(N_p, \theta, m)$ to be the space of entire functions f on N_p satisfying the condition:

$$\|f\|_{\theta,p,m} = \sup_{x\in N_p} |f(x)|e^{-\theta(m|x|_p)} < \infty.$$

Then the spaces $\mathcal{F}_\theta(N')$ and $\mathcal{G}_\theta(N)$ can be represented as

$$\mathcal{F}_\theta(N') = \bigcap_{p\in\mathbb{N}_0,\, m>0} \mathrm{Exp}(N_{-p}, \theta, m),$$

$$\mathcal{G}_\theta(N) = \bigcup_{p\in\mathbb{N}_0,\, m>0} \mathrm{Exp}(N_p, \theta, m),$$

and are equipped with the projective limit topology and the inductive limit topology, respectively.

The space $\mathcal{F}_\theta(N')$ is called the space of *test functions* on N'. Its dual space $\mathcal{F}'_\theta(N')$, equipped with the strong topology, is called the space of *distributions* on N'.

1.3 Formal Power Series Spaces

For $p \in \mathbb{N}_0$ and $m > 0$, we define the Hilbert spaces

$$F_{\theta,m}(N_p) = \left\{\varphi = (\varphi_n)_{n=0}^{\infty}\ ;\ \varphi_n \in N_p^{\widehat{\otimes} n},\ \sum_{n\geq 0} \theta_n^{-2} m^{-n} |\varphi_n|_p^2 < \infty\right\},$$

$$G_{\theta,m}(N_{-p}) = \left\{\boldsymbol{\Phi} = (\Phi_n)_{n=0}^{\infty}\ ;\ \Phi_n \in N_{-p}^{\widehat{\otimes} n},\ \sum_{n\geq 0} (n!\theta_n)^2 m^n |\Phi_n|_{-p}^2 < \infty\right\},$$

where $\theta_n = \inf_{r>0} e^{\theta(r)}/r^n,\ n \in \mathbb{N}_0$ Put

$$F_\theta(N) = \bigcap_{p\in\mathbb{N}_0, m>0} F_{\theta,m}(N_p),$$

$$G_\theta(N') = \bigcup_{p\in\mathbb{N}_0, m>0} G_{\theta,m}(N_{-p}).$$

Proposition 1 *The space $F_\theta(N)$ equipped with the projective limit topology is a nuclear Frechét space [4].*

The space $G_\theta(N')$ carries the dual topology of $F_\theta(N)$ with respect to the $\mathbb{C}$-bilinear pairing given by

$$\langle\langle \boldsymbol{\Phi}, \boldsymbol{\varphi} \rangle\rangle = \sum_{n\geq 0} n! \langle \Phi_n, \varphi_n \rangle, \tag{2}$$

where

$$\boldsymbol{\Phi} = (\Phi_n)_{n=0}^{\infty} \in G_\theta(N')$$

and

$$\boldsymbol{\varphi} = (\varphi_n)_{n=0}^{\infty} \in F_\theta(N)$$

1.4 Taylor Map

It was proved in [4] that the Taylor map defined by

$$T:\ \varphi \longmapsto \left(\frac{1}{n!}\varphi^{(n)}(0)\right)_{n=0}^{\infty}$$

is a topological isomorphism :

$$\mathcal{F}_\theta(N') \mapsto F_\theta(N)$$

Proposition 2 *The Taylor map T establish a topological isomorphism from $\mathcal{G}_{\theta^*}(N)$ onto $G_\theta(N')$.*

The action of a distribution $\Phi \in \mathcal{F}'_\theta(N')$ on a test function $\varphi \in \mathcal{F}_\theta(N')$ can be expressed in terms of the Taylor map as follows:

$$\langle\langle \Phi, \varphi \rangle\rangle = \langle\langle \boldsymbol{\Phi}, \boldsymbol{\varphi} \rangle\rangle,$$

where $\boldsymbol{\Phi} = (T^*)^{-1}\Phi$ and $\boldsymbol{\varphi} = T\varphi$

1.5 Laplace Transform

It is easy to see that for each $\xi \in N$, the exponential function $e_\xi(z) = e^{\langle z, \xi\rangle}, \quad z \in N'$, is a test function in the space $\mathcal{F}_\theta(N')$ for any Young function θ.

Thus we can define the Laplace transform of a distribution $\Phi \in \mathcal{F}'_\theta(N')$ by

$$\widehat{\Phi}(\xi) = \langle\langle \Phi, e_\xi \rangle\rangle, \quad \xi \in N. \tag{3}$$

From the paper [4], we have the following duality theorem (or characterization theorem of generalized functions)

Theorem 1 *The Laplace transform is a topological isomorphism from $\mathcal{F}'_\theta(N')$ onto $\mathcal{G}_{\theta^*}(N)$.*

2 Convolution Calculus

For $\varphi \in \mathcal{F}_\theta(N')$, the *translation* $\mathrm{t}_x\varphi$ of φ by $x \in N'$ is defined by

$$\mathrm{t}_x\varphi(y) = \varphi(y - x), \quad y \in N'.$$

The translation operator t_x is a continuous linear operator from $\mathcal{F}_\theta(N')$ into itself for any $x \in N'$.

By a convolution operator on the space $\mathcal{F}_\theta(N')$ of test functions we mean a continuous linear operator from $\mathcal{F}_\theta(N')$ into itself which commutes with translation operators t_x for all $x \in N'$.

Definition 1 We define the *convolution* $\Phi * \varphi$ of a distribution $\Phi \in \mathcal{F}'_\theta(N')$ and a test function $\varphi \in \mathcal{F}_\theta(N')$ to be the function

$$(\Phi * \varphi)(x) = \langle\langle \Phi, \mathrm{t}_{-x}\varphi \rangle\rangle, \quad x \in N'.$$

Lemma 1 *Let $\Phi \in \mathcal{F}'_\theta(N')$ arbitrarily fixed.Then for any $\varphi \in \mathcal{F}_\theta(N')$, $\Phi * \varphi \in \mathcal{F}_\theta(N')$ and the mapping T_Φ defined by*

$$T_\Phi \colon \varphi \longmapsto \Phi * \varphi, \quad \varphi \in \mathcal{F}_\theta(N'),$$

is a convolution linear operator on $\mathcal{F}_\theta(N')$.

Conversely, it was proved in [2] that all convolution operators on $\mathcal{F}_\theta(N')$ occur this way, i.e., if T is a convolution operator on $\mathcal{F}_\theta(N')$, then there exists a unique $\Phi \in \mathcal{F}'_\theta(N')$ such that $T = T_\Phi$ or equivalently,

$$T(\varphi) = T_\Phi(\varphi) = \Phi * \varphi, \quad \varphi \in \mathcal{F}_\theta(N'). \tag{4}$$

2.1 Convolution of Two Distributions

Suppose $\Phi_1, \Phi_2 \in \mathcal{F}'_\theta(N')$. Let T_{Φ_1} and T_{Φ_2} be the convolution operators given by Φ_1 and Φ_2, respectively, as in Equation (4). It is clear that the composition $T_{\Phi_1} \circ T_{\Phi_2}$ is also a convolution operator on $\mathcal{F}_\theta(N')$. Hence there exists a unique distribution, denoted by $\Phi_1 * \Phi_2$ in $\mathcal{F}'_\theta(N')$ such that

$$T_{\Phi_1} \circ T_{\Phi_2} = T_{\Phi_1 * \Phi_2}. \tag{5}$$

Proposition 3 ([2]) *For any $\Phi_1, \Phi_2 \in \mathcal{F}'_\theta(N')$, the distribution $\Phi_1 * \Phi_2$ in Eq. (5) is called the convolution of Φ_1 and Φ_2, and we have the following equality via the Laplace transform:*

$$\widehat{\Phi_1 * \Phi_2} = \widehat{\Phi}_1 \, \widehat{\Phi}_2. \tag{6}$$

Remark 1 One should keep in mind that $\Phi * \varphi$ defined in (4) is not consistency with the convolution product of two distributions in (6); see Remark 5.8 in [13].

2.2 White Noise Gel'fand Triple

Let γ be the standard Gaussian measure on the dual space X' of the real nuclear space X, namely, its characteristic function is given by

$$\int_{X'} e^{i\langle y,\xi\rangle}\, d\gamma(y) = e^{-|\xi|_0^2/2}, \quad \xi \in X,$$

where $|\cdot|_0$ is the norm $|\cdot|_p$ on X for $p = 0$.

Theorem 2 ([4]) *Suppose that the Young function θ satisfies the condition: $\lim_{r\to+\infty} \frac{\theta(r)}{r^2} < +\infty$. Then we obtain the Gel'fand triple*

$$\mathcal{F}_\theta(N') \hookrightarrow L^2(X', \gamma) \hookrightarrow \mathcal{F}'_\theta(N').$$

Remark 2 In the White noise theory, i.e., in the particular case of $X = S(\mathbb{R})$, and $\theta(x) = x^2$) we use usually the S-transform to characterize Hida distributions in term of analytic functionals with growth conditions (see [4, 6, 8–10, 12]) and the S-transform, defined on $\mathcal{F}'_\theta(N')$, is related to the Laplace transform by

$$S(\Phi)(\xi) = \widehat{\Phi}(\xi)e^{-\frac{\langle\xi,\xi\rangle}{2}}, \quad \xi \in N, \ \Phi \in \mathcal{F}'_\theta(N'). \tag{7}$$

2.3 Convolution Functionals

Let β be a continuous, convex, and increasing function on $\mathbb{R}^+$. Suppose f is a function in $\mathrm{Exp}(\mathbb{C}, \beta, m)$ for some $m > 0$. For each distribution Φ in $\mathcal{F}'_\theta(N')$, we define the *convolution composition* $f^*(\Phi)$ of f and Φ by

$$\widehat{f^*(\Phi)} = f(\widehat{\Phi}). \tag{8}$$

It was proved in [2] that $f^*(\Phi)$ belongs to $\mathcal{F}'_\lambda(N')$ with

$$\lambda = (\beta \circ e^{\theta^*})^*$$

In particular, when $\beta(x) = x$, $x \in \mathbb{R}_+$, and $f(z) = e^z$, $z \in \mathbb{C}$ we get a distribution $e^{*\Phi}$ in the space $\mathcal{F}'_{(e^{\theta^*})^*}(N')$ for each $\Phi \in \mathcal{F}'_\theta(N')$. Moreover, by Eq. (8), we have

$$\widehat{e^{*\Phi}} = e^{\widehat{\Phi}}. \tag{9}$$

Proposition 4 *The distribution $e^{*\Phi}$ has the following series expansion*

$$e^{*\Phi} = \sum_{n=0}^{\infty} \frac{1}{n!} \Phi^{*n},$$

where $\Phi^{*n} = \Phi * \Phi * \cdots * \Phi$ *(n times) and the convergence is in* $\mathcal{F}'_{(e^{\theta^*})_*}(N')$ *with respect to the strong topology.*

3 Initial-Valued Evolution Equation

Let $I \subset \mathbb{R}$ be an interval containing the origin. Consider a family $\{\Phi_t\,;\, t \in I\}$ of distributions in $\mathcal{F}'_\theta(N')$. We assume that the function $t \mapsto \Phi_t$ is continuous from I into $\mathcal{F}'_\theta(N')$. Then the function $t \mapsto \widehat{\Phi}_t$ is continuous from I into $\mathcal{G}_{\theta^*}(N)$. Thus for each $t \in I$, the set $\{\widehat{\Phi_s}\,;\, s \in [0, t]\}$ is a compact subset of $\mathcal{G}_{\theta^*}(N)$. In particular, it is bounded in $\mathcal{G}_{\theta^*}(N)$. Hence there exist constants $p \in \mathbb{N}_0$, $m > 0$, and $C_t > 0$ such that

$$|\widehat{\Phi_s}(\xi)| \le C_t\, e^{\theta^*(m|\xi|_p)}, \quad \forall s \in [0, t] \text{ and } \xi \in N_p.$$

This inequality shows that the function

$$\xi \mapsto \int_0^t \widehat{\Phi_s}(\xi)\, ds$$

belongs to the space $\mathcal{G}_{\theta^*}(N)$. So we obtain the following lemma:

Lemma 2 *There exists a unique distribution, denoted by* $\int_0^t \Phi_s\, ds$, *in* $\mathcal{F}'_\theta(N')$ *satisfying*

$$\left(\widehat{\int_0^t \Phi_s\, ds}\right)(\xi) = \int_0^t \widehat{\Phi_s}(\xi)\, ds, \quad \xi \in N.$$

Moreover, the process $E_t = \int_0^t \Phi_s\, ds$, $t \in I$ *is differentiable in* $\mathcal{F}'_\theta(N')$ *and satisfies the equation* $\frac{\partial}{\partial t} E_t = \Phi_t$.

3.1 Main Theorem

Let $\{\Phi_t\}$ and $\{M_t\}$ be two continuous $\mathcal{F}'_\theta(N')$-processes. Consider the initial value problem

$$\frac{dX_t}{dt} = \Phi_t * X_t + M_t\,, \quad X_0 = F \in \mathcal{F}'_\theta(N'). \tag{10}$$

The next theorem is from Theorem 4 of the paper [2].

Theorem 3 *The stochastic differential equation* (10) *has a unique solution in* $\mathcal{F}'_{(e^{\theta^*})^*}(N')$ *given by*

$$X_t = F * e^{* \int_0^t \Phi_s \, ds} + \int_0^t e^{* \int_s^t \Phi_u \, du} * M_s \, ds. \tag{11}$$

We can apply the above Theorem 3 to study an evolution equation for a power of the adjoint of the Gross Laplacian and a generalized potential function with the initial condition being a generalized function.

4 Gross Laplacian

Let $\varphi \in \mathcal{F}_\theta(N')$ be represented by $\varphi(x) = \sum_{n \geq 0} \langle x^{\otimes n}, \varphi^{(n)} \rangle$. The Gross Laplacian $(\Delta_G \varphi)(x)$ of φ at $x \in N'$ is defined to be

$$(\Delta_G \varphi)(x) = \sum_{n \geq 0} (n+2)(n+1) \langle x^{\otimes n}, \langle \tau, \varphi^{(n+2)} \rangle \rangle,$$

where τ is the trace operator, namely, $\langle \tau, \xi \otimes \eta \rangle = \langle \xi, \eta \rangle, \quad \xi, \eta \in N$. We denote by Δ_G^* the adjoint of the Gross Laplacian. For more details and obtained results on the Gross Laplacian and its adjoint, see [3, 5, 6, 10–12, 14].

4.1 Gross Laplacian as a Convolution Operator

It turns out that the adjoint Δ_G^* of the Gross Laplacian can be also defined to be a continuous linear operator from $\mathcal{F}'_\theta(N')$ into itself by

$$\Psi \in \mathcal{F}'_\theta(N') \to \mathcal{T} * \Psi,$$

where $\mathcal{T}$ is the generalized function in $\mathcal{F}'_\theta(N')$ with the Formal power series is given by

$$\mathcal{T} = (0, 0, \tau, 0, \cdots) \in G_\theta(N') \tag{12}$$

as in Eq. (2).

Theorem 4 *For any* $\Psi \in \mathcal{F}'_\theta(N')$,

$$\Delta_G^* \Psi = \mathcal{T} * \Psi, \tag{13}$$

where $\mathcal{T}$ *is given by* (12).

Proof For any $\Psi \in \mathcal{F}'_\theta(N')$ and $\eta \in N$, the Laplace transform of $\Delta_G^* \Psi$ is given by

$$\begin{aligned}\widehat{\Delta_G^* \Psi}(\eta) &:= \langle \Delta_G^* \Psi, e^\eta \rangle \\ &= \langle \psi, \Delta_G e^\eta \rangle \\ &= \langle \eta, \eta \rangle \widehat{\Psi}(\eta).\end{aligned}$$

On the other hand,

$$\widehat{\mathcal{T} * \Psi}(\eta) = \widehat{\mathcal{T}}(\eta)\widehat{\Psi}(\eta) = \langle \eta, \eta \rangle \widehat{\Psi}(\eta).$$

So we have $\Delta_G^* \Psi = \mathcal{T} * \Psi$.

Proposition 5 *For every positive integer p we have*

$$(\Delta_G^*)^p \Psi = (\mathcal{T}^{*p}) * \Psi, \quad \Psi \in \mathcal{F}'_\theta(N'). \tag{14}$$

Moreover, the generalized function associated with $(\Delta_G^*)^p$ *is given by*

$$\overrightarrow{\mathcal{T}^{*p}} = (0, 0, \ldots, \tau^{\otimes p}, 0, \ldots). \tag{15}$$

Proof Using Equations (5) and (13), we obtain

$$(\Delta_G^*)^p \Psi = \mathcal{T}^{*p} * \Psi.$$

But the Laplace transform of $\mathcal{T}$ is given by

$$\widehat{\mathcal{T}}(\xi) = \langle \tau, \xi^{\otimes 2} \rangle = \langle \xi, \xi \rangle = |\xi|_0^2.$$

Hence we have

$$\widehat{(\mathcal{T}^{*p})}(\xi) = \langle \tau, \xi^{\otimes 2} \rangle^p = \langle \xi, \xi \rangle^p = |\xi|_0^{2p}.$$

For any positive integer p, let $S = \mathcal{T}^{*p}$ and let the formal power series associated with S be given by $\widehat{S} = (S_0, S_1, \ldots, S_n, \ldots)$. Then we can use the definition of the Laplace transform and the bilinear pairing between test functions and distributions in Eq. (2) to deduce the following relationship

$$\widehat{(\mathcal{T}^{*p})}(\xi) = \langle \mathcal{T}^{*p}, e^\xi \rangle = \sum_{n \geq 0} n! \langle S_n, \frac{\xi^{\otimes n}}{n!} \rangle = \langle \xi, \xi \rangle^p,$$

which implies that $S_n = 0$ for all $n \neq 2p$ and $S_{2p} = \tau^{\otimes p}$. Therefore,

$$\overrightarrow{S} = \overrightarrow{\mathcal{T}^{*p}} = (0, 0, \ldots, \tau^{\otimes p}, 0, \ldots).$$

This proves Eq. (15).

Theorem 5 *Let θ be a Young function such that*

$$\lim_{r\to\infty} \theta(r)/r^2 < \infty$$

and $F \in \mathcal{F}'_\theta(N')$. Then the following evolution equation associated with the p-th power of the adjoint of the Gross Laplacian and a continuous $\mathcal{F}'_\theta(N')$-valued potential function V_t

$$\frac{\partial U_t}{\partial t} = (-1)^{p+1}\frac{1}{2}(\Delta_G^*)^p U_t + V_t, \quad U_0 = F, \tag{16}$$

has a unique solution in the space $\mathcal{F}'_\theta(N')$ given by

$$U_t = F * e^{*\frac{t}{2}(-1)^{p+1}\mathcal{T}^{*p}} + \int_0^t e^{*\frac{t-s}{2}(-1)^{p+1}\mathcal{T}^{*p}} * V_s\, ds, \tag{17}$$

where $\mathcal{T}$ is the generalized function given by Equation (13).

Proof Use Eq. (13) to rewrite Eq. (16) as

$$\frac{\partial U_t}{\partial t} = (-1)^{p+1}\frac{1}{2}\mathcal{T}^{*p} * U_t + V_t, \quad U_0 = F.$$

Then we can apply Theorem 3 to this equation to get the unique solution in Eq. (17).

5 Concluding Remarks

5.1 Interpretation of the Solutions of the Evolution Equation

We can further rewrite the solution in Equation (17) in another form. For $t > 0$, define the distribution $\mu_{t,p}$ by its Laplace transform

$$\widehat{\mu_{t,p}}(\xi) = \exp\Big[\frac{(-1)^{p+1}t}{2}\langle\xi,\xi\rangle^p\Big], \quad \xi \in N. \tag{18}$$

Recall the duality theorem [4] which states that the Laplace transform is a topological isomorphism from $\mathcal{F}'_\theta(N')$ onto $\mathcal{G}_{\theta^*}(N)$. Hence Eq. (18) implies that $\mu_{t,p}$, $t > 0$, are generalized functions in the space $\mathcal{F}'_\theta(N')$ with the Young function given by

$$\theta(x) = x^{\frac{2p}{2p-1}}, \quad x \geq 0.$$

Therefore, the solution U_t in Eq. (17) can be rewritten as

$$U_t = F * \mu_{t,p} + \int_0^t \mu_{t-s,p} * V_s \, ds.$$

In particular, when $V_t = 0$, we have the evolution equation

$$\frac{\partial U_t}{\partial t} = (-1)^{p+1} \frac{1}{2} (\Delta_G^*)^p U_t, \quad U_0 = F, \tag{19}$$

which has a unique solution given by

$$U_t = F * \mu_{t,p}. \tag{20}$$

Hochberg [7] has studied the one-dimensional case of Eq. (19) and showed that the fundamental solution defines a finitely additive measure with unbounded total variation. Using the white noise theory, we can now interpret this "finitely additive measure with unbounded total variation" as a generalized function in the space $\mathcal{F}'_\theta(N')$, which is given by Eq. (20). This phenomenon is very much like the case of Feynman integral, which had been regarded as a finitely additive measure with unbounded total variation before the theory of white noise was introduced by T. Hida in 1975. It is a well-known fact that the Feynman integral is a generalized function [6, 10].

5.2 Heat Equation Associated to the Gross Laplacian

When $p = 1$, Eq. (18) gives the equality

$$\widehat{\mu_{t,1}}(\xi) = \exp\Big[-\frac{t}{2} |\xi|_0^2 \Big], \quad \xi \in X,$$

which shows that $\mu_{t,1}$ is the standard Gaussian measure on X' with variance t, i.e., $\mu_{t,1} = \gamma_t$ with γ_t defined by $\gamma_t(\cdot) = \gamma\Big(\frac{\cdot}{\sqrt{t}}\Big)$. Note that the probability measure $\mu_{t,1}$ induces a positive distribution in the space $\mathcal{F}'_\theta(N')$ given by

$$\langle\langle \mu_{t,1} \varphi \rangle\rangle = \int_{X'} \varphi(x) \, d\mu_{t,1}(x) = \int_{X'} \varphi(\sqrt{t}\, x) \, d\gamma(x), \quad \varphi \in \mathcal{F}_\theta(N').$$

Moreover, if $p = 1$ and the potential function is given by $V_t = \alpha \dot{W}_t$ with $\alpha \in \mathbb{R}$ and $\dot{W}_t$ a white noise, then the solution in Eq. (17) reduces to the one obtained in [1], see also [14].

References

1. Barhoumi, A., Kuo, H.H., Ouerdiane, H.: Generalized Gross Heat equation with noises. Soochow J. Math. **32**, 113–125 (2006)
2. Ben Chrouda, M., El Oued, M., Ouerdiane, H.: Convolution calculus and applications to stochastic differential equations. Soochow J. Math. **28**, 375–388 (2002)
3. Chung, D.M., Ji, U.C.: Some Cauchy problems in white noise analysis and associated semi-groups of operators. (English summary) Stoch. Anal. Appl. **17**(1), 1–22 (1999)
4. Gannoun, R., Hachaichi, R., Ouerdiane, H., Rezgui, A.: Un théorème de dualité entre espace de fonctions Holomorphes à croissance exponentielle. J. Funct. Anal. **171**, 1–14 (2000)
5. Gross, L.: Potential theory on Hilbert space. J. Funct. Anal. **1**, 123–181 (1967)
6. Hida, T., Kuo, H.-H., Potthoff, J., Streit, L.: White Noise: An Infinite Dimensional Calculus. Kluwer Academic Publishers, Dordrecht (1993)
7. Hochberg, K.J.: A signed measure on path space related to Wiener measure. Ann. Probab. **6**, 433–458 (1978)
8. Horrigue, S., Ouerdiane, H.: Gross Laplacian acting on operators. Acta Appl. Math. **105**, 227–239 (2009)
9. Ji, U.C., Obata, N., Ouerdiane, H.: Quantum Levy Laplacian and associated heat equation. J. Funct. Anal. **249**, 3154 (2007)
10. Kuo, H.-H.: White Noise Distribution Theory. CRC Press, Boca Raton (1996)
11. Lee, Y.J.: Applications of the Fourier-Wiener transform to differential equations on infinite-dimensional spaces. I. Trans. Am. Math. Soc. **262**, 259–283 (1980)
12. Obata, N.: White Noise Calculus and Fock Space. Lecture Notes in Mathematics, vol. 1577. Springer, Berlin (1994)
13. Obata, N., Ouerdiane, H.: A note on convolution operators in white noise calculus. Infin. Dimens. Anal. Quantum. Proba. Relat. Top. **14**, 661–674 (2011)
14. Ouerdiane, H.: Solutions of evolution equations associated to infinite-dimensional Laplacian. Int. J. Quantum Inf. **14**, 16 (2016)

On Some Properties of Solution Sets of Discontinuous Quantum Stochastic Differential Inclusions

Dauda Alani Dikko

Abstract We establish properties of solution sets for upper semicontinuous quantum stochastic differential inclusions whose existence of solution satisfy a noncommutative generalisation of Kakutani fixed point theorem. The result here complement earlier results.

Keywords Connectedness · Contractibility · Noncommutative stochastic processes · Lower semicontinuous · Upper semicontinuous

AMS Classifications 81S25, 34F05, 34A60, 60H99

1 Introduction

There are numerous established qualitative theories which describes the existence of solution of quantum stochastic differential inclusions (QSDI) with each of these possessing different regularity conditions for the existence of solution [see [2, 6, 10, 12]], for some of these existence results. This necessitates establishing topological properties for the respective existence of solution results. Recent studies show that for QSDI with Lipschitzian coefficient [6] the solution set is arcwise connected [4] both locally and globally and similarly for QSDI with coefficients that are discontinuous and whose existence rely on the continuous selection of a lower semicontinuous coefficient and also satisfy a noncommutative Micheal selection property [9] , the solution set is arcwise connected . This present paper consider the case where the solutions

---

WWW home page: https://sci.ui.edu.ng/content/dadikko

D. A. Dikko (✉)
Department of Mathematics, University of Ibadan, Ibadan, Federal Republic of Nigeria
e-mail: da.dikko@mail.ui.edu.ng

L. Accardi et al. (eds.), *Infinite Dimensional Analysis, Quantum Probability and Applications*, Springer Proceedings in Mathematics & Statistics 390,
https://doi.org/10.1007/978-3-031-06170-7_15

exist subject to a noncommutative Kakutani fixed point property [10]. These results which are valid within the framework of the well known Hudson and Parthasarathy (H-P) formulation [7] of Boson fock space of noncommutative stochastic calculus and as extended to noncommutative stochastic inclusions by Ekhaguere [6], were shown to be connected and contractible . These can be employed in prescribed applications for example in studying noncommutative stochastic control [6, 9, 11] in which they sometimes occur as regularization of noncommutative stochastic differential equations having discontinuous coefficients.

Studying properties such as connectedness, path connectedness, arcwise connectedness and contractibility of solution sets in the classical or noncommutative settings for differential inclusions include, the need to determine the stability of approximate solutions for variational problems, fixed point theory and optimization theory which in turn provides conditions under which the stability of such a system hold. In the classical settings, differential inclusions on finite or infinite dimensional spaces are useful tools for solving control problems [1, 13] and optimization problems [14]. Some studies on the topological properties of solution sets for the noncommutative cases can be found in [3–5]. Our result here is a generalisation of some classical results [8] to the present quantum settings. These concerns solution sets for upper semicontinuous noncommutative stochastic differential inclusions which in integral form has the representation :

$$\begin{aligned} x(t) \in x_0 + \int_0^t \big(E(s, x(s)) d \wedge_\pi (s) + F(s, x(s)) d A_f(s) \\ + G(s, x(s)) d A_g^+(s) + H(s, x(s)) ds \big), \quad x(t_0) = x_0, \quad \text{almost all} \quad t \in [0, T]. \end{aligned} \tag{1}$$

In (1), the coefficients $\{E, F, G, H\}$ lie in a class of stochastic processes in which the noncommutative stochastic integrals against the creation, annihilation and guage processes A_f^+, A_g, $\bigwedge_\pi$ and the Lebesgue measure are defined. Equation (1) involves unbounded linear operators on a Hilbert space and it is a quantum generalization of the classical stochastic integral equations of the form

$$x(t, w) = x_o + \int_{t_0}^T H(t, x) dt + \int_{t_0}^T F(t, x) d Q(t), \quad x(t_0) = x_0, \quad t \in [t_0, T] \tag{2}$$

to a set valued form. The driving process $Q(t)$ is a martingale while H and F are smooth functions. We make use of the various spaces of the noncommutative stochastic processes introduced in [3, 6]. The paper shall be put in the following order; In Sect. 2, we state some of the necessary notations and key fundamental structures. We state some required definitions, results and relevant assumptions in Sect. 3 while in Sect. 4 we establish the main result.

2 Notations and Fundamental Structures

In the following, N is a given topological space with clos(N), as the collection of all non-empty closed subsets of N. To each pair (D, H) which consists of a pre - Hilbert space D with completion H , we associate $L_w^+(D, H)$ the set of all linear maps x from a pre-Hilbert space D to its completion H. With the property that the domain of the operator adjoint x^* of x contains D. The members of $L_w^+(D, H)$ are densely-defined linear operators on H which do not necessarily leave D invariant and $L_w^+(D, H)$ is a linear space when equipped with the usual notions of addition and scalar multiplication. To H also corresponds a Hilbert space $\Gamma(H)$, called the Boson Fock space determined by H. A natural dense subset of $\Gamma(H)$ consists of the linear space generated by the set of exponential vectors in $\Gamma(H)$ of the form

$$e(f) = \bigoplus_{n=0}^{\infty}(n!)^{-\frac{1}{2}}\bigotimes^{n} f, \quad f \in H$$

where $\otimes^0 f = 1$ and $\otimes^n f$ is the n-fold tensor product of f with itself for $n \geq 1$. In what follows, $I\!D$ is some pre-Hilbert space whose completion is $\mathcal{R}$ and γ is a fixed Hilbert space. We shall write $L^2_\gamma(I\!R_+)$ (resp. $L^2_\gamma([0, t]))$' resp. $L^2_\gamma([t, \infty)), t \in I\!R_+ \equiv [0, \infty)$ for the Hilbert space of square integrable, γ-valued maps on $I\!R_+ = [0, \infty)$ (resp. on $[0, t)$; resp. on $[t, \infty)$; $t \in I\!R_+$ The noncommutative stochastic processes discussed in what follows are densely-defined linear operators on $\mathcal{R} \otimes \Gamma(L^2_\gamma(I\!R_+))$; the inner product of this Hilbert space will be denoted by $\langle \cdot, \cdot \rangle$. For each $t > 0$, the direct sum decomposition

$$L^2_\gamma(I\!R_+) = L^2_\gamma([0, t)) \oplus L^2_\gamma([t, \infty))$$

induces a factorization

$$\Gamma(L^2_\gamma(I\!R_+)) = \Gamma(L^2_\gamma([0, t))) \otimes \Gamma(L^2_\gamma([t, \infty)))$$

of Fock space.

Let $I\!E$, $I\!E_t$ and $I\!E^t$, $t > 0$ be the linear spaces generated by the exponential vectors in $\Gamma(L^2_\gamma(I\!R_+))$, $\Gamma(L^2_\gamma([0, t)))$ and $\Gamma(L^2_\gamma([t, \infty)))$, $t > 0$ respectively. Then we define

$$\begin{aligned}
\mathcal{A} &\equiv L_w^+(I\!D \underline{\otimes} I\!E,\ \mathcal{R} \otimes \Gamma(L^2_\gamma(I\!R_+))) \\
\mathcal{A}^t &\equiv L_w^+(I\!D \underline{\otimes} I\!E_t,\ \mathcal{R} \otimes \Gamma(L^2_\gamma([0, t)))) \otimes 1^t \\
\mathcal{A}_t &\equiv 1_t \otimes L_w^+(I\!E^t, \Gamma(L^2_\gamma([t, \infty)))) \qquad t > 0
\end{aligned}$$

where $\underline{\otimes}$ denotes algebraic tensor product and 1_t (resp. 1^t) denotes the identify map on $\mathcal{R} \otimes \Gamma(L^2_\gamma([0, t)))$ (resp. $\Gamma(L^2_\gamma([t, \infty)))$ $t > 0$. We note that the spaces $\mathcal{A}_t$ and $\mathcal{A}^t, t > 0$, may be naturally identified with subspaces of $\mathcal{A}$. For $\eta, \xi \in I\!D \underline{\otimes} I\!E$, define $\|\cdot\|_{\eta\xi}$ by $\|x\|_{\eta\xi} = |\langle \eta, x\xi \rangle|, \quad x \in \mathcal{A}$ Then, $\{\|\cdot\|_{\eta\xi},\ \eta, \xi \in I\!D \underline{\otimes} I\!E\}$ is a family of locally convex seminorms on $\mathcal{A}$; we write τ_w for the locally convex topology on $\mathcal{A}$ determined by this family. In the foregoing $\tilde{\mathcal{A}}$, $\tilde{\mathcal{A}}_t$ and $\tilde{\mathcal{A}}^t$ denote the completions of the locally convex spaces $(\mathcal{A}, \tau_w)$, $(\mathcal{A}_t, \tau_w)$, $\mathcal{A}^t, \tau_w)$, $t > 0$ respectively we then note that $\{\tilde{\mathcal{A}}_t,\ t \in I\!R_+\}$ is a filtration of $\tilde{\mathcal{A}}$.

Hausdorff topology: If A is a topological space, then $Clos(A)$ [resp. $Comp(A)$] denotes the collection of all nonvoid closed (resp. Compact) Subsets of A. We shall employ the Hausdorff topology on $Clos(\tilde{\mathcal{A}})$ which is defined as follows.

For $x \in \tilde{\mathcal{A}}$, $\mathcal{M}, \mathcal{N} \in Clos(\tilde{\mathcal{A}})$, and $\eta, \xi \in I\!D \underline{\otimes} I\!E$, set

$$d_{\eta\xi}(x, \mathcal{N}) = \inf_{y \in \mathcal{N}} \|x - y\|_{\eta\xi}$$

$$\delta_{\eta\xi}(\mathcal{M}, \mathcal{N}) = \sup_{x \in \mathcal{M}} d_{\eta\xi}(x, \mathcal{N})$$

and

$$\rho_{\eta\xi}(\mathcal{M}, \mathcal{N}) = \max(\delta_{\eta\xi}(\mathcal{M}, \mathcal{N}), \delta_{\eta\xi}(\mathcal{N}, \mathcal{M}))$$

Then $\{\rho_{\eta\xi}(\cdot) : \eta, \xi \in I\!D \underline{\otimes} I\!E\}$ is a family of pseudometrics which determines a Hausdorff topology on $Clos(\tilde{A})$ denoted in the sequel by τ_H. If $\mathcal{M} \in Clos(\tilde{\mathcal{A}})$, then $\|\mathcal{M}\|_{\eta\xi}$ is defined by $\|\mathcal{M}\|_{\eta\xi} \equiv \rho_{\eta\xi}(\mathcal{M}, \{0\})$, for arbitrary $\eta, \xi \in I\!D \underline{\otimes} I\!E$.

For $A, B \in clos(\mathbb{C})$ and $x \in \mathbb{C}$, a complex number, define

$$d(x, B) \equiv \inf_{y \in B} |x - y|$$

$$\delta(A, B) \equiv \sup_{x \in A} d(x, B)$$

and

$$\rho(A, B) \equiv max(\delta(A, B), \delta(B, A)).$$

Then ρ is a metric on $clos(\mathbb{C})$ and induces a metric topology on the space. Let $I \subseteq I\!R_+$. A stochastic process indexed by I is an $\tilde{\mathcal{A}}$-valued map on I. A stochastic process X is called adapted if $X(t) \in \tilde{\mathcal{A}}_t$ for each $t \in I$. We write $Ad(\tilde{\mathcal{A}})$ for the set of all adapted stochastic processes indexed by I.

Definition 1 A member X of $Ad(\tilde{\mathcal{A}})$ is called

(i) Weakly absolutely continuous if the map $t \longmapsto \langle \eta, X(t)\xi \rangle$, $t \in I$, is absolutely continuous for arbitrary $\eta, \xi \in I\!D \underline{\otimes} I\!E$.
(ii) Locally absolutely p-integrable if $\|X(\cdot)\|_{\eta\xi}^p$ is Lebesgue-measurable and integrable on $[0, t) \subseteq I$ for each $t \in I$ and arbitrary $\eta, \xi \in I\!D \underline{\otimes} I\!E$.

Notation We write $Ad(\tilde{\mathcal{A}})_{wac}$ [resp. $L^p_{loc}(\tilde{\mathcal{A}})$] for the set of all weakly absolutely continuous (resp. locally absolutely p-integrable) members of $Ad(\tilde{\mathcal{A}})$.

Stochastic Integrators

Let $L^{\infty}_{\gamma,loc}(I\!R_+)$ [resp. $L^{\infty}_{B(\gamma),loc}(I\!R_+)$] be the linear space of all measurable, locally bounded functions from $I\!R_+$ to γ [resp. to $B(\gamma)$, the Banach space of Bounded maps of γ]. If $f \in L^{\infty}_{\gamma,loc}(I\!R_+)$ and $\pi \in L^{\infty}_{B(\gamma),loc}(I\!R_+)$, then πf is the member of $L^{\infty}_{\gamma,loc}(I\!R_+)$ given by $(\pi f)(t) = \pi(t) f(t)$, $t \in I\!R_+$. For $f \in L^2_{\gamma}(I\!R_+)$ and $\pi \in L^{\infty}_{B(\gamma),loc}(I\!R_+)$, define the operators $a(f)$, $a^+(f)$ and $\lambda(\pi)$ in $L^+_w(I\!D, \Gamma(L^2_{\gamma}(I\!R_+)))$ as follows;

$$a(f)e(g) = \langle f, g\rangle_{L^2_{\gamma}(I\!R_+)} e(g)$$

$$a^+(f)e(g) = \frac{d}{d\sigma} e(g + \sigma f)\Big|_{\sigma=0}$$

$$\lambda(\pi)e(g) = \frac{d}{d\sigma} e(e^{\sigma\pi} f)\Big|_{\sigma=0}$$

for $g \in L^2_{\gamma}(I\!R_+)$.

These are the annihilation, creation and gauge operators of quantum field theory.

For arbitrary $f \in L^{\infty}_{\gamma,loc}(I\!R_+)$ and $\pi \in L^{\infty}_{B(\gamma),loc}(I\!R_+)$, they give rise to the operator-valued maps A_f, A^+_f, and A_π defined by

$$A_f(t) \equiv a(f\chi_{[0,t)})$$

$$A^+_f(t) \equiv a^+(f\chi_{[0,t)})$$

$$\wedge_\pi(t) \equiv \lambda(\pi\chi_{[0,t)})$$

$t \in I\!R_+$, where χ_I denotes the indicator function of the Borel set $I \subseteq I\!R_+$. The maps A_f, A^+_f and $\wedge_\pi$ are stochastic processes, called the annihilation, creation and gauge processes, respectively, when their values are identified with their ampliations on $\mathcal{R} \otimes \Gamma(L^2_{\gamma}(I\!R_+))$. These are the stochastic integrators in the Hudson and Parthasarathy [10] formulation of Boson quantum stochastic integration, which we adopt in the following. If given $p, q, u, v \in L^2_{loc}(\tilde{\mathcal{A}})$, $f, g \in L^{\infty}_{\gamma,loc}(I\!R_+)$ and $\pi \in L^{\infty}_{B(\gamma),loc}(I\!R_+)$ then the integral.

$$\int_{t_0}^{t} p(s)d \wedge_\pi (s) + q(s)dA_f(s) + u(s)dA^+_g(s) + v(s)ds; \;\; t_0, t \in I\!R_+$$

can be interpreted as it is in the Hudson and Parthasarathy [7] formulation.

Stochastic Differential Inclusions

Definition 2 (i) By a multivalued stochastic process indexed by $I \subseteq I\!R_+$, we mean a multifunction on I with values in $Clos(\tilde{\mathcal{A}})$.

(ii) If Φ is a multivalued stochastic process indexed by $I \subseteq I\!R_+$, then a selection of Φ is a stochastic process $X : I \to \tilde{\mathcal{A}}$ with the property that $X(t) \in \Phi(t)$ for almost all $t \in I$.

(iii) A multivalued stochastic process Φ will be called

(a) adapted if $\Phi(t) \subseteq \tilde{\mathcal{A}}$, for each $t \in I\!R^+$;

(b) measurable if $t \longmapsto d_{\eta\xi}(x, \Phi(t))$ is measurable for arbitrary $x \in \tilde{\mathcal{A}}$, $\eta, \xi \in I\!D \underline{\otimes} I\!E$

(c) locally absolutely p-integrable if $t \mapsto \|\phi(t)\|_{\eta\xi}$, $t \in I\!R_+$, lies in $L^p_{loc}(I)$ for arbitrary $\eta, \xi \in I\!D \underline{\otimes} I\!E$.

Notation

(1) the set of all locally absolutely p-integrable multivalued stochastic processes will be denoted by $L^p_{loc}(\tilde{\mathcal{A}})_{mvs}$

(2) For $p \in (0, \infty)$ and $I \subseteq I\!R_+$, $L^p_{loc}(I \times \tilde{\mathcal{A}})_{mvs}$ is the set of maps $\Phi : I \times \tilde{\mathcal{A}} \longrightarrow Clos(\tilde{\mathcal{A}})$, such that $(t, X) \longmapsto \Phi(t, X(t))$, $t \in I$, lies in $L^p_{loc}(\tilde{\mathcal{A}})_{mvs}$ for every $X \in L^p_{loc}(\tilde{\mathcal{A}})$,

Remark 1 (i) If $\Phi \in L^p_{loc}(I \times \tilde{\mathcal{A}})_{mvs}$, then $L_p(\Phi) \equiv \{\varphi \in L^p_{loc}(\tilde{\mathcal{A}}) : \varphi\}$ is a selection of Φ.

(ii) For $f, g \in L^\infty_{\gamma,loc}(I\!R_+)$, $\pi \in L^\infty_{B(Y),loc}(I\!R_+)$, 1 is the identity map on $\mathcal{R} \otimes \Gamma(L^2_\gamma(I\!R_+))$, and M is any of the stochastic processes A_f, A_g^+, A_π and $s \longmapsto s1$, $s \in I\!R_+$.

Thus, we introduce stochastic integral (resp. differential) expressions as follows.

If $\Phi \in L^2_{loc}(I \times \tilde{\mathcal{A}})_{mvs}$ and $(t, X) \in I \times L^2_{loc}(\tilde{\mathcal{A}})$, then we make the definition

$$\int_{t_0}^t \Phi(s, X(s))dM(s) \equiv \left\{\int_{t_0}^t \varphi(s)dM(s) : \varphi \in L_2(\Phi)\right\}$$

This leads to the following notion.

Definition 3 Let $E, F, G, H \in L^2_{loc}(I \times \tilde{\mathcal{A}})_{mvs}$ and (t_0, x_0) be a fixed point of $I \times \tilde{\mathcal{A}}$, then a relation of the form

$$X(t) \in x_0 + \int_{t_0}^t (E(s, X(s))d\wedge_\pi + F(s, X(s))dA_f(s) \\ + G(s, X(s))dA_g^+(s) + H(s, X(s))ds); t \in I$$

is called a stochastic integral inclusion with coefficient E, F, G, H and initial data (t_0, x_0). We shall sometimes write the foregoing inclusion as follows;

$$dX(t) \in E(t, X(t))d\wedge_\pi(t) + F(t, X(t))dA_f(t) \\ + G(t, X(t))dA_g^+(t) + H(t, X(t))dt \qquad (3)$$

for almost all $t \in I$, $X(t_0) = x_0$.

This we refer to as stochastic differential inclusions with coefficients E, F, G, H and initial data (t_0, X_0).

Definition 4 By a solution of (3) we mean a weakly absolutely continuous stochastic process $\varphi \in L^2_{loc}(\tilde{A})$ such that

$$\begin{aligned} d\varphi(t) \in\ & E(t, \varphi(t))d \wedge_\pi (t) + F(t, \varphi(t))dA_f(t) \\ & + G(t, \varphi(t))dA_g^+(t) + H(t, \varphi(t))dt \end{aligned}$$

almost all $t \in I$, $\varphi(t_0) = x_0$.

Remark 2 (i) The existence of solution to a stochastic differential inclusion with Lipschitzian coefficients has been proved in [6].

(ii) If $\mathcal{M}$ is a subset of $\tilde{A}$, we write $co\mathcal{M}$ for the closed convex hull of $\mathcal{M}$ and if $\Phi : I \times \tilde{A} \longrightarrow Clos(\tilde{A})$, we define
$co\ \Phi : I \times \tilde{A} \longrightarrow Clos(\tilde{A})$ by

$$(co\ \Phi)(t, X) = co\ \Phi(t, X), \quad (t, X) \in I \times \tilde{A}$$

(iii) Related to (3) is the following stochastic differential inclusion:

$$dX(t) \in co\ E(t, X(t))dA_\pi(t) + co\ F(t, X(t))dA_f(t)$$

$$+co\ G(t, X(t))dA_g^+(t) + co\ H(t, X(t))dt$$

almost all $t \in I$

$$X(t_0) = X_0 \tag{4}$$

In [6], Ekhaguere established equivalent form of (1) and (2.4) as follows; suppose $E, F, G, H \in L^2_{loc}(I \times \tilde{A})_{mus}$ and (t_0, x_0) is some fixed point of $I \times \tilde{A}$. Taking Theorems 4.1 and 4.4 of Hudson and Parthasarathy which describes the matrix elements of the quantum stochastic integral.
For $\eta, \xi \in I\!D\underline{\otimes} I\!E$, with $\eta = c \otimes e(\alpha)$ and $\xi = d \otimes c(\beta)$, define $\mu_{\alpha\beta}, \upsilon_\beta, \sigma_\alpha$: $I \to \mathbb{C}$, $I \subset I\!R_+$, by

$$\begin{aligned} \mu_{\alpha\beta}(t) &= \langle \alpha(t), \pi(t)\beta(t) \rangle_\gamma \\ \upsilon_\beta(t) &= \langle f(t), \beta(t) \rangle_\gamma \\ \sigma_\alpha(t) &= \langle \alpha(t), g(t) \rangle_\gamma \qquad t \in I. \end{aligned}$$

To these functions, are associated the maps $\mu E, \upsilon F, \sigma G, P$ and coP from $I \times \tilde{A}$ into the set of multivalued sesquilinear forms on $I\!D\underline{\otimes} I\!E$ defined by

$$
\begin{aligned}
(\mu E)(t,x)(\eta,\xi) &= \{\langle \eta, \mu_{\alpha\beta}(t)p(t,x)\xi\rangle : p(t,x) \in E(t,x)\} \\
(\nu F)(t,x)(\eta,\xi) &= \{\langle \eta, \nu_{\beta}(t)q(t,x)\xi\rangle : q(t,x) \in F(t,x)\} \\
(\sigma G)(t,x)(\eta,\xi) &= \{\langle \eta, \sigma_{\alpha}(t)u(t,x)\xi\rangle : u(t,x) \in G(t,x)\} \\
P(t,x)(\eta,\xi) &= (\mu E)(t,x)(\eta,\xi) + (\nu F)(t,x)(\eta,\xi) \\
&\quad + (\sigma G)(t,x)(\eta,\xi) + H(t,x)(\eta,\xi) \\
(coP)(t,x)(\eta,\xi) &= \text{closed convex/hull of } P(t,x)(\eta,\xi)
\end{aligned} \tag{5}
$$

$$\eta, \xi \in I\!D \underline{\otimes} I\!E, \ (t,x) \in I \times \tilde{\mathcal{A}}$$

where

$$H(t,x)(\eta,\xi) = \{v(t,x)(\eta,\xi) : v(\cdot, X(\cdot))$$

is a selection of $H(\cdot, X(\cdot))\ \forall\, X \in L^2_{loc}(\tilde{\mathcal{A}})$
$\eta, \xi \in I\!D \underline{\otimes} I\!E,\ (t,x) \in I \times \tilde{\mathcal{A}}$.
As in [6] we shall consider the equivalent form of (1) given by

$$
\begin{aligned}
\frac{d}{dt}\langle \eta, X(t)\xi\rangle &\in P(t, X(t))(\eta,\xi) \\
X(0) &= a, \quad t \in [0,T].
\end{aligned} \tag{6}
$$

Inclusion (2.4) is a nonclassical ordinary differential inclusion and the map $(\eta,\xi) \to P(t,x)(\eta,\xi)$ is a multivalued sesquilinear form on $(I\!D \underline{\otimes} I\!E)^2$ for $(t,x) \in [0,T] \times \tilde{\mathcal{A}}$. We refer the reader to [6] for the explicit forms of the map and the existence results for solutions of QSDI (1) of Lipschitzian coefficient.

3 Preliminary Results and Assumptions

As employed in [3, 4], let clos(N) denote the family of all nonempty closed subsets of a topological space N. For $N \in \{\tilde{\mathcal{A}}, \mathbb{C}\}$. We denote by $d(x, A)$, the distance from a point $x \in \mathbb{C}$ to a set $A \subseteq \mathbb{C}$. For $A, B \in$ clos($\mathbb{C}$), $\rho(A, B)$ denote the Hausdorff distance between the sets.

As it is the references [3, 4, 6], we shall employ the space $\mathbf{wac}(\tilde{\mathcal{A}})$ which is the completion of the locally convex topological space $(Ad(\tilde{\mathcal{A}})_{wac}, \tau)$ of adapted weakly absolutely continuous stochastic processes $\Phi : [0,T] \to \tilde{\mathcal{A}}$ whose topology τ is generated by the family of seminorms given by :

$$|\Phi|_{\eta\xi} := \|\Phi(0)\|_{\eta\xi} + \int_0^T |\frac{d}{dt}\langle \eta, \Phi(t)\xi\rangle| dt, \quad \text{for } \eta,\ \xi \in I\!D \underline{\otimes} I\!E. \tag{7}$$

Associated with the space $\mathbf{wac}(\tilde{\mathcal{A}})$, we shall employ the space $\mathbf{wac}(\tilde{\mathcal{A}})(\eta,\xi)$ consisting of absolutely continuous complex valued functions $\langle \eta, \Phi(\cdot)\xi\rangle$, where $\Phi \in$

wac$(\tilde{A})$ for arbitrary pair of points $\eta, \xi \in I\!D\underline{\otimes}I\!E$. By a solution of QSDI (1) or its equivalent form (2.4), we mean a stochastic process $\Phi : [0, T] \to \tilde{A}$ lying in the space $Ad(\tilde{A})_{wac} \bigcap L^2_{loc}(\tilde{A})$ satisfying QSDI (1) or its equivalent form (2).

Associated with the space $\tilde{A}$, we define the space of complex numbers $\tilde{A}(\eta, \xi) := \{\langle \eta, a\xi \rangle : a \in \tilde{A}\}$. We shall denote by $S^{(T)}(a)$, the subset of **wac**$(\tilde{A})$ consisting of the set of solutions of QSDI (1) corresponding to the initial value $a \in \tilde{A}$ and write $S^{(T)}(a)(\eta, \xi) = \{\langle \eta, \Phi(\cdot)\xi \rangle : \Phi \in S^{(T)}(a)\}$. Moreover, $S^{(T)}(P)(\eta, \xi) := \bigcup_{a \in \tilde{A}} S^{(T)}(a)(\eta, \xi)$. In what follows, $a \to S^{(T)}(a)$ is the multivalued solution map of QSDI (1) corresponding to the initial value $x = a$. It is well known that the set $S^{(T)}(a)$ is not empty for arbitrary $a \in \tilde{A}$ (see [3, 6]).

Definition 5 (1) $A(\eta, \xi) \subset \mathbb{C}$ is called an absolute retract if for any $\omega(\eta, \xi) \subset A(\eta, \xi)$, closed ball $\bar{B}(\eta, \xi) \subset \omega(\eta, \xi)$ and a continuous map $\bar{f} : \omega(\eta, \xi) \mapsto A(\eta, \xi)$.
There exists a continuous extension $\bar{f} : \omega(\eta, \xi) \mapsto A(\eta, \xi)$ of f.

(2) $A(\eta, \xi) \subset \mathbb{C}$ is said to be a compact R_δ, if $A(\eta, \xi)$ is compact and $A(\eta, \xi) = \bigcap_{n\geq 1} (A_n(\eta, \xi))$ for a decreasing sequence $(A_n(\eta, \xi))$ of compact absolute retracts, $(A_n(\eta, \xi)) \subset \mathbb{C}$.

(3) Let $f : B(\eta, \xi) \mapsto A(\eta, \xi)$ be a continuous map, a set $C(\eta, \xi) \subset A(\eta, \xi)$ is said to be contractible if there exists a continuous map $h(\cdot, \cdot) : [0, 1] \times C(\eta, \xi) \mapsto C(\eta, \xi)$ and a fixed point $x_{0,\eta\xi} \in C(\eta, \xi)$ such that $h(0, x_{\eta\xi}) = x_{\eta\xi}$ and $h(1, x_{\eta\xi}) = x_{0,\eta\xi}$ on $C(\eta, \xi)$.
Therefore, an absolute retract $A_\eta(\eta, \xi)$ is contractible.

Definition 6 The following definitions will be employed in what follows for the Upper Semicontinuous Multivalued Maps

(1) Let $\mathcal{N} \subset \tilde{A}$ and $I \subset I\!R_+$. For arbitrary $\eta, \xi \in I\!D\underline{\otimes}I\!E$, $(t, x), (t_0, x_0) \in I \times \mathcal{N}$ and real numbers $\epsilon, \delta_{\eta\xi} > 0$, we define the map $d_{\eta\xi} : I \times \mathcal{N} \to I\!R_+$ by

$$d_{\eta\xi}((t, x), (t_0, x_0)) = \max \left\{ |t - t_0| \, , \, \|x - x_0\|_{\eta,\xi} \right\}.$$

(2) For arbitrary $\eta, \xi \in I\!D\underline{\otimes}I\!E$ we define

$$\tilde{A}(\eta, \xi) = \{x_{\eta\xi} = \langle \eta, x\xi \rangle, x \in \tilde{A}\}$$

$$B(\eta, \xi) := B_{\eta\xi,\epsilon}(0) = \{x_{\eta\xi} \in \tilde{A}(\eta, \xi) : |x_{\eta\xi}| < \delta_{\eta\xi}\}.$$

$$B_{\delta_{\eta\xi}}(t_0, x_0) = \{(t, x) \in I \times \mathcal{N} : d_{\eta\xi}((t, x), (t_0, x_0)) < \delta_{\eta\xi}\}.$$

(3) A map $\phi : I \times \mathcal{N} \to 2^{sesq(I\!D\underline{\otimes}I\!E)^2}$ is said to be upper semicontinuous at a point $(t_0, x_0) \in I \times \mathcal{N}$, if for each pair $\eta, \xi \in I\!D\underline{\otimes}I\!E$ and $\epsilon > 0$ there exists $\delta_{\eta\xi} = \delta_{\eta\xi}((t_0, x_0), \epsilon) > 0$ such that

$$\phi(t, x)(\eta, \xi) \subset \phi(t_0, x_0)(\eta, \xi) + B_{\eta\xi}(0)$$

on $B_{\delta_{\eta\xi}}(t_0, x_0)$.

The map ϕ is said to be upper semicontinuous if it is upper semicontinuous at every point $(t, x) \in I \times \mathcal{N}$. Furthermore, for a sesquilinear formed valued map we define

$$B_{P,\epsilon}(0) = \{\varphi(t, x)(\eta, \xi) \in P(t, x)(\eta, \xi) : |\phi(t, x)(\eta, \xi)| < \epsilon\}$$

Remark 3 For example, if we consider $\omega(\eta, \xi) = [0, 1] \times A_n(\eta, \xi)$, $B(\eta, \xi) = \{0, 1\} \times A_n(\eta, \xi)$, $f(0, x_{\eta\xi}) = x_{\eta\xi} \in A_n(\eta, \xi)$, $f(1, x_{\eta\xi}) = x_{0,\eta\xi} \in A_n(\eta, \xi)$.

Therefore, a compact R_δ is the intersection of a decreasing sequence of compact contractible sets. Hence an absolute retract is contractible.

As employed in [6] we consider the space $wac(\tilde{A})$ the completion of the locally convex space $(Ad(\tilde{A})_{wac}, \tau_{wac})$ where the topology τ_{wac} is generated by the family of seminorms $\{| \cdot |_{\eta\xi} : \eta, \xi \in I\!\!D\underline{\otimes} I\!\!E\}$ defined for each $\phi \in Ad(\tilde{A})_{wac}$ by

$$|\Phi|_{\eta\xi} = \|\Phi(0)\|_{\eta\xi} + \int_0^k \left| \frac{d}{dx}\langle \eta, \Phi(s)\xi\rangle \right| ds$$

Associated with $wac(\tilde{A})$, we define for arbitrary $\eta, \xi \in I\!\!D\underline{\otimes} I\!\!E$, the space of complex value functions

$$wac(\tilde{A})(\eta, \xi) = \{\langle \eta, \Phi(\cdot)\xi\rangle | \Phi \in wac(\tilde{A})\}$$

and that each element $\Phi_{\eta\xi}(\cdot) := \langle \eta, \phi(\cdot)\xi\rangle$ of $wac(\tilde{A})(\eta, \xi)$ is an absolutely continuous complex valued function on the interval $[0, k]$. We assume that $A(\eta, \xi)$ is a subset of $\tilde{A}$ such that the set of complex numbers $A(\eta, \xi) = \{\langle \eta, a\xi\rangle | a \in A\}$ is compact in $\mathbb{C}$ with diameter

$$D_{\eta\xi} = \sup_{x,y \in A(\eta,\xi)} |x - y|.$$

Let $A(\eta, \xi)$ be the space of all continuous functions $x : [0, \infty) \longrightarrow \mathbb{C}$ such that there exists $\frac{d}{dt}\langle \eta, x\xi\rangle \in L^1_{loc}([0, \infty), \mathbb{C})$.

Next we state noncommutative generalization of the Kakutani fixed point theorem and Tangent cone which shall be employed in the main result.

Lemma 1 ([10]) *Assume that the maps $E, F, G, H \in L^2_{loc}(I \times \tilde{A})_{mvs}$ and $P : I \times \tilde{A} \to 2^{sesq(D\otimes E)^2}$, a sesquilinear form valued map with closed and convex values such that*

(a) $t \mapsto P(t, X(t))(\eta, \xi)$ *has a measurable selection.*
(b) $X \mapsto P(t, X(t))(\eta, \xi)$ *is upper semicontinuous*
(c) $\rho(P(t, X(t)))((\eta, \xi), \{0\}) \leq C(t)\left(1 + ||X||_{\eta\xi}\right)$ *on* $I \times \tilde{A}$, *with* $c \in L'_{loc}(I)$

(d) $\lim_{\tau\to 0^+}\alpha_{\eta\xi}\left(P(I_{t,\tau}\times B)(\eta,\xi)\right) \le |K(t)\alpha_{\eta\xi}(B)|$ *where* $P(I_{t,\tau}\times B)(\eta,\xi) = \{P(t, X(t)(\eta,\xi)) : (t, X) \in I_{t,\tau}\times B\}$, $I_{t,\tau} = [t-\tau, t+\tau]\cap I$, *for* $B \in \mathcal{B}$, $\eta, \xi \in \mathbb{D}\underline{\otimes}\mathbb{E}$ *and* $K \in L^1_{loc}(I)$. *Then the quantum stochastic differential Inclusion* $\frac{d}{dt}\langle\eta, X(t)\xi\rangle \in P(t, X(t))(\eta,\xi)$ $X(t_0) = x_0$ *a.e. on* I *has a solution on* I

Lemma 2 ([2]) *The set* $\mathcal{T}_{k(\eta,\xi)}(x_{\eta\xi})$ *of all vectors which are tangent to the set* $K(\eta,\xi)$ *at the point* $x_{\eta\xi}$ *is a closed cone.*

Assumptions $\mathcal{S}_{(1)}$ The coefficients E, F, G, H appearing in QSDI (1) are upper semi-continuous.
$\mathcal{S}_{(2)}$ The multivalued map $(t,x)\to P(t,x)(\eta,\xi)$ has nonempty and closed values as subsets of the field $\mathbb{C}$ of complex numbers.
$\mathcal{S}_{(3)}$ For each $x\in\tilde{\mathcal{A}}$, the map $t\to P(t,x)(\eta,\xi)$ is measurable.
$\mathcal{S}_{(4)}$ There exists a map $K^P_{\eta\xi} : [0,\infty]\to \mathbb{R}_+$ lying in $L^1_{loc}([0,T])$ such that

$$\rho(P(t,x)(\eta,\xi), P(t,y)(\eta,\xi)) \le K^P_{\eta\xi}(t)\|x-y\|_{\eta\xi} \tag{8}$$

for $t\in[0,T]$, and for each pair $x,\ y\in\tilde{\mathcal{A}}$.
$\mathcal{S}_{(5)}$ There exist a stochastic process $Y : [0,T]\to\tilde{\mathcal{A}}$ lying in $Ad(\tilde{\mathcal{A}})_{wac}$ such that for each pair $\eta,\xi\in\mathbb{D}\underline{\otimes}\mathbb{E}$,

$$d\left(\{0\}, P(t, Y(t))(\eta,\xi)\right) \le \beta_{\eta\xi}(t), \tag{9}$$

for almost all $t\in[0,\infty]$ and for some locally integrable map $\beta_{\eta\xi} : [0,\infty]\to R_+$.

4 Main Result

Theorem 1 *Suppose*

(a) *The maps* $E, F, G, H \in L^2_{loc}(I\times\tilde{\mathcal{A}})_{mvs}$
(b) $P : I\times\tilde{\mathcal{A}} \to 2^{sesq(D\otimes E)^2}$, *is a sesquilinear form valued map with closed and convex values.*
(c) *Conditions (a-d)of Lemma 1 are satisfied such that on* $I\times\tilde{\mathcal{A}}$, *with* $c\in L'_{loc}(I)$ $A(\eta,\xi)\subset\tilde{\mathcal{A}}$, $P(t, X(t))(\eta,\xi)\subset T_{A(\eta,\xi)}(x_{\eta\xi})$ *on* $(t_0,T)\times A(\eta,\xi)$.

Then for $x_{\eta\xi}\in A(\eta,\xi)$, *there exist a decreasing Sequence of Compact Contractible*

$$S_{T,n}(a)(\eta,\xi)\subset Ad(\tilde{\mathcal{A}})_{wac}\bigcap L^2_{loc}(\tilde{\mathcal{A}})$$

Such that $S_T(x)(\eta,\xi)\subset S_{T,n}(a)(\eta,\xi)$ *and* $d\left(S_{T,n}(a)(\eta,\xi), S_T(x)(\eta,\xi)\right)\to 0$ *as* $n\to\infty$. *In particular* $S_T(x)(\eta,\xi)$ *is connected.*

Proof Assume on $A(\eta,\xi)$, $\|P(t,X(t))(\eta,\xi)\|\le 1$ on $[t_0,T]\times A(\eta,\xi)$ and consider the approximations

$$P_n(t,X(t))(\eta,\xi)=\sum_{\lambda\in\Lambda} Q_\lambda(x_{\eta,\xi})C_\lambda(t)(\eta,\xi)$$

with $C_\lambda(t)(\eta,\xi)=\overline{conv}P\left(t,B_{2r_n}(x_\lambda)\right)(\eta,\xi)$. We recall that $(Q_\Lambda(\cdot)(\eta,\xi))_{\lambda\in\Lambda}$ is a locally Lipschitz partition of unity subordinate to some locally finite refinement [10] $(U_\lambda(\cdot)(\eta,\xi))_{\lambda\in\Lambda}$ of $\left\{B_{r_n}(X_1)(\eta,\xi):X_{\eta\xi}\in A(\eta,\xi)\right\}$ with $r_n=3^{-n}$, and $x_{\lambda,\eta\xi}$ is such that $U_\lambda(\cdot)(\eta,\xi)\subset B_{r_n}(x_\lambda)(\eta,\xi)$, then $P(t,X)(\eta,\xi)\subset P_{n+1}(t,X)(\eta,\xi)\subset P_n(t,X)(\eta,\xi)\subset\overline{Conv}P\left(t,B_{3r_n}(X)\right)(\eta,\xi)$...(*) on $(t_0,T]\times A(\eta,\xi)$. Therefore, we let $S_{T,n}(\cdot)(\eta,\xi)$ be the solution set of $\frac{d}{dt}\langle\eta,X(t)\xi\rangle\in P(t,X(t))(\eta,\xi)$ on $[t_0,T)$, $X(0)=x_0\in A(\eta,\xi)$ with $P_n(t,X(t))(\eta,\xi)$ instead of $P(t,X(t))(\eta,\xi)$ and then we have $S_T(x)(\eta,\xi)\subset S_{T,n+1}(\cdot)(\eta,\xi)\subset S_{T,n}(\cdot)(\eta,\xi)$. Since $P_n(t,\cdot)(\eta,\xi)$ satisfy (*) then $S_{T,n}(\cdot)(\eta,\xi)$ is compact. To show that $S_{T,n}(\cdot)(\eta,\xi)$ is contractible, let $g_{\lambda,\eta\xi}\in P(\cdot,X_\lambda)(\eta,\xi)$ be a measurable selection and we define

$$f:[t_0,T]\times A(\eta,\xi)\to\overline{B_1}(0)(\eta,\xi)$$

by

$$f(t,x)(\eta,\xi)=\sum_\lambda Q_\lambda(x_{\eta\xi})g_{\lambda,\eta\xi}(t)\in P_n(t,X)(\eta,\xi)$$

on $[t_0,T)\times A(\eta,\xi)$ Since $(U_\lambda(\cdot)(\eta,\xi))$ is locally finite, $f(\cdot,x)(\eta,\xi)$ is measurable and $f(t,\cdot)(\eta,\xi)$ is locally lipschitz. Consequently, the non-classical initial value problem

$$\frac{d}{dt}\langle\eta,Z\xi\rangle=f(t,Z(t))(\eta,\xi)$$

a.e on $[t_0,T)$, $Z(0)=y$ has a unique solution $z(\cdot,s,y_{\eta\xi})$ which depend continuously on $(s,y_{\eta\xi})\in[t_0,T)\times A(\eta,\xi)$. Hence

$$h_n(\tau,u)(t)(\eta,\xi)=\begin{cases}U_{\eta\xi}(t) & \text{in } [0,\tau a]\\ Z_{\eta\xi}\left(t;\tau a,u_{\eta\xi}(\tau a)\right) & \text{in } (\tau a,a]\end{cases}$$

defines a continuous

$$h_n:[0,1]\times S_{T,n}(\cdot)(\eta,\xi)\to S_{T,n}(\cdot)(\eta,\xi)$$

with $h_n(0,U_{\eta\xi})=Z_\eta(\cdot:0,x_{0,\eta\xi})$ and $h_n(1,U_{\eta\xi})=U_{\eta\xi}$ and therefore $S_{T,n}(\cdot)(\eta,\xi)$ is contractible. By $P(t,X)(\eta,\xi)\subset P_{n+1}(t,X)(\eta,\xi)\subset P_n(t,X)(\eta,\xi)\subset\overline{Conv}P\left(t,B_{3r_n}(X)\right)(\eta,\xi)$ on $(t_0,T]\times A(\eta,\xi)$ a sequence $(U_{n,\eta\xi})$ with $U_{n,\eta\xi}\in S_{T,n}(\cdot)(\eta,\xi)$ for $n\ge 1$ has a Uniformly Convergent Subsequence with limits in $S_T(X)(\eta,\xi)$ and we have

$$d\left(S_{T,n}(\cdot)(\eta,\xi), S_T(x)(\eta,\xi)\right) \to 0$$

as $n \to \infty$. Since $S_T(X)(\eta,\xi) \subseteq S_{T,n}(\cdot)(\eta,\xi)$ is compact. As a result, $S_T(X)(\eta,\xi)$ is not a union of two disjoint nonempty compact sets, because the $S_{T,n}(\cdot)(\eta,\xi)$ is arcwise connected. Hence $S_T(X)(\eta,\xi)$ is connected.

References

1. Aubin, J.P., Cellina, A.: Differential Inclusions. Springer, Berlin (1984)
2. Akinwunmi, T.O. and Ayoola, E.O.: Viable Solutions of Lower Semicontinuous Quantum Stochastic Differential Inclusions, Analysis and Mathematical Physics (2020)
3. Ayoola, E.O.: Topological properties of solution sets of Lipschitzian quantum stochastic differential inclusions. Acta Appl. Math. **100**(1), 15–37 (2008)
4. Dikko, D.A., Ayoola, E.O.: Arcwise connectedness of solution sets of Lipschitzian quantum stochastic differential inclusions. J. Phys: Conf. Ser. **819**(3012005) (2017)
5. Dikko, D.A.: Further results on the arcwise connectedness of solution sets of discontinuous quantum stochastic differential inclusions. Inf. Dimens. Anal. Quantum Prob. Rel. Top. **23**(3)(2050022) (2020)
6. Ekhaguere, G.O.S.: Lipschitzian quantum stochastic differential inclusions. Int. J. Theoret. Phys. **31**(11), 2003–2034 (1992)
7. Hudson, R.L., Parthasarathy, K.R.: Quantum Ito's formulae and stochastic evolutions. Commun. Math. Phys. **93**, 301–324 (1984)
8. Deimling, K., Mohana Rao, M.R.: On solution sets of multivalued differential equations. Appl. Anal. **30**, 129–135 (1988)
9. Ogundiran, M.O., Ayoola, E.O.: An application of Michael selection theorems. Int. J. Func. Anal. Oper. Theory Appl. **21**, 65–79 (2012)
10. Ogundiran, M.O., Ayoola, E.O.: Upper semicontinuous quantum stochastic differential inclusions via Kakutani fixed point theorem. Dyn. Syst. Appl. **21**(3), 121–132 (2012)
11. Ogundiran, M.O., Ayoola, E.O.: Mayer problem for quantum stochastic control. J. Math. Phys. **51**, 023521 (2010)
12. Ogundiran, M.O., Ayoola, E.O.: Lower semicontinuous quantum stochastic differential inclusions. Eur. J. Math. Sci. **2**(1), 1–16 (2012)
13. Winship Heath, R.: Arcwise connectedness in semi metric spaces. Pac. J. Math. **12**(4), 1301–1319 (1962)
14. Suneja, S.K., Aggarwal, S.: Arcwise cone connected functions and optimality. Opsearch **37**(3), 1–16 (2000)

Fractional Operators from Vanishing Morrey to Vanishing Campanato Spaces in the Variable Exponent Setting on Quasi-metric Measure Spaces

Humberto Rafeiro and Stefan Samko

Abstract In the framework of bounded quasi-metric measure spaces (X, d, μ), we prove that the fractional operator of variable order $\alpha(x)$ is bounded from vanishing variable exponent Morrey space $VL^{p(\cdot),\lambda(\cdot)}(X)$ to vanishing variable exponent Campanato space $\mathrm{V}\mathcal{L}^{p(\cdot),\lambda(\cdot)}(X)$, when $\gamma \leqslant \alpha(x)p(x) + \lambda(x) \leqslant \gamma + \theta p(x)$. We do not assumed that the space (X, d, μ) is homogeneous and impose only the growth condition $\mu B(x, r) \leqslant Cr^\gamma$ on the measure and suppose that the distance is θ-regular in the sense of R. Macías and C. Segovia.

Keywords Fractional operator · Morrey space · Campanato space · Quasi-metric measure space

1 Introduction

To study the action of the fractional operator

$$I^{\alpha(\cdot)} f(x) = \int_\Omega f(y)|x - y|^{\alpha(x)-n} dy, \quad x \in \Omega \subset \mathbb{R}^n,$$

in variable exponent Morrey space $L^{p(\cdot),\lambda(\cdot)}(\Omega)$, see (10) for the definition of such space, one needs to consider three distinct cases,

H. Rafeiro (✉)
Department of Mathematical Sciences, College of Science, United Arab Emirates University, P.O. Box 15551, Al Ain, United Arab Emirates
e-mail: rafeiro@uaeu.ac.ae

S. Samko
Department of Mathematics, University of Algarve, FCT Faro, Portugal
e-mail: ssamko@ualg.pt

Kh. Ibragimov Complex Institute of the Russian Academy of Science, Grozny, Russia

L. Accardi et al. (eds.), *Infinite Dimensional Analysis, Quantum Probability and Applications*, Springer Proceedings in Mathematics & Statistics 390,
https://doi.org/10.1007/978-3-031-06170-7_16

$$1)\ \alpha(x)-\frac{n-\lambda(x)}{p(x)}<0,\quad 2)\ \alpha(x)=\frac{n-\lambda(x)}{p(x)},\quad \text{and}\quad 3)\ \alpha(x)-\frac{n-\lambda(x)}{p(x)}>0,$$

the so-called *pre-borderline*, *borderline*, and *post-borderline case*, respectively.

The result for the pre-borderline case when α, λ, and p are all constants is due to D. Adams [1], see [2] for the case of variable exponents. The extension to the borderline and post-borderline cases was done in variable exponent setting, see [14–16]. The borderline case when $\lambda(x) \equiv 0$ may be found in [20, 21]. Mapping properties of the fractional operator, in the borderline case when all the exponents are constant, from Morrey spaces were characterized in [17], not via BMO spaces but via grand Lebesgue spaces corresponding to $p = \infty$.

From [14, 16], it became clear that the proper language to treat the borderline and post-borderline case is via Campanato spaces, which allows to obtain such cases in a unified way. Recently, in [12], this approach was generalized to the setting of quasi-metric measure spaces. Namely, for the operator

$$I^{\alpha(\cdot)} f(y) = \int_X \frac{f(z)}{d(y,z)^{\gamma-\alpha(y)}} d\mu(z),$$

the authors studied Morrey to Campanato mapping properties over bounded quasi-metric measure space (X, d, μ) with the growth condition $\mu B(x,r) \leqslant C r^\gamma$, $\gamma > 0$, and the distance being θ-regular in the sense of R. Macías and C. Segovia, see Definition 1. It has been proved in [12], under the assumptions of Theorem 2, that

$$I^{\alpha(\cdot)} : L^{p(\cdot),\lambda(\cdot)}(X) \hookrightarrow \mathfrak{L}^{p(\cdot),\lambda(\cdot)+\alpha(\cdot)p(\cdot)}_{\mathfrak{c}(\cdot,\cdot)}(X). \tag{1}$$

Such boundedness was obtained when $\gamma \leqslant \alpha(x)p(x) + \lambda(x) \leqslant \gamma + \theta p(x)$, and special attention was needed to included the endpoints of the inequalities. In the case of the right-hand side inequality, a certain "correcting factor" of logarithmic type was introduced in the target Campanato space.

The vanishing Morrey space with constant characteristics was introduced in [23] related with applications to PDEs, see also [3, 18] for some properties of such spaces. In [19] it was studied the boundedness of maximal, potential, and singular operators in vanishing generalized Morrey spaces. It is worth pointing out that the boundedness of Calderón–Zygmund operators on generalized variable exponent vanishing Morrey spaces was obtained in [7].

In this paper, we show that the result similar to (1) holds, see Theorem 2, whenever the source and target space are replaced with their corresponding vanishing variable exponent versions. Extension to vanishing spaces is based on estimation of Campanato modular of fractional integral $I^{\alpha(\cdot)} f$ via Morrey modular of the function f. This result seems new even in the Euclidean case and even for constant exponents, where the previous known case, see [14], was just for the borderline case in the Euclidean setting. Our result covers both borderline and post-bordeline case and in the setting of bounded quasi-metric measure spaces satisfying only the growth condition.

The paper is organized as follows: in Sect. 2 we provide main definitions and some auxiliary results and in Sect. 3 we prove the boundedness result.

2 Preliminaries

2.1 Quasi-metric Measure Spaces

By a quasi-metric measure space we mean a triplet (X, d, μ), where X is a bounded set, μ a regular Borel measure, and d a quasi-distance, i.e., it satisfies the conditions: $d(x, y) \geqslant 0$, $d(x, y) = 0 \Leftrightarrow x = y$, $d(x, y) = d(y, x)$, and

$$d(x, y) \leqslant \varkappa\left(d(x, z) + d(z, y)\right), \quad \varkappa \geqslant 1. \tag{2}$$

We denote $\operatorname{diam} X =: \mathfrak{D} < \infty$, $B(x, r) := \{y \in X : d(x, y) < r\}$, and $S(x, r) := \{y \in X : d(x, y) = r\}$. We assume that all $B(x, r)$ are measurable sets and $\mu S(x, r) = 0$ for all $x \in X$ and $r \geqslant 0$.

The measure μ satisfies the so-called *growth condition*, also known as the *upper Ahlfors condition*, if for some $\gamma > 0$, we have

$$\mu B(x, r) \leqslant C r^{\gamma}. \tag{3}$$

From [10], we know that every quasi-metric space (X, d) admits an equivalent quasi-distance d_* with Lipschitz property; i.e., there exists $\theta \in (0, 1]$ such that

$$|d_*(x, z) - d_*(y, z)| \leqslant M d_*^{\theta}(x, y)\, [d_*(x, z) + d_*(y, z)]^{1-\theta} \tag{4}$$

and the quasi-distance d_* is a power of a distance ρ; i.e., $d_*(x, y) = \rho(x, y)^{\frac{1}{\theta}}$.

Definition 1 We say that the quasi-distance d is regular of order $\theta \in (0, 1]$ if it itself satisfies the inequality (4).

When (X, d) is a quasi-metric space, regular of order $\theta \in (0, 1]$, we have

$$|d(x, z)^{-\beta} - d(y, z)^{-\beta}| \leqslant \frac{|\beta|}{\theta} d(x, y)^{\theta} \min\{d(x, z), d(y, z)\}^{-\beta-\theta}, \; \beta + \theta > 0, \tag{5}$$

which follows from the inequality $|a^{-\beta} - b^{-\beta}| \leqslant |\beta||a - b| \max\{a^{-\beta-1}, b^{-\beta-1}\}$, $\beta > -1$ and the fact that there exists a distance ρ such that $d(x, y) = \rho(x, y)^{\frac{1}{\theta}}$, see [22].

Recall that $\widetilde{M_N} f(x) = \sup_{r>0} \frac{1}{\mu B(x,Nr)} \int_{B(x,r)} |f(y)| d\mu(y)$ is the so-called *modified maximal operator*, see e.g. [8, Sect. 13.5.2].

For future use, we recall the following result proved in [12].

Lemma 1 *Let (X, d, μ) be a quasi-metric measure space, μ satisfies* (3), *and* $\inf_{x\in X} \alpha(x) > 0$. *Then, for $N > 0$, the estimate*

$$\left|\int_{d(x,y)<r} \frac{f(y)}{d(x,y)^{\gamma-\alpha(x)}} d\mu(y)\right| \leqslant C_N r^{\alpha(x)} \widetilde{M_N} f(x) \tag{6}$$

holds, where the constant C_N does not depend on f, r, and x.

2.2 Variable Exponent Spaces

The variable exponent Lebesgue space $L^{p(\cdot)}(X)$ is defined as the set of measurable functions such that

$$\|f\|_{L^{p(\cdot)}(X)} := \inf\left\{\eta > 0 : \int_X \left|\frac{f(x)}{\eta}\right|^{p(x)} d\mu(x) \leqslant 1\right\} < \infty.$$

We say that $p \in \mathbb{P}^{\log}(X)$ if $1 < p_- \leqslant p(x) \leqslant p_+ < \infty$ and

$$|p(x) - p(y)| \leqslant \frac{C}{-\ln d(x,y)}, \quad d(x,y) \leqslant \frac{1}{2}, \quad x, y \in X, \tag{7}$$

where C does not depend on x, y. Likewise, we say that $p \in \mathbb{P}_\mu^{\log}(X)$ if

$$|p(x) - p(y)| \leqslant \frac{C}{-\ln \mu B(x, d(x,y))}, \quad \mu B(x, d(x,y)) < 1/2, \quad x, y \in X.$$

If μ satisfies (3) then $\mathbb{P}_\mu^{\log}(X) \subset \mathbb{P}^{\log}(X)$, see [9, Lemma 2.56].

Lemma 2 ([13, Lemma 2.5]) *Let (X, d, μ) be a quasi-metric measure space with finite measure and $p \in \mathbb{P}_\mu^{\log}(X)$. Then $\|\chi_{B(x,r)}\|_{L^{p(\cdot)}(X)} \leqslant C(\mu B(x,r))^{\frac{1}{p(x)}}$.*

Lemma 3 ([12, Lemma 2.3]) *Let (X, d, μ) be a bounded quasi-metric measure space, $p \in \mathbb{P}^{\log}(X)$, μ satisfy the condition* (3), *and $\sup_{x\in X} \beta(x) < \infty$. Then*

$$\|d(x,\cdot)^{\beta(x)} \chi_{B(x,2r)\setminus B(x,r)}(\cdot)\|_{L^{p(\cdot)}(X)} \leqslant C r^{\beta(x)+\frac{\gamma}{p(x)}}, \quad 0 < r < \mathfrak{D}/2, \tag{8}$$

where C does not depend on x and r.

A modified formulation of [8, Theorem 13.73] is given in Theorem 1.

Theorem 1 *Let (X, d, μ) be a quasi-metric measure space, $\mu X < \infty$, and $1 < p_- \leqslant p_+ < \infty$. If exists a constant $C > 0$ such that, for all $x \in X$ and $r > 0$,*

$$\sup_{y\in B(x,r)} [p(x) - p(y)] \leqslant \frac{C}{-\ln \mu B(x, 3\varkappa r)}, \quad \mu B(x, 3\varkappa r) < 1/2, \tag{9}$$

then $\widetilde{M_{3\varkappa}}$ *is bounded in* $L^{p(\cdot)}(X)$, *where* $\varkappa$ *is the constant from* (2).

We refer the reader to [4, 5] for more details regarding variable exponent Lebesgue spaces in the Euclidean setting, and to [9] for several results in the setting of quasi-metric measure spaces.

The variable exponent Morrey space $L^{p(\cdot),\lambda(\cdot)}(X)$ is defined as the set of measurable functions for which

$$\|f\|_{L^{p(\cdot),\lambda(\cdot)}(X)} = \sup_{x\in X, r>0} \mathfrak{M}_{p(\cdot),\lambda(\cdot)}(f; x, r) < \infty, \tag{10}$$

where $\mathfrak{M}_{p(\cdot),\lambda(\cdot)}(f; x, r) := r^{-\frac{\lambda(x)}{p(x)}} \|f\chi_{B(x,r)}\|_{L^{p(\cdot)}(X)}$. See [2, 8] for other definitions of $L^{p(\cdot),\lambda(\cdot)}$ and [11] for several generalizations of Morrey spaces.

By $\mathfrak{L}^{p(\cdot),\lambda(\cdot)}(X)$ and $\mathcal{L}^{p(\cdot),\lambda(\cdot)}(X)$ we denote two versions of variable exponent Campanato space defined by the norms

$$\|f\|_{\mathfrak{L}^{p(\cdot),\lambda(\cdot)}(X)} := \|f\|_{L^1(X)} + [f]_{\mathfrak{L}^{p(\cdot),\lambda(\cdot)}(X)},$$
$$\|f\|_{\mathcal{L}^{p(\cdot),\lambda(\cdot)}(X)} := \|f\|_{L^1(X)} + \lceil f \rfloor_{\mathcal{L}^{p(\cdot),\lambda(\cdot)}(X)},$$

where

$$[f]_{\mathfrak{L}^{p(\cdot),\lambda(\cdot)}(X)} := \sup_{x\in X, r>0} \inf_{c\in\mathbb{R}} \mathfrak{M}_{p(\cdot),\lambda(\cdot)}(f - c; x, r),$$
$$\lceil f \rfloor_{\mathcal{L}^{p(\cdot),\lambda(\cdot)}(X)} := \sup_{x\in X, r>0} \mathfrak{M}_{p(\cdot),\lambda(\cdot)}(f - \langle f\rangle_{B(x,r)}; x, r),$$

and $\langle f\rangle_E := \frac{1}{\mu E}\int_E f(y)d\mu(y)$. When $p \in \mathbb{P}^{\log}_{\mu}(X)$, the norms of $\mathfrak{L}^{p(\cdot),\lambda(\cdot)}(X)$ and $\mathcal{L}^{p(\cdot),\lambda(\cdot)}(X)$ are equivalent, see [12, Theorem 2.2].

The semi-norm

$$[f]_{\mathcal{L}^{p(\cdot),\lambda(\cdot)}_{\mathfrak{c}(\cdot,\cdot)}(X)} = \sup_{x\in X, r>0} \inf_{c\in\mathbb{R}} \mathfrak{c}(x, r)\mathfrak{M}_{p(\cdot),\lambda(\cdot)}(f - c; x, r),$$

allows us to introduce, in a similar way, the "weighted" variable exponent Campanato space $\mathcal{L}^{p(\cdot),\lambda(\cdot)}_{\mathfrak{c}(\cdot,\cdot)}(X)$, where $\mathfrak{c}(x, r)$ is a non-negative function which will be specified in Theorem 2.

The vanishing variable exponent Morrey space, denoted by $VL^{p(\cdot),\lambda(\cdot)}(X)$, is introduced as the space of all functions $f \in L^{p(\cdot),\lambda(\cdot)}(X)$ satisfying additionally the vanishing condition

$$\lim_{r\to 0} \sup_{x\in X} \mathfrak{M}_{p(\cdot),\lambda(\cdot)}(f; x, r) = 0.$$

Similarly, by $V\mathcal{L}^{p(\cdot),\lambda(\cdot)}_{\mathfrak{c}(\cdot,\cdot)}(X)$ we denote the set of all functions $f \in \mathcal{L}^{p(\cdot),\lambda(\cdot)}_{\mathfrak{c}(\cdot,\cdot)}(X)$ satisfying the additional vanishing condition

$$\lim_{r\to 0} \sup_{x\in X} \mathfrak{M}_{p(\cdot),\lambda(\cdot)}(f - \langle f\rangle_{B(x,r)}; x, r) = 0.$$

We define the *generalized variable Hölder space* $H^{\omega(\cdot,\cdot)}$ by

$$[f]_{H^{\omega(\cdot,\cdot)}(X)} = \sup_{r>0} \sup_{x,y\in X} \frac{|f(x)-f(y)|}{\omega(x,d(x,y))} < \infty,$$

where ω, defined on $X \times [0, \operatorname{diam} X]$, is a continuity modulus for every $x \in X$. The latter means that $\omega(x,\cdot) \subset C([0,\operatorname{diam} X])$, $\omega(x,0)=0$ and $\omega(x,t)>0$, $0<t\leqslant \operatorname{diam} X$, $x \in X$, and $\omega(x,t)$ is increasing in t.

3 Boundedness Result

We denote $a_{p,\lambda,\alpha}(x) := \alpha(x) - \frac{\gamma-\lambda(x)}{p(x)}$ and assume that

$$0 \leqslant a_{p,\lambda,\alpha}(x) \leqslant \theta \tag{11}$$

and

$$\sup_{x\in X} \lambda(x) < \gamma. \tag{12}$$

In the proof of the following theorem we based ourselves on the inequality

$$\begin{aligned}&\|g-\langle g\rangle_{B(x,r)}\|_{L^{p(\cdot)}(B(x,r))}\\ &\qquad \leqslant 2(\|g_1(\cdot,r)-c_1\|_{L^{p(\cdot)}(B(x,r))} + \|g_2(\cdot,r)-c_2\|_{L^{p(\cdot)}(B(x,r))})\end{aligned}$$

for any splitting $g(y) = g_1(y,r) + g_2(y,r)$, where $c_i = c_i(x,r,g)$ are arbitrary constants. The inequality is valid for $p \in \mathbb{P}^{\log}_\mu(X)$, see [12, Eq. (3.4)].

Theorem 2 *Let (X,d,μ) be a bounded quasi-metric measure space, μ satisfy* (3), *$p \in \mathbb{P}^{\log}_\mu(X)$, and λ satisfy* (12) *and* (7). *Assume also that*

$$\alpha \in H^{\omega(\cdot,\cdot)}(X) \quad \textit{and} \quad \omega(x,r) \leqslant C\frac{r^{a_{p,\lambda,\alpha}(x)}}{\left(\ln \frac{2\mathfrak{D}}{r}\right)^2}. \tag{13}$$

Then, under the condition (11),

$$I^{\alpha(\cdot)} : VL^{p(\cdot),\lambda(\cdot)}(X) \to V\mathcal{L}^{p(\cdot),\lambda(\cdot)+\alpha(\cdot)p(\cdot)}_{\mathfrak{c}(\cdot,\cdot)}(X),$$

is bounded, where

$$\mathfrak{c}(x,r) := \left(r^{\theta-a_{p,\lambda,\alpha}(x)}\int_r^{2\mathfrak{D}} \frac{dt}{t^{1+\theta-a_{p,\lambda,\alpha}(x)}}\right)^{-1}.$$

Proof It suffices to prove that the vanishing property is preserved under the operator $I^{\alpha(\cdot)}$, since (1) is already known under the assumed conditions, see [12].

Fix $r > 0$, $x \in X$, and suppose that $f \geqslant 0$. We decompose the function f as $f(z) = f(z)\chi_{B(x,2\varkappa r)}(z) + f(z)\chi_{X\setminus B(x,2\varkappa r)}(z) =: f_1(z) + f_2(z)$, where $\varkappa$ is the constant from (2). From the previous splitting, it follows $I^{\alpha(\cdot)} f(y) = I^{\alpha(\cdot)} f_1(y) + I^{\alpha(\cdot)} f_2(y)$.

ESTIMATION OF $I^{\alpha(\cdot)} f_1$: From (13) we have that α is log-Hölder continuous which, together with (6), entails the inequality $I^{\alpha(\cdot)} f_1(y) \leqslant Cr^{\alpha(x)}\widetilde{M_N} f_1(y)$.

Since $p \in \mathbb{P}^{\log}_{\mu}(X)$ implies that p satisfies condition (9), then from Theorem 1 we have $\|I^{\alpha(\cdot)} f_1\|_{L^{p(\cdot)}(B(x,r))} \leqslant Cr^{\alpha(x)}\|f\|_{L^{p(\cdot)}(B(x,r))}$, yielding immediately the inequality $r^{-\frac{\lambda(x)}{p(x)}-\alpha(x)}\|I^{\alpha(\cdot)} f_1\|_{L^{p(\cdot)}(B(x,r))} \leqslant Cr^{-\frac{\lambda(x)}{p(x)}}\|f\|_{L^{p(\cdot)}(B(x,r))}$, from which, by the fact that $f \in VL^{p(\cdot),\lambda(\cdot)}(X)$, it follows

$$\limsup_{r\to 0}\sup_{x\in X} \frac{\mathfrak{c}(x,r)}{r^{\frac{\lambda(x)}{p(x)}+\alpha(x)}}\|I^{\alpha(\cdot)} f_1\|_{L^{p(\cdot)}(B(x,r))} = 0. \tag{14}$$

ESTIMATION OF $I^{\alpha(\cdot)} f_2$: Let $y \in B(x,r)$ and define the constant $c_F := F_2(x) = \int_{X\setminus B(x,2\varkappa r)} \frac{f(z)}{d(x,z)^{\gamma-\alpha(x)}}d\mu(z)$, where $\varkappa$ comes from (2). We have

$$\begin{aligned}|I^{\alpha(\cdot)} f_2(y) - c_F| &\leqslant \int_{X\setminus B(x,2\varkappa r)} f(z)\left|\frac{1}{d(y,z)^{\gamma-\alpha(x)}} - \frac{1}{d(x,z)^{\gamma-\alpha(x)}}\right| d\mu(z)\\ &+ \int_{X\setminus B(x,2\varkappa r)} f(z)\left|\frac{1}{d(y,z)^{\gamma-\alpha(y)}} - \frac{1}{d(y,z)^{\gamma-\alpha(x)}}\right| d\mu(z) =: G_1 + G_2.\end{aligned}$$

From (5), inequality $d(x,z) \leqslant 2\varkappa d(y,z)$, and dyadic decomposition, we obtain

$$\begin{aligned}G_1 &\leqslant C\frac{\gamma-\alpha(x)}{\theta}d(x,y)^{\theta}\int_{X\setminus B(x,2\varkappa r)} \frac{f(z)}{d(x,z)^{\gamma-\alpha(x)+\theta}}d\mu(z)\\ &\leqslant C\frac{\gamma-\alpha(x)}{\theta}d(x,y)^{\theta}\sum_{1\leqslant k<\log_2(\frac{2\mathfrak{D}}{r})}\int_{2^k\varkappa r\leqslant d(x,z)<2^{k+1}\varkappa r} \frac{f(z)}{d(x,z)^{\gamma-\alpha(x)+\theta}}d\mu(z).\end{aligned}$$

Applying Hölder's inequality for variable exponent, the estimate (8), and the fact that μ satisfies the growth condition, we obtain

$$G_1 \leqslant Cr^{\theta}\sum_{1\leqslant k<\log_2(\frac{2\mathfrak{D}}{r})}\|f\|_{L^{p(\cdot)}(B(x,2^{k+1}\varkappa r))}(2^k\varkappa r)^{\frac{\alpha(x)p(x)-\gamma}{p(x)}-\theta}. \tag{15}$$

Recall that, for an increasing function φ and a constant $\varrho \geqslant 0$, we have

$$\int_r^{4\mathfrak{D}} \frac{\varphi(t)}{t^{\varrho}}\frac{dt}{t} \geqslant C\sum_{0\leqslant k<\log_2(\frac{2\mathfrak{D}}{r})} \frac{\varphi(2^k r)}{(2^{k+1}r)^{\varrho}}. \tag{16}$$

Consequently, by (15), (16), and the fact that $\varkappa \geqslant 1$, we get

$$G_1 \leqslant Cr^\theta \int_{2\varkappa r}^{4\mathfrak{D}} \frac{\|f\|_{L^{p(\cdot)}(B(x,t))}}{t^{\theta+\frac{\gamma}{p(x)}-\alpha(x)}}\frac{dt}{t} \leqslant Cr^\theta \int_r^{4\mathfrak{D}} \frac{\mathfrak{M}_{p(\cdot),\lambda(\cdot)}(f;x,t)}{t^{\theta-a_{p,\lambda,\alpha}(x)}}\frac{dt}{t}. \tag{17}$$

From (17), Lemma 2, and the growth condition, it follows

$$\begin{aligned}
&\frac{\mathfrak{c}(x,r)}{r^{\frac{\lambda(x)}{p(x)}+\alpha(x)}}\|G_1\|_{L^{p(\cdot)}(B(x,r))}\\
&\leqslant C\frac{\mathfrak{c}(x,r)r^\theta}{r^{\frac{\lambda(x)}{p(x)}+\alpha(x)}}\int_r^{4\mathfrak{D}} \frac{\mathfrak{M}_{p(\cdot),\lambda(\cdot)}(f;x,t)}{t^{\theta-a_{p,\lambda,\alpha}(x)}}\frac{dt}{t}\|1\|_{L^{p(\cdot)}(B(x,r))} \qquad (18)\\
&\leqslant Cr^{\theta-a_{p,\lambda,\alpha}(x)}\mathfrak{c}(x,r)\int_r^{4\mathfrak{D}} \frac{\mathfrak{M}_{p(\cdot),\lambda(\cdot)}(f;x,t)}{t^{\theta-a_{p,\lambda,\alpha}(x)}}\frac{dt}{t} =: I(x,r).
\end{aligned}$$

From the vanishing condition on f, for a fixed $\varepsilon > 0$, there exists $\delta = \delta(\varepsilon) > 0$ such that $\sup_{x\in X}\mathfrak{M}_{p(\cdot),\lambda(\cdot)}(f;x,t) < \varepsilon$, for all $0 < t < \delta$. Thus, for $0 < r < \delta$, we have

$$\begin{aligned}
I(x,r) &\leqslant r^{\theta-a_{p,\lambda,\alpha}(x)}\mathfrak{c}(x,r)\left(\varepsilon\int_r^{\min\{\delta,2\mathfrak{D}\}}\frac{dt}{t^{1+\theta-a_{p,\lambda,\alpha}(x)}} + C_\delta\|f\|_{L^{p(\cdot),\lambda(\cdot)}(X)}\right)\\
&\leqslant \varepsilon + C_\delta\cdot r^{\theta-a_{p,\lambda,\alpha}(x)}\mathfrak{c}(x,r)\|f\|_{L^{p(\cdot),\lambda(\cdot)}(X)}.
\end{aligned} \tag{19}$$

From (19), we obtain that $\lim_{r\to 0}\sup_{x\in X} I(x,r) \leqslant \varepsilon$ which, together with (18) and the arbitrariness of $\varepsilon > 0$, yields

$$\lim_{r\to 0}\sup_{x\in X}\frac{\mathfrak{c}(x,r)}{r^{\frac{\lambda(x)}{p(x)}+\alpha(x)}}\|G_1\|_{L^{p(\cdot)}(B(x,r))} = 0.$$

Using the elementary inequalities

$$|t^a - t^b| \leqslant |\ln(t)||a-b|\begin{cases} t^{\min\{a,b\}}, & 0 < t \leqslant 1;\\ t^{\max\{a,b\}}, & t \geqslant 1,\end{cases}$$

and $|\ln d(y,z)| \leqslant \max\{1, |\log_2\mathfrak{D}|\}\ln\frac{2\mathfrak{D}}{d(y,z)}$, we estimate G_2 as

$$\begin{aligned}
G_2 &\leqslant C|\alpha(x)-\alpha(y)|\times\\
&\times\int_{X\setminus B(x,2\varkappa r)} f(z)\left(\frac{1}{d(y,z)^{\gamma-\alpha(y)}}+\frac{1}{d(y,z)^{\gamma-\alpha(x)}}\right)\ln\left(\frac{2\mathfrak{D}}{d(y,z)}\right)d\mu(z).
\end{aligned} \tag{20}$$

Taking into account that $2\varkappa d(y,z) \geqslant d(x,z)$ and $X\setminus B(x,2\varkappa r) \subset X\setminus B(y,r)$ we obtain, by the estimate (20), that

$$G_2 \leqslant C|\alpha(x)-\alpha(y)| \left(\int_{X\setminus B(y,r)} \frac{f(z)}{d(y,z)^{\gamma-\alpha(y)}} \ln\left(\frac{2\mathfrak{D}}{d(y,z)}\right) d\mu(z) + \right.$$
$$\left. + \int_{X\setminus B(x,2\varkappa r)} \frac{f(z)}{d(x,z)^{\gamma-\alpha(x)}} \ln\left(\frac{4\varkappa\mathfrak{D}}{d(x,z)}\right) d\mu(z) \right)$$
$$=: C|\alpha(x)-\alpha(y)|[K(y)+\widetilde{K}(x)].$$

Let us estimate $K(y)$ as

$$K(y) = \sum_{0\leqslant k<\log_2(\frac{2\mathfrak{D}}{r})} \int_{2^k r\leqslant d(y,z)\leqslant 2^{k+1}r} \frac{f(z)}{d(y,z)^{\gamma-\alpha(y)}} \ln\left(\frac{2\mathfrak{D}}{d(y,z)}\right) d\mu(z)$$
$$\leqslant \sum_{0\leqslant k<\log_2(\frac{2\mathfrak{D}}{r})} \ln\left(\frac{2\mathfrak{D}}{2^k r}\right) \int_{2^k r\leqslant d(y,z)\leqslant 2^{k+1}r} \frac{f(z)}{d(y,z)^{\gamma-\alpha(y)}} d\mu(z).$$

Defining $\chi_k := \chi_{B(y,r2^{k+1})\setminus B(y,r2^k)}$ and applying Hölder's inequality together with estimates (8) and (16), we have

$$K(y) \leqslant C\ln\left(\frac{2\mathfrak{D}}{r}\right) \sum_{0\leqslant k<\log_2(\frac{2\mathfrak{D}}{r})} \|f\|_{L^{p(\cdot)}(B(y,2^{k+1}r))} \left\| \frac{\chi_k}{d(y,\cdot)^{\gamma-\alpha(y)}} \right\|_{p'(\cdot)}$$
$$\leqslant C\ln\left(\frac{2\mathfrak{D}}{r}\right) \int_r^{4\mathfrak{D}} \frac{\|f\|_{L^{p(\cdot)}(B(y,t))}}{t^{1-\alpha(y)+\frac{\gamma}{p(y)}}} dt \tag{21}$$
$$\leqslant C\ln\left(\frac{2\mathfrak{D}}{r}\right) \int_r^{4\mathfrak{D}} \frac{\mathfrak{M}_{p(\cdot),\lambda(\cdot)}(f;y,t)}{t^{1-a_{p,\lambda,\alpha}(y)}} dt.$$

Fixing $\varepsilon>0$ and taking $\delta=\delta(\varepsilon)$ as in (19), we have

$$|\alpha(x)-\alpha(y)|K(y) \leqslant C\omega(x,r)\ln\left(\frac{2\mathfrak{D}}{r}\right)\left(\varepsilon\int_r^\delta \frac{dt}{t^{1-\alpha(y)}} + C_\delta\|f\|_{L^{p(\cdot),\lambda(\cdot)}(X)}\right). \tag{22}$$

From (22), Lemma 2, and the growth condition, it follows

$$r^{-\frac{\lambda(x)}{p(x)}-\alpha(x)} \| |\alpha(x)-\alpha(\cdot)|K(\cdot)\|_{L^{p(\cdot)}(B(x,r))}$$
$$\leqslant Cr^{-a_{p,\lambda,\alpha}(x)}\omega(x,r)\ln\frac{2\mathfrak{D}}{r}\left(\varepsilon\int_r^\delta \frac{dt}{t^{1-\alpha(y)}} + C_\delta\|f\|_{L^{p(\cdot),\lambda(\cdot)}(X)}\right) \tag{23}$$
$$\leqslant C\varepsilon + \left(\ln\frac{2\mathfrak{D}}{r}\right)^{-1} \|f\|_{L^{p(\cdot),\lambda(\cdot)}(X)}.$$

Since $\varepsilon>0$ is arbitrary, from (23) we obtain

$$\limsup_{r\to 0}\sup_{x\in X} r^{-\frac{\lambda(x)}{p(x)}-\alpha(x)} \| |\alpha(x)-\alpha(\cdot)|K(\cdot)\|_{L^{p(\cdot)}(B(x,r))} = 0.$$

Since $\widetilde{K}(x) \leqslant (2 + \log_2 \varkappa) \, K(x)$, the estimate for $\widetilde{K}(x)$ follows from the estimate for $K(y)$, $y \in B(x, r)$.

It remains to collect all the estimates. □

Acknowledgements The research of H. Rafeiro was supported by a Start-up Grant of United Arab Emirates University, Al Ain, United Arab Emirates via Grant No. G00002994. The research of S. Samko was supported by: (a) Russian Foundation for Basic Research under the grant No. 19-01-00223, and (b) TUBITAK and Russian Foundation for Basic research under the grant No. 20-51-46003.

References

1. Adams, D.R.: A note on Riesz potentials. Duke Math. J. **42**(4), 765–778 (1975)
2. Almeida, A., Hasanov, J., Samko, S.: Maximal and potential operators in variable exponent Morrey spaces. Georgian Math. J. **15**(2), 195–208 (2008)
3. Chiarenza, F., Frasca, M.: Morrey spaces and Hardy-Littlewood maximal function. Rend. Math. **7**, 273–279 (1987)
4. Cruz-Uribe, D., Fiorenza, A.: Variable Lebesgue spaces. In: Foundations and Harmonic Analysis. Birkhäuser, New York (2013)
5. Diening, L., Harjulehto, P., Hästö, P., Růžička, M.: Lebesgue and Sobolev Spaces with Variable Exponents. Springer, Berlin (2011)
6. Genebashvili, I., Gogatishvili, A., Kokilashvili, V., Krbec, M.: Weight Theory for Integral Transforms on Spaces of Homogeneous Type. Chapman and Hall/CRC (1997)
7. Karapetyants, A.N., Rafeiro, H., Samko, S.G.: On singular operators in vanishing generalized variable-exponent Morrey spaces and applications to Bergman-type spaces. Math. Notes **106**(5), 727–739 (2019)
8. Kokilashvili, V., Meskhi, A., Rafeiro, H., Samko, S.: Integral Operators in Non-standard Function Spaces. Volume 2: Variable Exponent Hölder, Morrey-Campanato and Grand spaces. Birkhäuser, Basel (2016)
9. Kokilashvili, V., Meskhi, A., Rafeiro, H., Samko, S.: Integral Operators in Non-standard Function Spaces. Volume 1: Variable Exponent Lebesgue and Amalgam Spaces. Birkhäuser, Basel (2016)
10. Macías, R.A., Segovia, C.: Lipschitz functions on spaces of homogeneous type. Adv. Math. **33**(3), 257–270 (1979)
11. Rafeiro, H., Samko, N., Samko, S.: Morrey-Campanato spaces: an overview. In: Karlovich, Y., Rodino, L., Silbermann, B., Spitkovsky, I. (eds.) Operator Theory, Pseudo-Differential Equations, and Mathematical Physics. Operator Theory: Advances and Applications, vol. 228, pp. 293–323. Birkhäuser, Basel (2013)
12. Rafeiro, H., Samko, S.: Fractional operators of variable order from variable exponent Morrey spaces to variable exponent Campanato spaces on quasi-metric measure spaces with growth condition. Ric. Mat. (2021). https://doi.org/10.1007/s11587-021-00639-4
13. Rafeiro, H., Samko, S.: Variable exponent Campanato spaces. J. Math. Sci. **172**, 143–164 (2011)
14. Rafeiro, H., Samko, S.: BMO-VMO results for fractional integrals in variable exponent Morrey spaces. Nonlinear Anal. **184**, 35–43 (2019)
15. Rafeiro, H., Samko, S.: A note on vanishing Morrey → VMO result for fractional integrals of variable order. Fract. Calc. Appl. Anal. **23**(1), 298–302 (2020)
16. Rafeiro, H. Samko., S.: Addendum to "On the Riesz potential operator of variable order from variable exponent Morrey space to variable exponent Campanato space". Math. Meth. Appl. Sci. 1–8 (2020). Math. Meth. Appl. Sci. (2021). https://doi.org/10.1002/mma.7796

17. Rafeiro, H., Samko, S., Umarkhadzhiev, S.: Grand Lebesgue space for $p = \infty$ and its application to Sobolev-Adams embedding theorems in borderline cases. Math. Nachr. (2021). https://doi.org/10.1002/mana.202000347
18. Ragusa, M.A.: Commutators of fractional integral operators on vanishing-Morrey spaces. J. Global Optim. **40**(1–3), 361–368 (2008)
19. Samko, N.: Maximal, potential and singular operators in vanishing generalized Morrey spaces. J. Glob. Optim. **57**, 1385–1399 (2013)
20. Samko, S.: A note on Riesz fractional integrals in the limiting case $\alpha(x)p(x) \equiv n$. Fract. Calc. Appl. Anal. **16**(2), 370–377 (2013)
21. Samko, S.: Remark to the paper of S. Samko, "A note on Riesz fractional integrals in the limiting case $\alpha(x)p(x) \equiv n$". FCAA **16**(2) (2013). Fract. Calc. Appl. Anal. **17**(1)277–278 (2014)
22. Samko, S.: Variable exponent fractional integrals in the limiting case $\alpha(x)p(x) \equiv n$ on quasimetric measure spaces. Georgian Math. J. **27**(1), 157–164 (2020)
23. Vitanza, C.: Functions with vanishing Morrey norm and elliptic partial differential equations. In: Proceedings of Methods of Real Analysis and Partial Differential Equations, Capri, pp. 147–150. Springer (1990)

Part V
Operator Algebras

Characterization of Certain Traces on von Neumann Algebras

Airat Bikchentaev

Abstract Consider a unital C^*-algebra $\mathcal{A}$. Let $n \geq 2$ and let $P_1, \cdots, P_n$ be projections in $\mathcal{A}$ such that $P_1 + \ldots + P_n = I$. We costruct $\mathcal{P}_n \colon \mathcal{A} \to \mathcal{A}$ being a block projection operator given by the formula $\mathcal{P}_n(X) = \sum_{k=1}^{n} P_k X P_k$ for all $X \in \mathcal{A}$. For a weight φ on a von Neumann algebra $\mathcal{A}$, we prove that φ is a trace if and only if $\varphi(\mathcal{P}_2(A)) = \varphi(A)$ for all $A \in \mathcal{A}^+$. We also prove that if $\mathcal{A}$ is a von Neumann algebra then for a normal semifinite weight φ on $\mathcal{A}$ the following conditions are equivalent: (i) φ is a trace; (ii) $\varphi((A^{m/2} B^m A^{m/2})^k) \leq \varphi((A^{k/2} B^k A^{k/2})^m)$ for all $A, B \in \mathcal{A}^+$ and some numbers $k, m \in \mathbb{R}$ such that $k > m > 0$; (iii) $\varphi(|\mathcal{P}_n(A)|) \leq \varphi(|A|)$ for all $A \in \mathcal{A}$ and for all projections $P_1, \ldots, P_n \in \mathcal{A}$. As a consequence, we obtain a criterions for commutativity of von Neumann algebras and C^*-algebras.

Keywords Hilbert space · Linear operator · von Neumann algebra · C^*-algebra · Block projection operator · Weight · Trace · Tracial inequality · Commutativity

1 Introduction

Traces and weights on C^*-algebras are basic tools in the operator theory and its applications. So it seems important to characterize traces in different classes of weights on C^*-algebras, see [1–10].

Consider a tracial positive normal linear functional φ on a von Neumann algebra $\mathcal{A}$, and positive numbers p, q such that $1/p + 1/q = 1$, then we have:

- Hölder's inequality [10, Theorem 5], [11, Chap. IX, Theorem 2.13]:

$$\varphi(|XY|) \leq \varphi(X^p)^{1/p} \varphi(Y^q)^{1/q} \quad \text{for all} \quad X, Y \in \mathcal{A}^+;$$

A. Bikchentaev (✉)
N.I. Lobachevskii Institute of Mathematics and Mechanics, Kazan (Volga Region) Federal University, Kremlevskaya ul. 18, 420008 Kazan, Tatarstan, Russia
e-mail: Airat.Bikchentaev@kpfu.ru

L. Accardi et al. (eds.), *Infinite Dimensional Analysis, Quantum Probability and Applications*, Springer Proceedings in Mathematics & Statistics 390,
https://doi.org/10.1007/978-3-031-06170-7_17

- Cauchy–Schwarz–Buniakowski inequality [12, Theorem 4.21]:

$$\varphi(|XY|^{1/2}) \le \varphi(X)^{1/2}\varphi(Y)^{1/2} \quad \text{for all} \quad X, Y \in \mathcal{A}^+;$$

- Golden–Thompson inequality [13, Theorem 4]:

$$\varphi(e^{X+Y}) \le \varphi(e^{X/2}e^Y e^{X/2}) \quad \text{for all} \quad X, Y \in \mathcal{A}^{\text{sa}};$$

- Peierls–Bogoliubov inequality [13, Theorem 7]:

$$\varphi(e^X) \exp \frac{\varphi(e^{X/2}Ye^{X/2})}{\varphi(e^X)} \le \varphi(e^{X+Y}) \quad \text{for all} \quad X, Y \in \mathcal{A}^+.$$

Also inequality

$$\operatorname{tr}((X^{1/2}YX^{1/2})^{rp}) \le \operatorname{tr}((X^{r/2}Y^r X^{r/2})^p), \quad r \ge 1, \ p > 0,$$

holds true for positive operators X, Y on a Hilbert space $\mathcal{H}$ [14]. This inequality generalizes the inequalities of Lieb and Thirring, and resembles the Golden–Thompson inequality (see [15, Sect. 8]).

It can be said that any of the given trace inequalities is sharp in the following sense: only the trace of all positive linear functionals satisfies the inequality. It is known that if we limit ourselves only to projections of a von Neumann algebra $\mathcal{A}$ then each and every inequality of Hölder, Cauchy–Schwarz–Buniakowski, Golden–Tompson, Peierls–Bogoliubov, Araki–Lieb–Thirring etc., characterizes the tracial functionals among all positive normal functionals, see [16–22].

Gohberg and Krein had begun to study the block projection operators in [23]. These operators admit a natural extension to the setting of quasi-normed ideals and noncommutative integration. Consider a number $n \ge 2$ and let $P_1, \dots, P_n$ be such projections in a unital C^*-algebra $\mathcal{A}$ that $P_1 + \cdots + P_n = I$. Introduce a block projection operator $\mathcal{P}_n \colon \mathcal{A} \to \mathcal{A}$ as follows: $\mathcal{P}_n(X) = \sum_{k=1}^n P_k X P_k$ for all $X \in \mathcal{A}$.

Consider a weight φ on a von Neumann algebra $\mathcal{A}$. Here we prove that the following conditions are equivalent: (i) φ is a trace; (ii) $\varphi(\mathcal{P}_2(A)) = \varphi(A)$ for all $A \in \mathcal{A}^+$ (Theorem 1). Note that the block projection operators on certain algebras (von Neumann algebras and algebras of operators measurable with respect to semifinite normal traces) already appeared in [24, 25]. We also proved several uniform submajorization inequalities for block projection operators [26]. Here we show that the following conditions are equivalent for a normal semifinite weight φ on a von Neumann algebra $\mathcal{A}$: (i) φ is a trace; (ii) $\varphi((A^{m/2}B^m A^{m/2})^k) \le \varphi((A^{k/2}B^k A^{k/2})^m)$ for all $A, B \in \mathcal{A}^+$ and some numbers $k, m \in \mathbb{R}$ such that $k > m > 0$; (iii) $\varphi(|\mathcal{P}_n(A)|) \le \varphi(|A|)$ for all $A \in \mathcal{A}$ and for all $P_1, \dots, P_n \in \mathcal{A}^{\text{pr}}$ with $P_1 + \ldots + P_n = I$, where $\mathcal{P}_n(A) = \sum_{k=1}^n P_k A P_k$ (Theorem 2). As a consequence, we obtain certain criterions for commutativity of von Neumann algebras and C^*-algebras (Corollaries 1, 3).

2 Definitions and Notation

The basic notion here is a C^*-algebra, being a complex Banach $*$-algebra $\mathcal{A}$ such that $\|A^*A\| = \|A\|^2$ for every $A \in \mathcal{A}$. For a C^*-algebra $\mathcal{A}$ by $\mathcal{A}^{\rm pr}$, $\mathcal{A}^{\rm sa}$ and $\mathcal{A}^+$ we denote its subsets of projections ($A = A^* = A^2$), self-adjoint elements ($A^* = A$) and positive elements, respectively. For any $A \in \mathcal{A}$ we have $|A| = \sqrt{A^*A} \in \mathcal{A}^+$. If I is the unit of the algebra $\mathcal{A}$ and $P \in \mathcal{A}^{\rm pr}$, then $P^\perp = I - P$.

We say that a mapping $\varphi : \mathcal{A}^+ \to [0, +\infty]$ is *a weight* on a C^*-algebra $\mathcal{A}$, if $\varphi(X + Y) = \varphi(X) + \varphi(Y)$, $\varphi(\lambda X) = \lambda\varphi(X)$ for all $X, Y \in \mathcal{A}^+$, $\lambda \geq 0$ (moreover, $0 \cdot (+\infty) \equiv 0$). Introduce the set

$$\mathfrak{M}_\varphi^+ = \{X \in \mathcal{A}^+ \colon \ \varphi(X) < +\infty\}, \quad \mathfrak{M}_\varphi = \mathrm{lin}_{\mathbb{C}} \mathfrak{M}_\varphi^+$$

for a weight φ.

We can always extend by linearity the restriction $\varphi|_{\mathfrak{M}_\varphi^+}$ to a functional on $\mathfrak{M}_\varphi$. This extension is denoted by the same letter φ. Such an extension tells us that finite weights (i.e., $\varphi(X) < +\infty$ for all $X \in \mathcal{A}^+$) are virtually the same with positive functionals on $\mathcal{A}$. We call a positive linear functional φ on $\mathcal{A}$ with $\|\varphi\| = 1$ *a state*. A weight φ is said to be *faithful*, if $\varphi(X) = 0$ $(X \in \mathcal{A}^+)$ implies that $X = 0$; *a trace*, if $\varphi(Z^*Z) = \varphi(ZZ^*)$ for every $Z \in \mathcal{A}$. If $\varphi(A) = \sup\{\varphi(B) : \ B \in \mathcal{A}^+, \ B \leq A, \ \varphi(B) < +\infty\}$, for every $A \in \mathcal{A}^+$ then a trace φ on a C^*-algebra $\mathcal{A}$ is *semifinite*. *A subadditive weight* on a C^*-algebra $\mathcal{A}$ is a mapping $\varphi : \mathcal{A}^+ \to [0, +\infty]$ such that $\varphi(X + Y) \leq \varphi(X) + \varphi(Y)$, $\varphi(\lambda X) = \lambda\varphi(X)$ for all $X, Y \in \mathcal{A}^+$, $\lambda \geq 0$ (here $0 \cdot (+\infty) \equiv 0$), see [27–30]. A subadditive weight φ is called *a subadditive trace*, if $\varphi(Z^*Z) = \varphi(ZZ^*)$ for all $Z \in \mathcal{A}$.

Let $\mathcal{B}(\mathcal{H})$ be the $*$-algebra of all bounded linear operators on a Hilbert space $\mathcal{H}$ over the field $\mathbb{C}$. Gelfand–Naimark theorem states that every C^*-algebra is isometrically isomorphic to a concrete C^*-algebra of operators on a Hilbert space $\mathcal{H}$ [31, II.6.4.10]. For any set $\mathcal{X} \subset \mathcal{B}(\mathcal{H})$ we construct the commutant

$$\mathcal{X}' = \{Y \in \mathcal{B}(\mathcal{H}) \colon \ XY = YX \ \text{ for all } \ X \in \mathcal{X}\}.$$

A von Neumann algebra $\mathcal{A}$ acting on a Hilbert space $\mathcal{H}$ is a $*$-subalgebra of the algebra $\mathcal{B}(\mathcal{H})$ such that $\mathcal{A} = \mathcal{A}''$. For $P, Q \in \mathcal{A}^{\rm pr}$ we write $P \sim Q$ (the Murray–von Neumann equivalence), if $P = U^*U$ and $Q = UU^*$ for some $U \in \mathcal{A}$.

A *normal* weight φ on von Neumann algebra $\mathcal{A}$ is a weight such that $\varphi(\sup X_i) = \sup \varphi(X_i)$ for every bounded increasing net $\{X_i\}$ in $\mathcal{A}^+$; a weight φ is *semifinite*, if the set $\mathfrak{M}_\varphi$ is ultraweakly dense in $\mathcal{A}$ (see [32, Definition VII.1.1]).

Consider a von Neumann algebra $\mathcal{A}$, let $U \in \mathcal{A}$ be a unitary operator, i.e., $U^*U = UU^* = I$. There exists an automorphism α of a von Neumann algebra $\mathcal{A}$, defined by the formula $\alpha(A) = U^*AU$ for all $A \in \mathcal{A}$. By [33, Theorem 1.4] it follows that the automorphism α can be represented as a finite product of involutions $U_S : \ A \mapsto SAS$, here S is a symmetry in $\mathcal{A}$, i.e., $U^*AU = S_1 \cdots S_m A S_m \cdots S_1$ with unitaries

$S_1, \ldots, S_m \in \mathcal{A}^{sa}$. Moreover, if $\mathcal{A}$ possesses no type I_{fin} direct summands then the unitary operator U by itself is a finite product of symmetries from $\mathcal{A}^{sa}$ [33, Theorem 1.6].

The universal representation of a C^*-algebra $\mathcal{A}$ is the pair

$$\{\pi, \mathfrak{H}\} = \sum_{\varphi \in \mathcal{S}(\mathcal{A})}^{\oplus} \{\pi_\varphi, \mathfrak{H}_\varphi\},$$

where $\mathcal{S}(\mathcal{A})$ is the set of all states on $\mathcal{A}$, $(\pi_\varphi, \mathfrak{H}_\varphi)$ is the Gelfand–Naimark–Segal representation of a C^*-algebra $\mathcal{A}$, assosiated with φ. Here we say that the von Neumann algebra $\mathcal{M} = \pi(\mathcal{A})''$, generated by $\pi(\mathcal{A})$, is *the universal enveloping von Neumann algebra* of C^*-algebra $\mathcal{A}$ [11, Chap. III, Definition 2.3].

Consider a C^*-algebra $\mathcal{A}$. Let φ be a positive linear functional on $\mathcal{A}$ and π be the universal representation of $\mathcal{A}$. Then arbitrary state on $\mathcal{A}$ by construction of π turns into a vector state on $\pi(\mathcal{A})$, hence it extends to a normal state on the universal enveloping algebra $\mathcal{M} = \pi(\mathcal{A})''$. Then φ yields such a positive normal functional $\widehat{\varphi}$ on the universal enveloping von Neumann algebra that $\widehat{\varphi}(\pi(A)) = \varphi(A)$ $(A \in \mathcal{A}^+)$.

Recall the finite dimensional Spectral Theorem: every normal matrix $A \in \mathbb{M}_n(\mathbb{C})$ reduces to the sum $A = \sum_{i=1}^m \lambda_i P_i$, where $\lambda_i \in \mathbb{C}$, and the projections $P_i \in \mathbb{M}_n(\mathbb{C})^{pr}$ with $P_i P_j = 0$ for $i \neq j$ and $i, j = 1, \ldots, m$, $m \leq n$.

3 Trace Characterization on C^*-Algebras

Lemma 1 *Let $\mathcal{A}$ be a unital C^*-algebra. Let $n \geq 2$ and let $P_1, \ldots, P_n \in \mathcal{A}^{pr}$ be such that $P_1 + \ldots + P_n = I$, $\mathcal{P}_n(A) = \sum_{k=1}^n P_k A P_k$ for $A \in \mathcal{A}$. Then*

(i) *If φ is a trace on $\mathcal{A}$, then $\varphi(\mathcal{P}_n(A)) = \varphi(A)$ for all $A \in \mathcal{A}^+$.*
(ii) *If φ is a subadditive trace on $\mathcal{A}$, then $\varphi(\mathcal{P}_n(A)) \leq \varphi(A)$ for all $A \in \mathcal{A}^+$.*

Proof (i) For any $A \in \mathcal{A}^+$ and $n \geq 2$ we have

$$\varphi(\mathcal{P}_n(A)) = \sum_{k=1}^n \varphi(P_k A P_k) = \sum_{k=1}^n \varphi(P_k A^{1/2} \cdot A^{1/2} P_k) = \sum_{k=1}^n \varphi((A^{1/2} P_k)^* A^{1/2} P_k) =$$

$$= \sum_{k=1}^n \varphi(A^{1/2} P_k (A^{1/2} P_k)^*) = \sum_{k=1}^n \varphi(A^{1/2} P_k A^{1/2}) = \varphi\Big(A^{1/2}\Big(\sum_{k=1}^n P_k\Big)A^{1/2}\Big) = \varphi(A).$$

(ii) For any $A \in \mathcal{A}$ and $n \geq 2$ by [25, Lemma 2] we have the representation

$$\mathcal{P}_n(A) = \frac{1}{2^{n-1}} \sum_{k=1}^{2^{n-1}} S_k A S_k,$$

where the unitaries $S_k \in \mathcal{A}^{sa}$, $k = 1, \ldots, 2^{n-1}$, have the form

$$P_1 \pm P_2 \pm \cdots \pm P_n.$$

Consider $A \in \mathcal{A}^+$. We have $\varphi(S_k A S_k) = \varphi(A)$ for all $k = 1, \ldots, 2^{n-1}$ and

$$\varphi(\mathcal{P}_n(A)) = \varphi\Big(\frac{1}{2^{n-1}} \sum_{k=1}^{2^{n-1}} S_k A S_k\Big) = \frac{1}{2^{n-1}} \varphi\Big(\sum_{k=1}^{2^{n-1}} S_k A S_k\Big) \leq$$

$$\leq \frac{1}{2^{n-1}} \sum_{k=1}^{2^{n-1}} \varphi(S_k A S_k) = \varphi(A).$$

Theorem 1 *For a weight φ on a von Neumann algebra $\mathcal{A}$ the following conditions are equivalent:*

(i) *φ is a trace;*
(ii) *$\varphi(\mathcal{P}_2(A)) = \varphi(A)$ for all $A \in \mathcal{A}^+$.*

Proof For (i) $\Rightarrow$ (ii) see item (i) of Lemma 1.

(ii) $\Rightarrow$ (i). By applying Upmeier's results [33], one can see that [34, Theorem 1.4.2] actually shows us that a weight φ on a von Neumann algebra $\mathcal{A}$ is a trace if and only if $\varphi(SAS) = \varphi(A)$ for any positive operator $A \in \mathcal{A}^+$ and any symmetry $S \in \mathcal{A}^{\rm sa}$, see Sect. 2. Let $A \in \mathcal{A}^+$, a symmetry $S \in \mathcal{A}^{\rm sa}$ and $P_1 \in \mathcal{A}^{\rm pr}$ be such that $S = 2P_1 - I$, i.e., $P_1 = (S + I)/2$. For $P_2 = P_1^{\perp}$ we have $\mathcal{P}_2(A) = \frac{1}{2}(A + SAS)$ and

$$\varphi(A) = \varphi(\mathcal{P}_2(A)) = \varphi\Big(\frac{1}{2}(A + SAS)\Big) = \frac{1}{2}\varphi(A) + \frac{1}{2}\varphi(SAS).$$

For $\varphi(A) < +\infty$, or $\varphi(SAS) = \varphi(A) = +\infty$ we obtain $\varphi(SAS) = \varphi(A)$, the theorem is thereby established. Assume that $\varphi(SAS) < \varphi(A) = +\infty$. By repeating the above argument for the operator $A_1 = SAS$ instead of A (then $SA_1S = A$), we conclude that

$$+\infty > \varphi(SAS) = \frac{1}{2}\varphi(SAS) + \frac{1}{2}\varphi(A) = +\infty.$$

This is a contradiction, hence the theorem holds.

Recall Taylor's formula with Peano's remainder. Then we have

Lemma 2 *If $a \in \mathbb{R}$, then*

$$(1+t)^a = 1 + at + \frac{1}{2!}a(a-1)t^2 + \cdots + \frac{1}{n!}a(a-1)\cdots(a-n+1)t^n + o(t^n) \quad \text{as } t \to 0.$$

Theorem 2 *For a normal semifinite weight φ on a von Neumann algebra $\mathcal{A}$ the following conditions are equivalent:*

(i) *φ is a trace;*
(ii) *$\varphi((A^{m/2} B^m A^{m/2})^k) \leq \varphi((A^{k/2} B^k A^{k/2})^m)$ for all $A, B \in \mathcal{A}^+$ and some numbers $k, m \in \mathbb{R}$ with $k > m > 0$;*

(iii) $\varphi(|\mathcal{P}_n(A)|) \leq \varphi(|A|)$ *for all* $A \in \mathcal{A}$ *and for all* $P_1, \ldots, P_n \in \mathcal{A}^{\mathrm{pr}}$ *with* $P_1 + \cdots + P_n = I$, *where* $\mathcal{P}_n(A) = \sum_{k=1}^{n} P_k A P_k$.

Proof (i) $\Rightarrow$ (ii). Consider $A, B \in \mathcal{A}^+$ and $k > m > 0$. Put

$$X = A^m, \quad Y = B^m, \quad r = \frac{k}{m}, \quad p = m. \tag{1}$$

Then $r > 1$ and

$$(A^{m/2} B^m A^{m/2})^k = (X^{1/2} Y X^{1/2})^{rp}, \quad (A^{k/2} B^k A^{k/2})^m = (X^{r/2} Y^r X^{r/2})^p.$$

The inequality

$$\varphi((X^{1/2} Y X^{1/2})^{rp}) \leq \varphi((X^{r/2} Y^r X^{r/2})^p), \quad r \geq 1, \ p > 0 \tag{2}$$

was proved in [35].

(ii) $\Rightarrow$ (i). Again we rewrite the inequality of item (ii) in the form of relation (2) applying transformations (1). It follows by Lemma 2 of [20] that for every projection $P \in \mathcal{A}^{\mathrm{pr}}$ with $\varphi(P) < +\infty$ the reduced weight $\varphi|_P$ on the reduced von Neumann algebra $P\mathcal{A}P$ is a trace. Therefore φ is a trace by Lemma 2 of [36].

(i) $\Rightarrow$ (iii). For $n = 2$ see [37, Lemma 13]; for the general case see [24, Lemma 3.2].

(iii) $\Rightarrow$ (i). *Step 1.* Consider a positive normal functional φ on a von Neumann algebra $\mathcal{A}$. Then the proof of implication (iii) $\Rightarrow$ (i) for arbitrary von Neumann algebra is reducible to the case of the algebra $\mathbb{M}_2(\mathbb{C})$ as in similar situations (see [7] or [38]).

A positive normal linear functional φ on a von Neumann algebra $\mathcal{A}$ is tracial if and only if $\varphi(P) = \varphi(Q)$ for all $P, Q \in \mathcal{A}^{\mathrm{pr}}$ with $PQ = 0$ and $P \sim Q$ (see [7], [38, Lemma 2]). Consider a *-algebra $\mathcal{N}$ in the reduced algebra $(P + Q)\mathcal{A}(P + Q)$ generated by a partial isometry $V \in \mathcal{A}$ that realizes the equivalence between P and Q. The algebra $\mathcal{N}$ is *-isomorphic to $\mathbb{M}_2(\mathbb{C})$. Inequality (iii) holds for operators of $\mathcal{N}$ and the restricted functional $\varphi|\mathcal{N}$. Let us show that this restriction is tracial on $\mathcal{N}$; henceforce, $\varphi(P) = \varphi(Q)$.

Recall that every linear functional φ on $\mathbb{M}_2(\mathbb{C})$ possesses the form $\varphi(\cdot) = \mathrm{tr}(S_\varphi \,\cdot)$. The matrix $S_\varphi \in \mathbb{M}_2(\mathbb{C})$ is the so-called density matrix of φ. Without loss of generality assume that

$$S_\varphi = \mathrm{diag}\Big(\frac{1}{2} - s, \frac{1}{2} + s\Big), \quad 0 \leq s \leq \frac{1}{2}.$$

Thus $\varphi(X)$ equals $(1/2 - s)x_{11} + (1/2 + s)x_{22}$ for $X = [x_{ij}]_{i,j=1}^2 \in \mathbb{M}_2(\mathbb{C})$.

Consider a complex $\sigma \in \mathbb{C}$ with $|\sigma| = 1$ and a real $t \in [0, 1]$. These numbers define the projection

$$R^{(t,\sigma)} = \begin{pmatrix} t & \sigma\sqrt{t - t^2} \\ \bar{\sigma}\sqrt{t - t^2} & 1 - t \end{pmatrix} \in \mathbb{M}_2(\mathbb{C}).$$

Put $P_1 = R^{(1/2,1)}$, $P_2 = P_1^{\perp} = R^{(1/2,-1)}$ and

$$A = \begin{pmatrix} 1 & -\varepsilon \\ \varepsilon & \varepsilon^2 \end{pmatrix}$$

for $0 < \varepsilon < 1/2$. Then $\mathcal{P}_2(A) = \frac{1+\varepsilon^2}{2} I$ and $\varphi(|\mathcal{P}_2(A)|) = \mathrm{tr}(S_\varphi \mathcal{P}_2(A)) = \frac{1+\varepsilon^2}{2}$.

The matrix

$$|A|^2 = \begin{pmatrix} 1+\varepsilon^2 & -\varepsilon+\varepsilon^3 \\ -\varepsilon+\varepsilon^3 & \varepsilon^2+\varepsilon^4 \end{pmatrix}$$

has the characteristic equation $\lambda^2 - (1+\varepsilon^2)^2\lambda + 4\varepsilon^4 = 0$. Therefore,

$$\lambda_1 = \frac{(1+\varepsilon^2)^2 + \sqrt{(1+\varepsilon^2)^4 - 16\varepsilon^4}}{2} \quad \text{and} \quad \lambda_2 = \frac{(1+\varepsilon^2)^2 - \sqrt{(1+\varepsilon^2)^4 - 16\varepsilon^4}}{2}.$$

Taylor's formula with Peano remainder (cf Lemma 2 with $a = 1/2$) implies that

$$\sqrt{(1+\varepsilon^2)^4 - 16\varepsilon^4} = 1 + 2\varepsilon^2 - 7\varepsilon^4 + o(\varepsilon^5), \quad (\varepsilon \to 0+).$$

Hence

$$\lambda_1 = 1 + 2\varepsilon^2 - 3\varepsilon^4 + o(\varepsilon^5), \quad \lambda_2 = 4\varepsilon^4 + o(\varepsilon^5), \quad (\varepsilon \to 0+).$$

The finite-dimensional Spectral Theorem yields the representation

$$|A|^2 = \lambda_1 R^{(t,1)} + \lambda_2 R^{(t,1)\perp} = \lambda_1 R^{(t,1)} + \lambda_2 R^{(1-t,-1)}.$$

We determine the parameter $t \in [0,1]$ from the equation

$$1 + \varepsilon^2 = \lambda_1 t + \lambda_2(1-t), \quad \text{i.e.,}$$

$$1 + \varepsilon^2 = (1 + 2\varepsilon^2 - 3\varepsilon^4 + o(\varepsilon^5))t + (4\varepsilon^4 + o(\varepsilon^5))(1-t), \quad (\varepsilon \to 0+).$$

Hence

$$t = \frac{1 + \varepsilon^2 - 4\varepsilon^4}{1 + 2\varepsilon^2 - 7\varepsilon^4 + o(\varepsilon^5)}, \quad (\varepsilon \to 0+).$$

By Lemma 2 with $a = -1$ we find

$$t = (1 + \varepsilon^2 - 4\varepsilon^4)(1 - 2\varepsilon^2 + 11\varepsilon^4 + o(\varepsilon^5)) = 1 - \varepsilon^2 + 5\varepsilon^4 + o(\varepsilon^5), \quad (\varepsilon \to 0+).$$

Finite-dimensional Spectral Theorem yields $|A| = \sqrt{\lambda_1} R^{(t,1)} + \sqrt{\lambda_2} R^{(1-t,-1)}$, where

$$\sqrt{\lambda_1} = 1 + \varepsilon^2 - 2\varepsilon^4 + o(\varepsilon^5), \quad \sqrt{\lambda_2} = 2\varepsilon^2\sqrt{1 + o(\varepsilon)} = 2\varepsilon^2 + o(\varepsilon^3), \qquad (\varepsilon \to 0+)$$

for a=1/2 in Lemma 2 and thanks to the relation $\varepsilon^k o(\varepsilon^m)$=$o(\varepsilon^{k+m})$, ($\varepsilon$→0+) for all $k, m \in \mathbb{N}$. Hence

$$\varphi(|A|) = \mathrm{tr}(S_\varphi|A|) = \sqrt{\lambda_1}\,\mathrm{tr}(S_\varphi R^{(t,1)}) + \sqrt{\lambda_2}\,\mathrm{tr}(S_\varphi R^{(1-t,-1)}) =$$
$$= \sqrt{\lambda_1}\Big(\Big(\frac{1}{2} - s\Big)t + \Big(\frac{1}{2} + s\Big)(1-t)\Big) + \sqrt{\lambda_2}\Big(\Big(\frac{1}{2} - s\Big)(1-t) + \Big(\frac{1}{2} + s\Big)t\Big) =$$
$$= \frac{1}{2} - s + \frac{3}{2}\varepsilon^2 + 3s\varepsilon^2 + o(\varepsilon^3) \quad \text{as} \quad \varepsilon \to 0+.$$

The inequality $\varphi(|\mathcal{P}_2(A)|) \le \varphi(|A|)$ then takes the form

$$\frac{1}{2} + \frac{\varepsilon^2}{2} \le \frac{1}{2} - s + \frac{3}{2}\varepsilon^2 + 3s\varepsilon^2 + o(\varepsilon^3), \qquad (\varepsilon \to 0+).$$

It holds true only for $s = 0$ for all ε, $0 < \varepsilon < \frac{1}{2}$.

Step 2. Let a normal semifinite weight φ on a von Neumann algebra $\mathcal{A}$ meet condition (iii). It follows by Step 1 that for every projection $P \in \mathcal{A}^{\mathrm{pr}}$ with $\varphi(P) < +\infty$ the reduced weight $\varphi|_P$ on the reduced von Neumann algebra $P\mathcal{A}P$ is a trace. Therefore φ is a trace by Lemma 2 of [36]. The assertion is proved.

Corollary 1 *For a von Neumann algebra $\mathcal{A}$ the following conditions are equivalent:*

(i) *the algebra $\mathcal{A}$ is commutative;*
(ii) $\varphi(\mathcal{P}_2(A)) = \varphi(A)$ *for all normal states φ on $\mathcal{A}$, for all operators $A \in \mathcal{A}^+$, for all $P_1 \in \mathcal{A}^{\mathrm{pr}}$ and $P_2 = P_1^\perp$;*
(iii) $\varphi(|\mathcal{P}_2(A)|) \le \varphi(|A|)$ *for all normal states φ on $\mathcal{A}$, for all operators $A \in \mathcal{A}$, for all $P_1 \in \mathcal{A}^{\mathrm{pr}}$ and $P_2 = P_1^\perp$;*
(iv) $\varphi((A^{m/2}B^mA^{m/2})^k) \le \varphi((A^{k/2}B^kA^{k/2})^m)$ *for all normal states φ on $\mathcal{A}$, for all operators $A, B \in \mathcal{A}^+$ and some numbers $k, m \in \mathbb{R}$ with $k > m > 0$.*

Proof (iv) ⇒ (i). Every normal state on $\mathcal{A}$ is tracial by Theorem 2, i.e., $\varphi(XY) = \varphi(YX)$ for all $X, Y \in \mathcal{A}$. The set of all normal states separates elements of the algebra $\mathcal{A}$ [31, Chap. III, Theorem 2.4.5]. This fact implies that $XY = YX$ $(X, Y \in \mathcal{A})$. So the von Neumann algebra $\mathcal{A}$ is commutative.

Corollary 2 *Let φ be a positive functional on C^*-algebra $\mathcal{A}$ such that the inequality $\varphi((A^{m/2}B^mA^{m/2})^k) \le \varphi((A^{k/2}B^kA^{k/2})^m)$ holds for any $A, B \in \mathcal{A}^+$ and some numbers $k, m \in \mathbb{R}$ such that $k > m > 0$. Then φ is a tracial functional.*

Proof Consider the universal enveloping von Neumann algebra $\mathcal{M}$ [11, III. 2]. Assume that π is the corresponding universal representation of the C^*-algebra $\mathcal{A}$, and $\widehat{\varphi}$ is the positive normal functional on $\mathcal{M}$ with $\widehat{\varphi}(\pi(A)) = \varphi(A)$ for $A \in \mathcal{A}$, see Sect. 2. Fix positive operators $\widehat{A}, \widehat{B} \in \mathcal{M}$. Then by Kaplansky density theorem there exist bounded positive nets $\{A_\alpha\}$ and $\{B_\alpha\}$ in $\mathcal{A}$ such that $\pi(A_\alpha) \to \widehat{A}$ and $\pi(B_\alpha) \to$

$\widehat{B}$ in the strong operator topology. Fix $k, m \in \mathbb{R}$ such that $k > m > 0$. We take into account inequality (ii) of Theorem 1, apply continuity of the operations in the strong operator topology and conclude that $\widehat{\varphi}((\widehat{A}^{m/2}\widehat{B}^m\widehat{A}^{m/2})^k) \le \widehat{\varphi}((\widehat{A}^{k/2}\widehat{B}^k\widehat{A}^{k/2})^m)$. By Theorem 2 $\widehat{\varphi}$ is a tracial functional on $\mathcal{M}$, hence φ is a tracial functional on $\mathcal{A}$.

Corollary 3 *Let $\mathcal{A}$ be a C^*-algebra such that $(A^{m/2}B^m A^{m/2})^k \le (A^{k/2}B^k A^{k/2})^m$ for all $A, B \in \mathcal{A}^+$ and some numbers $k, m \in \mathbb{R}$ with $k > m > 0$. Then $\mathcal{A}$ is commutative.*

Proof The inequality of Corollary 2 holds for every positive functional on $\mathcal{A}$. Then every positive functional on $\mathcal{A}$ is tracial, and $\mathcal{A}$ is commutative.

For other trace characterizations see also [38–44].

Acknowledgements The author was supported by the development program of Volga Region Mathematical Center (agreement no. 075-02-2021-1393).

References

1. Abed, S.A.: An inequality for projections and convex functions. Lobachevskii J. Math. **39**(9), 1287–1292 (2018). https://doi.org/10.1134/S1995080218090214
2. Bikchentaev, A.M.: Commutation of projections and characterization of traces on von Neumann algebras III. Int. J. Theor. Phys. **54**(12), 4482–4493 (2015). https://doi.org/10.1007/s10773-015-2639-6
3. Bikchentaev, A.M.: Inequality for a trace on a unital C^*-algebra. Math. Notes **99**(4), 487–491 (2016). https://doi.org/10.1134/S0001434616030214
4. Bikchentaev, A.M.: Differences of idempotents in C^*-algebras. Sib. Math. J. **58**(2), 183–189 (2017). https://doi.org/10.1134/S003744661702001X
5. Bikchentaev, A.M.: Differences of idempotents in C^*-algebras and the quantum Hall effect. Theor. Math. Phys. **195**(1), 557–562 (2018). https://doi.org/10.1134/S0040577918040074
6. Bikchentaev, A.M.: Trace and differences of idempotents in C^*-algebras. Math. Notes **105**(5–6), 641–648 (2019). https://doi.org/10.1134/S0001434619050018
7. Gardner, L.T.: An inequality characterizes the trace. Canad. J. Math. **31**(6), 1322–1328 (1979). https://doi.org/10.4153/CJM-1979-109-9
8. Hoa, D.T., Tikhonov, O.E.: Weighted monotonicity inequalities for traces on operator algebras. Math. Notes **88**(1–2), 177–182 (2010). https://doi.org/10.1134/S0001434610070175
9. Hoa, D.T., Osaka, H., Toan, H.M.: On generalized Powers-Størmer's inequality. Linear Algebra Appl. **438**(1), 242–249 (2013). https://doi.org/10.1016/j.laa.2012.07.053
10. Petz, D., Zemánek, J.: Characterizations of the trace. Linear Algebra Appl. **111**, 43–52 (1988). https://doi.org/10.1016/0024-3795(88)90050-X
11. Takesaki, M.: Theory of operator algebras, vol. I. In: Operator Algebras and Non-commutative Geometry, 5. Springer-Verlag, Berlin (2002)
12. Manjegani, S.M.: Inequalities in operator algebras. A thesis for the Ph.D. in mathematics degree of Regina University. Canada, Regina, 95 pp (2004)
13. Ruskai, M.: Inequalities for traces on von Neumann algebras. Commun. Math. Phys. **26**(4), 280–289 (1972). https://doi.org/10.1007/BF01645523
14. Araki, H.: On an inequality of Lieb and Thirring. Lett. Math. Phys. **19**(2), 167–170 (1990). https://doi.org/10.1007/BF01045887
15. Simon, B.: Trace Ideals and Their Applications, 2nd edn. Mathematical Surveys and Monographs, vol. 120. American Mathematical Society, Providence, RI, (2005)
16. Bikchentaev, A.M., Tikhonov, O.E.: Characterization of the trace by Young's inequality. JIPAM. J. Inequal. Pure Appl. Math. **6**(2), Article 49, 3 pp (2005)

17. Bikchentaev, A.M., Tikhonov, O.E.: Characterization of the trace by monotonicity inequalities. Linear Algebra Appl. **422**(1), 274–278 (2007). https://doi.org/10.1016/j.laa.2006.10.005
18. Bikchentaev, A.M.: Commutativity of projections and characterization of traces on von Neumann algebras. Sib. Math. J. **51**(6), 971–977 (2010). https://doi.org/10.1007/s11202-010-0096-2
19. Bikchentaev, A.M.: Commutation of projections and trace characterization on von Neumann algebras II. Math. Notes **89**(3–4), 461–471 (2011). https://doi.org/10.1134/S0001434611030175
20. Bikchentaev, A.M.: The Peierls-Bogoliubov inequality in C^*-algebras and characterization of tracial functionals. Lobachevskii J. Math. **32**(3), 175–179 (2011). https://doi.org/10.1134/S1995080211030061
21. Bikchentaev, A.M.: Commutativity of operators and characterization of traces on C^*-algebras. Dokl. Math. **87**(1), 79–82 (2013). https://doi.org/10.1134/S1064562413010298
22. Cho, K., Sano, T.: Young's inequality and trace. Linear Algebra Appl. **431**(8), 1218–1222 (2009). https://doi.org/10.1016/j.laa.2009.04.016
23. Gohberg, I.C., Kreĭn, M.G.: Introduction to the Theory of Linear Nonselfadjoint Operators. Translations of Mathematical Monographs, vol. 18. American Mathematical Society, Providence, RI (1969)
24. Chilin, V., Krygin, A., Sukochev, F.: Extreme points of convex fully symmetric sets of measurable operators. Integr. Equ. Oper. Theory **15**(2), 186–226 (1992). https://doi.org/10.1007/BF01204237
25. Bikchentaev, A.M.: Block projection operator in normed ideal spaces of measurable operators. Russ. Math. (Iz. VUZ) **56**(2), 75–79 (2012). https://doi.org/10.3103/S1066369X12020107
26. Bikchentaev, A.M., Sukochev, F.: Inequalities for the block projection operators. J. Funct. Anal. **280**(7), Article 108851, 18 pp (2021). https://doi.org/10.1016/j.jfa.2020.108851
27. Haagerup, U.: Normal weights on W^*-algebras. J. Funct. Anal. **19**(3), 302–317 (1975). https://doi.org/10.1016/0022-1236(75)90060-9
28. Bikchentaev, A.M.: On the Haagerup problem on subadditive weights on W^*-algebras. Russ. Math. (Iz. VUZ) **55**(10), 82–85 (2011). https://doi.org/10.3103/S1066369X11100112
29. Bikchentaev, A.M.: The Haagerup problem on subadditive weights on W^*-algebras. II. Russ. Math. (Iz. VUZ) **57**(12), 66–69 (2013). https://doi.org/10.3103/S1066369X13120074
30. Bikchentaev, A.M.: Seminorms associated with subadditive weights on C^*-algebras. Math. Notes **107**(3–4), 383–391 (2020). https://doi.org/10.1134/S0001434620030025
31. Blackadar, B.: Operator algebras, theory of C^*-algebras and von Neumann algebras. In: Operator Algebras and Non-commutative Geometry, III. Encyclopaedia of Mathematical Sciences, vol. 122. Springer-Verlag, Berlin (2006)
32. Takesaki, M.: Theory of Operator Algebras, Operator Algebras and Non-Commutative Geometry, 6. v. II. Encyclopaedia of Mathematical Sciences, vol. 125. Springer-Verlag, New York (2003)
33. Upmeier, H.: Automorphism groups of Jordan C^*-algebras. Math. Z. **176**(1), 21–34 (1981). https://doi.org/10.1007/BF01258901
34. Ayupov, Sh.A.: Classification and Representation of Ordered Jordan Algebras (Russian). "Fan", Tashkent (1986)
35. Kosaki, H.: On an inequality of Araki-Lieb-Thirring (von Neumann algebra case). Proc. Am. Math. Soc. **114**(2), 477–481 (1992). https://doi.org/10.1090/S0002-9939-1992-1065951-1
36. Stolyarov, A.I., Tikhonov, O.E., Sherstnev, A.N.: Characterization of normal traces on von Neumann algebras by inequalities for the modulus. Math. Notes **72**(3–4), 411–416 (2002). https://doi.org/10.1023/A:1020559623287
37. Kaftal, V., Weiss, G.: Compact derivations relative to semifinite von Neumann algebras. J. Funct. Anal. **62**(2), 202–220 (1985). https://doi.org/10.1016/0022-1236(85)90003-5
38. Tikhonov, O.E.: Subadditivity inequalities in von Neumann algebras and characterization of tracial functionals. Positivity **9**(2), 259–264 (2005). https://doi.org/10.1007/s11117-005-2711-1

39. Bikchentaev, A.M.: On a property of L_p-spaces on semifinite von Neumann algebras. Math. Notes **64**(1–2), 159–163 (1998). https://doi.org/10.1007/bf02310299
40. Bikchentaev, A.M.: Metrics on projections of the von Neumann algebra associated with tracial functionals. Sib. Math. J. **60**(6), 952–956 (2019). https://doi.org/10.1134/S003744661906003X
41. Bikchentaev, A.M., Abed, S.A.: Projections and traces on von Neumann algebras. Lobachevskii J. Math. **40**(9), 1260–1267 (2019). https://doi.org/10.1134/S1995080219090051
42. Bikchentaev, A.M.: Inequalities for determinants and characterization of the trace. Sib. Math. J. **61**(2), 248–254 (2020). https://doi.org/10.1134/S0037446620020068
43. Bikchentaev, A.M., Sherstnev, A.N.: Studies on noncommutative measure theory in Kazan University (1968–2018). Int. J. Theor. Phys. **60**(2), 585–596 (2021). https://doi.org/10.1007/s10773-019-04156-x
44. Hoa, D.T., Tikhonov, O.E.: Weighted trace inequalities of monotonicity. Lobachevskii J. Math. **26**, 63–67 (2007)

Actions of $*$-Morphisms on Certain Projections of C^*-Matrix Algebras

Fouzia Shaheen and Ahmed Al-Rawashdeh

Abstract H. Dye defined certain projections $P_{i,j}(a)$ of a C^*-matrix algebra $\mathbb{M}_n(A)$, and he used it to prove that any isomorphism between unitary groups of von Neumann factors is implemented by a $*$-isomorphism. In this paper we study actions of some $*$-isomorphism on the projections $P_{i,j}(a)$. Indeed, we deduce results related to the Cuntz algebras and its unitary groups.

1 Introduction

Let A be a unital C^*-algebra. The set of projections, the unitary groups, and the self-adjoint unitaries of A are denoted by $\mathcal{P}(A)$, $\mathcal{U}(A)$ and $\mathcal{U}_{sa}(A)$, respectively. The C^*-matrix algebra over A, which is denoted by $\mathbb{M}_n(A)$, is the algebra of all $n \times n$-matrices $(a_{i,j})$ over A, with the usual addition, scalar multiplication and multiplication of matrices, and the involution (adjoint) is $(a_{i,j})^* = (a^*_{j,i})$.

Recall that (see [6, p. 74]) a system of matrix units of a unital C^*-algebra A is a subset $\{e^r_{i,j}\}$, $1 \le i, j \le n$ and $1 \le r \le m$ of A, such that

$$e^r_{i,j}e^r_{j,k} = e^r_{i,k},\ e^r_{i,j}e^s_{k,l} = 0 \text{ if } r \neq s \text{ or } j \neq k,\ (e^r_{i,j})^* = e^r_{j,i},\ \sum_{i,r}^{n,m} e^r_{i,i} = 1$$

and for every i, $e^r_{i,i} \in \mathcal{P}(A)$.

Dye in [6] associated to each $a \in A$, the projections $P_{i,j}(a)$ in $\mathbb{M}_n(A)$ as follows:

F. Shaheen (✉) · A. Al-Rawashdeh
Department of Mathematical Sciences, UAE University, Al Ain, United Arab Emirates
e-mail: 201990173@uaeu.ac.ae

A. Al-Rawashdeh
e-mail: aalrawashdeh@uaeu.ac.ae

L. Accardi et al. (eds.), *Infinite Dimensional Analysis, Quantum Probability and Applications*, Springer Proceedings in Mathematics & Statistics 390,
https://doi.org/10.1007/978-3-031-06170-7_18

$$P_{i,j}(a) = \begin{pmatrix} 0 \cdots & \cdots & \cdots & \cdots & \cdots 0 \\ & \cdots\cdots\cdots\cdots\cdots\cdots\cdots\cdots & & & \\ 0 \cdots & (1+aa^*)^{-1} & \cdots & (1+aa^*)^{-1}a & \cdots 0 \\ & \cdots\cdots\cdots\cdots\cdots\cdots\cdots\cdots & & & \\ 0 \cdots & a^*(1+aa^*)^{-1} & \cdots & a^*(1+aa^*)^{-1}a & \cdots 0 \\ & \cdots\cdots\cdots\cdots\cdots\cdots\cdots\cdots & & & \\ 0 \cdots & \cdots & \cdots & \cdots & \cdots 0 \end{pmatrix}$$

Dye used these projections as a main tool to prove that an isomorphism between the discrete unitary groups of von Neumann factors not of type I_n, is implemented by a $*$-isomorphism between the factors themselves [6, Theorem 2]. Also, these projections were used in the extension problem in the case of UHF-algebras by Al-Rawashdeh in [2]. More properties of the projections $P_{i,j}(a)$ were discussed by Al-Rawashdeh in [3].

Let A and B be two unital C^*-algebras and let $\varphi : \mathcal{U}(A) \to \mathcal{U}(B)$ be an isomorphism. As φ preserves self-adjoint unitaries, it induces a natural bijection $\theta_\varphi : \mathcal{P}(A) \to \mathcal{P}(B)$ between the sets of projections of A and B given by

$$1 - 2\theta_\varphi(p) = \varphi(1 - 2p), \;\; p \in \mathcal{P}(A).$$

A bijective map $\theta : \mathcal{P}(A) \to \mathcal{P}(B)$ is called a projection orthoisomorphism, if it preserves orthogonality, i.e. $pq = 0$ iff $\theta(p)\theta(q) = 0$; $\forall p, q \in \mathcal{P}(A)$.

Let us recall the following results of Dye.

Theorem 1 ([6, Lemma 8]) *Let θ be a projection orthoisomorphism on a von Neumann matrix algebra $\mathbb{M}_n(M)$ to a von Neumann matrix algebra $\mathbb{M}_n(N)$, under which diagonal matrix units corresponds. Then each projection $\theta(P_{i,j}(a))$ has the form $P_{i,j}(a')$.*

Theorem 2 ([6, Lemma 6]) *Let θ be a projections orthoisomorphism between the C^*-matrix algerbas M_n and N_n $(n \geq 3)$, under which the diagonal matrix units and more generally, projections of the form $P_{i,j}(a)$ correspond. Then there exists a linear mapping ψ of M_n and N_n, ψ being the direct sum of a $*$-antiisomorphism and a $*$-isomorphism, with the property that $\psi(P_{i,j}(a)) = \theta(P_{i,j}(a))$, for all $a \in M$ and all i, j.*

Remark 1 The main idea of the proof of [6, Lemma 6] is that, the hypothesis $\theta(P_{i,j}(a)) = P_{i,j}(a')$ induces a $*$-isomoprhism ψ from M_n and N_n. In [6, Lemma 7] Dye proved that if θ leaves all $P_{i,j}(a)$ fixed, then it is the identity mapping on $\mathcal{P}(A)$.

In Sect. 2, we recall some results of the Cuntz algebra $\mathcal{O}_n$, and C^*-matrix algebras $\mathbb{M}_n(A)$, which will be used to prove our main results. In Sect. 3, we recall known results about the images of $P_{i,j}(a) \in \mathcal{P}(\mathbb{M}_n(A))$ under orthoisomorphisms. The case of von Neumann factors were studied by Dye in [6], and by Al-Rawashdeh in [1] for the case of unital C^*-algebras. In a connection to these results, we raise some questions and we discuss the action of some $*$-isomoprhisms on $P_{i,j}(a)$ for general unital C^*-algberas, and we deduce some results in the case of Cuntz algebra $\mathcal{O}_n$, as in Theorem 3, Theorem 4 and Proposition 4. In Theorem 5 and Corollary 2, we give a characterization of the condition, under which a group automorphism φ on $\mathcal{U}(\mathcal{O}_n)$ is implemented by a ∗-automorphism on $\mathcal{O}_n$.

2 Preliminaries

Let A be a unital C^*-algebra having a system of matrix units $\{e_{i,j}\}_{i,j=1}^n$, for some $n \geq 3$. Recall that $e_{1,1}Ae_{1,1}$ is a C^*-algebra (corner algebra) which has $e_{1,1}$ as a unit. This system of matrix units implements a ∗-isomorphism between A and $\mathbb{M}_n(e_{1,1}Ae_{1,1})$. Indeed, let the isomorphism

$$\eta_e : \mathbb{M}_n(e_{1,1}Ae_{1,1}) \to A$$

be defined by

$$\eta_e((a_{i,j})^n) = \sum_{i,j=1}^{n} e_{i,1}a_{i,j}e_{1,j}.$$

Moreover, if $e_{1,1}$ is equivalent to 1 (i.e. A is assumed to be an infinite C^*-algebra), then there exists a partial isometry v of A such that $v^*v = e_{1,1}$ and $vv^* = 1$, and this defines the ∗-isomorphism $\Delta_v : A \to e_{1,1}Ae_{1,1}$ by $\Delta_v(x) = v^*xv$.

For two unital C^*-algebras A and B, if $\alpha : A \to B$ is a ∗-isomorphism, then α induces a ∗-isomorphism $\widehat{\alpha} : \mathbb{M}_n(A) \to \mathbb{M}_n(B)$, which is defined by $(a_{i,j}) \mapsto (\alpha(a_{i,j}))$. Then we have the following result.

Proposition 1 *Let A be an infinite unital C^*-algebra having a system of matrix units $\{e_{i,j}\}_{i,j=1}^n$. If $e_{1,1}$ is equivalent to 1, then $\mathbb{M}_n(A)$ is ∗-isomorphic to A.*

Proof Let $\Delta_v : A \to e_{1,1}Ae_{1,1}$ and $\eta_1 : \mathbb{M}_n(e_{1,1}Ae_{1,1}) \to A$ be defined as above. Then the mapping $\eta = \eta_e \circ \widehat{\Delta_v}$ is a ∗-isomorphism from $\mathbb{M}_n(A)$ onto A. Moreover,

$$\eta(a_{i,j})^n = \sum_{i,j}^{n} e_{i,1}v^*a_{i,j}ve_{1,j}, \text{ and}$$

$$\eta^{-1}(x) = (ve_{1,i}xe_{j,1}v^*)_{i,j}^n.$$

As a main example of such C^*-algebras, let us recall the Cuntz algebra:

Definition 1 The Cuntz algebra $\mathcal{O}_n$; $n \geq 2$, is the universal C^*-algebra which is generated by isometries $s_1, s_2, \ldots, s_n$, such that $\sum_{i=1}^n s_is_i^* = 1$ with $s_i^*s_j = 0$, when $i \neq j$ and $s_i^*s_i = 1$ (for more details, see [4], [5, p. 149]).

For every $1 \leq i \leq n$, let p_i denote the diagonal matrix unit $e_{i,i}$, and u_i denote the associated self-adjoint unitary $1 - 2p_i$. Put

$$e_{i,j} = s_i s_j^*, \qquad 1 \leq i, j \leq n . \tag{1}$$

Then $\{e_{i,j}\}_{i,j=1}^n$ forms a system of matrix units for $\mathcal{O}_n$.

As s_1^* is a partial isometry between $e_{1,1}$ and the unity. Proposition 1 shows that the mapping

$$\eta : \mathbb{M}_n(\mathcal{O}_n) \to (\mathcal{O}_n), \ (a_{i,j})_{i,j} \mapsto \sum_{i,j=1}^{n} s_i a_{i,j} s_j^* \tag{2}$$

is a $*$-isomorphism. Moreover, for $x \in \mathcal{O}_n$, $\eta^{-1}(x) = (s_i^* x s_j)_{i,j} \in \mathbb{M}_n(\mathcal{O}_n)$. The isomorphism η is associated with the isometries $s_1, s_2, \ldots, s_n$. This proves that the Cuntz algebra $\mathcal{O}_n$ is isomorphic to the C^*-algebra $\mathbb{M}_n(\mathcal{O}_n)$. Then for $a \in \mathcal{O}_n$ and applying the mapping η, the projections $P_{i,j}(a)$ are seen inside $\mathcal{O}_n$ as:

$$s_i(1+aa^*)^{-1}s_i^* + s_i(1+aa^*)^{-1}as_j^* + s_ja^*(1+aa^*)^{-1}s_i^* + s_ja^*(1+aa^*)^{-1}as_j^*.$$

3 Actions of $*$-Morphisms on $P_{i,j}(a)$

Consider a unital C^*-algebra A, and a mapping α defined on A. Let us define a lifting map of α on a matrix algebra over A, as follows:

Definition 2 (*Lifting-Map*) Let A be a unital C^*-algebra, $\{e_{i,j}\}_{i,j=1}^n$ and $\{f_{i,j}\}_{i,j=1}^n$ be two system of matrix units of A. If α is a mapping on A, then the lifting-map of α on $\mathbb{M}_n(e_{1,1}Ae_{1,1})$ w.r.t. the matrix units is defined as:
$\widetilde{\alpha}_{e,f} = \eta_f^{-1}\alpha\eta_e$.

Remark 2 Consider the Cuntz algebra $\mathcal{O}_n$, and fix $s_1, s_2, \ldots, s_n$ as partial isometries generating $\mathcal{O}_n$ and consider s_1^* as a partial isometry between $e_{1,1}$ and the unity. The lifting map of α on $\mathbb{M}_n(\mathcal{O}_n)$ is $\widetilde{\alpha} = \eta^{-1}\alpha\eta$.

Remark 3 It is easy to check the followings

1. α is $*$-automorphism on A iff $\tilde{\alpha}$ is $*$-automorphism $\mathbb{M}_n(e_{1,1}Ae_{1,1})$;
2. The restriction of η on $\mathcal{P}(\mathbb{M}_n(e_{1,1}Ae_{1,1}))$ is an orthoisomorphism, onto $\mathcal{P}(A)$, which is denoted by $\eta|_{\mathcal{P}(A)}$;
3. The restriction of η on $\mathcal{U}(\mathbb{M}_n(e_{1,1}Ae_{1,1}))$ is a group isomorphism onto $\mathcal{U}(A)$;
4. θ is an orthoisomorphism on $\mathcal{P}(A)$ iff $\tilde{\theta}$ is an orthoisomorphism from $\mathcal{P}(\mathbb{M}_n(e_{1,1}Ae_{1,1}))$ onto $\mathcal{P}(\mathbb{M}_n(f_{1,1}Af_{1,1}))$;
5. φ is group-automorphism on $\mathcal{U}(A)$ iff $\tilde{\varphi}$ is group-isomorphism from $\mathcal{U}(\mathbb{M}_n(e_{1,1}Ae_{1,1}))$ onto $\mathcal{U}(\mathbb{M}_n(f_{1,1}Af_{1,1}))$.

Proposition 2 *Let A be a unital C^*-algebras having a system of matrix units $\{e_{i,j}\}_{i,j=1}^n$ and let φ be an automorphism on $\mathcal{U}(A)$. Then $\theta_{\tilde{\varphi}} = \tilde{\theta}_{\varphi}$.*

Proof Let $p \in \mathcal{P}(\mathbb{M}_n(A))$ and $u = I - 2p, where\ u \in \mathcal{U}_{sa}(\mathbb{M}_n(A))$. Then

$$\begin{aligned}
\theta_{\tilde{\varphi}}(p) &= \frac{I_n - \tilde{\varphi}(u)}{2} \\
&= \frac{I - \eta^{-1}\varphi\eta(u)}{2} \\
&= \frac{I - \eta^{-1}\varphi(1 - 2\eta(p))}{2} \\
&= \frac{I - \eta^{-1}(1 - 2\theta_\varphi(\eta(p)))}{2} \\
&= \frac{I - (I - 2\eta^{-1}\theta_\varphi(\eta(p)))}{2} \\
&= \eta^{-1}\theta_\varphi(\eta(p)) \\
&= \tilde{\theta}_\varphi(p),
\end{aligned}$$

hence, the proof is completed.

Let us recall the following main results about the images of $P_{i,j}(a)$ under orthoisomorphisms, in the case of unital C^*-algebras:

Lemma 1 ([1, Lemma 6.2.0.3]) *Let* $\{e_{i,j}\}_{i,j=1}^n$ *and* $\{f_{i,j}\}_{i,j=1}^n$ *be two systems of matrix units in* A *and* B *respectively such that* $\theta_\varphi(e_{i,i}) = f_{i,i}$. *If* θ_φ *is an orthoisomorphism, then for every* $a \in e_{1,1}Ae_{1,1}$ *and* $1 \leq i \neq j \leq n$, $\widetilde{\theta_\varphi}(P_{i,j}(a)) = P_{i,j}(b)$, *for some* $b \in f_{1,1}Bf_{1,1}$.

Remark 4 Al-Rawashdeh in [1, Lemma 6.2.0.3] proved that if A is an unital C^*-algebra having the system of matrix units $\{e_{i,j}\}_{i,j=1}^n$ and φ is a group isomorphism on $\mathcal{U}(A)$ such that θ_φ is an orthoisomorphism, then $\tilde{\theta}_\varphi(P_{i,j}(u)) = P_{i,j}(u')$, hence $\tilde{\theta}_\varphi(P_{i,j}(a)) = P_{i,j}(a')$. There is no direct relation between u' and $\varphi(u)$.

The projections $P_{i,j}(a)$ w.r.t. a system of matrix units $\{e_{i,j}\}_{i,j=1}^n$ are denoted by $P_{i,j}^e(a)$.

Lemma 2 ([1, Lemma 6.1.0.4]) *The set* $\{P_{i,j}(a);\ a \in A,\ 1 \leq i, j \leq n\}$ *depends only on the partition of the unity. If* $\{e_{i,j}\}_{i,j=1}^n$ *and* $\{f_{i,j}\}_{i,j=1}^n$ *are two systems of matrix units in* A, *such that* $e_{i,i} = f_{i,i}$ *for all* $1 \leq i \leq n$ *and* $v \in \mathcal{U}(e_{1,1}Ae_{1,1})$, *then there exists* $u \in \mathcal{U}(e_{1,1}Ae_{1,1})$ *such that* $P_{i,j}^f(v) = P_{i,j}^e(u)$ *and* $P_{i,j}^e(a) = P_{i,j}^f(f_{1,i}e_{i,1}ae_{1,j}f_{j,1})$.

Now, let us ask the following questions:

Question 1 Let A be a C^*-algebra having a system of matrix units $\{e_{i,j}\}_{i,j=1}^n$ and let ψ be a $*$-automorphism of A. Then for $a \in A$, what is $\tilde{\psi}(P_{i,j}(a))$?

Question 2 Let ψ be a $*$-automorphism on A. For all $a \in A$, is

$$\tilde{\psi}(P_{i,j}(a)) = P_{i,j}(\psi(a))?$$

Question 3 If φ is a group automorphism on $\mathcal{U}(A)$, then for which C^*-algebra A and $u \in \mathcal{U}(A)$, do we have that

$$\tilde{\theta_\varphi}(P_{i,j}(u)) = P_{i,j}(\varphi(u))?$$

As partial solutions of Questions 1, we have the following answer in the case of the Cuntz algebras $\mathcal{O}_n$, on which a $*$-automorphism fixes all diagonal matrix.

Theorem 3 *Consider the Cuntz algebra $\mathcal{O}_n$, which is generated by the isometries $s_1, s_2, \ldots, s_n$. If ψ is a $*$-automorphism of $\mathcal{O}_n$ such that $\psi(s_i s_i^*) = s_i s_i^*$, for all $1 \le i \le n$, then*

$$\tilde{\psi}(P^e_{k,l}(a)) = P^e_{k,l}(s_k^* \psi(s_k s_1^*) s_1 \psi(a) s_1^* \psi(s_1 s_l^*) s_l),$$

for every $a \in \mathcal{O}_n$.

Proof Recall the system of matrix units $\{e_{i,j}\}_{i,j=1}^n$ of $\mathcal{O}_n$ in Eq. (1). Suppose that $\psi(s_i) = t_i$, for $1 \le i \le n$. Then $\{t_1, t_2, \ldots t_n\}$ is a set of partial isometries generates $\mathcal{O}_n$, hence we define another system of matrix units by $f_{i,j} = t_i t_j^*$, $1 \le i, j \le n$. By assumption, the systems $\{e_{i,j}\}$, $\{f_{i,j}\}$ have the same diagonal matrix units. Then

$$\begin{aligned}
\tilde{\psi}(P^e_{k,l}(a)) &= \eta^{-1}\psi(s_k(I+aa^*)^{-1}s_k^* + s_k(I+aa^*)^{-1}as_l^* \\
&\quad + s_l a^*(I+aa^*)^{-1}s_k^* + s_l a^*(I+aa^*)^{-1}as_l^*) \\
&= \eta^{-1}(s_k(I+\psi(a)\psi(a^*))^{-1}s_k^* + s_k(I+\psi(a)\psi(a^*))^{-1}\psi(a)s_l^* \\
&\quad + s_l\psi(a^*)(I+\psi(a)\psi(a^*))^{-1}s_k^* + s_l\psi(a^*)(I+\psi(a)\psi(a^*))^{-1}\psi(a)s_l^*) \\
&= P^f_{k,l}(s_1\psi(a)s_1^*) && (\text{using } \Delta_{s_1^*}) \\
&= P^e_{k,l}(e_{1,k} f_{k,1} s_1 \psi(a) s_1^* f_{1,l} e_{l,1}) && (\text{by Lemma 2}) \\
&= P^e_{k,l}(s_1 s_k^* \psi(s_k)\psi(s_1)^* s_1 \psi(a) s_1^* \psi(s_1)\psi(s_l)^* s_l s_1^*) \\
&= P^e_{k,l}(s_k^* \psi(s_k s_1^*) s_1 \psi(a) s_1^* \psi(s_1 s_l^*) s_l) && (\text{using } \Delta^{-1}_{s_1^*}).
\end{aligned}$$

Corollary 1 *Let ψ be an automorphism of $\mathcal{O}_n$ such that $\psi(s_i s_i^*) = s_i s_i^*$, for all $1 \le i \le n$. If $u \in \mathcal{U}(\mathcal{O}_n)$, then $\tilde{\psi}(P^e_{k,l}(u)) = P^e_{k,l}(\omega)$, for some unitary $\omega \in \mathcal{U}(\mathcal{O}_n)$.*

Proof The proof follows from the previous theorem as now

$$s_k^* \psi(s_k s_1^*) s_1 \psi(u) s_1^* \psi(s_1 s_l^*) s_l \in \mathcal{U}(\mathcal{O}_n).$$

If A is a unital C^*-algebra having a system of matrix unit $\{e_{i,j}\}_{i,j=1}^n$ and ψ is a $*$-automorphism on A, then ψ induces a system of matrix units $\{f_{i,j}\}_{i,j=1}^n$ of A with $f_{i,j} = \psi(e_{i,j})$. Then we have the following results as a partial solution for Question 2.

Proposition 3 *Let A be a unital C^*-algebra, having a system of matrix units $\{e_{i,j}\}_{i,j=1}^n$ and ψ be a $*$-automorphism of A. If $\{f_{i,j}\}_{i,j=1}^n$ is the induced system* *of matrix units, then for all* $u \in \mathcal{U}(e_{1,1}Ae_{1,1})$, *we have*

$$\widetilde{\psi}(P^{e}_{i,j}(u)) = P^{f}_{i,j}(\psi(u)).$$

Proof For $i = 1,\ j = 2$, let $u \in \mathcal{U}(e_{1,1}Ae_{1,1})$. Then

$$P_{1,2}(u) = \frac{1}{2}\begin{pmatrix} e_{1,1} & u \\ u^* & e_{1,1}\end{pmatrix} \quad \in \mathcal{P}(\mathbb{M}_n(e_{1,1}Ae_{1,1}))$$

where $uu^* = u^*u = e_{1,1}$. Now,

$$\begin{aligned}\tilde{\psi}_{e,f}(P^{e}_{1,2}(u)) &= \eta_f^{-1}\psi(\frac{1}{2}[e_{1,1} + e_{1,1}ue_{1,2} + e_{2,1}u^*e_{1,1} + e_{2,2}]) \\ &= \eta_f^{-1}\frac{1}{2}[f_{1,1} + f_{1,1}\psi(u)f_{1,2} + f_{2,1}\psi(u^*)f_{1,1} + f_{2,2}] \\ &= \frac{1}{2}\begin{pmatrix} f_{1,1} & f_{1,1}\psi(u)f_{1,1} \\ f_{1,1}\psi(u^*)f_{1,1} & f_{1,1}\end{pmatrix} \\ &= P^{f}_{1,2}(\psi(u)),\end{aligned}$$

which completes the proof.

Theorem 4 *Consider the Cuntz algebra* $\mathcal{O}_n$, *and its generating partial isometries* $s_1, s_2, \ldots, s_n$. *If* α *is a* ∗*-automorphism of* $\mathcal{O}_n$, *such that* $\alpha(p_j) = p_j$ *and* $w \in \mathcal{U}(\mathcal{O}_n)$, *then there exists a* ∗*-automorphism* β_α *of* $\mathcal{O}_n$ *such that:*

$$\beta_\alpha(p_i) = p_i, \ \text{ and } \ \tilde{\beta_\alpha}(P_{i,j}(w)) = P_{i,j}(\alpha(w)).$$

Proof Let $s_1, s_2, \ldots, s_n$ be partial isometries generating $\mathcal{O}_n$ and consider the ∗-isomorphism η associated with the partial isometries. Let α be a given ∗-automorphism of the algebra. Define the mapping β_α on $\mathcal{O}_n$ by

$$\beta_\alpha(x) = \sum_{i,j=1}^{n} s_i\alpha(s_i^*)\alpha(x)\alpha(s_j)s_j^*.$$

We claim that β_α is a ∗-automorphism of $\mathcal{O}_n$: It is clear that β_α is linear map and bijection, indeed $(\beta_\alpha)^{-1}(x) = \sum_{i,j=1}^{n} s_i\beta_\alpha^{-1}(s_i^*)\alpha^{-1}(x)\alpha^{-1}(s_j)s_j^*$. Also, β_α preserves the adjoint of elements as:

$$\begin{aligned}\beta_\alpha(x^*) &= \sum_{i,j=1}^{n} s_i\alpha(s_i^*)\alpha(x)^*\alpha(s_j)s_j^* \\ &= (\sum_{i,j=1}^{n} s_i\alpha(s_i^*)\alpha(x)\alpha(s_j)s_j^*)^* \\ &= (\beta_\alpha(x))^*.\end{aligned}$$

It preserves the product as:

$$\begin{aligned}
\beta_\alpha(x)\beta_\alpha(y) &= \sum_{i,j=1}^{n} s_i\alpha(s_i^*)\alpha(x)\alpha(s_j)s_j^* \sum_{k,l=1}^{n} s_k\alpha(s_k^*)\alpha(y)\alpha(s_l)s_l^* \\
&= \sum_{i,j,l=1}^{n} s_i\alpha(s_i^*)\alpha(x)\alpha(s_j s_j^*)\alpha(y)\alpha(s_l)s_l^* \\
&= \sum_{i,j,l=1}^{n} s_i\alpha(s_i^*)\alpha(x)\alpha(p_j)\alpha(y)\alpha(s_l)s_l^* \\
&= \sum_{i,j,l=1}^{n} s_i\alpha(s_i^*)\alpha(x)(p_j)\alpha(y)\alpha(s_l)s_l^* \qquad \text{(using } \alpha(p_j) = p_j\text{)} \\
&= \sum_{i,l=1}^{n} s_i\alpha(s_i^*)\alpha(xy)\alpha(s_l)s_l^* \qquad \text{(using } \sum_{i=1}^{n} p_i = 1\text{)} \\
&= \beta_\alpha(xy).
\end{aligned}$$

As $\alpha(p_i) = p_i$, for all $k = 1, \ldots n$, we have

$$\begin{aligned}
\beta_\alpha(p_k) &= \sum_{i,j=1}^{n} s_i\alpha(s_i^*)\alpha(p_k)\alpha(s_j)s_j^* \\
&= \sum_{i,j,s=1}^{n} s_i\alpha(s_i^* s_k s_k^* s_j)s_j^* \qquad i = j = k \\
&= s_k\alpha(1)s_k^* \\
&= s_k s_k^* \\
&= p_k.
\end{aligned}$$

Moreover, for all $w \in \mathcal{U}(\mathcal{O}_n)$

$$\begin{aligned}
\tilde{\beta_\alpha}(P_{i,j}(w)) &= \eta^{-1}\beta_\alpha(\frac{1}{2}(s_i s_i^* + s_i w s_j^* + s_j w^* s_i^* + s_j s_j^*)) \\
&= \eta^{-1}\frac{1}{2}(\{s_i\alpha(s_i^*) + s_i\alpha(w s_j^*) + s_j\alpha(w^* s_i^*) + s_j\alpha(s_j^*)\}\{\alpha(s_i)s_i^* + \alpha(s_j)s_j^*\}) \\
&= \eta^{-1}(\frac{1}{2}(s_i s_i^* + s_j\alpha(w^*)s_i^* + s_i\alpha(w)s_j^* + s_j s_j^*)) \\
&= P_{i,j}(\alpha(w)),
\end{aligned}$$

hence the proof is now completed.

Recall that if A is a unital C^*-algebra and $v \in \mathcal{U}(A)$, then Ad_v which is defined by $x \mapsto vxv^*$ is a $*$-automoprhism on A.

Proposition 4 *Consider the Cuntz algebra $\mathcal{O}_n$, and its generating partial isometries $s_1, s_2, \ldots, s_n$. If $v \in \mathcal{U}(O_n)$, then*

$$\widetilde{Ad_v}(P_{i,j}(u)) = P_{i,j}(Ad_v(u)).$$

Proof Given $v \in \mathcal{U}(O_n)$ and $v = \sum_{i=1}^{n} s_i u s_i^* \in \mathcal{U}(O_n)$.

$$\begin{aligned} Ad_v\eta(a_{i,j}) &= v(\sum_{i=1}^{n} s_i a_{i,j} s_j^*)v^* \\ &= \sum_{i,j=1}^{n} s_i u a_{i,j} u^* s_j^* \in \mathcal{P}(O_n). \end{aligned}$$

Therefore, $\eta^{-1}Ad_u\eta(a_{i,j}) = (ua_{i,j}u^*)_{i,j}$.

For $i = 1$, $j = 2$, we have

$$\begin{aligned} \widetilde{Ad_v}(P_{1,2}(v)) &= \frac{1}{2}\begin{pmatrix} 1 & uvu^* \\ uv^*u^* & 1 \end{pmatrix} \\ &= P_{1,2}(uvu^*) \\ &= P_{1,2}(Ad_u(v)). \end{aligned}$$

Hence the result holds for any i and j.

Proposition 5 *Let α be a ∗-automorphism of a unital C^*-algebra A having a system of matrix units $\{e_{i,j}\}_{i,j=1}^{n}$. Then for all $a \in A$, we have*

$$\widehat{\alpha}(P_{i,j}(a)) = P_{i,j}(\alpha(a)).$$

Proof If α is a ∗-automorphism on A, then α induces the ∗-automorphism $\widehat{\alpha}$ on $\mathbb{M}_n(A)$, which is defined by $(a_{i,j}) \mapsto (\alpha(a_{i,j}))$. Let us fix $i = 1$ *and* $j = 2$.
For every $a \in A$, we have

$$P_{1,2}(a) = \begin{pmatrix} (1+aa^*)^{-1} & (1+aa^*)^{-1}a \\ a^*(1+aa^*)^{-1} & a^*(1+aa^*)^{-1}a \end{pmatrix}. \text{ Then}$$

$$\begin{aligned} \widehat{\alpha}(P_{1,2}(a)) &= \begin{pmatrix} (1+\alpha(a)\alpha(a^*))^{-1} & (1+\alpha(a)\alpha(a^*))^{-1}\alpha(a) \\ \alpha(a^*)(1+\alpha(a)\alpha(a^*))^{-1} & \alpha(a^*)(1+\alpha(a)\alpha(a^*))^{-1}\alpha(a) \end{pmatrix} \\ &= P_{1,2}(\alpha(a)), \end{aligned}$$

which ends the proof.

Recall Theorem 1 and Lemma 1, which proved that the image of $P_{i,j}(u)$ ($u \in \mathcal{U}(A)$) under orthoisomorphisms is $P_{i.j}(u')$; with $u' \in \mathcal{U}(A)$.
The following result gives a characterization of Question 3 in the case of Cuntz algebra $\mathcal{O}_n$, and the induced map θ_φ on $\mathcal{P}(\mathcal{O}_n)$.

Theorem 5 *Consider the Cuntz algebra $\mathcal{O}_n$, with generating partial isometries $s_1, s_2, \ldots, s_n$, and let φ be an automorphism on $\mathcal{U}(\mathcal{O}_n)$. Then for all $u \in \mathcal{U}(\mathcal{O}_n)$, we have*

$$\tilde{\theta}_\varphi(P_{i,j}(u)) = P_{i,j}(\varphi(u))$$

if and only if

$$\theta_\varphi(\frac{1}{2}(s_is_i^* + s_js_j^* + s_ju^*s_i^* + s_ius_j^*)) = \frac{1}{2}(s_is_i^* + s_js_j^* + s_j\varphi(u^*)s_i^* + s_i\varphi(u)s_j^*).$$

Proof For $i = 1$, $j = 2$, assume that

$$\theta_\varphi(\frac{1}{2}(s_1s_1^* + s_2s_2^* + s_2u^*s_1^* + s_1us_2^*)) = \frac{1}{2}(s_1s_1^* + s_2s_2^* + s_1\varphi(u)s_2^* + s_2\varphi(u^*)s_1^*).$$

Recall that $P_{1,2}(u) = \frac{1}{2}\begin{pmatrix} 1 & u \\ u^* & 1 \end{pmatrix} \in \mathcal{P}(\mathbb{M}_n(\mathcal{O}_n))$. Then

$$\begin{aligned} \eta(P_{1,2}(u)) &= \frac{1}{2}(s_1s_1^* + s_1us_2^* + s_2u^*s_1^* + s_2s_2^*) \\ &= Q_{1,2}(u) \in \mathcal{P}(\mathcal{O}_n). \end{aligned}$$

Now let $x = 1 - 2Q_{1,2}(u)$. Therefore, $\varphi(x) = 1 - 2\theta_\varphi(Q_{1,2}(u)) \in \mathcal{U}_{sa}(\mathcal{O}_n)$. Put $\varphi^{'}(x) = \eta^{-1}(\varphi(x)) \in \mathcal{U}_{sa}(\mathbb{M}_n(\mathcal{O}_n))$. Then we get

$$\varphi^{'}(x) = I_n - 2\eta^{-1}\theta_\varphi(Q_{1,2}(u)) \tag{3}$$

which implies that

$$\begin{aligned} \varphi^{'}(x) &= I_n - 2\eta^{-1}(s_1s_1^* + s_2s_2^* + s_1\varphi(u)s_2^* + s_2\varphi(u^*)s_1^*) \\ &= I_n - \begin{pmatrix} 1 & \varphi(u) & \ldots & 0 \\ \varphi(u^*) & 1 & \ldots & 0 \\ \vdots & & & \\ 0 & 0 & \ldots & 0 \end{pmatrix} \\ &= \begin{pmatrix} 0 & -\varphi(u) & \ldots & 0 \\ -\varphi(u^*) & 0 & \ldots & 0 \\ \vdots & & & \\ 0 & 0 & \ldots & 1 \end{pmatrix}. \end{aligned}$$

Equation (3) implies that

$$\frac{I_n - \varphi'(x)}{2} = \eta^{-1}\theta_\varphi \eta(P_{1,2}(u))$$
$$= \tilde{\theta}_\varphi(P_{1,2}(u)).$$

Hence,

$$\tilde{\theta}_\varphi(P_{1,2}(u)) = \frac{1}{2}\begin{pmatrix} 1 & \varphi(u) & \dots 0 \\ \varphi(u^*) & 1 & \dots 0 \\ \vdots & & \\ 0 & 0 & \dots 0 \end{pmatrix}$$

$$= P_{1,2}(\varphi(u)).$$

Conversely, assume $\tilde{\theta}_\varphi(P_{i,j}(u)) = P_{i,j}(\varphi(u))$. For $i = 1$, $j = 2$, let $\eta(P_{1,2}(u)) = Q_{1,2}(u)$, and $x = 1 - 2Q_{1,2}(u) \in \mathcal{U}(\mathcal{O}_n)$. Then

$$Q_{1,2} = \frac{1-x}{2} \in \mathcal{P}(\mathcal{O}_n).$$

Therefore, $\theta_\varphi(Q_{1,2}(u)) = \dfrac{I - \varphi(x)}{2}$, which implies that

$$\eta^{-1}\theta_\varphi(Q_{1,2}(u)) = \frac{1 - \eta^{-1}\varphi(x)}{2}.$$

Now by assumption,

$$\frac{1 - \eta^{-1}\varphi(x)}{2} = \frac{1}{2}\begin{pmatrix} 1 & \varphi(u) & \dots 0 \\ \varphi(u^*) & 1 & \dots 0 \\ \vdots & & \\ 0 & 0 & \dots 0 \end{pmatrix}. \quad \text{Therefore,}$$

$$\eta^{-1}\varphi(x) = I_n - \begin{pmatrix} 1 & \varphi(u) & \dots 0 \\ \varphi(u^*) & 1 & \dots 0 \\ \vdots & & \\ 0 & 0 & \dots 0 \end{pmatrix} \in \mathcal{U}_{sa}(\mathbb{M}_n(\mathcal{O}_n)), \ \textit{and}$$

$$\varphi(x) = 1 - \eta\begin{pmatrix} 1 & \varphi(u) & \dots 0 \\ \varphi(u^*) & 1 & \dots 0 \\ \vdots & & \\ 0 & 0 & \dots 0 \end{pmatrix}$$
$$= 1 - (s_1 s_1^* + s_1\varphi(u)s_2^* + s_2\varphi(u^*)s_1^* + s_2 s_2^*)$$
$$= 1 - 2\theta_\varphi(Q_{1,2}(u)),$$

which implies that

$$\theta_\varphi(Q_{1,2}(u)) = \frac{1}{2}(s_1s_1^* + s_1\varphi(u)s_2^* + s_2\varphi(u^*)s_1^* + s_2s_2^*),$$

hence the proof is completed.

Consequently, we have the following partial result for $\mathcal{O}_n$:

Corollary 2 *Given φ be an automorphism on $\mathcal{U}(\mathcal{O}_n)$. If for each $u \in \mathcal{U}(\mathcal{O}_n)$ and for all $1 \le i, j \le n$, one has*

$$\theta_\varphi(\frac{1}{2}(s_is_i^* + s_js_j^* + s_ju^*s_i^* + s_ius_j^*)) = \frac{1}{2}(s_is_i^* + s_js_j^* + s_j\varphi(u^*)s_i^* + s_i\varphi(u)s_j^*),$$

then there exists a $*$-automorphism Φ(linear or conjugate linear) on $\mathcal{O}_n$, such that

1. $\Phi(u) = \phi(u)$; $\forall\, u \in \mathcal{U}(\mathcal{O}_n)$;
2. $\widehat{\Phi}(P_{i,j}(a)) = \widetilde{\theta}_\varphi(P_{i,j}(a))$; $\forall\, 1 \le i, j \le n$ *and* $\forall a \in \mathcal{O}_n$.

Proof Assume that

$$\theta_\varphi(\frac{1}{2}(s_is_i^* + s_js_j^* + s_ju^*s_i^* + s_ius_j^*)) = \frac{1}{2}(s_is_i^* + s_js_j^* + s_j\varphi(u^*)s_i^* + s_i\varphi(u)s_j^*).$$

By Theorem 3.17, we have that $\tilde{\theta_\varphi}(P_{i,j}(u)) = P_{i,j}(\varphi(u))$. Therefore, using Dye's technique in the first part of Theorem 1.1 ([6], Lemma 8), we have for all $a \in \mathcal{O}_n$, $\tilde{\theta_\varphi}(P_{i,j}(a)) = P_{i,j}(a')$ for some $a' \in \mathcal{O}_n$. By Theorem 5.2.2.2 of [1], the induced map θ_φ is an orthoisomorphism on $\mathcal{P}(\mathcal{O}_n)$. Therefore, $\tilde{\theta}_\varphi$ is an orthoisomorphism on $\mathcal{P}(\mathbb{M}_n(\mathcal{O}_n))$. Then following Dye's result in Theorem 1.2 ([6], Lemma 6), and as $\mathcal{O}_n$ is a simple C^*-algebra, there exists a $*$ automorphism $\Phi : \mathcal{O}_n \to \mathcal{O}_n$ defined by $a \mapsto a'$, where a and a' satisfy the relation

$$\tilde{\theta}_\varphi(P_{i,j}(a)) = P_{i,j}(a') \;\forall\, 1 \le i, j \le n.$$

Also, Φ is a bijective, additive, preserves product and the adjoints, which is either linear or conjugate linear. Indeed, $\Phi(u) = u' = \varphi(u)$, for all $u \in \mathcal{O}_n$, which means that φ is implemented by the $*$-automorphism Φ on $\mathcal{O}_n$. Moreover, by the last part of the proof of [6, Lemma 6], we have $\widehat{\Phi}(P_{i,j}(a)) = \widetilde{\theta}_\varphi(P_{i,j}(a))$.

Acknowledgements The authors gratefully acknowledge and thank the Department of Research Affairs at UAE University, as this article is financially supported by the Grant: UPAR (2019), Fund No. 31S397(COS), and Post Doc Grant (2019), Fund no. 31s404.

References

1. Al-Rawashdeh, A.: The Unitary Group as an Invariant of a Simple Unital C^*-Algebra. Ph.D. Thesis, University of Ottawa, Canada (2003)
2. Al-Rawashdeh, A.: On the extension of unitary group isomorphisms of unital UHF-algebras. Inter. J. Math. **26**(8) (2015)
3. Al-Rawashdeh, A.: On the certain projections of C^*-matrix algberas. Ann. Funct. Anal. **3**(2), 144–154 (2012)
4. Cuntz, J.: Simple C^*-algebras generated by isometries. Commun. Math. Phys. **57**, 173–185 (1977)
5. Davidson, K.R.: C^*-Algebras by Example, Fields Institute Monographs, vol. 6. American Mathematical Society, Providencs, RI (1996)
6. Dye, H.: On the geometry of projections in certain operator algebras. Ann. Math. **61**, 73–89 (1955)

Part VI
Stochastic Operators

Compatible Linear Lypunov Function for Infinite Dimensional Volterra Quadratic Stochastic Operators

Ahmad Fadillah bin Embong

Abstract The simplest non-linear operator is the quadratic ones. Most of the researches in this direction were investigating on finite set of all probability distributions. However, there are models where the probability distributions are countably infinite, which means that the considered operators are defined on infinite-dimensional spaces. We restrict ourselves to Quadratic Stochastic Operators (QSOs) define on infinite dimension, specifically a class of QSOs called Volterra. In this paper, we construct a linear Lyapunov function for infinite dimensional Volterra QSOs by means of finite dimensional ones.

Keywords Infinite dimensional · Volterra operator · Lyapunov function

1 Introduction

The space of ℓ^∞ is defined as follows:

$$\ell^\infty = \{\mathbf{x} = (x_1, x_2, \dots),\ \ x_k \in \mathbb{R},\ \ \sup |x_k| < \infty\}.$$

For a subset $\ell^1 \subset \ell^\infty$ denotes the space of all *absolutely summable sequences* i.e.,

$$\ell^1 = \{\mathbf{x} = (x_1, x_2, \dots),\ \ \ \|\mathbf{x}\| < \infty\}, \tag{1}$$

with the norm $\|\mathbf{x}\| = \sum\limits_{k=1}^{\infty} |x_k|$. Let $E \subset \mathbb{N}$ and denote

$$S_r^E = \left\{\mathbf{x} = (x_i)_{i\in E} \in \mathbb{R}^E,\ \ x_i \geq 0,\ \ \|\mathbf{x}\| = r\right\},\ \ \text{for } r \in [0, 1]. \tag{2}$$

A. F. bin Embong (✉)
Universiti Teknologi Malaysia, 81300 Skudai, Johor Bahru, Johor, Malaysia
e-mail: ahmadfadillah@utm.my

L. Accardi et al. (eds.), *Infinite Dimensional Analysis, Quantum Probability and Applications*, Springer Proceedings in Mathematics & Statistics 390,
https://doi.org/10.1007/978-3-031-06170-7_19

The notation S_1^E is simply written as S^E. By face of S_r^E we mean that subset $\Gamma \subset S_r^E$ and for any $\mathbf{x} = (x_i)_{i\in E} \in \Gamma$, then there exists $x_{i_0} = 0$ for some $i_0 \in E$. Define a mapping V on S^E by

$$V(\mathbf{x}) = \sum_{i,j\in E} P_{ij,k} x_i x_j, \tag{3}$$

where $P_{ij,k}$ are known as *heredity coefficients* which are satisfying,

$$P_{ij,k} \geq 0, \quad P_{ij,k} = P_{ji,k}, \quad \text{and} \quad \sum_{k\in E} P_{ij,k} = 1. \tag{4}$$

One may check that the mapping V is well-defined and it maps from S^E into itself i.e., S^E is an invariant with respect to V. The mapping is known as *Quadratic Stochastic Operators (QSOs)* where the they originally appeared from the problems of population genetics [1]. In case of the cardinality $|E| < \infty$, a lot of literatures were devoted to investigate such operators (for examples, see [3, 6–8, 18] and the references therein). Without the loss of generality, we consider $E = \{1, 2, \ldots, n\}$ to represent the case $|E| < \infty$. Instead of S_r^E, we simply write S_r^{n-1}. Denote

$$\mathbf{B}_r^{n-1} = \bigcup_{i\in[0,r]} S_i^{n-1} \quad \text{for } r \in [0, 1].$$

On the other hand, when $|E| = \infty$, then a lot of well-known facts on finite case will demolish. For example, infinite dimensional simplex, S^E is not compact with respect to ℓ^1-norm, consequently the limiting point of the trajectory of $\mathbf{x}_0 \in S^E$ with respect to V may not exist. It worth to note that, some studies of infinite dimensional QSOs have been done in the following papers [9, 14–16]. The book presented in [10] serves a good introduction to the theory of QSOs and the associated processes. Some of the fascinating application related to QSOs were given in [20–23].

For the sake of simplicity, we are going to consider $E = \mathbb{N}$ and $S_r^{\mathbb{N}}$ is written as S_r. We notice that S_1 (denoted by S) is a convex subset of the classical ℓ^1-space. Analogously, we write

$$\mathbf{B}_r = \bigcup_{t\in[0,r]} S_t \quad \text{for } r \in [0, 1].$$

A QSO is the simplest nonlinear operator, yet many challenging open problems need to be solved [3]. Therefore, researchers are likely to introduce classes of QSOs. Note, the most well-studied class of QSOs is known as Volterra [2, 19] in which the biological means is that each individual can inherit only the species (or trait) from their parents. Recall that, a QSO $V : S \to S$ is called *Volterra* if one has

$$P_{ij,k} = 0 \text{ if } k \notin \{i, j\} \text{ for all } i, j, k \in \mathbb{N}. \tag{5}$$

In [9] it was given an alternative definition of Volterra operators in terms of extremal elements of S. Moreover, using (5) any Volterra QSOs can be written as follows:

$$(V(\mathbf{x}))_k = x_k \left(1 + \sum_{i=1}^{\infty} a_{ki} x_i\right), \quad k \in \mathbb{N}, \tag{6}$$

where $a_{ki} = 2P_{ik,k} - 1$ and $i, k \in \mathbb{N}$. One can see that

$$a_{ki} = -a_{ik}, \ |a_{ki}| \leq 1 \ \text{ for every } k, i \in \mathbb{N}. \tag{7}$$

Hence, a Volterra QSO can be uniquely determined by a skew-symmetric matrix (a_{ij}) that satisfies (7).

The main problem in the theory of nonlinear operators is that to determine the asymptotic behavior (or dynamics). This problem is not an easy task, even for the simplest nonlinear case i.e., quadratic operators it is not fully solved. Nevertheless, the dynamics of finite dimensional Volterra QSOs were completely investigated by using the theory of Lyapunov function and tournament [2]. In case of infinite dimension, certain dynamical behavior of Volterra QSOs have been investigated in [16, 17], however in general setting, it still an open problem. Furthermore, the study in [9] constructed the infinite dimensional Volterra QSOs by extending them from the finite ones. The result is crucial since it gives information on how to investigate the limiting behavior of infinite dimensional Volterra QSOs by means of finite dimensional ones. Therefore, in the present paper, we are going to construct the Lyapunov functions of infinite dimensional Volterra QSOs from the finite case.

We recall the definition of Lyapunov function as follows:

Definition 1 An ℓ^1-continuous function $\varphi : S \to \mathbb{R}$ is called a *Lyapunov function* for a QSO V acting on S if the limit $\lim_{n\to\infty} \varphi(V^{(n)}(\mathbf{x}))$ exists for any initial point $\mathbf{x} \in S$.

We notice that, a bounded function $\varphi : S \to \mathbb{R}$ with $\varphi(V(\mathbf{x})) \leq (\geq)\varphi(\mathbf{x})$ for any $\mathbf{x} \in S$, then φ is a Lyapunov function for V. Next, we list out some of the known results related to Lypunov functions for Volterra QSOs defined on S^{n-1}.

Theorem 1 ([2, 5]) *Let V be a finite dimensional Volterra operator and let $A = (a_{ki})$ be its corresponding skew-symmetric matrix. Then the function $\varphi : S^{n-1} \to \mathbb{R}$ defined by*

(i) $\varphi_{\mathbf{b}}(\mathbf{x}) = \sum_{k=1}^{n} b_k x_k$ *is a Lyapunov function if* $b_k a_{ki} \leq 0$ *for all* $i, k \in \{1, 2, \ldots, n\}$.

(ii) $\varphi_{\mathbf{b}_\uparrow}(\mathbf{x}) = \sum_{k=1}^{n} b_k x_k$ *is a Lyapunov function if* $a_{ki} \geq 0$ *for all* $k \leq i$ *where* $i, j \in \{1, 2, \ldots, n\}$.

(iii) $\varphi_{\mathbf{b}_\downarrow}(\mathbf{x}) = \sum_{k=1}^{n} b_k x_k$ *is a Lyapunov function if* $a_{ki} \leq 0$ *for all* $k \leq i$ *where* $i, j \in \{1, 2, \ldots, n\}$.

(iv) $\varphi_p(\mathbf{x}) = \prod_{k=1}^{n} x_k^{p_k}$ *is a Lyapunov function for arbitrary* V *where* $\mathbf{x} \in ri\,S^{n-1}$ *and* $p \in \mathbb{P}_A^{n-1}$.

(v) $\varphi(\mathbf{x}) = \frac{x_i}{x_j}, i \neq j$ *is a Lyapunov function for any* V *where* $\mathbf{x} \in ri\,S^{n-1}$.

(vi) $\varphi(\mathbf{x}) = \sum_{i>r+1}^{n} x_i, \mathbf{x} \in S^{n-1}$ *is a Lyapunov function if there is* $r \in \{1, 2, \ldots, n\}$ *such that* $a_{ij} < 0$ *for all* $i \leq r, j > r$,

where $\mathbb{P}_A^{n-1} = \left\{\mathbf{x} \in S^{n-1}; \sum_{i=1}^{n} a_{ki}x_i \geq 0, k \in \{1, 2, \ldots, n\}\right\}$.

The construction of Lyapunov functions for infinite dimensional Volterra QSOs and their dynamics were investigated in [12, 13, 16, 17], but in general setting, these are still open problems.

2 Main Results

Next, we are going to construct Lyapunov function of infinite dimensional Volterra QSOs by means of finite dimensional ones. Let us recall some notions from [9]. A sequence of Volterra $\{V_{n]}\}$ operators is called *compatible* if

$$V_{n+1} \upharpoonright S^{n-1} = V_n, \tag{8}$$

for every $n \in \mathbb{N}$. Here, $V \upharpoonright S^{n-1}$ stands for the restriction for V to the simplex S^{n-1}. The compatibility condition (8) together with (6) imply that

$$a_{ki}^{n+1} = a_{ki}^n, \quad \forall k, i \in \{1, \cdots, n\}, \tag{9}$$

where a_{ki}^{n+1} and a_{ki}^n are associated skew-symmetric matrices to V_{n+1} and V_n, respectively.

Theorem 2 ([9]) *The sequence* $\{V_n\}$ *of compatible Volterra operators weakly converges to a Volterra operator* V.

Next remark gives some observation on the relationship between the Lyapunov functions on the simplex S with its restriction to the face of S.

Remark 1 If the functional φ defined on S is a Lyapunov function for a Volterra QSO V, then from the definition of Lyapunov function one concludes that restriction of φ to any faces Γ of S is again a Lyapunov function for $V \upharpoonright \Gamma$.

Now, a natural question arises: how can we extend Lyapunov functions given for finite-dimensional Volterra operators to infinite dimensional ones. So here, we introduce a notion of compatible sequence of functions. Let $\{\varphi^{n]}\}$ be a sequence of functions and we say that the sequence $\{\varphi^{n]}\}$ is *compatible* if

$$\varphi^{n+1]} \restriction \mathbf{B}_1{}^{n-1} = \varphi^{n]},$$

for any $n \in \mathbb{N}$. Again, $\varphi^{n+1]} \restriction \mathbf{B}_1{}^{n-1}$ represents the restriction of functional $\varphi^{n+1]}$ to the ball $\mathbf{B}_1{}^{n-1}$.

In particular, let $\varphi^{n]}_{\mathbf{b}^{n]}}(\mathbf{x}) = \sum_{k=1}^{n} b^{n]}_k x_k$ where $\mathbf{x} \in S^{n-1}$ and $\mathbf{b}^{n]} = (b^{n]}_1, \dots, b^{n]}_n)$. Thus, if the sequence $\{\varphi^{n]}_{\mathbf{b}^{n]}}(\mathbf{x})\}$ is compatible, then this means,

$$b^{n+1]}_k = b^{n]}_k \text{ for any } k \in \{1, 2, \dots n\}. \tag{10}$$

Remark 2 Let V be a finite dimensional Volterra QSOs defined on S^{n-1}. If for any $\mathbf{x} \in S^{n-1}$ one has

$$\varphi_{\mathbf{b}}(V(\mathbf{x})) \le \varphi_{\mathbf{b}}(\mathbf{x}), \tag{11}$$

(where the functional $\varphi_{\mathbf{b}} = \sum_{k=1}^{n} b_k x_k$), then for any $\mathbf{y} \in S^{n-1}_r$ and $r \in (0, 1)$ we obtain

$$\varphi_{\mathbf{b}}(V(\mathbf{y})) \le \varphi_{\mathbf{b}}(\mathbf{y}).$$

This fact can be seen as follows: For a fix $r \in (0, 1)$, we take any $\mathbf{y} \in S^{n-1}_r$. Then

$$\begin{aligned}
\varphi_{\mathbf{b}}(V(\mathbf{y})) - \varphi_{\mathbf{b}}(\mathbf{y}) &= \sum_{k=1}^{n} b_k y_k \left(1 + \sum_{i=1}^{n} a_{ki} y_i\right) - \sum_{k=1}^{n} b_k y_k \\
&= \sum_{k=1}^{n} b_k y_k \left(\sum_{i=1}^{n} a_{ki} y_i\right) \\
&= r^2 \sum_{k=1}^{n} b_k \frac{y_k}{r} \left(\sum_{i=1}^{n} a_{ki} \frac{y_i}{r}\right) \\
&= r^2 \left(\sum_{k=1}^{n} b_k \frac{y_k}{r} \left(1 + \sum_{i=1}^{n} a_{ki} \frac{y_i}{r}\right) - \sum_{k=1}^{n} b_k \frac{y_k}{r}\right) \\
&= r^2 \left(\varphi_{\mathbf{b}}\left(V\left(\frac{\mathbf{y}}{r}\right)\right) - \varphi_{\mathbf{b}}\left(\frac{\mathbf{y}}{r}\right)\right).
\end{aligned}$$

Clearly $\frac{\mathbf{y}}{r} \in S^{n-1}$, hence from (11) it follows that $\varphi_{\mathbf{b}}(V(\mathbf{y})) - \varphi_{\mathbf{b}}(\mathbf{y}) \le 0$ which implies the desired statement.

The following theorem is the main result in this paper.

Theorem 3 *Let $\{V_{n]}\}$ be a compatible sequence of Volterra QSOs. For each $n \in \mathbb{N}$, we assume that $V_{n]}$ has a Lyapunov function in the form*

$$\varphi^{n]}_{\mathbf{b}^{n]}}(\mathbf{x}) = \sum_{k=1}^{n} b^{n]}_k x_k. \tag{12}$$

If $\varphi^{n]}_{\mathbf{b}^{n]}}(V_{n]}(\mathbf{x})) \leq \varphi^{n]}_{\mathbf{b}^{n]}}(\mathbf{x})$ *is satisfied for any* $n \in \mathbb{N}$ *and* $\mathbf{x} \in S^{n-1}$*, then there exists a Lyapunov function for* V*, where* $\{V_{n]}\}$ *weakly converges to* V*.*

Proof First we show that from the assumption (12), then there is a compatible sequence of Lypunov function. Take any $n_0 \in \mathbb{N}$, then from (12) we infer that $\varphi^{n_0]}_{\mathbf{b}^{n_0]}}$ is a Lyapunov function for $V_{n_0]}$. Due to Remark 1 for any $m < n_0$ one has

$$\varphi^{n_0]}_{\mathbf{b}^{n_0]}} \upharpoonright S^{m-1} = \varphi^{m]}_{\mathbf{b}^{m]}}, \tag{13}$$

which yields that $\varphi^{m]}_{\mathbf{b}^{m]}}$ a Lyapunov function for $V_{m]} = V_{n_0]} \upharpoonright S^{m-1}$. Since n_0 is an arbitrary element of $\mathbb{N}$, then there exists a compatible sequence of Lypunov functions. We denote this sequence by $\{\varphi^{n]}_{\mathbf{b}^{n]}}\}$. Taking into account (10), we infer that $\mathbf{b}^{n]} = (b_1^{n]}, \ldots, b_n^{n]})$ weakly converges to $\mathbf{b}^* = (b_1^*, \ldots, b_n^*, \ldots)$. Indeed, $\mathbf{b}^* \in \ell^\infty$. If it is not so, then there is i_0 such that $|b^*_{i_0}| > M$ for any $M > 0$. Without the loss of generality, we assume that $b_i^{i_0]} \geq 0$ for all $i \in \{1, \ldots, i_0\}$, then for $\mathbf{x} \in S^{i_0-1}$ such that $x_{i_0} > 0$ one has

$$\varphi^{i_0]}_{\mathbf{b}^{i_0]}}(\mathbf{x}) = \sum_{k=1}^{i_0} b_k^{i_0]} x_k \geq b_{i_0}^{i_0]} x_{i_0} \geq M x_{i_0}. \tag{14}$$

Since x_{i_0} is a fix number and M is an arbitrary positive number, therefore the functional $\varphi^{i_0]}_{\mathbf{b}^{i_0]}}(\mathbf{x})$ cannot be finite. Taking into account any Volterra QSOs satisfies $V(ri\Gamma) \subset ri\Gamma$, we conclude that $\varphi^{i_0]}_{\mathbf{b}^{i_0]}}(\mathbf{x})$ cannot be a Lypunov function for $V_{i_0]}$. However, this contradicts to the fact that $\varphi^{i_0]}_{\mathbf{b}^{i_0]}}$ is a Lypunov function for the operator $V_{i_0]}$. Thus $|b_i^*| \leq M$ for some $M \in \mathbb{R}$.

Next we want to show that $\{\varphi^{n]}_{\mathbf{b}^{n]}}(\mathbf{x})\}$ forms a Cauchy sequence. Take any $\mathbf{x} \in S$, then for every $\epsilon > 0$, there exist n_0 such that $\sum_{k=n_0+1}^{\infty} x_k < \frac{\epsilon}{M}$. Now take $m > n \geq n_0$. Therefore,

$$\begin{aligned}\left|\varphi^{m]}_{\mathbf{b}^{m]}}(\mathbf{x}) - \varphi^{n]}_{\mathbf{b}^{n]}}(\mathbf{x})\right| &= \left|\sum_{k=n+1}^{m} b_k^{m]} x_k\right| \leq \sum_{k=n+1}^{m} |b_k^{m]}| x_k \\ &\leq M \sum_{k=n+1}^{m} x_k \leq M \sum_{k=n_0+1}^{\infty} x_k < M\frac{\epsilon}{M} = \epsilon. \end{aligned} \tag{15}$$

Hence, the sequence $\{\varphi^{n]}_{\mathbf{b}^{n]}}(\mathbf{x})\}$ is Cauchy, therefore $\varphi^{n]}_{\mathbf{b}^{n]}}(\mathbf{x})$ converges to $\varphi_{\mathbf{b}^*}(\mathbf{x})$. Let us show $\varphi_{\mathbf{b}^*}(\mathbf{x})$ is indeed a Lyapunov function for V where $V_{n]}$ weakly converges to V. Indeed, we have

$$\begin{aligned}
\varphi_{\mathbf{b}^*}(V(\mathbf{x})) - \varphi_{\mathbf{b}^*}(\mathbf{x}) &= \sum_{k=1}^{\infty} b_k^* x_k \left(1 + \sum_{i=1}^{\infty} a_{ki} x_i\right) - \sum_{k=1}^{\infty} b_k^* x_k \\
&= \sum_{k=1}^{\infty} b_k^* x_k \sum_{i=1}^{\infty} a_{ki} x_i \\
&= \sum_{k=1}^{n_0} b_k^* x_k \sum_{i=1}^{\infty} a_{ki} x_i + \sum_{k=n_0+1}^{\infty} b_k^* x_k \sum_{i=1}^{\infty} a_{ki} x_i \\
&= \sum_{k=1}^{n_0} b_k^* x_k \sum_{i=1}^{n_0} a_{ki} x_i + \sum_{k=1}^{n_0} b_k^* x_k \sum_{i=n_0+1}^{\infty} a_{ki} x_i + \\
&\quad \sum_{k=n_0+1}^{\infty} b_k^* x_k \sum_{i=1}^{\infty} a_{ki} x_i \\
&= \varphi_{\mathbf{b}^{n_0]}}(V_{n_0]}(\mathbf{x})) - \varphi_{\mathbf{b}^{n_0]}}(\mathbf{x}) + \sum_{k=1}^{n_0} b_k^* x_k \sum_{i=n_0+1}^{\infty} a_{ki} x_i + \\
&\quad \sum_{k=n_0+1}^{\infty} b_k^* x_k \sum_{i=1}^{\infty} a_{ki} x_i .
\end{aligned} \tag{16}$$

From the assumption $\varphi_{\mathbf{b}^{n_0]}}(V_{n_0]}(\mathbf{x})) \leq \varphi_{\mathbf{b}^{n_0]}}(\mathbf{x})$ and Remark 2 we conclude that,

$$\varphi_{\mathbf{b}^{n_0]}}(V_{n_0]}(\mathbf{x})) - \varphi_{\mathbf{b}^{n_0]}}(\mathbf{x}) \leq 0. \tag{17}$$

Further, we will show that $\left| \sum\limits_{k=1}^{n_0} b_k^* x_k \sum\limits_{i=n_0+1}^{\infty} a_{ki} x_i \right| < \epsilon$ for arbitrary small $\epsilon > 0$. Clearly one has

$$\begin{aligned}
\left| \sum_{k=1}^{n_0} b_k^* x_k \sum_{i=n_0+1}^{\infty} a_{ki} x_i \right| &\leq \sum_{k=1}^{n_0} \left| b_k^* x_k \sum_{i=n_0+1}^{\infty} a_{ki} x_i \right| \\
&\leq \sum_{k=1}^{n_0} |b_k^*| \, x_k \sum_{i=n_0+1}^{\infty} |a_{ki}| x_i \\
&\leq M \sum_{k=1}^{n_0} x_k \sum_{i=n_0+1}^{\infty} x_i \\
&< M \frac{\epsilon}{M} = \epsilon.
\end{aligned} \tag{18}$$

Next, we get

$$\begin{aligned}\left|\sum_{k=n_0+1}^{\infty} b_k^* x_k \sum_{i=1}^{\infty} a_{ki} x_i\right| &\le \sum_{k=n_0+1}^{\infty} \left|b_k^* x_k \sum_{i=1}^{\infty} a_{ki} x_i\right| \\ &\le \sum_{k=n_0+1}^{\infty} \left|b_k^*\right| x_k \sum_{i=1}^{\infty} |a_{ki}| x_i \\ &\le M \sum_{k=n_0+1}^{\infty} x_k \sum_{i=1}^{\infty} x_i < \epsilon. \end{aligned} \tag{19}$$

Hence, thanks to (16), (18) and (19) one finds

$$\varphi_{\mathbf{b}^*}(V(\mathbf{x})) - \varphi_{\mathbf{b}^*}(\mathbf{x}) < \varphi_{\mathbf{b}^{n_0]}}(V_{n_0]}(\mathbf{x})) - \varphi_{\mathbf{b}^{n_0]}}(\mathbf{x}) + 2\epsilon.$$

The arbitrariness of ϵ implies $\varphi_{\mathbf{b}^*}(V(\mathbf{x})) \le \varphi_{\mathbf{b}^*}(\mathbf{x})$ for all $\mathbf{x} \in S$ which yields that the limit $\lim_{m\to\infty} \varphi_{\mathbf{b}}\left(V^{(n)}(\mathbf{x})\right)$ exists. Consequently, $\varphi_{\mathbf{b}^*}(\mathbf{x})$ is a Lyapunov function for V. This completes the proof.

Immediately from the last theorem, we can use the same procedures to prove the following theorem.

Theorem 4 *Let $\{V_{n]}\}$ be a compatible sequence of Volterra QSOs and for each $n \in \mathbb{N}$, we assume that $V_{n]}$ has Lyapunov function in the form*

$$\varphi^{n]}_{\mathbf{b}^{n]}}(\mathbf{x}) = \sum_{k=1}^{n} b_k^{n]} x_k. \tag{20}$$

If $\varphi^{n]}_{\mathbf{b}^{n]}}(V_{n]}(\mathbf{x})) \ge \varphi^{n]}_{\mathbf{b}^{n]}}(\mathbf{x})$ for any $n \in \mathbb{N}$ and $\mathbf{x} \in S^{n-1}$, then there exists a Lyapunov function for V.

Denote

$$\varphi_{\mathbf{b}}(\mathbf{x}) = \sum_{k=1}^{\infty} b_k x_k, \quad \mathbf{x} \in \ell^1. \tag{21}$$

Let us list out concrete examples of linear Lyapunov function for infinite dimensional Volterra QSOs that have been proven in [12, 17] as follows:

1. Assume that $\mathbf{b} \in \ell^\infty$ and for any pair $(k, i) \in \mathbb{N}^2$ one has $b_k a_{ki} \le 0$ (resp. $b_k a_{ki} \ge 0$). Then the functional $\varphi_{\mathbf{b}}$ given by (21) defined on S is a Lyapunov function for V.
2. Let $a_{ki} \ge 0$ for all $k < i$ and $\mathbf{b}_\uparrow \in \ell^\infty$. Then a functional $\varphi_{\mathbf{b}_\uparrow}$ given by (21) defined on S is a Lyapunov function for V.
3. Let $a_{ki} \le 0$ for all $k < i$ and $\mathbf{b}_\downarrow \in \ell^\infty$. Then the functional $\varphi_{\mathbf{b}_\downarrow}$ given by (21) defined on S is a Lyapunov function for V.

where,

$$\begin{array}{ll} \mathbf{b}_{\uparrow} = (b_1, \ldots, b_n, \ldots), & \text{such that } b_1 \leq \cdots \leq b_n \leq \cdots \\ \mathbf{b}_{\downarrow} = (b_1, \ldots, b_n, \ldots), & \text{such that } b_1 \geq \cdots \geq b_n \geq \cdots \end{array}.$$

One can see that, the constructed Lyapunov functions above are the extension of Lyapunov function in Theorem 1 (i), (ii) and (iii). The Lyapunov functions open a path to describe the limiting set of the associated Volterra QSOs and here we provide brief results that have been proven in [12, 17]. Let us define a skew-symmetric matrix as follows:

$$\begin{pmatrix} A & O \\ O & B \end{pmatrix},$$

where A is a $(k_0 - 1) \times (k_0 - 1)$-skew-symmetric matrix, O is a null matrix and B is an infinite dimensional skew-symmetric matrix. Moreover, all the entries of matrices A and B satisfy (7). If the entries of matrix $B = (a_{ki})_{i,k \geq k_0}$ are $a_{ki} \geq 0$ (respectively, $a_{ki} \leq 0$) for all $k < i$, then the correspondence set of all Volterra QSOs is denoted by $\tilde{\mathcal{V}}^{+}$ (respectively, $\tilde{\mathcal{V}}^{-}$).

The authors in [17] use the linear Lyapunov functions to study the dynamics of V taken from the class $\tilde{\mathcal{V}}^{+} \cup \tilde{\mathcal{V}}^{-}$. It was proven that the omega limiting sets of $V \in \tilde{\mathcal{V}}^{+}$ are not empty and belong to the boundary of the simplex S. However, the set could be empty if $V \in \tilde{\mathcal{V}}^{-}$. The results are summerized in the following theorem.

Theorem 5 ([17]) *Let $V \in \tilde{\mathcal{V}}^{+} \cup \tilde{\mathcal{V}}^{-}$ and $V \neq Id$, then the following statements hold:*

(i) if $V \in \tilde{\mathcal{V}}^{+}$ then $\omega_V(\mathbf{x}_0) = \omega_V^{(w)}(\mathbf{x}_0)$. Moreover, $\omega_V^{(w)}(\mathbf{x}_0) \in \partial S$ for all $\mathbf{x}_0 \in S \setminus Fix(V)$;

(ii) if $V \in \tilde{\mathcal{V}}^{-}$ then for any $\mathbf{x}_0 \in S \setminus Fix(V)$ one has:

$(ii)_a$ $\omega_V(\mathbf{x}_0) = \omega_V^{(w)}(\mathbf{x}_0) \in \partial S$ if and only if $\omega_V(\mathbf{x}_0) \neq \emptyset$;

$(ii)_b$ if $\omega_V(\mathbf{x}_0) = \emptyset$ then there exists $r < 1$ such that $\omega_V^{(w)}(\mathbf{x}_0) \in \partial S_r$,

where $\omega_V(\mathbf{x}_0)$ (respectively, $\omega_V^{(w)}(\mathbf{x}_0)$) denotes the set of limit points of $\{V^{(n)}(\mathbf{x}_0)\}_{n=1}^{\infty}$ with respect to ℓ^1-norm. (respectively, pointwise convergence). Moreover, the cardinality of the omega limiting sets is described as follows:

(i) if $\omega_V(\mathbf{x}_0) \neq \emptyset$ then $|\omega_V(\mathbf{x}_0)| = 1 \vee \infty$;

(ii) $|\omega_V^{(w)}(\mathbf{x}_0)| = 1 \vee \infty$,

for any $\mathbf{x}_0 \in S$.

From the last theorem we can conclude that any operators taken from the class $\tilde{\mathcal{V}}^{+} \cup \tilde{\mathcal{V}}^{-}$ do not have periodic orbit. This statement is an analogous result like finite dimensional case that was well-establish in [2]. Furthermore, the same study was done for more general operators called Lotkha-Volterra operators [16].

Acknowledgements The author acknowledges the Ministry of Higher Education (MOHE) and Research Management Centre-UTM, Universiti Teknologi Malaysia (UTM) for the financial support through Fundamental Research Grant Scheme (FRGS/1/2021/STG06/UTM/02/5).

References

1. Bernstein, S.N.: The solution of a mathematical problem concerning the theory of heredity. Ucheniye-Zapiski N.-I. Kaf. Ukr. Otd. Mat. **1**, 83–115 (1924) (in Russian)
2. Ganikhodzhaev, R.N.: Quadratic stochastic operators, Lyapunov functions, and tournaments. Russ. Acad. Sci. Sb. Math. **76**, 489–506 (1993)
3. Ganikhodzhaev, R., Mukhamedov, F., Rozikov, U.: Quadratic stochastic operators and processes: results and open problems. Infin. Dimens. Anal. Quan. Probab. Relat. Top. **14**, 279–335 (2011)
4. Ganikhodjaev, N.N., Ganikhodjaev, R.N., Jamilov, U.U.: Quadratic stochastic operators and zero-sum game dynamics. Ergod. Theory Dyn. Syst. **35**, 1443–1473 (2015)
5. Jamilov, U.U.: Linear Lyapunov functions for Volterra quadratic stochastic operators. TWMS J. Pure Appl. Math. **3**(1), 28–34 (2012)
6. Kesten, H.: Quadratic transformations: a model for population growth I. Adv. Appl. Probab. **2**, 1–82 (1970)
7. Karim, S.N., Hamzah, N.Z.A., Ganikhodjaev, N.: New class of Lebesgue quadratic stochastic operators on continual state space. Int. J. Allied Health Sci. **5**(1) (2021)
8. Lyubich Y.I.: Mathematical Structures in Population Genetics. Biomathematics, vol. 22. Springer-Verlag (1992)
9. Mukhamedov, F., Akin, H., Temir, S.: On infinite dimensional quadratic Volterra operators. J. Math. Anal. Appl. **310**(2), 533–556 (2005)
10. Mukhamedov, F., Ganikhodjaev, N.: Quantum Quadratic Operators and Processes. Lecture Notes in Mathematics, vol. 2133. Springer (2015)
11. Mukhamedov, F., Embong, A.F.: On b-bistochastic quadratic stochastic operators. J. Inequal. Appl. **1**, 1–16 (2015)
12. Mukhamedov, F., Embong, A.F.: Linear Lyapunov functions of infinite dimensional Volterra operators. Malays. J. Math. Sci. **13**, 201–210 (2019)
13. Mukhamedov, F., Embong, A.F.: On Lyapunov functions for infinite dimensional Volterra quadratic stochastic operators. J. Phys: Conf. Ser. **949**, 12–22 (2017)
14. Mukhamedov, F., Embong, A.F.: Infinite dimensional orthogonality preserving nonlinear Markov operators. Linear Multilinear Algebr. **69**(3), 526–550 (2021)
15. Mukhamedov, F., Khakimov, O., Embong, A.F.: Projective surjectivity of quadratic stochastic operators on L1 and its application. Chaos Solitons Fract. **148**, 111034 (2021)
16. Mukhamedov, F., Khakimov, O., Embong, A.F.: Ergodicities of infinite dimensional nonlinear stochastic operators. Qual. Theory Dyn. Syst. **19**(3), 1–20 (2020)
17. Mukhamedov, F., Khakimov, O., Embong, A.F.: On omega limiting sets of infinite dimensional Volterra operators. Nonlinearity **33**(11), 5875–5904 (2020)
18. Mukhamedov, F., Saburov, M., Qaralleh, I.: On $\xi^{(s)}$-quadratic stochastic operators on two dimensional simplex and their behavior. Abst. Appl. Anal. **12** (2013). Article ID 942038
19. Mukhamedov, F., Saburov, M.: On Homotopy of Volterrian quadratic stochastic operators. Appl. Math. Inform. Sci. **4**, 47–62 (2010)
20. Narendra, S.G., Samaresh, C.M., Elliott, W.M.: On the Volterra and other nonlinear models of interacting populations. Rev. Mod. Phys. **43**, 231–276 (1971)
21. Takeuchi, Y.: Global Dynamicsal Properties of Lotka-Volterra Systems. World Scientific. (1996)

22. Rozikov, U.A., Shoyimardonov, S.K.: Ocean ecosystem discrete time dynamics generated by ℓ-Volterra operators. Int. J. Biomath. **12**(02), 1950015 (2019)
23. Rozikov, U.A., Velasco, M.V.: A discrete-time dynamical system and an evolution algebra of mosquito population. J. Math. Biol. **78**(4), 1225–1244 (2019)

Bijectivity of a Class of Lotka-Volterra Operators Defined on 2D-Simplex

Chin Hee Pah and Azizi Rosli

Abstract It is known that if Lotka-Volterra operators, defined on the simplex S^{m-1}, satisfy $\underline{\mathbf{f}}$-monotonicity condition, then they are bijection of S^{m-1}. In the present paper, we consider a class of Lotka-Volterra operators defined on two dimensional simplex which are bijective, however, they are not $\underline{\mathbf{f}}$-monotone.

Keywords Bijective · Lotka-volterra · Stochastic operator

1 Introduction

It is well-known that Lotka-Volterra (LV) systems typically model the time evolution of conflicting species in biology [1, 15]. For more natural applications of LV systems we refer to [13]. Discrete time LV systems are well-known subject of applied mathematics [7]. Since then, LV systems have proved to be a rich source of analysis for the investigation of dynamical properties and modeling in different domains of natural sciences (see for example, [3, 6]).

The mentioned papers show importance the study of limiting behavior of discrete LV type operators. In [10], it has been proved that any LV stochastic operator is surjective. Furthermore, it is proved that $\underline{\mathbf{f}}$-monotone LV operators are bijections. In that paper, it has been provided an example of LV operator which is not bijective and not $\underline{\mathbf{f}}$-monotone as well. Therefore, it is natural to ask: is there bijective but not $\underline{\mathbf{f}}$-monotone LV operators. In this paper, we are going to provide a class of bijective Lotka-Volterra operators defined on 2-simplex which are not $\underline{\mathbf{f}}$-monotone.

C. H. Pah (✉) · A. Rosli
International Islamic University Malaysia Kuantan, Pahang, Malaysia
e-mail: pahchinhee@iium.edu.my

L. Accardi et al. (eds.), *Infinite Dimensional Analysis, Quantum Probability and Applications*, Springer Proceedings in Mathematics & Statistics 390,
https://doi.org/10.1007/978-3-031-06170-7_20

2 Preliminaries

Let $E = \{1, 2, \ldots, m\}$, and

$$S^{m-1} = \left\{ \mathbf{x} = (x_1, x_2, \ldots, x_m) \in \mathbb{R}^m : x_i \geq 0, \sum_{i=1}^{m} x_i = 1 \right\} \tag{1}$$

be the $(m-1)$-dimensional simplex. Let $\text{int}S^{m-1} = \{\mathbf{x} \in S^{m-1} : x_1 x_2 \cdots x_m > 0\}$ be the interior of S^{m-1}, $\Gamma_\alpha = \left\{\mathbf{x} \in S^{m-1} : x_i = 0, i \in \alpha \subset E \cup \{0\}\right\}$ be the faces of S^{m-1}, $\text{int}\Gamma_\alpha = \{\mathbf{x} \in \Gamma_\alpha : \prod_{k \notin \alpha} x_k > 0\}$.

A mapping $U : S^{m-1} \to S^{m-1}$ defined on the simplex given by

$$U(\mathbf{x})_k = \sum_{i,j=1}^{m} p_{ij,k} x_i x_j, \quad k \in E \tag{2}$$

where $p_{ij,k}$ is a coefficient satisfying

$$p_{ij,k} \geq 0, \quad p_{ij,k} = p_{ji,k}, \quad \sum_{k=1}^{m} p_{ij,k} = 1, \quad i, j, k \in E \tag{3}$$

is known as *Quadratic Stochastic Operator (QSO)*.

A QSO is called *Volterra* if it satisfies

$$P_{ij,k} = 0 \text{ if } k \notin \{i, j\}. \tag{4}$$

One can easily check that any operator (2) satisfying (4) is equivalent to

$$U(\mathbf{x})_k = x_k \left(1 + \sum_{i=1}^{m} a_{ki} x_i \right), \tag{5}$$

where $a_{ki} = 1 - 2p_{ik,k}$ [4].

In present paper, we consider a class of Lotka-Volterra stochastic Operators. A mapping $V : S^{m-1} \to S^{m-1}$ is called a *Lotka-Volterra operator* (for short LV operator) if every face of the simplex is invariant under V, i.e. $V(\Gamma_\alpha) \subset \Gamma_\alpha$ for any $\alpha \in E$ [9]. By $DV(\mathbf{x})$ we denote the Jacobian of V at the point $\mathbf{x}$.

Given a mapping $\mathbf{f} : \mathbf{x} \in S^{m-1} \to (f_1(\mathbf{x}), f_2(\mathbf{x}), \ldots, f_m(\mathbf{x})) \in \mathbb{R}^m$ in what follows, we are interested in the following operator defined by

$$(V\mathbf{x})_k = x_k(1 + f_k(\mathbf{x})), \; k = \overline{1, m} \; x \in S^{m-1}. \tag{6}$$

Theorem 1 ([9]) *Let V be an operator given by* (6). *The following conditions are equivalent:*

(i) The operator V is Lotka-Volterra;
(ii) The mapping $\underline{f} \equiv (f_1, f_2, \ldots, f_m) : S^{m-1} \to \mathbb{R}^m$ *satisfies the following conditions:*

1^0 $\underline{f}$ *is continuous in* S^{m-1};
2^0 *for every* $\mathbf{x} \in S^{m-1}$ *one has* $f_k(\mathbf{x}) \geq -1$, *for all* $k = \overline{1, m}$;
3^0 *for every* $\mathbf{x} \in S^{m-1}$ *one has* $\sum_{k=1}^{m} x_k f_k(\mathbf{x}) = 0$;
4^0 *for every* $\alpha \subset I$ *one holds* $f_k(\mathbf{x}) > -1$ *for all* $\mathbf{x} \in ri\,\Gamma_\alpha$ *and* $k \in \alpha$.

Theorem 2 ([8]) *Let* $f \equiv (f_1, f_2, \ldots, f_m) : \mathbb{R}^m \to \mathbb{R}^m$ *be a linear mapping. Then* f *satisfies the conditions* $1^0 - 4^0$ *if and only if one has*

$$f_k(\mathbf{x}) = \sum_{i=1}^{m} a_{ki} x_i, \quad k = \overline{1, m} \tag{7}$$

with

$$a_{ki} = -a_{ik}, \quad |a_{ki}| \leq 1 \quad \forall\, k, i = \overline{1, m}.$$

Remark 1 Note that LV type operator $V : S^{m-1} \to S^{m-1}$ with generating mapping given by (7), then it is Volterra operator. In [4, 8, 11] limiting behaviors of quadratic Volterra operators were studied. We refer a reader to [5] for review on quadratic stochastic operators.

Theorem 3 ([10]) *Any LV type operator given by* (6) *maps simplex* S^{m-1} *onto itself. Namely, V is a surjection of* S^{m-1}.

We recall that a mapping $\mathbf{f} : S^{m-1} \to \mathbb{R}^m$ is called *monotone* on the simplex S^{m-1} if for every points $x, y \in S^{m-1}$ the following condition is satisfied

$$\langle \mathbf{f}(\mathbf{x}) - \mathbf{f}(\mathbf{y}), \mathbf{x} - \mathbf{y} \rangle \geq 0, \tag{8}$$

where $\langle \mathbf{x}, \mathbf{y} \rangle$ stands for the standard scalar product in $\mathbb{R}^m$.

Observe that for a generating function $\mathbf{f}$ of LV type operator, its monotonicity can be replaced by

$$\langle \mathbf{f}(\mathbf{x}), \mathbf{y} \rangle + \langle \mathbf{x}, \mathbf{f}(\mathbf{y}) \rangle \leq 0, \qquad \mathbf{x}, \mathbf{y} \in S^{m-1}.$$

Definition 1 An LV type operator (6) is called $\mathbf{f}$-monotone if its generating mapping $\mathbf{f} : S^{m-1} \to \mathbb{R}^m$ is monotone on S^{m-1}.

Theorem 4 ([10]) *Let V be a* $\underline{f}$*-monotone LV operator on* S^{m-1} *then V is a homeomorphism of the simplex.*

In [9], it has been provided an example of LV operator which is not bijective and not $\underline{\mathbf{f}}$-monotone as well. Therefore, it is natural to ask: is there bijective but not

$\underline{\mathbf{f}}$-monotone LV operators. In this paper, we are going to provide a class of bijective Lotka-Volterra operators defined on 2-simplex which are not $\underline{\mathbf{f}}$-monotone.

Definition 2 A continuous mapping $F : X \to Y$ is a local homeomorphism if for every $\mathbf{x} \in X$ there exist an open set U containing $\mathbf{x}$ such that the image $F(U)$ is open in Y, and the restriction $\overline{F} : U \to F(U)$ is a homeomorphism.

Theorem 5 (Inverse Function Theorem) *Let the mapping $F : \mathbb{R}^n \to \mathbb{R}^n$ be continuously differentiable on some open set containing $\mathbf{x}$, and suppose that the determinant of the Jacobian of F at $\mathbf{x}$ is non-zero, i.e. $|DF(\mathbf{x})| \neq 0$. Then there is an open set P containing $\mathbf{x}$, and an open set Q containing $F(\mathbf{x})$ such that $F : P \to Q$ has a continuous inverse $F^{-1} : Q \to P$ which is differentiable for all $\mathbf{y} \in Q$.*

3 Main Result

In this section, we consider the following LV operator given by

$$V(\mathbf{x}) = \begin{cases} x_1' = x_1\left(1 + \gamma_1 x_1^r x_2 - \gamma_2 x_3^{r+1}\right), \\ x_2' = x_2\left(1 + \gamma_3 x_2^r x_3 - \gamma_1 x_1^{r+1}\right), \\ x_3' = x_3\left(1 + \gamma_2 x_3^r x_1 - \gamma_3 x_2^{r+1}\right). \end{cases} \tag{9}$$

where $\gamma_k \in (0, 1]$ and $r \geq 0$ are any real numbers. Such kind of operator has been considered in [12], where limiting behaviour of such operator was investigated. We point out that if $r = 0$ then (9) reduced to the Stein-Ulam operator [2, 14]. In this section, we show that LV operator given by (9) is bijective.

Before to formulate and prove main result, we need some auxiliary facts.

Let us consider the following matrix:

$$M_c = \begin{bmatrix} a_{11} & a_{12} & -ca_{31} \\ -ca_{12} & a_{22} & a_{23} \\ a_{31} & -ca_{23} & a_{33} \end{bmatrix}, \tag{10}$$

where

$$a_{11} = b_1 + a_{12}(c-1)\frac{x_2}{x_1},$$
$$a_{22} = b_2 + a_{23}(c-1)\frac{x_3}{x_2},$$
$$a_{33} = b_3 + a_{31}(c-1)\frac{x_1}{x_3}.$$

We have next result.

Corollary 1 *If $x_1, x_2, x_3 \in S^2$, $a_{ij} > 0$ for $i \neq j$, and $b_1, b_2, b_3 > 0$ then*

$$\det |M_c| > 0$$

for any constant $c \geq 1$.

Proof Now, expanding $\det |M_c|$, one finds

$$\det |M_c| = a_{11}a_{22}a_{33} + a_{12}a_{23}a_{31} - c^3 a_{12}a_{23}a_{31} + ca_{23}^2 a_{11} + ca_{12}^2 a_{33} + ca_{31}^2 a_{22}. \tag{11}$$

If $c = 1$, it is obvious that $\det |M_c| > 0$. Now suppose that $c > 1$. Since $b_1 b_2 b_3 > 0$, by AM-GM inequality, from (11) it follows that

$$\begin{aligned}
\det |M_c| &= \left(b_1 + a_{12}(c-1)\frac{y}{x}\right)\left(b_2 + a_{23}(c-1)\frac{z}{y}\right)\left(b_3 + a_{31}(c-1)\frac{x}{z}\right) + a_{12}a_{23}a_{31} \\
&\quad - c^3 a_{12}a_{23}a_{31} + ca_{23}^2\left(b_1 + a_{12}(c-1)\frac{y}{x}\right) + ca_{12}^2\left(b_3 + a_{31}(c-1)\frac{x}{z}\right) \\
&\quad + ca_{31}^2\left(b_2 + a_{23}(c-1)\frac{z}{y}\right) \\
&> b_1 b_2 b_3 + c^3 a_{12}a_{23}a_{31} - 3c(c-1)a_{12}a_{23}a_{31} - a_{12}a_{23}a_{31} + a_{12}a_{23}a_{31} \\
&\quad - c^3 a_{12}a_{23}a_{31} + ca_{23}^2 b_1 + ca_{12}^2 b_3 + ca_{31}^2 b_2 + c(c-1)\left[a_{23}^2 a_{12}\frac{y}{x}\right. \\
&\quad \left. + a_{12}^2 a_{31}\frac{x}{z} + a_{31}^2 a_{23}\frac{z}{y}\right] \\
&\geq b_1 b_2 b_3 - 3c(c-1)a_{12}a_{23}a_{31} + ca_{23}^2 b_1 + ca_{12}^2 b_3 + ca_{31}^2 b_2 \\
&\quad + 3c(c-1)\left(a_{23}^3 a_{31}^3 a_{12}^3 \frac{xyz}{xyz}\right)^{\frac{1}{3}} \\
&= b_1 b_2 b_3 + ca_{23}^2 b_1 + ca_{12}^2 b_3 + ca_{31}^2 b_2 \\
&> 0.
\end{aligned}$$

Hence, $\det |M_c| > 0$ for $c \geq 1$. This completes the proof.

Remark 2 If $c = 1$, then M_1 is skew-symmetric, and $\det |M_1| > 0$ was proven in [4].

Theorem 6 *The mapping* $V : S^2 \to S^2$ *defined by* (9) *is a homeomorphism.*

Proof It is known that by Theorem 3 the LV operator (9) is surjection.

One could easily check that the restriction of V on each face $\overline{V} : S^1 \to S^1$ maps the edges to itself. Note that the Jacobian of such restriction on the edge has the following form:

$$J_{\overline{V}} = \begin{bmatrix} a_{ii} & a_{ij} \\ -ha_{ij} & a_{jj} \end{bmatrix}, \qquad i, j \notin \alpha,$$

where $a_{ii}, a_{jj} > 0$, and $h > 0$. One can check it is positive, hence $\overline{V}$ is homeomorphism.

Consequently, V maps any faces of S^2 homeomorphically onto itself. Hence, by the compactness of S^2, it suffices to show that the mapping V is a local homeomorphism at each interior point of S^2.

Now let us compute the Jacobian of V at $\mathbf{x} = (x_1, x_2, x_3) \in \text{int}S^2$:

$$J_V = \begin{bmatrix} 1+\gamma_1(r+1)x_1^r x_2 - \gamma_2 x_3^{r+1} & \gamma_1 x_1^{r+1} & -(r+1)x_1\gamma_2 x_3^r \\ -(r+1)\gamma_1 x_1^r x_2 & 1+\gamma_3(r+1)x_2^r x_3 - \gamma_1 x_1^{r+1} & \gamma_3 x_2^{r+1} \\ \gamma_2 x_3^{r+1} & -(r+1)\gamma_3 x_2^r x_3 & 1+\gamma_2(r+1)x_3^r x_1 - \gamma_3 x_2^{r+1} \end{bmatrix}.$$

From the Jacobian we yield the determinant

$$|J_V| = x_1x_2x_3 \begin{vmatrix} B_1 + r\gamma_1 x_1^r \dfrac{x_2}{x_1} & \gamma_1 x_1^r & -(r+1)\gamma_2 x_3^r \\ -(r+1)\gamma_1 x_1^r & B_2 + r\gamma_3 x_2^r \dfrac{x_3}{x_2} & \gamma_3 x_2^r \\ \gamma_2 x_3^r & -(r+1)\gamma_3 x_2^r & B_3 + r\gamma_2 x_3^r \dfrac{x_1}{x_3} \end{vmatrix}$$

where

$$B_1 = \frac{1+\gamma_1 x_1^r x_2 - \gamma_2 x_3^{r+1}}{x_1},$$
$$B_2 = \frac{1+\gamma_3 x_2^r x_3 - \gamma_1 x_1^{r+1}}{x_2},$$
$$B_3 = \frac{1+\gamma_2 x_3^r x_1 - \gamma_3 x_2^{r+1}}{x_3}.$$

Let us denote $c = r+1$, $a_{12} = \gamma_1 x_1^r$, $a_{23} = \gamma_3 x_2^r$, $a_{31} = \gamma_2 x_3^r$. Then J_V reduces to M_c (10). Since $x_1x_2x_3 \in \text{int}S^2$, it follows that $x_1, x_2, x_3 > 0$, and $B_1, B_2, B_3 > 0$. Therefore, Corollary 1 implies that $\det|J_V| > 0$, i.e. the mapping V is a local homeomorphism at each point $\mathbf{x} \in \text{int}S^2$. The theorem is proved.

Now, we show that considered LV-operator is not $\underline{\mathbf{f}}$-monotone. Indeed, from (9) with (8), one finds

$$\begin{aligned}
\langle \mathbf{f}(x) - \mathbf{f}(y), x - y \rangle &= -x_1 \left(\gamma_1 y_1^r y_2 - \gamma_2 y_3^{r+1}\right) - x_2 \left(\gamma_3 y_2^r y_3 - \gamma_1 y_1^{r+1}\right) \\
&\quad - x_3 \left(\gamma_2 y_3^r y_1 - \gamma_3 y_2^{r+1}\right) - y_1 \left(\gamma_1 x_1^r x_2 - \gamma_2 x_3^{r+1}\right) \\
&\quad - y_2 \left(\gamma_3 x_2^r x_3 - \gamma_1 x_1^{r+1}\right) - y_3 \left(\gamma_2 x_3^r x_1 - \gamma_3 x_2^{r+1}\right) \\
&= -x_1 \gamma_1 y_2 (y_1^r - x_1^r) - x_2 \gamma_3 y_3 (y_2^r - x_2^r) - x_3 \gamma_2 y_1 (y_3^r - x_3^r) \\
&\quad + x_1 \gamma_2 y_3 (y_3^r - x_3^r) + x_2 \gamma_1 y_1 (y_1^r - x_1^r) + x_3 \gamma_3 y_2 (y_2^r - x_2^r) \\
&= \gamma_1 (y_1^r - x_1^r)(x_2 y_1 - x_1 y_2) + \gamma_3 (y_2^r - x_2^r)(x_3 y_2 - x_2 y_3) + \\
&\quad + \gamma_2 (y_3^r - x_3^r)(x_1 y_3 - x_3 y_1).
\end{aligned}$$

Let us take $\mathbf{x} = (\beta, 1 - \beta, 0)$ and $\mathbf{y} = (1 - \beta, 0, \beta)$, then

$$\begin{aligned}
\langle \mathbf{f}(\mathbf{x}) - \mathbf{f}(\mathbf{y}), \mathbf{x} - \mathbf{y} \rangle &= \gamma_1 ((1 - \beta)^r - \beta^r)(1 - \beta)^2 + \gamma_3 \beta (1 - \beta)^{r+1} + \gamma_2 \beta^{r+2} \\
&= \gamma_1 (1 - \beta)^{r+2} - \gamma_1 \beta^r (1 - \beta)^2 + \gamma_3 \beta (1 - \beta)^{r+1} + \gamma_2 \beta^{r+2} \\
&= (1 - \beta)^{r+1} [\gamma_1 (1 - \beta) + \gamma_3 \beta] + \beta^r [\gamma_2 \beta^2 - \gamma_1 (1 - \beta)^2].
\end{aligned}$$

From here, we may deduce that given $r > 0$, there exist $\beta > \frac{1}{2}$ such that

$$\gamma_1 > \frac{\gamma_3 \beta (1 - \beta)^{r+1} + \gamma_2 \beta^{r+2}}{(\beta^r - (1 - \beta)^r)(1 - \beta)^2},$$

where

$$\gamma_3 < -\frac{\beta^{r+2}}{\beta (1 - \beta)^{r+1}} \gamma_2 + \frac{(\beta^r - (1 - \beta)^r)(1 - \beta)^2}{\beta (1 - \beta)^{r+1}},$$

and

$$\gamma_2 < \frac{(\beta^r - (1 - \beta)^r)(1 - \beta)^2}{\beta^{r+2}}$$

satisfying $\gamma_1, \gamma_2, \gamma_3 \in (0, 1]$. This yields that $\langle \mathbf{f}(\mathbf{x}) - \mathbf{f}(\mathbf{y}), \mathbf{x} - \mathbf{y} \rangle < 0$. Hence, LV operator given by (9) is not $\underline{\mathbf{f}}$-monotone.

Let us provide a more concrete example.

Example 1 Let $\varepsilon_1, \varepsilon_2, \varepsilon_3 > 0$ and $r > 0$. Take $\beta = \frac{2}{3}$, we choose

$$\gamma_2 < \frac{1}{4} \left(1 - \frac{1}{2^r}\right).$$

Take $\gamma_2 = \frac{1}{4} \left(1 - \frac{1}{2^r}\right) - \varepsilon_1 > 0$, then

$$\gamma_3 < 2^{r+1} \varepsilon_1.$$

So, $\gamma_3 = 2^{r+1} \varepsilon_1 - \varepsilon_2 > 0$, then

$$\gamma_1 > \frac{2^r - 1 - 2\varepsilon_2}{2^r - 1}.$$

For $x = (\beta, 1-\beta, 0)$, and $y = (1-\beta, 0, \beta)$, we obtain

$$\langle \mathbf{f}(x) - \mathbf{f}(y), x - y \rangle = \frac{\varepsilon_3(2^{r+1} - 2^{2r} - 1)}{9(6^r - 3^r)} < 0.$$

Hence, the corresponding LV operator is not $\underline{\mathbf{f}}$-monotone.

As consequence to Theorem 6, we get the following corollary:

Corollary 2 *For any initial point* $\mathbf{x}^{(0)} \in S^2$, *the negative trajectory*

$$\mathbf{x}^{(0)}, V^{-1}\mathbf{x}^{(0)}, V^{-2}\mathbf{x}^{(0)}, \ldots$$

exists.

4 Conclusion

In this paper, we have established that the operator (9) is bijective. Now, let us consider the following LV operator:

$$V(\mathbf{x}) = \begin{cases} x_1' = x_1\left[1 + (a\varphi(x_1)x_2 - b\varphi(x_3)x_3)f(\mathbf{x})\right], \\ x_2' = x_2\left[1 + (c\varphi(x_2)x_3 - a\varphi(x_1)x_1)f(\mathbf{x})\right], \\ x_3' = x_3\left[1 + (b\varphi(x_3)x_1 - c\varphi(x_2)x_2)f(\mathbf{x})\right], \end{cases} \tag{12}$$

where $\varphi : [0, 1] \to [0, 1]$ is a strictly increasing function, and $f(\mathbf{x}) : S^2 \to (0, 1]$. Clearly, if $\varphi(x) = x^r$ and $f(\mathbf{x}) = 1$, then the last operator reduces to (9), hence it is more general one. Consequently, the bijectivity of (12) is still an open problem.

Acknowledgements The authors are grateful to the MOHE grant FRGS21-230-0839.

References

1. Bernstein, S.N.: The solution of a mathematical problem concerning the theory of heredity. Ucheniye-Zapiski N.-I. Kaf. Ukr. Otd. Mat. **1**, 83–115 (1924) (in Russian)
2. Baranski, K., Misiurewicz, M.: Omega-limit sets for the Stein-Ulam spiral map. Top. Proc. **36**, 145–172 (2010)
3. Dohtani, A.: Occurrence of chaos in higher-dimensional discrete-time systems. SIAM J. Appl. Math. **52**, 1707–1721 (1992)

4. Ganikhodjaev, R.N.: Quadratic stochastic operators, Lyapunov functions, and tournaments. Russ. Acad. Sci. Sb. Math. **76**(2), 489–506 (1993). https://doi.org/10.1070/SM1993v076n02ABEH003423
5. Ganikhodzhaev, R.N., Mukhamedov, F., Rozikov, U.: Quadratic stochastic operators and processes: results and open problems. Infin. Dimens. Anal. Quantum Probab. Relat. Top. **14**, 279–335 (2011)
6. Hofbauer, J., Hutson, V., Jansen, W.: Coexistence for systems governed by difference equations of Lotka-Volterra type. J. Math. Biol. **25**, 553–570 (1987)
7. Lyubich, Y.I.: Mathematical Structures in Population Genetics. Springer-Verlag, Berlin (1992)
8. Mukhamedov, F., Saburov, M.: On homotopy of Volterrian quadratic stochastic operator. Appl. Math. Inform. Sci. **4** 47–62 (2010)
9. Mukhamedov, F., Saburov, M.: On dynamics of Lotka-Volterra type operators. Bull. Malays. Math. Sci. Soc. **37**(2), 59–64 (2014)
10. Mukhamedov, F., Saburov, M.: Stability and monotonicity of Lotka-Volterra type operators. Qual. Theory Dyn. Syst. **16**(2), 249–267 (2017)
11. Mukhamedov, F., Saburov, M., Qaralleh, I.: On $\xi^{(s)}$-quadratic stochastic operators on two-dimensional simplex and their behavior. Abstr. Appl. Anal. **2013**, Art. ID 942038 (2013)
12. Saburov, M.: The discrete-time Kolmogorov systems with historic behavior. Math. Meth. Appl. Sci. **44**, 813–819 (2021)
13. Takeuchi, Y.: Global Dynamical Properties of Lotka–Volterra Systems. World Scientific (1996)
14. Ulam, S.M.: A Collection of Mathematical Problems. Interscience Publishers, New York-London (1960)
15. Volterra, V.: Lois de fluctuation de la population de plusieurs espèces coexistant dans le même milieu. Association Franc. Lyon **1926**, 96–98 (1927)

Dynamics of Stochastic Cesaro Operators

Otabek Khakimov

Abstract For a given linear operator A on $\mathbb{R}^m$ we define a nonlinear operator T_A by renormalizing A. Necessary and sufficient conditions for the stochasticity of T_A are found. A new class of nonlinear stochastic operators which is called Riesz Stochastic Operators (RSO) is introduced and its some basic properties are studied. The dynamics of the Cesaro operator (from the class RSOs) is fully investigated. It is shown that the stochastic Cesaro operator has exactly m fixed points and these fixed points are attracting. Moreover, it is established that the attractors of these points form a partition of the simplex.

Keywords Stochastic analysis · Nonlinear operator · Fixed points

1 Introduction

It is known that nonlinear dynamical systems appear in some problems of physics, biology, chemistry, economics, etc. [3]. Unlike linear dynamical systems, there is still no general approach to the study of nonlinear dynamical systems. For example, quadratic stochastic operators on a simplex which have applications on population genetics and economy have been studied for almost a hundred years [1, 3]. However, even in this class, full description of the dynamics is not studied yet. Because, the dynamics of such kind operators involves several polynomials with degree ≥ 2 at each step. On the other hand, there are another nonlinear stochastic operators between linear and quadratic stochastic operators, which have several applications in many areas of natural sciences. In the present paper, we are going to deal with such types of operators. Namely, we introduce nonlinear operators, which are generated by a linear ones. At first glance it might seem that the studying such kind of operators is straightforward. However, in reality, they are tricky. Main motivation for the studying

O. Khakimov (✉)
Institute of Mathematics Named After V. I. Romanovski, Tashkent 100125, Uzbekistan
e-mail: o.khakimov@mathinst.uz

AKFA University, 1st Deadlock 10, Kukcha Darvoza, Tashkent 100095, Uzbekistan

L. Accardi et al. (eds.), *Infinite Dimensional Analysis, Quantum Probability and Applications*, Springer Proceedings in Mathematics & Statistics 390,
https://doi.org/10.1007/978-3-031-06170-7_21

such kind of operators came from [2, 5]. In [5] were presented several non-stochastic dynamical systems which naturally arise from biological systems. Moreover, such types of operators normalization gives a class of nonlinear stochastic dynamical systems. Another class (called Lotka-Volterra type operators) of non-linear stochastic operators has been recently investigated in [4].

In the present paper, a given linear operator A on $\mathbb{R}^m$ we are going to define a nonlinear operator T_A by renormalizing the operator A. Furthermore, in Sect. 2, the stochasticity of T_A is investigated. In Sect. 3, a new class of of nonlinear stochastic operators (called *Riesz Stochastic Operators (RSO)*) is introduced and its some basic properties are studied. Furthermore, in Sect. 4, we fully describe the dynamics of the Cesaro operator (from the class RSOs).

2 Nonlinear Stochastic Operators Generated by Linear Ones

Let $m \geq 2$ be an integer. Then the set of all probability distributions on $\{1, 2, \ldots, m\}$ is given by the following set

$$S^{m-1} = \left\{ (x_1, x_2, \ldots, x_m) \in \mathbb{R}^m : \sum_{k=1}^{m} x_k = 1 \text{ and } x_i \geq 0, \ \forall i = \overline{1, m} \right\},$$

which is an $(m-1)$-dimensional simplex. The following sets

$$\begin{aligned} int S^{m-1} &= \left\{ \mathbf{x} \in S^{m-1} : x_i > 0 \text{ for every } 1 \leq i \leq m \right\}, \\ \partial S^{m-1} &= S^{m-1} \setminus int S^{m-1} \end{aligned}$$

are called an *interior* and a *boundary* of the simplex, respectively. By $\mathbf{c}_m = \left(\frac{1}{m}, \frac{1}{m}, \ldots, \frac{1}{m}\right)$ we denote the center of S^{m-1}. The vertices of the simplex are given by the vectors $\mathbf{e}_k = (\delta_{k1}, \delta_{k2}, \ldots, \delta_{km})$, $k = \overline{1, m}$, where δ_{ij} is a Kronecker symbol.

Let $A : \mathbb{R}^m \to \mathbb{R}^m$ be a linear operator. Let define the following operator

$$(T_A(\mathbf{x}))_i = \frac{(A\mathbf{x})_i}{\sum_{k=1}^{m} (A\mathbf{x})_k}, \qquad i = \overline{1, m}, \tag{1}$$

where $\mathbf{x} \in \mathbb{R}^m$.

We are interested in the following two problems:

- Classification of all linear operators A for which $T_A(S^{m-1}) \subset S^{m-1}$, i.e. the stochasticity of T_A.
- Investigate asymptotical behavior of the trajectory of $\mathbf{x}_{(0)} \in S^{m-1}$ w.r.t. T_A.

For a given matrix $A = (a_{ij})_{i,j=1}^{m}$ we denote the following set

$$\mathcal{I}_A = \{(i, j) : a_{ij} \neq 0\}.$$

Lemma 1 *Let T_A be an operator given by (1). If $T_A(S^{m-1}) \subset S^{m-1}$ then for any $j \in \{1, 2, \dots, m\}$ there exists an index $i \in \{1, 2, \dots, m\}$ such that $(i, j) \in \mathcal{I}_A$.*

Proof Let us pick arbitrary $j \in \{1, 2, \dots, m\}$. Then

$$A\mathbf{e}_j = (a_{1j}, a_{2j}, \dots, a_{mj}).$$

Assume that $T_A(S^{m-1}) \subset S^{m-1}$. Hence, $T_A(\mathbf{e}_j) \in S^{m-1}$ yields that

$$(a_{1j}, a_{2j}, \dots, a_{mj}) \neq \mathbf{0}.$$

This means that there exists an index $i \in \{1, 2, \dots, m\}$ such that $a_{ij} \neq 0$. □

Lemma 2 *Let T_A be an operator given by (1) and $rank(A) = 1$. If $T_A(S^{m-1}) \subset S^{m-1}$ then T_A is a constant.*

Proof Let $rank(A) = 1$. For the sake of convenience, we write $(A\mathbf{x})_k = (A_k, \mathbf{x}), k = \overline{1, m}$, where $(\cdot, \cdot)$ is the standard inner product in $\mathbb{R}^m$, and $A_k = (a_{k1}, a_{k2}, \dots, a_{mk})$. Since $rank(A) = 1$, without loss of generality, we may assume that $A_1 \neq \mathbf{0}$. Then there exist $\lambda_2, \lambda_3, \dots, \lambda_m \in \mathbb{R}$such that $A_k = \lambda_k A_1$, $k = \overline{2, m}$. Hence, we have $(A_k, \mathbf{x}) = \lambda_k(A_1, \mathbf{x})$. This implies that

$$T_A(\mathbf{x}) = \left(\frac{1}{1 + \lambda_2 + \dots + \lambda_m}, \frac{\lambda_2}{1 + \lambda_2 + \dots + \lambda_m}, \dots, \frac{\lambda_m}{1 + \lambda_2 + \dots + \lambda_m}\right).$$

The last one together with the arbitrariness of $\mathbf{x} \in S^{m-1}$ implies that T_A is constant.

Lemma 3 *Let $A = (a_{ij})_{i,j=1}^m$ be a matrix with $rank(A) > 1$. Then the following statements are equivalent:*

(i) $T_A(S^{m-1}) \subset S^{m-1}$;
(ii) $sgn(a_{ij}) = sgn(a_{kl})$ *for any* $(i, j), (k, l) \in \mathcal{I}_A$.

Proof $(i) \Rightarrow (ii)$ First we show that $a_{ij}a_{kj} \geq 0$ for any $i, j, k \in \{1, 2, \dots, m\}$. Let us pick any $j \in \{1, 2, \dots, m\}$. One can see that $A\mathbf{e}_j = (a_{1j}, a_{2j}, \dots, a_{mj})$. Then by $T_A(\mathbf{e}_j) \in S^{m-1}$ we obtain

$$0 \leq (T_A(\mathbf{e}_j))_i (T_A(\mathbf{e}_j))_k = \frac{a_{ij}a_{kj}}{(a_{1j} + a_{2j} + \dots + a_{mj})^2}, \quad \forall i, k \in \{1, 2, \dots, m\}.$$

So, $a_{ij}a_{kj} \geq 0$ for any $i, j, k \in \{1, 2, \dots, m\}$.

Now, we prove that $sgn(a_{ij}) = sgn(a_{kl})$ for any $(i, j), (k, l) \in \mathcal{I}_A$. Let us assume that it is not true. Then according to Lemma 1 without loss of generality we may suppose that there exists an integer $1 \leq n < m$ such that $a_{ij} > 0$ for all $i \leq n$, $(i, j) \in \mathcal{I}_A$ and $a_{ij} < 0$ for all $n < i \leq m$, $(i, j) \in \mathcal{I}_A$. By $rank(A) > 1$ one can finds

$i_1, i_2 \in \{1, 2, \ldots, m\}$ and $j_1 \le n < j_2$ such that $i_1 \ne i_2$, $a_{i_1 j_1} > 0$ and $a_{i_1 j_1}a_{i_2 j_2} - a_{i_1 j_1}a_{i_2 j_1} \ne 0$. Let us choose λ as follows

$$\begin{array}{ll}
0 < \lambda < -\frac{a_{i_2 j_2}}{a_{i_2 j_1} - a_{i_2 j_2}}, & \text{if } a_{i_2 j_1}a_{i_2 j_2} \ne 0,\ a_{i_1 j_2} = 0;\\
0 < -\frac{a_{i_1 j_2}}{a_{i_1 j_1} - a_{i_1 j_2}}\lambda < 1, & \text{if } a_{i_1 j_2}a_{i_2 j_2} \ne 0,\ a_{i_2 j_1} = 0;\\
0 < \lambda < -\frac{a_{i_1 j_2}}{a_{i_1 j_1} - a_{i_1 j_2}}, & \text{if } a_{i_1 j_2}a_{i_2 j_1} \ne 0,\ a_{i_2 j_2} = 0;\\
0 < \lambda < 1, & \text{if } a_{i_2 j_2} \ne 0,\ a_{i_1 j_2} = a_{i_2 j_1} = 0;\\
-\frac{a_{i_1 j_2}}{a_{i_1 j_1} - a_{i_1 j_2}} < \lambda < -\frac{a_{i_2 j_2}}{a_{i_2 j_1} - a_{i_2 j_2}}, & \text{if } \frac{a_{i_1 j_1}}{a_{i_1 j_2}} < \frac{a_{i_2 j_2}}{a_{i_2 j_1}};\\
-\frac{a_{i_2 j_2}}{a_{i_2 j_1} - a_{i_2 j_2}} < \lambda < -\frac{a_{i_1 j_2}}{a_{i_1 j_1} - a_{i_1 j_2}}, & \text{if } \frac{a_{i_1 j_1}}{a_{i_1 j_2}} > \frac{a_{i_2 j_2}}{a_{i_2 j_1}}.
\end{array}$$

Then for $\mathbf{x}_\lambda = \lambda \mathbf{e}_{j_1} + (1 - \lambda)\mathbf{e}_{j_2} \in S^{m-1}$ we obtain

$$(T_A(\mathbf{x}_\lambda))_{i_1}(T_A(\mathbf{x}_\lambda))_{i_2} < 0,$$

which contradicts $T_A(\mathbf{x}_\lambda) \in S^{m-1}$. So, we conclude that $sgn(a_{ij}) = sgn(a_{kl})$ holds for any $(i, j), (k, l) \in \mathcal{I}_A$.

$(ii) \Rightarrow (i)$ Without loss of generality, let us assume that $a_{ij} > 0$ for any $(i, j) \in \mathcal{I}_A$. Thanks to Lemma 1 for any $k \in \{1, 2, \ldots, m\}$ one has $A\mathbf{e}_k \ne \mathbf{0}$. We notice that for any $\mathbf{y} \in S^{m-1}$ there exists an element $(\alpha_1, \alpha_2, \ldots, \alpha_m) \in S^{m-1}$ such that $\mathbf{y} = \sum_{k=1}^m \alpha_k \mathbf{e}_k$. Denote $\mathbf{y}_\alpha := \sum_{k=1}^m \alpha_k \mathbf{e}_k$ then $A\mathbf{y}_\alpha \ne \mathbf{0}$. Hence, by $a_{ij} > 0$ for any $(i, j) \in \mathcal{I}_A$ one gets

$$\left(T_A\left(\mathbf{y}_\alpha\right)\right)_i \ge 0, \quad \forall i \in \{1, 2, \ldots, m\}.$$

This means that $T_A\left(\mathbf{y}_\alpha\right) \in S^{m-1}$. The arbitrariness of $\mathbf{y}_\alpha$ implies that $T_A(S^{m-1}) \subset S^{m-1}$.

Remark 1 Keeping in mind $T_{-A} = T_A$, due to Lemmas 1 and 3 without loss of generality we may conclude the following: $T_A(S^{m-1}) \subset S^{m-1}$ if and only if the following statements hold

(A_1) for any $j \in \{1, 2, \ldots, m\}$ there exists $i \in \{1, 2, \ldots, m\}$ such that $(i, j) \in \mathcal{I}_A$;
(A_2) $a_{ij} > 0$ for any $(i, j) \in \mathcal{I}_A$.

Remark 2 We notice that T_A is a linear if $\sum_{i=1}^m a_{ij} = 1$ for any $j \in \{1, 2, \ldots, m\}$.

3 Riesz Stochastic Operators

Definition 1 Let a matrix $A = (a_{ij})_{i,j=1}^m$ have the following properties:

(A1) $a_{ij} \ge 0$ for every $i, j \in \{1, 2, \ldots, m\}$;
(A2) $\sum_{j=1}^i a_{ij} = 1$ for all $i \in \{1, 2, \ldots, m\}$;

(A3) $\sum_{i=1}^{m} a_{ii} < m - i + 1$ for any $i \in \{1, 2, \dots, m\}$.

Then, the associated operator T_A is called Riesz type stochastic operator (in short RSO). Moreover, if $a_{ij} \neq 0$ for any $j \leq i$ then T_A is called positive RSO.

By $Fix(T_A)$ we denote the set of all fixed points (belonging to S^{m-1}) of RSO T_A, i.e. $\mathbf{x} \in Fix(T_A)$ if and only if $T_A(\mathbf{x}) = \mathbf{x}$, $\mathbf{x} \in S^{m-1}$.

Remark 3 We notice that the center $\mathbf{c}_m$ is a fixed point for every RSO. Indeed, for any matrix $A = (a_{ij})_{i,j=1}^{m}$ with $\sum_{j=1}^{m} a_{ij} = 1$, $i = \overline{1,m}$ one has $A\mathbf{c}_m = \mathbf{c}_m$. So, $T_A(\mathbf{c}_m) = \mathbf{c}_m$.

Proposition 1 *Let T_A be a positive RSO. Then T_A has the following properties:*

(i) For every $k \in \{1, 2, \dots, m\}$ a set $M_k = \{\mathbf{x} \in S^{m-1} : x_1 = x_2 = \cdots = x_k \neq x_{k+1}\}$ is invariant w.r.t. T_A.
(ii) $T_A(x) \in int S^{m-1}$ if and only if $x_1 \neq 0$.
(iii) For every $k \in \{1, 2, \dots, m\}$ a set $\Gamma_k = \{\mathbf{x} \in S^{m-1} : x_i = 0$ for every $i < k$ and $x_k \neq 0\}$ is invariant w.r.t. T_A. Moreover, $\{\Gamma_k\}_{k=1}^{m}$ is a partition of S^{m-1}.

Proof (*i*) Take an arbitrary $k \in \{1, 2, \dots, m\}$. Let us show that $T_A(\mathbf{x}) \in M_k$ for any $\mathbf{x} \in M_k$. Pick $\mathbf{x} \in M_k$, this means that $x_1 = x_2 = \cdots = x_i$ for any $i \leq k$. Hence, from (1) it easily follows that

$$(T_A(\mathbf{x}))_i = \frac{x_1}{\sum_{l=1}^{m} a_{lj} \sum_{j=1}^{l} x_j}, \qquad \forall i \leq k.$$

We notice that $M_m = \{\mathbf{c}_m\}$. Thanks to Remark 3 the center of the simplex is a fixed point of T_A. Therefore, it is enough to consider a case $k < m$. From $x_1 \neq x_{k+1}$ we obtain

$$(T_A(\mathbf{x}))_{k+1} = \frac{(1 - a_{k+1,k+1})x_1 + a_{k+1,k+1}x_{k+1}}{\sum_{l=1}^{m} a_{lj} \sum_{j=1}^{l} x_j} \neq (T_A(\mathbf{x}))_k,$$

which means $T_A(\mathbf{x}) \in M_k$.

(*ii*) Let us take $\mathbf{x} \in S^{m-1}$ with $x_1 \neq 0$. Then $\sum_{i=1}^{k} a_{ki}x_i \neq 0$ for every $k \leq m$. Hence, we infer that $T_A(\mathbf{x}) \in int S^{m-1}$.

Now we assume that $T_A(\mathbf{x}) \in int S^{m-1}$. It yields that

$$0 \neq (T_A(\mathbf{x}))_1 = \frac{x_1}{\sum_{i=1}^{m} a_{ij} \sum_{j=1}^{i} x_j}.$$

Consequently, $x_1 \neq 0$.

(*iii*) It is obvious that $\Gamma_i \cap \Gamma_j = \emptyset$ for $i \neq j$. Moreover, $\bigcup_{k=1}^{m} \Gamma_k = S^{m-1}$. This means that a collection $\{\Gamma_k\}_{k=1}^{m}$ is a partition of the simplex.

Let us take an arbitrary $x \in \Gamma_1$. Then due to (*ii*) we have $T_A(\mathbf{x}) \in int S^{m-1}$. Since $int S^{m-1} \subset \Gamma_1$ and the arbitrariness of $\mathbf{x}$, we conclude that $T_A(\Gamma_1) \subset \Gamma_1$.

Now we assume that $1 < k \leq m$. Then one has $\Gamma_k \subset M_{k-1}$. According to (*i*) we get $T_A(\Gamma_k) \subset M_{k-1}$. On the other hand, for any $\mathbf{x} \in S^{m-1}$ with $x_1 = 0$ one finds $(T_A(\mathbf{x}))_1 = 0$. This fact together with $\Gamma_k = \{\mathbf{y} \in M_{k-1} : y_1 = 0\}$ implies that $T_A(\Gamma_k) \subset \Gamma_k$.

Now we are going to study a trajectory of initial point w.r.t positive RSO for $m = 2$ and $m = 3$, respectively.

Example 1 Let $m = 2$ and $A = \begin{pmatrix} 1 & 0 \\ a & 1-a \end{pmatrix}$ with $0 < a < 1$. Then a positive RSO T_A has two fixed points: $\mathbf{c}_2$ and $\mathbf{e}_2$. Moreover, $\lim_{n\to\infty} T_A^n(\mathbf{x}_0) = \mathbf{c}_2$ for every $\mathbf{x}_0 \in S^1 \setminus \{\mathbf{e}_2\}$.

Example 2 Let $m = 3$ and $A = \begin{pmatrix} 1 & 0 & 0 \\ a & 1-a & 0 \\ b & c & 1-b-c \end{pmatrix}$ with $0 < a, b, c < 1$ and $b + c < 1$. Then for positive RSO we the following results:

(a) if $b + c \le a$ then

$$\lim_{n\to\infty} T_A^n(\mathbf{x}_0) = \begin{cases} \mathbf{c}_3, & \text{if } \mathbf{x}_0 \in \Gamma_1, \\ \mathbf{e}_3, & \text{if } \mathbf{x}_0 \in S^2 \setminus \Gamma_1. \end{cases}$$

(b) if $b + c > a$ then

$$\lim_{n\to\infty} T_A^n(\mathbf{x}_0) = \begin{cases} \mathbf{c}_3, & \text{if } \mathbf{x}_0 \in \Gamma_1, \\ \mathbf{x}_*, & \text{if } \mathbf{x}_0 \in \Gamma_2, \\ \mathbf{e}_3, & \text{if } \mathbf{x}_0 \in \Gamma_3, \end{cases}$$

where $\mathbf{x}_* = \left(0, \frac{b+c-a}{b+2c-a}, \frac{c}{b+2c-a}\right)$.

4 Cesaro Operators

In this section we consider a special case of positive RSO which is called *Cesaro operator*. Th Cesaro operator on S^{m-1} is defined as follows

$$(T(\mathbf{x}))_k = \frac{\frac{1}{k}\sum_{i=1}^{k} x_i}{\sum_{i=1}^{m} \frac{1}{i}\sum_{j=1}^{i} x_j}, \quad k = \overline{1, m}. \tag{2}$$

It is easy to see that (2) is associated to the following matrix

$$\begin{pmatrix} 1 & 0 & 0 & \ldots & 0 \\ \frac{1}{2} & \frac{1}{2} & 0 & \ldots & 0 \\ \ldots & \ldots & \ldots & \ldots & \ldots \\ \frac{1}{m} & \frac{1}{m} & \frac{1}{m} & \ldots & \frac{1}{m} \end{pmatrix}.$$

Since the Cesaro operator is a positive RSO, then all sets given in Proposition 1 are invariant w.r.t T. Moreover, the Cesaro operator has the following invariant sets

$$D_{\downarrow} = \{\mathbf{x} \in S^{m-1} : x_1 \geq x_2 \geq \cdots \geq x_m\}$$

and

$$D_{\uparrow} = \{\mathbf{x} \in S^{m-1} : x_1 \leq x_2 \leq \cdots \leq x_m\}.$$

Indeed, if we take $\mathbf{x} \in D_{\downarrow}$, then

$$\frac{\sum_{i=1}^{k} x_i}{k} \geq \frac{\sum_{i=1}^{k+1} x_i}{k+1}, \qquad \forall k \in \{1, 2, \ldots, m-1\}.$$

Hence, we have $T(\mathbf{x}) \in D_{\downarrow}$. By the same way, we can show that $T(\mathbf{x}) \in D_{\uparrow}$.

The following Theorem describes the set of all fixed points of the Cesaro operator on S^{m-1}.

Theorem 1 *Let T be an operator given by (2). Then $Fix(T) = \{\mathbf{p}_1, \mathbf{p}_2, \ldots, \mathbf{p}_m\}$, where for any $k \in \{1, 2, \ldots, m\}$ a vector $\mathbf{p}_k = (p_{k1}, p_{k2}, \ldots, p_{km})$ has the following form*

$$p_{kl} = \begin{cases} 0, & l < k; \\ \frac{C_{l-1}^{k-1}}{C_m^k}, & k \leq l \leq m. \end{cases} \tag{3}$$

Proof Due to the statement (iv) of Proposition 1 we consider an equation $T(\mathbf{x}) = \mathbf{x}$ in Γ_k for any $k \leq m$.

Let $\mathbf{x} \in \Gamma_k$ be a solution of $T(\mathbf{x}) = \mathbf{x}$. This means that

$$\frac{\frac{1}{l}\sum_{i=1}^{l} x_i}{\sum_{i=1}^{m} \frac{1}{i}\sum_{j=1}^{i} x_j} = x_l, \qquad \forall l \in \{1, 2, \ldots, m\}. \tag{4}$$

Keeping in mind $0 = x_1 = x_2 = \cdots = x_{k-1} \neq x_k$, we rewrite (4) for $l = k$ and obtain

$$\sum_{i=k}^{m} x_i \sum_{j=i}^{m} \frac{1}{j} = \frac{1}{k}. \tag{5}$$

Putting (6) into (4) yields

$$\frac{k}{l}\sum_{i=k}^{l} x_i = x_l, \qquad \forall l \geq k. \tag{6}$$

From the last equalities, one finds

$$x_l = \frac{\prod_{j=k}^{l-1} j}{\prod_{j=1}^{l-k} j} x_k, \qquad \forall l > k. \tag{7}$$

After simplification RHS of (8) we arrive at

$$x_l = C_{l-1}^{k-1} x_k, \quad \forall l > k. \tag{8}$$

On the other hand, from $\sum_{i=k}^{m} x_i = 1$ and (6) it follows that $x_m = \frac{k}{m}$. Putting it into (8) for $l = m$ implies

$$x_k = \frac{1}{C_m^k}.$$

Consequently, after rewriting (8) we obtain

$$x_l = \frac{C_{l-1}^{k-1}}{C_m^k}, \quad \forall l \geq k.$$

It is easy to check that $\sum_{l=k}^{m} \frac{C_{l-1}^{k-1}}{C_m^k} = 1$. This means that a vector $\mathbf{p}_k = (p_{k1}, p_{k2}, \ldots, p_{mk})$ whose coordinates defined by (3) belong to S^{m-1}. Moreover, $\mathbf{p}_k \in \Gamma_k$. Thus, we have shown that $Fix(T) \cap \Gamma_k = \{\mathbf{p}_k\}$ for any $k \in \{1, 2, \ldots, m\}$. This completes the proof.

To investigate an asymptotic behaviour of the trajectory of an initial point w.r.t the Cesaro operator, we need some auxiliary results.

Lemma 4 *Let $k, m \in \mathbb{N}$with $k \leq m$. Then the following sequence*

$$a_n(m,k) = k^n \sum_{i_1=k}^{m} \frac{1}{i_1} \sum_{i_2=i_1}^{m} \frac{1}{i_2} \cdots \sum_{i_n=i_{n-1}}^{m} \frac{1}{i_n}, \quad \forall n \in \mathbb{N} \tag{9}$$

tends to C_m^k as $n \to \infty$.

Proof One can see that

$$a_{n+1}(m,k) > a_n(m,k), \quad \forall n \in \mathbb{N}. \tag{10}$$

By induction we prove $a_n(m,k) \to C_m^k$ as $n \to \infty$.

The first step of induction is not empty. Indeed, for $m = k$ we have $a_n(k,k) = k^n \cdot \frac{1}{k^n} = 1$ and $C_k^k = 1$. Assume that for $m > k$ it holds $a_n(m,k) \to C_m^k$ as $n \to \infty$. Now we show that $a_n(m+1,k) \to C_{m+1}^k$ as $n \to \infty$. First of all we notice that

$$\begin{aligned}\sum_{i_1=k}^{m+1} \frac{1}{i_1} \sum_{i_2=i_1}^{m+1} \frac{1}{i_2} \cdots \sum_{i_n=i_{n-1}}^{m+1} \frac{1}{i_n} &= \sum_{i_1=k}^{m} \frac{1}{i_1} \sum_{i_2=i_1}^{m} \frac{1}{i_2} \cdots \sum_{i_n=i_{n-1}}^{m} \frac{1}{i_n} \\ &+ \frac{1}{m+1} \sum_{i_1=k}^{m} \frac{1}{i_1} \sum_{i_2=i_1}^{m} \frac{1}{i_2} \cdots \sum_{i_n-1=i_{n-2}}^{m} \frac{1}{i_{n-1}} \\ &+ \cdots + \frac{1}{(m+1)^n}.\end{aligned}$$

Hence,

$$a_n(m+1,1) = a_n(m,k) + \frac{k}{m+1}a_{n-1}(m,k) + \cdots + \left(\frac{k}{m+1}\right)^{n-1} a_1(m,k) + \left(\frac{k}{m+1}\right)^n. \tag{11}$$

Keeping in mind $0 < \frac{k}{m+1} < 1$, from $a_n(m,k) \to C_m^k$ as $n \to \infty$, for an arbitrary positive $\varepsilon < 2C_{m+1}^k$ one can find a positive integer n_0 such that

$$C_m^k - \frac{\varepsilon(m+k-1)}{2(m+1)} < a_n(m,k) < C_m^k + \frac{\varepsilon(m+k-1)}{2(m+1)} \tag{12}$$

and

$$\left(\frac{k}{m+1}\right)^n < \frac{\varepsilon}{2C_{m+1}^k - \varepsilon} \tag{13}$$

for all $n > n_0$.

Then plugging (10) into (11) and using the second inequality of (12), we obtain

$$\begin{aligned} a_n(m+1,k) &< a_n(m,k)\left(1 + \frac{k}{m+1} + \left(\frac{k}{m+1}\right)^n\right) \\ &= a_n(m,k)\frac{m+1}{m+1-k}\left(1 - \left(\frac{k}{m+1}\right)^{n+1}\right) \\ &< a_n(m,k)\frac{m+1}{m+1-k} \\ &< \left(C_m^k + \frac{\varepsilon(m+1-k)}{m+1}\right)\frac{m+1}{m+1-k} \\ &= \frac{m+1}{m+1-k}C_m^k + \varepsilon \\ &= C_{m+1}^k + \varepsilon. \end{aligned} \tag{14}$$

Using the first inequality of (12) and (13) from (11) it follows that

$$\begin{aligned} a_n(m+1,k) &> a_n(m,k) + \frac{k}{m+1}a_{n-1}(m,k) + \cdots + \left(\frac{k}{m+1}\right)^{n-n_0-1} a_{n_0+1}(m,k) \\ &> \left(C_m^k - \frac{\varepsilon(m+1-k)}{2(m+1)}\right)\left(1 + \frac{k}{m+1} + \cdots + \left(\frac{k}{m+1}\right)^{n-n_0-1}\right) \\ &> \frac{m+1}{m+1-k}\left(C_m^k - \frac{\varepsilon(m+1-k)}{2(m+1)}\right)\left(1 - \left(\frac{k}{m+1}\right)^{n+1}\right) \end{aligned}$$

$$\begin{aligned}
&= \left(C^k_{m+1} - \frac{\varepsilon}{2}\right)\left(1 - \left(\frac{k}{m+1}\right)^{n+1}\right) \\
&> C^k_{m+1} - \frac{\varepsilon}{2} - \frac{\varepsilon}{2} \\
&= C^k_{m+1} - \varepsilon. \qquad (15)
\end{aligned}$$

By (14) and (15) we arrive at $a_n(m+1, k) \to C^k_{m+1}$ as $n \to \infty$.

Corollary 1 *Let $k, m \in \mathbb{N}$ with $k \le m$. Then for any integer $1 \le j \le m-k$ the following sequence*

$$b_n(m,k,j) = k^n \sum_{i_1=k+j}^{m} \frac{1}{i_1} \sum_{i_2=i_1}^{m} \frac{1}{i_2} \cdots \sum_{i_n=i_{n-1}}^{m} \frac{1}{i_n}, \quad \forall n \in \mathbb{N} \qquad (16)$$

tends to 0 as $n \to \infty$.

Proof It is easy to check that $b_n(m,k,j) = \left(\frac{k}{k+j}\right)^n a_n(m, k+j)$, where $a_n(m, k+j)$ given by (11). Due to Lemma 4 we have $a_n(m, k+j) \to C^{k+j}_m$ as $n \to \infty$. On the other hand, it holds $\left(\frac{k}{k+j}\right)^n \to 0$ as $n \to \infty$. So, $b_n(m,k,j) \to 0$ as $n \to \infty$.

Theorem 2 *Let T be an operator given by (2) and $k \in \{1, 2, \ldots, m\}$. Then for any $\boldsymbol{x} \in \Gamma_k$ one has*

$$\lim_{n\to\infty} T^n(\boldsymbol{x}) = \boldsymbol{p}_k.$$

Proof Let $\mathbf{x} \in \Gamma_k$. Then

$$(T^n(\mathbf{x}))_l = \begin{cases} 0, & l < k, \\ \dfrac{\sum_{i_1=k}^{l} x_{i_1} \sum_{i_2=i_1}^{m} \frac{1}{i_2} \cdots \sum_{i_n=i_{n-1}}^{m} \frac{1}{i_n}}{l \sum_{i_0=k}^{m} x_{i_0} \sum_{i_1=i_0}^{m} \frac{1}{i_1} \cdots \sum_{i_n=i_{n-1}}^{m} \frac{1}{i_n}}, & k \le l \le m, \end{cases} \quad \forall n \in \mathbb{N}. \qquad (17)$$

By (17), to establish $T^n(\mathbf{x}) \to \mathbf{p}_k$ it is enough to prove $(T^n(\mathbf{x}))_l \to \frac{C^{k-1}_{l-1}}{C^k_m}$ for any $l \in \{k, k+1, \ldots, m\}$.

Let $l \in \{k, k+1, \ldots, m\}$. Then from (17) it follows that

$$(T^n(\mathbf{x}))_l = \frac{k x_k a_n(l,k) + \sum_{j=1}^{l-k} x_{k+j} b_n(l,k,j)}{l\left(k x_k a_n(m,k) + \sum_{j=1}^{m-k} x_{k+j} b_n(m,k,j)\right)}, \qquad (18)$$

where $a_n(m,k)$ and $b_n(m,k,j)$ are given by (11), (16), respectively. Due to Lemma 4 and Corollary 1 we get

$$\text{RHS of (18)} \to \frac{k x_k C^k_l}{l x_k C^k_m} = \frac{C^{k-1}_{l-1}}{C^k_m} \text{ as } n \to \infty.$$

From Examples 1 and 2 we can formulate the following conjecture.

Conjecture 1 Let T_A be a positive RSO. If $a_{ii} \geq a_{jj}$ for every $i < j$ then T_A is topologically conjugate to the Cesaro operator.

References

1. Ganikhodzhaev, R., Mukhamedov, F., Rozikov, U.: Quadratic stochastic operators and processes: results and open problems. Infin. Dimens. Anal. Quan. Probab. Relat. Top. **14**, 279–335 (2011). https://doi.org/10.1142/S0219025711004365
2. Hofbauer, J., Sigmund, K.: Evolutionary Games and Population Dynamics. Cambridge University Press, Cambridge (1998)
3. Lyubich, Yu.I.: Mathematical Structures in Population Genetics. Springer-Verlag, Berlin (1992)
4. Mukhamedov, F., Saburov, M.: Stability and monotonicity of Lotka-Volterra type operators. Qual. Theor. Dyn. Sys. **16**(2), 249–267 (2017)
5. Rozikov, U.: Population Dynamics, Algebraic and Probabilistic Approach. World Scientific, Singapoure (2020)

The Dynamics of a Volterra Cubic Operator

Uygun U. Jamilov and Elbek Kh. Ziyodullaev

Abstract In this paper, we consider a family of Volterra cubic stochastic operators defined on the two-dimensional simplex depending on two parameters. It is found all fixed points and their types. It is proved that depending on the parameters and an initial point a trajectory of such kind operator converges to a vertex of the two-dimensional simplex.

Keywords Cubic stochastic operator · Volterra operator · Regular operator

1 Introduction

We consider discrete-time dynamical systems generated by cubic stochastic operators. The motivations for such investigations can be seen in [12, 14, 20, 26, 27] and references therein. In [26], a new notion of a cubic stochastic operator was introduced and investigated.

Let $E = \{1, \ldots, m\}$ be a set, then the set of all probability distributions on E is

$$S^{m-1} = \left\{ \mathbf{x} = (x_1, x_2, \ldots, x_m) \in \mathbb{R}^m : x_i \geq 0, \text{ for any } i \text{ and } \sum_{i=1}^{m} x_i = 1 \right\}$$

which is an $(m-1)$-dimensional simplex.

U. U. Jamilov (✉)
V. I. Romanovskiy Institute of Mathematics, Uzbekistan Academy of Sciences, 9, University Str., 100174 Tashkent, Uzbekistan
e-mail: uygun.jamilov@mathinst.uz; jamilovu@yandex.ru

AKFA University, 264, National Park Str., 111221 Qibray district, Tashkent region, Uzbekistan

National University of Uzbekistan, 4, University Str., 100174 Tashkent, Uzbekistan

E. Kh. Ziyodullaev
Navoi State Pedagogical Institute, 45, Ibn Sino street, Navoi, Uzbekistan

L. Accardi et al. (eds.), *Infinite Dimensional Analysis, Quantum Probability and Applications*, Springer Proceedings in Mathematics & Statistics 390,
https://doi.org/10.1007/978-3-031-06170-7_22

A *cubic stochastic operator* (CSO) is a map $W: S^{m-1} \to S^{m-1}$ given by

$$W: x_l' = \sum_{i,j,k \in E} p_{ijk,l} x_i x_j x_k, \quad l \in E, \tag{1}$$

where $p_{ijk,l}$ are the heredity coefficients such that

$$p_{ijk,l} \geq 0, \ \sum_{l \in E} p_{ijk,l} = 1, \tag{2}$$

and the coefficients $p_{ijk,l}$ do not change for any permutation of i, j and k.

Let $p_{ijk,l} = p_{ij,l}$ for any $i, j, k, l \in E$, then (1) has the form

$$x_l' = \sum_{i,j,k \in E} p_{ijk,l} x_i x_j x_k = \sum_{k \in E} x_k \sum_{i,j \in E} p_{ij,l} x_i x_j = \sum_{i,j \in E} p_{ij,l} x_i x_j, \quad l \in E,$$

that is, in this case, the cubic stochastic operator (1) coincides with a quadratic stochastic operator defined on a finite-dimensional simplex (see e.g. [4, 9–11, 15, 18, 22, 28, 29, 32]) for more classes of quadratic stochastic operators and their applications. For a recent review on the theory of quadratic stochastic operators see [6].

Note that W (defined by (1)) is a non-linear operator, and its dimension increases with m. Higher-dimensional dynamical systems are essential, but only relatively few dynamical systems have yet been analyzed (see e.g. [2]).

For a given $\mathbf{x}^{(0)} \in S^{m-1}$ the trajectory $\{\mathbf{x}^{(n)}\}$, $n = 0, 1, 2, \ldots$, of an initial point $\mathbf{x}^{(0)}$ under the action of the CSO (1) is defined by $\mathbf{x}^{(n+1)} = W(\mathbf{x}^{(n)})$, where $n = 0, 1, 2, \ldots$.

One of the main problems in mathematical biology is to study the asymptotical behaviour of the trajectories. In [20, 21, 26], this problem was considered for a class of Volterra CSOs. A Volterra CSO is defined by (1), (2) and with the additional assumption

$$p_{ijk,l} = 0 \ \text{if} \ l \notin \{i, j, k\}. \tag{3}$$

The biological treatment of conditions (3) is rather precise: the offspring repeats the genotype of one of its parents. It is known that a Volterra CSO has the form

$$x_l' = x_l \left(x_l^2 + 3 \sum_{i \in E \setminus \{l\}} p_{lli,l} x_l x_i + 3 \sum_{i \in E \setminus \{l\}} p_{iil,l} x_i^2 + 6 \sum_{\substack{i \in E \setminus \{l\} \\ i<j}} p_{ijl,l} x_i x_j \right), \quad l \in E, \tag{4}$$

and such operator (4) can be considered as a discrete version of Kolmogorov systems for m interacting species (see e.g. [3]).

In [27], the authors gave a construction of a cubic stochastic operator. A family of Volterra cubic stochastic operators defined on a finite-dimensional simplex depending on a parameter was investigated in [12]. In [8] was considered a family of independent and identically distributed maps from the family and studied its long term dynamics, in particular its random point attractors. The problem connected with surjectivity of CSOs were considered in [23].

In the present paper, we consider a family of Volterra cubic stochastic operators defined on the two-dimensional simplex. The article is organized as follows. In Sect. 2, we recall the definitions and known results related to cubic stochastic operators. In Sect. 3, we study the asymptotical behaviour of trajectories of CSOs. We described the set fixed points and we found the types of the all fixed points. Therein we showed that any trajectory of such CSO converges to a vertex of the two-dimensional simplex.

2 Preliminaries

First, we recall some necessary definitions.

Definition 1 A point $\mathbf{x} \in S^{m-1}$ is called a fixed point of W if $W(\mathbf{x}) = \mathbf{x}$.

Denote the set of all fixed points by Fix (W). The fixed points of a nonlinear operator describe limit or long-run behaviour of the trajectories for any initial point.

Let $DW(\mathbf{x}^*) = \left(\partial W_i/\partial x_j(\mathbf{x}^*)\right)_{i,j=1}^{m}$ be the Jacobian of W at the point $\mathbf{x}^*$.

Definition 2 ([2]) A fixed point $\mathbf{x}^*$ is called hyperbolic if its Jacobian $DW(\mathbf{x}^*)$ has no eigenvalues on the unit sphere.

Definition 3 ([2]) A hyperbolic fixed point $\mathbf{x}^*$ is called:

(i) attracting if all the eigenvalues of the $DW(\mathbf{x}^*)$ are inside the unit ball;
(ii) repelling if all the eigenvalues of the $DW(\mathbf{x}^*)$ are outside the unit ball;
(iii) a saddle otherwise.

Definition 4 A CSO W is called regular if for any initial point $\mathbf{x} \in S^{m-1}$, the limit $\lim\limits_{n\to\infty} W^n(\mathbf{x})$ exists.

Definition 5 A CSO W is said to be ergodic if the limit

$$\lim_{n\to\infty} \frac{1}{n} \sum_{k=0}^{n-1} W^k(\mathbf{x})$$

exists for any $\mathbf{x} \in S^{m-1}$.

In [30] based on numerical calculations, Ulam conjectured that the ergodic theorem holds for any quadratic stochastic operator. In [31], Zakharevich proved that this conjecture is false in general. In [4, 13] the ergodicity problem of Volterra quadratic stochastic operators was studied.

The limit behaviour of trajectories and fixed points also play an important role in many applied problems connected with cubic stochastic operators. In [26], the authors proved that a Volterra CSO has the regular property. Similar results were proved in [12] for a class of Volterra CSO depending on a parameter. In [16] by authors the results of the articles [12] were generalised up to a class of constrained Volterra CSOs. A CSO which is a convex combination of a regular and non-ergodic operators was studied in [17]. The families of non-ergodic Volterra CSOs was considered in [19, 24].

The following notations will be used throughout this paper. Let the set $\partial S^{m-1} = \{\mathbf{x} \in S^{m-1} : x_i = 0 \quad \text{for at least one} \quad i \in E\}$ be the boundary of the simplex S^{m-1}; the set $\Gamma_\tau = \{\mathbf{x} \in S^{m-1} : x_i = 0, \quad i \in \tau \subset E\}$ a face of the simplex S^{m-1}; the set $\operatorname{int} S^{m-1} = \{\mathbf{x} \in S^{m-1} : x_1 x_2 \cdots x_m > 0\}$ the interior of S^{m-1}; and $\mathbf{e}_i = (\delta_{1i}, \delta_{2i}, \dots, \delta_{ni}) \in S^{m-1}$, $i = 1, \dots, n$, the vertices of the simplex S^{m-1}, where δ_{ij} is the Kronecker delta. Let $\omega(\mathbf{x}^0)$ be the set of limit points of the trajectory $\{W^n(\mathbf{x}^0) \in S^{m-1}\}_{n=1}^{\infty}$. Let 1_A be the characteristic (or indicator) function of a subset A of X. For an arbitrary set T, will denote by $1_A T$ the set

$$1_A T = \begin{cases} T, & \text{if } x \in A, \\ \emptyset, & \text{if } x \notin A. \end{cases}$$

3 Main Results

Consider the following cubic stochastic operator defined on the S^2

$$W : \begin{cases} x_1' = x_1(x_1^2 + 3\alpha x_2^2 + 3\beta x_3^2 + 3x_1x_2 + 3x_1x_3 + 6x_2x_3) \\ x_2' = x_2^2(x_2 + 3(1-\alpha)x_1 + 3x_3) \\ x_3' = x_3^2(x_3 + 3(1-\beta)x_1 + 3x_2). \end{cases} \tag{5}$$

where $\alpha, \beta \in [0, 1]$. Note that the CSO (5) can be considered as a discrete version of Kolmogorov system for three interacting species (see e.g. [3]) and as modified Lotka-Volterra type models (see e.g. [7, 25]) which are arise in problems of ecology.

To investigate the dynamics of the CSO (5) we need some certain auxiliary results from one-dimensional dynamical systems.

The dynamics of a function. Let us consider the following function

$$f_\theta(x) := (3\theta - 2)x^3 + 3(1-\theta)x^2, \quad x, \theta \in [0, 1]. \tag{6}$$

Denote $f_\theta^n = \underbrace{f_\theta \circ \cdots \circ f_\theta}_{n \text{ times}}$, the n-fold composition of f_θ with itself and

$$x_\theta := 1/(2 - 3\theta).$$

Proposition 1 *For the function f_θ the following statements are true:*

i) $\mathrm{Fix}(f_\theta) = \{0, 1\} \cup 1_{\{3\theta \in [0,1]\}}\{x_\theta\}$*;*
ii) The point $x = 0$ is an attracting point.

$$\text{The point } x = 1 \text{ has type } \begin{cases} \text{attracting}, & \text{if } 0 \le 3\theta < 1, \\ \text{non-hyperbolic}, & \text{if } 3\theta = 1, \\ \text{repelling}, & \text{if } 1 < 3\theta \le 3, \end{cases}$$

$$\text{and the point } x_\theta \text{ has type } \begin{cases} \text{non-hyperbolic}, & \text{if } 3\theta = 1, \\ \text{repelling}, & \text{if } 0 \le 3\theta < 1. \end{cases}$$

iii) if $1 \le 3\theta \le 3$ then $\lim_{n\to\infty} f_\theta^n(x) = 0$ for any $x \in [0, 1)$;
iv) if $0 \le 3\theta < 1$ then

$$\lim_{n\to\infty} f_\theta^n(x) = \begin{cases} 0, & \text{if } 0 < x < x_\theta, \\ 1, & \text{if } x_\theta < x < 1. \end{cases}$$

Proof (i) From the (6) we have $f_\theta(x) - x = x(x-1)(1-(2-3\theta)x)$. Consequently, it follows that $x = 0$ and $x = 1$ are solutions of the equation $f_\theta(x) = x$. Moreover, if $0 \le 3\theta < 1$ then also there is a solution x_θ of this equation.

(ii) One has that $f_\theta'(0) = 0$, $f_\theta'(1) = 3\theta$, and $f_\theta'(x_\theta) = 3(1-3\theta)/(2-3\theta)$. It immediately follows the statement of Proposition.

(iii) Let $1 < 3\theta \le 3$ then by virtue of the part (ii) the function $f_\theta(x)$ has an attracting point $x = 0$ and a repelling $x = 1$ point. It easy to verify that $f_\theta(x) < x$ for any $x \in (0, 1)$ and using the last one has that $\lim_{n\to\infty} f_\theta^n(x) = 0$.

(iv) Let $0 \le 3\theta < 1$ then by virtue of the part ii) the function has $x = 0$, $x = 1$ the attracting and $x = x_\theta$ a repelling fixed points. It easy to check that $f_\theta(x) < x$ for any $x \in (0, x_\theta)$ and $f_\theta(x) > x$ for any $x \in (x_\theta, 1)$. So we have

$$\lim_{n\to\infty} f_\theta^n(x) = \begin{cases} 0, & \text{if } 0 < x < x_\theta, \\ 1, & \text{if } x_\theta < x < 1. \end{cases}$$

The proof is completed.

Consider the following function

$$g_\tau(x) := 2(3\tau - 1)x^3 + 3(1-\tau)x^2, \quad x \in [0, 1/2], \quad \tau \in [0, 1]. \tag{7}$$

We denote $g_\tau^n = \underbrace{g_\tau \circ \cdots \circ g_\tau}_{n \text{ times}}$, the n-fold composition of g_τ with itself.

Proposition 2 *For the function g_τ the following statements are true:*

(i) Fix $(g_\tau) = \{0, 1/2\}$*;*
(ii) The point $x = 0$ is an attracting and $x = 1/2$ is a repelling point;
(iii) for any $x \in [0, 1/2)$ we have $\lim_{n\to\infty} g_\tau^n(x) = 0$.

Proof (i) From the (7) we have $g_\tau(x) - x = x(2x-1)(1-(1-3\theta)x)$. Hence it follows that $x = 0$ and $x = 1/2$ are solutions of the equation $g_\tau(x) = x$ in the segment $[0, 1/2]$.

(ii) One can easily has that $g'_\tau(0) = 0, \quad g'_\tau(1/2) = 3(1+\tau)/2$. It immediately follows the statement of Proposition.

(iii) By assertion of the part (ii) the function g_τ has $x = 0$ an attracting and $x = 1/2$ a repelling fixed points. It easy to verify that $g_\tau(x) < x$ for any $x \in (0, 1/2)$ and using this one has that $\lim_{n\to\infty} g_\tau^n(x) = 0$. The proof is completed.

The dynamics of a Volterra CSO. Now using the above results study the dynamics of the Volterra CSO (5).

Proposition 3 *For the CSO W the following statements are true:*

(i) The faces $\Gamma_{\{i\}}, \ i = 1, 2, 3$ are invariant sets and the median $M_{23} = \{\mathbf{x} \in S^2 : x_2 = x_3\}$ is an invariant set whenever $\alpha = \beta$;
(ii) Fix $(W) = \{\mathbf{e}_1, \mathbf{e}_2, \mathbf{e}_3, \mathbf{c}\left(\Gamma_{\{1\}}\right)\} \cup 1_{\{3\alpha\in[0,1]\}}\{\mathbf{x}(\alpha)\} \cup 1_{\{3\beta\in[0,1]\}}\{\mathbf{x}(\beta)\}$, *where* $\mathbf{c}\left(\Gamma_{\{1\}}\right) = (0, 1/2, 1/2)$, $\mathbf{x}(\alpha) = (1 - x_\alpha, x_\alpha, 0)$, $\mathbf{x}(\beta) = \left(1 - x_\beta, 0, x_\beta\right)$*;*
(iii) The vertex $\mathbf{e}_1$ is an attracting point and the point $\mathbf{c}\left(\Gamma_{\{1\}}\right)$ is a repelling point. If $0 \le 3\alpha < 1$ (resp. $0 \le 3\beta < 1$) then the vertex $\mathbf{e}_2$ (resp. $\mathbf{e}_3$) is an attracting point, if $3\alpha = 1$ (resp. $3\beta = 1$) then it is a non-hyperbolic point and it is a saddle point when $1 < 3\alpha \le 3$ (resp. $1 < 3\beta \le 3$). If $3\alpha = 1$ (resp. $3\beta = 1$) then the fixed point $\mathbf{x}(\alpha)$ (resp. $\mathbf{x}(\beta)$) is a non-hyperbolic point and it is a saddle point when $3\alpha \in [0, 1)$ (resp. $3\beta \in [0, 1)$).

Proof (i) Let $\mathbf{x} \in \Gamma_{\{i\}}, \ i = 1, 2, 3$, that is $x_i = 0$. Then from (5) one can easily to see that $x'_i = 0$. Also it is easy to check that if $\alpha = \beta$ then for any $\mathbf{x} \in M_{23}$ we have $W(\mathbf{x}) \in M_{23}$.

(ii) A fixed point of the operator (5) is a solution of the equation $W(\mathbf{x}) = \mathbf{x}$ or, in other words, of the system

$$\begin{cases} x_1 = x_1(x_1^2 + 3\alpha x_2^2 + 3\beta x_3^2 + 3x_1x_2 + 3x_1x_3 + 6x_2x_3) \\ x_2 = x_2^2(x_2 + 3(1-\alpha)x_1 + 3x_3) \\ x_3 = x_3^2(x_3 + 3(1-\beta)x_1 + 3x_2). \end{cases} \tag{8}$$

We consider all possible cases.

Case $x_1 = 0$. In this case using $x_2 + x_3 = 1$ from the second equation of (8) we get $2x_2^3 - 3x_2 + x_2 = 0$. So from the last equation we have its solutions $x_2 = 0$, $x_2 = 1$ and $x_2 = 1/2$.

Let $x_2 = 0$ then we have $x_3 = 1$, that is we obtain the vertex $\mathbf{e}_3$.

Let $x_2 = 1$ then we have $x_3 = 0$, that is we obtain the vertex $\mathbf{e}_2$.

If $x_2 = 1/2$ then we have $x_3 = 1/2$, that is we obtain the center $\mathbf{c}\left(\Gamma_{\{1\}}\right)$.

Case $x_2 = 0$. Let $x_1 = 0$ then we have $x_3 = 1$, that is we obtain $\mathbf{e}_3 = (0, 0, 1)$. Let $x_1 > 0$ then using $x_1 + x_3 = 1$ from (8) one has

$$(3\beta - 2)x_1^2 + 3(1 - 2\beta)x_1 + 3\beta - 1 = 0. \tag{9}$$

So we have the solutions $x_1^{(1)} = 1 - x_\beta$ and $x_1^{(2)} = 1$. If we take $x_1^{(2)}$ then we have $\mathbf{e}_1 = (1, 0, 0)$. Let $3\beta > 2$ then one has that $x_1^{(1)} > 1$, that is $x_1^{(1)} \notin [0, 1]$. Let $3\beta = 2$ then from the Eq. (9) one has $x_1^{(1)} = -1 < 0$. Let $3\beta < 2$ then we have $x_1^{(1)} < 1$. It should be also $3\beta < 1$. Hence if $0 \leq 3\beta < 1$ then $x_1^{(1)} \in [0, 1]$. In this case one has $x_3^{(1)} = 1 - x_1^{(1)}$, i.e. $x_3^{(1)} = x_\beta$.

Thus we obtain the fixed points $\mathbf{e}_1 = (1, 0, 0)$, $\mathbf{e}_3 = (0, 0, 1)$ and

$$\mathbf{x}(\beta) = (1 - x_\beta, 0, x_\beta) = \left(\frac{1 - 3\beta}{2 - 3\beta}, 0, \frac{1}{2 - 3\beta}\right), \quad \text{where } 0 \leq 3\beta \leq 1.$$

Case $x_3 = 0$. In this case repeating the above algebra we have fixed points $\mathbf{e}_1 = (1, 0, 0)$, $\mathbf{e}_3 = (0, 0, 1)$ and

$$\mathbf{x}(\alpha) = (1 - x_\alpha, x_\alpha, 0) = \left(\frac{1 - 3\alpha}{2 - 3\alpha}, \frac{1}{2 - 3\alpha}, 0\right), \quad \text{where } 0 \leq 3\alpha \leq 1.$$

Case $x_1x_2x_3 > 0$. In this case the system (8) has form

$$\begin{cases} 1 = x_1^2 + 3\alpha x_2^2 + 3\beta x_3^2 + 3x_1x_2 + 3x_1x_3 + 6x_2x_3 \\ 1 = x_2(x_2 + 3(1 - \alpha)x_1 + 3x_3) \\ 1 = x_3(x_3 + 3(1 - \beta)x_1 + 3x_2). \end{cases} \tag{10}$$

(a) Let $\alpha = \beta$. Then from the second and third equations of the system (10) one has $0 = x_2^2 - x_3^2 + 3(1 - \alpha)x_1(x_2 - x_3) = (x_2 - x_3)(1 + (2 - 3\alpha)x_1)$. Therefore we get either $x_2 = x_3$ or $x_1 = x_\alpha$.

If we take $x_2 = x_3$ then from the first equation of (10) one has

$$1 = x_1^2 + 3\alpha x_2^2 + 3\alpha x_2^2 + 3x_1x_2 + 3x_1x_2 + 6x_2^2.$$

Using $x_1 = 1 - 2x_2$ we get

$$1 = (1 - 2x_2)^2 + (6 + 6\alpha)x_2^2 + 6(1 - 2x_2)x_2 \;\Leftrightarrow\; 0 = x_2(1 + (3\alpha - 1)x_2).$$

Therefore it follows either $x_2 = 0$ or $x_2 = 1/(1 - 3\alpha)$ and using $x_2 = x_3$ we have

$$x_1 = 1 - x_2 - x_3 = 1 \;\text{ or }\; x_1 = 1 - x_2 - x_3 = -\frac{1 + 3\alpha}{1 - 3\alpha}.$$

Since the following inequalities

$$x_2 = \frac{1}{1-3\alpha} > 0 \text{ and } x_1 = -\frac{1+3\alpha}{1-3\alpha} > 0$$

do not hold simultaneously it follows that if $\alpha = \beta$ then $\operatorname{int} S^2 \cap \operatorname{Fix}(W) = \emptyset$.

(b) Suppose $\alpha \neq \beta$. Then from the second equation of (10) we have

$$1 = x_2(x_2 + 3(1-\alpha)x_1 + 3x_3) \Rightarrow \alpha = -\frac{2x_2^2 - 3x_2 + 1}{3x_1x_2}.$$

Since $\alpha \in [0, 1]$ one has

$$\alpha = -\frac{2x_2^2 - 3x_2 + 1}{3x_1x_2} \geq 0 \Leftrightarrow 2x_2^2 - 3x_2 + 1 \leq 0 \Rightarrow x_2 \in \left[\frac{1}{2}, 1\right].$$

Similarly using $\beta \in [0, 1]$ from the third equation of (10) we have

$$\beta = -\frac{2x_3^2 - 3x_3 + 1}{3x_1x_3} \geq 0 \Leftrightarrow 2x_3^2 - 3x_3 + 1 \leq 0 \Rightarrow x_3 \in \left[\frac{1}{2}, 1\right].$$

Consequently using $\mathbf{x} \in S^2$ one has that

$$x_2 \in \left[\frac{1}{2}, 1\right], \quad x_3 \in \left[\frac{1}{2}, 1\right] \Rightarrow x_2 = x_3 = \frac{1}{2}, \quad x_1 = 0.$$

But the last contradicts to the condition $x_1x_2x_3 > 0$. Therefore it is follow that if $\alpha \neq \beta$ then $\operatorname{int} S^2 \cap \operatorname{Fix}(W) = \emptyset$.

(iii) The Jacobian $DW(\mathbf{x})$ at the fixed points have the following eigenvalues

$$
\begin{aligned}
&\mathbf{e}_1 : \lambda_1 = 0, \lambda_2 = 0;\\
&\mathbf{e}_2 : \lambda_1 = 0, \lambda_2 = 3\alpha;\\
&\mathbf{e}_3 : \lambda_1 = 0, \lambda_2 = 3\beta;\\
&\mathbf{c}\left(\Gamma_{\{1\}}\right) : \lambda_1 = \frac{3}{2}, \lambda_2 = \frac{3}{2} + \frac{3(\alpha+\beta)}{4};\\
&\mathbf{x}(\alpha) : \lambda_1 = 0, \lambda_2 = \frac{3(1-2\alpha)}{2-3\alpha};\\
&\mathbf{x}(\beta) : \lambda_1 = 0, \lambda_2 = \frac{3(1-2\beta)}{2-3\beta}.
\end{aligned}
\tag{11}
$$

It is easy to see that for any parameters the vertex $\mathbf{e}_1$ is an attracting point and the point $\mathbf{c}\left(\Gamma_{\{1\}}\right)$ is a repelling point. Since $3\alpha \in [0, 3]$ and $3\beta \in [0, 3]$ it follows that:

the vertex $\mathbf{e}_2$ is an attracting point for any $3\alpha \in [0, 1)$ and it is a non-hyperbolic point whenever $3\alpha = 1$ and it is a saddle point if $3\alpha > 1$; Similarly the vertex $\mathbf{e}_3$ is an attracting point for any $3\beta \in [0, 1)$ and it is a non-hyperbolic point whenever $3\beta = 1$ and it is a saddle point if $3\beta > 1$; if $3\alpha = 1$ then the fixed point $\mathbf{x}(\alpha)$ is a non-hyperbolic point and it is a saddle point when $3\alpha \in [0, 1)$; Similarly if $3\beta = 1$ then the fixed point $\mathbf{x}(\beta)$ is a non-hyperbolic point and it is a saddle point when $3\beta \in [0, 1)$.

The proof is completed.

Denote $S^2_> = \{\mathbf{x} \in S^2 : x_2 > x_3\}$ and $S^2_< = \{\mathbf{x} \in S^2 : x_2 < x_3\}$.

Theorem 1 *(i) For any* $\mathbf{x}^{(0)} \in \Gamma_{\{1\}}$ *we have* $W(\mathbf{x}^{(0)}) = \mathbf{x}^{(0)}$;
(ii) For any $\mathbf{x}^{(0)} \in \Gamma_{\{2\}}$ *we have*

(ii_a) *if* $1 \leq 3\beta \leq 3$ *then* $\lim_{n\to\infty} W^n(\mathbf{x}) = \mathbf{e}_1$;
(ii_b) *if* $0 \leq 3\beta < 1$ *then*

$$\lim_{n\to\infty} W^n(\mathbf{x}) = \begin{cases} \mathbf{e}_1, & \text{if } 0 < x_3^{(0)} < x_\beta, \\ \mathbf{e}_3, & \text{if } x_\beta < x_3^{(0)} < 1. \end{cases}$$

(iii) For any $\mathbf{x}^{(0)} \in \Gamma_{\{3\}}$ *we have*

(iii_a) *if* $1 \leq 3\alpha \leq 3$ *then* $\lim_{n\to\infty} W^n(\mathbf{x}) = \mathbf{e}_1$;
(iii_b) *if* $0 \leq 3\alpha < 1$ *then*

$$\lim_{n\to\infty} W^n(\mathbf{x}) = \begin{cases} \mathbf{e}_1, & \text{if } 0 < x_2^{(0)} < x_\alpha, \\ \mathbf{e}_2, & \text{if } x_\alpha < x_2^{(0)} < 1; \end{cases}$$

(iv) Let $\alpha = \beta$ *then*

(iv_a) *we have* $\lim_{n\to\infty} W^n(\mathbf{x}) = \mathbf{e}_1$ *for any* $\mathbf{x}^{(0)} \in \operatorname{int} S^2 \cap M_{23}$;
(iv_b) *for any* $1 \leq 3\alpha \leq 3$ *and for all* $\mathbf{x}^{(0)} \in (\operatorname{int} S^2 \setminus M_{23}) \setminus \operatorname{Fix}(W)$ *we have*

$$\lim_{n\to\infty} W^n(\mathbf{x}^{(0)}) = \mathbf{e}_1;$$

$iv_c)$ *for any* $0 \leq 3\alpha < 1$ *and* $\mathbf{x}^{(0)} \in (\operatorname{int} S^2 \cap S^2_>) \setminus \operatorname{Fix}(W)$ *we obtain*

$$\lim_{n\to\infty} W^n(\mathbf{x}^{(0)}) = \begin{cases} \mathbf{e}_1, & \text{if } 0 < x_2^{(0)} < x_\alpha, \\ \mathbf{e}_2, & \text{if } x_\alpha < x_2^{(0)} < 1. \end{cases}$$

(iv_d) *for any* $0 \leq 3\alpha < 1$ *and* $\mathbf{x}^{(0)} \in (\operatorname{int} S^2 \cap S^2_<) \setminus \operatorname{Fix}(W)$ *we obtain*

$$\lim_{n\to\infty} W^n\left(\mathbf{x}^{(0)}\right) = \begin{cases} \mathbf{e}_1, & if \quad 0 < x_3^{(0)} < x_\alpha, \\ \mathbf{e}_3, & if \quad x_\alpha < x_3^{(0)} < 1. \end{cases}$$

(v) Let $\alpha \neq \beta$ then

(v_a) *if $3\alpha > 3\beta \geq 1$ or $3\beta > 3\alpha \geq 1$ then there are curves γ and ν such that*

$$\lim_{n\to\infty} W^n\left(\mathbf{x}^{(0)}\right) = \begin{cases} \mathbf{e}_2, & if \quad \mathbf{x}^{(0)} \in \gamma, \\ \mathbf{e}_3, & if \quad \mathbf{x}^{(0)} \in \nu, \\ \mathbf{e}_1, & if \quad \mathbf{x}^{(0)} \in \mathrm{int} S^2 \setminus (\gamma \cup \nu)\,; \end{cases}$$

(v_b) *if $3\beta > 1 > 3\alpha$ then there are a neighbourhood $U(\mathbf{e}_2)$ and a curve γ such that*

$$\lim_{n\to\infty} W^n\left(\mathbf{x}^{(0)}\right) = \begin{cases} \mathbf{e}_2, & if \quad \mathbf{x}^{(0)} \in U(\mathbf{e}_2), \\ \mathbf{e}_3, & if \quad \mathbf{x}^{(0)} \in \gamma, \\ \mathbf{e}_1, & if \quad \mathbf{x}^{(0)} \in \mathrm{int}\, S^2 \setminus (U(\mathbf{e}_2) \cup \gamma)\,; \end{cases}$$

if $3\alpha > 1 > 3\beta$ then there are a neighbourhood $U(\mathbf{e}_3)$ and a curve ν such that

$$\lim_{n\to\infty} W^n\left(\mathbf{x}^{(0)}\right) = \begin{cases} \mathbf{e}_3, & if \quad \mathbf{x}^{(0)} \in U(\mathbf{e}_3), \\ \mathbf{e}_2, & if \quad \mathbf{x}^{(0)} \in \nu, \\ \mathbf{e}_1, & if \quad \mathbf{x}^{(0)} \in \mathrm{int}\, S^2 \setminus (U(\mathbf{e}_2) \cup \nu)\,; \end{cases}$$

(v_c) *if $1 > 3\beta > 3\alpha$ or $1 > 3\alpha > 3\beta$ then there are the neighbourhoods $U(\mathbf{e}_2)$ and $U(\mathbf{e}_3)$ such that*

$$\lim_{n\to\infty} W^n\left(\mathbf{x}^{(0)}\right) = \begin{cases} \mathbf{e}_2, & if \quad \mathbf{x}^{(0)} \in U(\mathbf{e}_2), \\ \mathbf{e}_3, & if \quad \mathbf{x}^{(0)} \in U(\mathbf{e}_3), \\ \mathbf{e}_1, & if \quad \mathbf{x}^{(0)} \in \mathrm{int}\, S^2 \setminus (U(\mathbf{e}_2) \cup U(\mathbf{e}_3))\,. \end{cases}$$

Proof (i) Let $\mathbf{x}^{(0)} \in \Gamma_{\{1\}}$ then from (5) one easily has $W\left(\mathbf{x}^{(0)}\right) = \mathbf{x}^{(0)}$.

(ii) Let $\mathbf{x}^{(0)} \in \Gamma_{\{2\}}$ then using $x_1 + x_3 = 1$ from the operator (5) one has

$$f_\beta(x) = x^2\big(3(1-\beta) - (2-3\beta)x\big), \quad x = x_3 \in [0, 1].$$

Due to Proposition 1 we have:
(ii_a) if $1 \leq 3\beta \leq 3$ then $\lim\limits_{n\to\infty} W^n(\mathbf{x}) = \mathbf{e}_1$ for any $\mathbf{x}^{(0)} \in \Gamma_{\{2\}}$;
(ii_b) if $0 \leq 3\beta < 1$ then

$$\lim_{n\to\infty} W^n(\mathbf{x}) = \begin{cases} \mathbf{e}_1, & \text{if } 0 < x_3^{(0)} < x_\beta, \\ \mathbf{e}_3, & \text{if } x_\beta < x_3^{(0)} < 1. \end{cases}$$

(iii) Let $\mathbf{x}^{(0)} \in \Gamma_{\{3\}}$ then using $x_1 + x_2 = 1$ from the operator (5) one has

$$f_\alpha(x) = x^2\big(3(1-\alpha) - (2-3\alpha)x\big), \quad x = x_2 \in [0, 1].$$

By virtue of Proposition 1 we have:
(iii_a) if $1 \le 3\alpha \le 3$ then $\lim\limits_{n\to\infty} W^n(\mathbf{x}) = \mathbf{e}_1$ for any $\mathbf{x}^{(0)} \in \Gamma_{\{3\}}$;
(iii_b) if $0 \le 3\alpha < 1$ then

$$\lim_{n\to\infty} W^n(\mathbf{x}) = \begin{cases} \mathbf{e}_1, & \text{if } 0 < x_2^{(0)} < x_\alpha, \\ \mathbf{e}_2, & \text{if } x_\alpha < x_2^{(0)} < 1. \end{cases}$$

(iv) Consider the case $\alpha = \beta$.
(iv_a) Suppose $\mathbf{x}^{(0)} \in \operatorname{int} S^2 \cap M_{23}$. Using $x_1 + 2x_2 = 1$ from (5) one has

$$g_\alpha(x) = x^2\big(3(1-\alpha) - 2(2-3\alpha)x\big), \quad x = x_2 \in [0, 1/2].$$

Due to Proposition 2 we have $\lim\limits_{n\to\infty} W^n(\mathbf{x}) = \mathbf{e}_1$ for any $\mathbf{x}^{(0)} \in \operatorname{int} S^2 \cap M_{23}$.

(iv_b) Let $1 \le 3\alpha \le 3$. It is easy to see that $W(S^2_>) \subset S^2_>$ and $W(S^2_<) \subset S^2_<$. From (5) one has

$$\begin{aligned} x_1' &= x_1(x_1^2 + 3\alpha x_2^2 + 3\alpha x_3^2 + 3x_1x_2 + 3x_1x_3 + 6x_2x_3) \\ &\ge x_1(x_1^2 + x_2^2 + x_3^2 + 3x_1x_2 + 3x_1x_3 + 6x_2x_3) \\ &\ge x_1(1 + x_1x_2 + x_1x_3 + 4x_2x_3). \end{aligned}$$

Since $x_1x_2 + x_1x_3 + 4x_2x_3 = 0$ if and only if when $\mathbf{x} \in \{\mathbf{e}_1, \mathbf{e}_2, \mathbf{e}_3\}$ it follows that for an initial $\mathbf{x}^{(0)} \in \big(S^2_> \cup S^2_<\big) \setminus \operatorname{Fix}(W)$ we have $x_1^{(n+1)} > x_1^{(n)}, \quad n = 0, 1, 2, \ldots$ Consequently $x_1^{(n)}$ is strictly increasing and with the upper bound 1. Hence its limit exists and equal to 1 as the supremum of $x_1^{(n)}$ in S^2. So in this case it is follow that $\lim\limits_{n\to\infty} W^n\big(\mathbf{x}^{(0)}\big) = \mathbf{e}_1$ for any $\mathbf{x}^{(0)} \in \big(\operatorname{int} S^2 \cap (S^2_> \cup S^2_<)\big) \setminus \operatorname{Fix}(W)$.

(iv_c) Let $0 \le 3\alpha < 1$ and let $\mathbf{x}^{(0)} \in \operatorname{int} S^2 \cap S^2_>$. Then from (5) one has

$$x_2' - x_3' = (x_2 - x_3)\left((x_2 + x_3)^2 + 3(1-\alpha)x_1(x_2 + x_3) + 2x_2x_3\right). \tag{12}$$

From the (12) easy to see that $x_2 > x_3 \;\; (x_2 < x_3) \Leftrightarrow x_2' > x_3' \;\; (x_2' < x_3')$. Moreover one has that

$$\frac{x_3'}{x_2'} = \frac{x_3}{x_2}\left(\frac{x_3}{x_2} \cdot \frac{x_2 + 3(1-\alpha)x_1 + x_3 + 2x_2}{x_2 + 3(1-\alpha)x_1 + x_3 + 2x_3}\right) < \frac{x_3}{x_2}, \quad \text{because}$$

$$\frac{x_3\,(x_2 + 3(1-\alpha)x_1 + x_3 + 2x_2)}{x_2\,(x_2 + 3(1-\alpha)x_1 + x_3 + 2x_3)} < 1 \Leftrightarrow (x_3 - x_2)(1 + (2 - -3\alpha)x_1) < 0 \Leftrightarrow x_3 - x_2 < 0.$$

Denote $Q_n := x_3^{(n)}/x_2^{(n)}, \quad n = 0, 1, 2, \ldots$ Therefore for an initial point with $x_2^{(0)} > x_3^{(0)}$ it is follow that $Q_{n+1} < Q_n,\ n = 0, 1, 2, \ldots$ Consequently Q_n is strictly decreasing and with the lower bound 1. Hence its limit exists and equal to 0 as the infimum of Q_n in $S^2_>$, that is $\lim_{n\to\infty} x_3^{(n)} = 0$. So for a small ε and for sufficiently enough large numbers n it is holds $-\varepsilon < x_3^{(n)} < \varepsilon$. Then from the second equation of (5) and (6) one has

$$f_\alpha\left(x_2^{(n)}\right) - 3\alpha\varepsilon\left(x_2^{(n)}\right)^2 < x_2^{(n+1)} < f_\alpha\left(x_2^{(n)}\right) + 3\alpha\varepsilon\left(x_2^{(n)}\right)^2.$$

Due to Proposition 1 for any $0 \leq 3\alpha < 1$ we get that if $0 < x_2^{(0)} < x_\alpha$ then $\lim_{n\to\infty} f_\alpha^n\left(x_2^{(0)}\right) = 0$ and $\lim_{n\to\infty} f_\alpha^n\left(x_2^{(0)}\right) = 1$ when $x_\alpha < x_2^{(0)} < 1$.

Thus for any $0 \leq 3\alpha < 1$ and $\mathbf{x}^{(0)} \in S^2_>\backslash\mathrm{Fix}\,(W)$ we obtain that if $0 < x_2^{(0)} < x_\alpha$ then $\lim_{n\to\infty} W^n\left(\mathbf{x}^{(0)}\right) = \mathbf{e}_1$ and $\lim_{n\to\infty} W^n\left(\mathbf{x}^{(0)}\right) = \mathbf{e}_2$ when $x_\alpha < x_2^{(0)} < 1$.

(iv_d) Similarly can be proved the following assertions: for any $0 \leq 3\alpha < 1$ and $\mathbf{x}^{(0)} \in \mathrm{int}\, S^2 \cap S^2_<\backslash\mathrm{Fix}\,(W)$ one has that if $0 < x_3^{(0)} < x_\alpha$ then $\lim_{n\to\infty} W^n\left(\mathbf{x}^{(0)}\right) = \mathbf{e}_1$ and $\lim_{n\to\infty} W^n\left(\mathbf{x}^{(0)}\right) = \mathbf{e}_3$ when $x_\alpha < x_3^{(0)} < 1$.

(v) Suppose $\alpha \neq \beta$ and we consider the following all possible cases.

(v_a) Let $3\beta > 3\alpha \geq 1$. The case $3\alpha > 3\beta \geq 1$ can be considered similarly. As $3\alpha > 1$ (resp. $3\beta > 1$) by Proposition 3 one has that the vertex $\mathbf{e}_2$ (resp. $\mathbf{e}_3$) is a saddle point. Therefore there is a curve γ (resp. ν) such that for any $\mathbf{x}^{(0)} \in \gamma$ (resp. $\mathbf{x}^{(0)} \in \nu$) it follows that $\lim_{n\to\infty} W^n\left(\mathbf{x}^{(0)}\right) = \mathbf{e}_2$ (resp. $\lim_{n\to\infty} W^n\left(\mathbf{x}^{(0)}\right) = \mathbf{e}_3$) (see Theorem 6.5 in [2] pp. 216–217).

Let $\mathbf{x}^{(0)} \in \mathrm{int}\, S^2 \setminus (\gamma \cup \nu)$ then from the operator (5) one has

$$\begin{aligned}
x_1' &= x_1(x_1^2 + 3\alpha x_2^2 + 3\beta x_3^2 + 3x_1x_2 + 3x_1x_3 + 6x_2x_3) \\
&\geq x_1(x_1^2 + 3\alpha x_2^2 + 3\beta x_3^2 + 3x_1x_2 + 3x_1x_3 + 6x_2x_3) \\
&\geq x_1(x_1^2 + x_2^2 + x_3^2 + 3x_1x_2 + 3x_1x_3 + 6x_2x_3) \\
&\geq x_1(1 + x_1x_2 + x_1x_3 + 4x_2x_3).
\end{aligned}$$

Since $x_1x_2 + x_1x_3 + 4x_2x_3 = 0$ if and only if when $\mathbf{x} \in \{\mathbf{e}_1, \mathbf{e}_2, \mathbf{e}_3\}$ it is follow that for an initial $\mathbf{x}^{(0)} \in S^2\backslash\mathrm{Fix}\,(W)$ we have $x_1^{(n+1)} > x_1^{(n)}, \quad n = 0, 1, 2, \ldots$ Consequently $x_1^{(n)}$ is strictly increasing and with the upper bound 1. Hence its limit exists and equal to 1 as the supremum of $x_1^{(n)}$ in S^2. So in this case it is follow that $\lim_{n\to\infty} W^n\left(\mathbf{x}^{(0)}\right) = \mathbf{e}_1$ for any $\mathbf{x}^{(0)} \in \mathrm{int}\, S^2\backslash\mathrm{Fix}\,(W)$.

(v_b) Let $3\beta > 1 > 3\alpha$. The case $3\alpha > 1 > 3\beta$ can be considered in a similar manner. Since $0 \leq 3\alpha < 1$ by Proposition 3 we have that the vertex $\mathbf{e}_2$ is an attracting point. Therefore there is an attraction basin $U(\mathbf{e}_2)$ of the vertex $\mathbf{e}_2$ such that for any $\mathbf{x}^{(0)} \in U(\mathbf{e}_2)$ it follows that $\lim_{n\to\infty} W^n\left(\mathbf{x}^{(0)}\right) = \mathbf{e}_2$ (see [2, Theorem 6.3, pp. 216–217]).

As $3\beta > 1$ by Proposition 3 one has that the vertex $\mathbf{e}_3$ is a saddle point. So there is a curve γ such that for any $\mathbf{x}^{(0)} \in \gamma$ it follows that $\lim_{n\to\infty} W^n\left(\mathbf{x}^{(0)}\right) = \mathbf{e}_3$ (see [2, Theorem 6.5, pp. 216–217]).

Let $\mathbf{x}^{(0)} \in \operatorname{int} S^2 \setminus (U(\mathbf{e}_2) \cup \gamma)$. From (5) we get

$$\begin{aligned} x_3' + x_2' = 1 - x_1' &= (1-x_1)^3 + 3x_1\left((1-\alpha)x_2^2 + (1-\beta)x_3^2\right) \\ &< (1-x_1)^3 + 3(1-\alpha)x_1\left(x_2^2 + x_3^2\right) \\ &\le (1-x_1)^3 + 3(1-\alpha)x_1(1-x_1)^2 \le 3(1-x_1)^2 . \end{aligned} \tag{13}$$

Iterating the last inequality (13) we have

$$1 - x_1^{(n+1)} < 3\left(1 - x_1^{(n)}\right)^2 < \cdots < 3^{2n+1}\left(1 - x_1^{(0)}\right)^{2^n} . \tag{14}$$

Using $0 < \left(1 - x_1^{(0)}\right) < 1$ from (14) one has $\lim_{n\to\infty} x_1^{(n)} = 1$.

Thus it follows that $\lim_{n\to\infty} W^n\left(\mathbf{x}^{(0)}\right) = \mathbf{e}_1$ for any $\mathbf{x}^{(0)} \in \operatorname{int} S^2 \setminus (U(\mathbf{e}_2) \cup \gamma)$.

(v_c) Let $1 > 3\beta > 3\alpha$. The case $1 > 3\alpha > 3\beta$ can be considered similarly. Since $0 \le 3\alpha < 1$ (resp. $0 \le 3\beta < 1$) by Proposition 3 we have that the vertex $\mathbf{e}_2$ (resp. $\mathbf{e}_3$) is an attracting point. Therefore there is an attraction basin $U(\mathbf{e}_2)$ (resp. $U(\mathbf{e}_3)$) of the vertex $\mathbf{e}_2$ (resp. $\mathbf{e}_3$) such that for any $\mathbf{x}^{(0)} \in U(\mathbf{e}_2)$ (resp. $\mathbf{x}^{(0)} \in U(\mathbf{e}_3)$) it follows that $\lim_{n\to\infty} W^n\left(\mathbf{x}^{(0)}\right) = \mathbf{e}_2$ (resp. $\lim_{n\to\infty} W^n\left(\mathbf{x}^{(0)}\right) = \mathbf{e}_3$)(see [2, Theorem 6.3, pp. 216-217]).

Let $\mathbf{x}^{(0)} \in \operatorname{int} S^2 \setminus (U(\mathbf{e}_2) \cup U(\mathbf{e}_3))$. Then using the techniques of the proof of part v_b) from (5), (13) and (14) one has $\lim_{n\to\infty} W^n\left(\mathbf{x}^{(0)}\right) = \mathbf{e}_1$.

The proof is completed.

Corollary 1 *For any parameters the CSO (5) is a regular operator. Therefore it is an ergodic transformation.*

References

1. Davronov, R.R., Jamilov (Zhamilov), U. U., Ladra, M.: Conditional cubic stochastic operator. J. Diff. Equ. Appl. **21**, 1163–1170 (2015). https://doi.org/10.1080/10236198.2015.1062481
2. Devaney, R.L.: An introduction to chaotic dynamical systems. In: Studies in Nonlinearity. Westview Press, Boulder, CO (2003)
3. Freedman, H.I., Waltman, P.: Persistence in models of three interacting predator-prey populations. Math. Biosci. **68**, 213–231 (1984). https://doi.org/10.1016/0025-5564(84)90032-4
4. Ganikhodjaev, N.N., Ganikhodjaev, R.N., Jamilov, U.U.: Quadratic stochastic operators and zero-sum game dynamics. Ergod. Theory Dyn. Syst. **35**, 1443–1473 (2015). https://doi.org/10.1017/etds.2013.109
5. Ganikhodzhaev, R.N.: Quadratic stochastic operators, Lyapunov functions and tournaments. Sb. Math. **76**, 489–506 (1993). https://doi.org/10.1070/SM1993v076n02ABEH003423
6. Ganikhodzhaev, R., Mukhamedov, F., Rozikov, U.: Quadratic stochastic operators and processes: results and open problems. Infin. Dimens. Anal. Quan. Probab. Relat. Top. **14**, 279–335 (2011). https://doi.org/10.1142/S0219025711004365

7. Gavid, C., Pokrovskii, A., Prentice, M., Sobolev, V.: Dynamics of a Lotka-Volterra type model with applications to marine phage population dynamics. J. Phys. Conf. Ser. **55**, 80–93 (2006). https://doi.org/10.1088/1742-6596/55/1/008
8. Homburg, A.J., Jamilov, U.U., Scheutzow, M.: Asymptotics for a class of iterated random cubic operators. Nonlinearity **32**, 3646–3660 (2019). https://doi.org/10.1088/1361-6544/ab1f24
9. Jamilov, U.U.: Quadratic stochastic operators corresponding to graphs. Lobachevskii J. Math. **34**(2), 148–151 (2013). https://doi.org/10.1134/S1995080213020042
10. Jamilov, U.U.: On a family of strictly non-Volterra quadratic stochastic operators. J. Phys. Conf. Ser. **697**, 012013 (2016). https://doi.org/10.1088/1742-6596/697/1/012013
11. Jamilov, U.U.: On symmetric strictly non-Volterra quadratic stochastic operators. Disc. Nonlin. Comp. **5**, 263–283 (2016). https://doi.org/10.5890/DNC.2016.09.006
12. Jamilov, U.U., Khamraev, A.Y., Ladra, M.: On a Volterra cubic stochastic operator. Bull. Math. Biol. **80**, 319–334 (2018). https://doi.org/10.1007/s11538-017-0376-0
13. Jamilov, U.U., Ladra, M.: Non-ergodicity of uniform quadratic stochastic operators. Qual. Theory Dyn. Syst. **15**, 257–271 (2016). https://doi.org/10.1007/s12346-015-0145-0
14. Jamilov, U.U., Ladra, M.: On identically distributed non-Volterra cubic stochastic operator. J. Appl. Nonlin. Dyn. **6**, 79–90 (2017). https://doi.org/10.5890/JAND.2017.03.006
15. Jamilov, U.U., Ladra, M., Mukhitdinov, R.T.: On the equiprobable strictly non-Volterra quadratic stochastic operators. Qual. Theory Dyn. Syst. **16**, 654–655 (2017). https://doi.org/10.1007/s12346-016-0209-9
16. Jamilov, U.U., Reinfelds, A.: On constrained Volterra cubic stochastic operators. J. Diff. Equ. App. **26**, 261–274 (2020). https://doi.org/10.1080/10236198.2020.1720664
17. Jamilov, U.U., Reinfelds, A.: A family of Volterra cubic stochastic operators. J. Convex Anal. **28**, 19–30 (2021). JCA/JCA28/JCA281/jca28003.htm
18. Jamilov, U.U., Scheutzow, M., Wilke-Berenguer, M.: On the random dynamics of Volterra quadratic operators. Ergod. Theory Dyn. Syst. **37**, 228–243 (2017). https://doi.org/10.1017/etds.2015.30
19. Jamilov, U.U., Scheutzow, M., Vorkastner, I.: A prey-predator model with three interacting species. Arxiv:1907.05100v1
20. Khamraev, A.Y.: On cubic operators of Volterra type. Uzbek. Math. J. **2**, 79–84 (2004). (Russian)
21. Khamraev, A.Y.: On a Volterra-type cubic operator. Uzbek. Math. J. **3**, 65–71 (2009). (Russian)
22. Lyubich, Y.I.: Mathematical Structures in Population Genetics. Biomathematics, vol. 22. Springer- Verlag, Berlin (1992)
23. Mukhamedov, F.M., Embong, A.F., Rosli, A.: Orthogonal preserving and surjective cubic stochastic operators. Ann. Func. Anal. **8**, 490–501 (2017). https://doi.org/10.1215/20088752-2017-0013
24. Mukhamedov, F.M., Pah, C.H., Rosli, A.: On non-ergodic Volterra cubic stochastic operators. Qual. Theory Dyn. Syst. **18**, 1225–1235 (2019). https://doi.org/10.1007/s12346-019-00334-8
25. Mukhamedov, F., Saburov, M.: Stability and monotonicity of Lotka-Volterra type operators. Qual. Theory Dyn. Syst. **16**, 249–267 (2017). https://doi.org/10.1007/s12346-016-0190-3
26. Rozikov, U.A., Khamraev, A.Y.: On cubic operators defined on finite-dimensional simplices. Ukr. Math. J. **56**, 1699–1711 (2004). https://doi.org/10.1007/s11253-005-0145-3
27. Rozikov, U.A., Khamraev, A.Yu.: On construction and a class of non-Volterra cubic stochastic operators. Nonlinear Dyn. Syst. Theory **14**, 92–100 (2014). www.e-ndst.kiev.ua/v14n1.htm
28. Rozikov, U.A., Zhamilov, U.: F-quadratic stochastic operators. Math. Notes **83**, 554–559 (2008). https://doi.org/10.1134/S0001434608030280
29. Rozikov, U.A., Zhamilov, U.U.: Volterra quadratic stochastic operators of a two-sex population. Ukr. Math. J. **63**, 1136–1153 (2011). https://doi.org/10.1007/s11253-011-0568-y
30. Ulam, S.M.: A Collection of Mathematical Problems. Interscience Tracts in Pure and Applied Mathematics, no. 8. Interscience Publishers, New York-London (1960)
31. Zakharevich, M.I.: On the behaviour of trajectories and the ergodic hypothesis for quadratic mappings of a simplex. Russ. Math. Surv. **33**, 265–266 (1978). https://doi.org/10.1070/RM1978v033n06ABEH003890

32. Zhamilov, U.U., Rozikov, U.A.: On the dynamics of strictly non-Volterra quadratic stochastic operators on the two-dimensional simplex. Russ. Acad. Sci. Sb. Math. **200**, 1339–1351 (2009). https://doi.org/10.1070/SM2009v200n09ABEH004039

The Dynamics of Superposition of Non-Volterra Quadratic Stochastic Operators on S^2

Uygun Jamilov and Kamola Aralova

Abstract In this paper, we study the dynamical systems generated by stochastic operators which are a superposition of non-Volterra quadratic stochastic operators defined on the two-dimensional simplex. We showed that such stochastic operator has two fixed points and there are no periodic points except fixed points. We found attractive basins for each fixed points. We constructed a Lyapunov function and using it we showed that such operator has the property being regular.

Keywords Quadratic stochastic operator · Volterra operator · Regular operator

1 Introduction

Let

$$S^{m-1} = \{\mathbf{x} = (x_1, x_2, \ldots, x_m) \in R^m : \text{ for any } i,\ x_i \geq 0, \text{ and } \sum_{i=1}^{m} x_i = 1\}$$

be the $(m-1)$-dimensional simplex.

A map $V : S^{m-1} \to S^{m-1}$ is called a stochastic operator (SO). The trajectory $\{\mathbf{x}^{(n)}\}_{n=0}^{\infty}$ of an operator V for an initial point $\mathbf{x}^{(0)} \in S^{m-1}$ is defined by

U. Jamilov (✉) · K. Aralova
V. I. Romanovskiy Institute of Mathematics, Uzbekistan Academy of Sciences, 9, University Str., Tashkent 100174, Uzbekistan
e-mail: uygun.jamilov@mathinst.uz; jamilovu@yandex.ru

K. Aralova
e-mail: aralovak.a@mail.ru

U. Jamilov
Akfa University, 264, National Park Str., 111221 Qibray District, Tashkent Region, Uzbekistan

National University of Uzbekistan, 9, Universitet Str., 100174 Tashkent, Uzbekistan

L. Accardi et al. (eds.), *Infinite Dimensional Analysis, Quantum Probability and Applications*, Springer Proceedings in Mathematics & Statistics 390,
https://doi.org/10.1007/978-3-031-06170-7_23

$$\mathbf{x}^{(n+1)} = V(\mathbf{x}^{(n)}) = V^{n+1}(\mathbf{x}^{(0)}), \ n = 0, 1, 2, \ldots$$

Denote by $\omega_V(\mathbf{x}^{(0)})$ the set of limit points of the trajectory $\{\mathbf{x}^{(n)}\}_{n=0}^{\infty}$.

Definition 1 A SO V is called

(i) regular if the limit $\lim_{n\to\infty} V^n(\mathbf{x})$ exits for any initial $\mathbf{x} \in S^{m-1}$;
(ii) ergodic if the limit

$$\lim_{n\to\infty} \frac{1}{n} \sum_{k=0}^{n-1} V^k(\mathbf{x})$$

exists for any initial $\mathbf{x} \in S^{m-1}$.

In the particular case a map $V : S^{m-1} \to S^{m-1}$ is called a quadratic stochastic operator (QSO) if

$$(V\mathbf{x})_k = \sum_{i,j=1}^{m} p_{ij,k} x_i x_j \tag{1}$$

for any $\mathbf{x} \in S^{m-1}$ and for all $k = 1, \ldots, m$, where

$$p_{ij,k} \geq 0, \quad p_{ij,k} = p_{ji,k} \ \text{for all} \ i, j, k; \quad \sum_{k=1}^{m} p_{ij,k} = 1. \tag{2}$$

Such operators frequently arises in many models of mathematical genetics, namely, theory of heredity and the theory of QSOs developed in many works (see e.g. [1, 4, 6, 8–11, 13, 19–21]). We refer a reader to [7, 15] for a recent review on quadratic stochastic operators.

The main problem in mathematical biology consists in the study of the asymptotical behaviour of the trajectories for a given QSO. In other words the main task is the description of the set $\omega_V\left(\mathbf{x}^{(0)}\right)$ for any initial point $\mathbf{x}^{(0)} \in S^{m-1}$ for a given QSO. In general, the main problem of the study of the asymptotic behaviour of a QSO is also a difficult problem, which remains open, even in the two-dimensional simplex case.

The main problem was more completely solved for the Volterra QSOs (see e.g. [3, 11, 14]). A Volterra QSO is defined by (1), (2) and with the additional assumption

$$p_{ij,k} = 0 \ \text{if} \ k \notin \{i, j\}, \ \text{for any} \ i, j, k \in E. \tag{3}$$

In [16, 18] general forms of Volterra operators have been explored.

But non-Volterra QSO's (i.e., not satisfying condition (3)) were not exhaustively studied, because there is no general theory that can be applied to the study of non-Volterra operators.

In the present paper, we consider a specific family of discrete-time dynamical systems generated by superposition of non-Volterra quadratic stochastic operators. The article is organized as follows. In Sect. 2, we recall the definitions and known results related to a quadratic stochastic operators. In Sect. 3, we study the asymptotical behaviour of superposition of non-Volterra QSOs defined on the two-dimensional simplex. It is described the set of all fixed points of the operator. We proved that for an initial point a trajectory of such operator converges to a fixed point.

2 Preliminaries

In this section we recall necessary definitions and known results. Let V be a stochastic operator.

Definition 2 A point $\mathbf{x} \in S^{m-1}$ is called a periodic point of V if there exists an n so that $V^n(\mathbf{x}) = \mathbf{x}$. The smallest positive integer n satisfying the above is called the prime period or least period of the point $\mathbf{x}$. A period-one point is called a fixed point of V.

Denote the set of all fixed points by $\mathrm{Fix}(V)$ and the set of all periodic points of (not necessarily prime) period n by $\mathrm{Per}_n(V)$. Evidently that the set of all iterates of a periodic point form a periodic trajectory (orbit).

Let $DV(\mathbf{x}^*) = (\partial V_i/\partial x_j)_{i,j=1}^m(\mathbf{x}^*)$ be a Jacobian of V at the point $\mathbf{x}^*$.

Definition 3 A fixed point $\mathbf{x}^*$ is called hyperbolic if its Jacobian $D_\mathbf{x}V(\mathbf{x}^*)$ has no eigenvalues on the unit circle. A hyperbolic fixed point $\mathbf{x}^*$ is called:

(i) attracting if all the eigenvalues of the Jacobian $D_\mathbf{x}V(\mathbf{x}^*)$ are less than 1 in absolute value;
(ii) repelling if all the eigenvalues of the Jacobian $D_\mathbf{x}V(\mathbf{x}^*)$ are greater than 1 in absolute value;
(iii) a saddle otherwise.

In [5] studied the following non-Volterra QSO defined on the two-dimensional simplex

$$V_0(\mathbf{x}) = \left(x_1^2 + 2x_2x_3,\ \ x_2^2 + 2x_1x_3,\ \ x_3^2 + 2x_1x_2\right). \tag{4}$$

Let $\mathbf{e}_1 = (1, 0, 0)$, $\mathbf{e}_2 = (0, 1, 0)$, $\mathbf{e}_3 = (0, 0, 1)$ be the vertices and $\mathbf{c} = (1/3, 1/3, 1/3)$ be the center of the simplex S^2.

Theorem 1 ([5]) *For the QSO (4) the following statements are true:*

(i) $Fix(V_0) = \{\mathbf{e}_1, \mathbf{e}_2, \mathbf{e}_3, \mathbf{c}\}$*;*
(ii) the center $\mathbf{c}$ *is an attracting point;*
(iii) for any $\mathbf{x}^{(0)} \in S^2 \setminus Fix(V_0)$ *the trajectory of non-Volterra QSO* V_0 *converges to the center.*

The following quasi strictly non-Volterra operator on the S^2

$$V_1(\mathbf{x}) = \left(2x_3x_2,\ 2x_3x_1,\ x_3^2 + (x_1 + x_2)^2\right) \tag{5}$$

was studied in [12].

Theorem 2 ([12]) *For the operator (5) the following statements are true:*

(i) *the vertex* $\mathbf{e}_3$ *and* $\bar{\mathbf{x}} = (1/4, 1/4, 1/2)$ *are fixed points;*
(ii) $V_1^2(\mathbf{x}) = \mathbf{x}$ *for any* $\mathbf{x} \in \{\mathbf{x} \in S^2 : x_1 + x_2 = 1/2, x_3 = 1/2\}$*;*
(iii) *the sets* $M_\nu = \{\mathbf{x} \in S^2 : x_1 = \nu x_2 \vee x_1 = (1/\nu)x_2\}$ *when* $\nu > 0$, *and* $M_0 = \{\mathbf{x} \in S^2 : x_1x_2 = 0\}$ *when* $\nu = 0$ *are invariant sets of* V*;*
(iv) *if* $\mathbf{x}^{(0)} \in \Theta = \{\mathbf{x} \in S^2 : x_3 = 0\} \cup \{\mathbf{e}_3\}$, *then* $\omega_{V_1}\left(\mathbf{x}^{(0)}\right) = \{\mathbf{e}_3\}$*;*
(v) $\omega_{V_1}\left(\mathbf{x}^{(0)}\right) = \{(1/2, 0, 1/2), (0, 1/2, 1/2)\}$ *for any initial* $\mathbf{x}^{(0)} \in M_0$*;*
(vi) *for any* $\mathbf{x}^{(0)} \in S^2 \setminus \Theta$ *there is* M_ν, $\nu \in (0, +\infty)$ *such that* $\mathbf{x}^{(0)} \in M_\nu$ *and* $\omega_{V_1}\left(\mathbf{x}^{(0)}\right) = \{\tilde{\mathbf{x}}, \hat{\mathbf{x}}\}$, *where*

$$\tilde{\mathbf{x}} = \left(\frac{1}{2}, \frac{\nu}{2(1+\nu)}, \frac{1}{2(1+\nu)}\right) \text{ and } \hat{\mathbf{x}} = \left(\frac{1}{2}, \frac{1}{2(1+\nu)}, \frac{\nu}{2(1+\nu)}\right).$$

Thus, we can see that the all trajectories of the QSO V_0 converge to the center of the two-dimensional simplex and the all trajectories of the QSO V_1 converge to a periodic trajectory with the period two. In [10, 17] these results were generalized and the dynamics of the convex combination of the general form operators also were studied.

3 Superposition of Operators

Let us consider the nonlinear operator $W_0 := V_0 \circ V_1$ defined by a superposition of the above given operators and it follows that this operator has the form

$$W_0 : \begin{cases} x_1' = 4x_3^2x_2^2 + 4x_3x_1(2x_3^2 - 2x_3 + 1), \\ x_2' = 4x_3^2x_1^2 + 4x_3x_2(2x_3^2 - 2x_3 + 1), \\ x_3' = (2x_3^2 - 2x_3 + 1)^2 + 8x_3^2x_1x_2. \end{cases} \tag{6}$$

First we need some axillary observations. Consider the function

$$f(x) = 6x^4 - 12x^3 + 10x^2 - 4x + 1, \quad x \in [0, 1]. \tag{7}$$

We define $f^n = \underbrace{f \circ \cdots \circ f}_{n \text{ times}}$ to be the n-fold composition of f with itself.

In the next Theorem we will give properties of the function $f(x)$.

Theorem 3 *For the function $f(x)$ the following statements are true:*

(i) $\mathrm{Fix}\,(f) = \{1, x^*\}$, *where* $x^* = \frac{\sqrt[3]{2+2\sqrt{17}}}{6} - \frac{2}{3\sqrt[3]{2+2\sqrt{17}}} + \frac{1}{3}$;
(ii) *the fixed point $x = 1$ is a repelling point and the fixed point x^* is attracting point;*
(iii) $\mathrm{Per}_n\,(f) \setminus \mathrm{Fix}\,(f) = \emptyset$ *for any value* $n \geq 2$;
(iv) $\lim_{n\to\infty} f^n(x) = x^*$ *for any* $0 < x < 1$ *and* $f(0) = f(1) = 1$.

Proof (i) It is easy to verify that $f(1) = 1$, that is $x = 1$ is a solution of the equation $f(x) = x$. One can write the last equation in the form

$$f(x) - x = (x - 1)(6x^3 - 6x^2 + 4x - 1).$$

Let $x \neq 1$ then some algebra shows that the equation $6x^3 - 6x^2 + 4x - 1 = 0$ has three solutions: a real number

$$x^* = \frac{\sqrt[3]{2+2\sqrt{17}}}{6} - \frac{2}{3\sqrt[3]{2+2\sqrt{17}}} + \frac{1}{3}$$

and another two complex numbers.

(ii) It is easy to check that $|f'(1)| = 4$ and we claim that $|f'(x^*)| < 1$. Indeed, since x^* is a solution of the equation $6x^3 - 6x^2 + 4x - 1 = 0$ it follows that

$$\begin{aligned} f'(x^*) &= 24x^{*3} - 36x^{*2} + 20x^* - 4 - 4(6x^{*3} - 6x^{*2} + 4x^* - 1) \\ &= 4x^*(1 - 3x^*) \approx -0.2567. \end{aligned}$$

Therefore it follows that $-1 < f'(x^*) < 0$.

Thus $x = 1$ is a repelling and x^* is an attracting points of the function $f(x)$.

(iii) Consider the function

$$h(x) = \frac{f^2(x) - x}{f(x) - x}$$

and it is easy to see that

$$\begin{aligned} h(x) = {} & 1296x^{12} - 7776x^{11} + 22032x^{10} - 38664x^9 + 46440x^8 - 39888x^7 + \\ & + 24900x^6 - 11280x^5 + 3660x^4 - 834x^3 + 130x^2 - 12x + 1 \end{aligned} \tag{8}$$

Simple graphical analysis shows that $h(x) > 0$ for any $x \in [0, 1]$, that is the last equation has no solution (Fig. 1).

Therefore, due to Sharkovsky's theorem (see e.g. [2]) it follows that the equation $f^n(x) = x$ has no solution for all $n \geq 2$.

(iv) Further one easily has that $f'(x) = 4(2x - 1)(3x^2 - 3x + 1)$ and the equation $f'(x) = 0$ has a unique real solution $x = 1/2$. Consequently it follows that

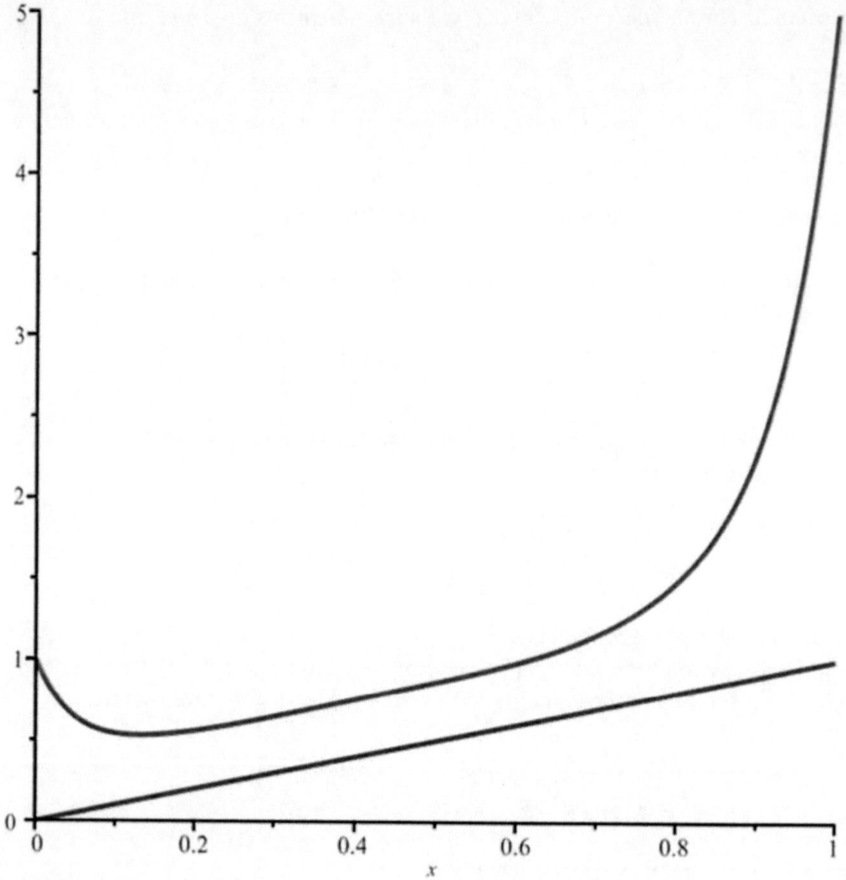

Fig. 1 The graph of the function $h(x)$

$$\min_{0<x<1} f(x) = \frac{3}{8} < x^* \approx 0.3884.$$

Suppose that $f^n(x) \geq x^*$. Then in a neighbourhood of the fixed point x^* of the function $f(x)$ we have $-1 < f'(x) < 0$ and it holds

$$f^{n+2k}(x) \leq f^n(x) \text{ and } f^{n+2k+1}(x) \geq f^n(x), \quad k = 0, 1, \ldots$$

Moreover, it follows that

$$\lim_{k\to\infty} f^{n+2k}(x) = \lim_{k\to\infty} f^{n+2k+1}(x) = x^*.$$

The case $f^n(x) \leq x^*$ can be considered similarly.

Thus for an arbitrary $x \in (0,1)$ we have $\lim_{n\to\infty} f^n(x) = x^*$. □

Theorem 4 *For the operator W_0 (6) the following statements are true:*

(i) the sets $M_1 = \{\mathbf{x} \in S^2 : x_1 = x_2\}$, $M_2 = \{\mathbf{x} \in S^2 : x_1 > x_2\}$ and $M_3 = \{\mathbf{x} \in S^2 : x_1 < x_2\}$ are invariant sets;
(ii) the vertex $\mathbf{e}_3$ and $\mathbf{x}^* = \left(\frac{1-x^*}{2}, \frac{1-x^*}{2}, x^*\right)$ are fixed points;
(iii) $W_0(\mathbf{x}^{(0)}) = \mathbf{e}_3$ for any $\mathbf{x}^{(0)} \in E_3 = \{\mathbf{x} \in S^2 : x_3 = 0\}$;
(iv) if $\mathbf{x}^{(0)} \in M_1 \setminus (E_3 \cup \{\mathbf{e}_3\})$, then $\omega_{W_0}\left(\mathbf{x}^{(0)}\right) = \{\mathbf{x}^*\}$
(v) the function $\psi(\mathbf{x}) = |x_1 - x_2|$ is a Lyapunov function;
(vi) if $\mathbf{x}^{(0)} \in S^2 \setminus (E_3 \cup \{\mathbf{e}_3\})$ then $\omega_{W_0}\left(\mathbf{x}^{(0)}\right) = \{\mathbf{x}^*\}$

Proof (i) From the first two equations of the (6) we have

$$\begin{aligned} x_1' - x_2' &= 4x_3^2(x_2^2 - x_1^2) + 4x_3(2x_3^2 - 2x_3 + 1)(x_1 - x_2) \\ &= 4x_3^2(x_2 - x_1)(1 - x_3) + 4x_3(2x_3^2 - 2x_3 + 1)(x_1 - x_2) \\ &= 4x_3(2x_3^2 - 2x_3 + 1 - x_3(1 - x_3))(x_1 - x_2) \\ &= 4x_3(3x_3^2 - 3x_3 + 1)(x_1 - x_2). \end{aligned} \tag{9}$$

Since $4x_3(3x_3^2 - 3x_3 + 1) > 0$, $\forall x_3 \in (0, 1]$ from the (9) it follows that

(a) if $\mathbf{x} \in M_1$, then by $x_1 - x_2 = 0$ it is holds $x_1' - x_2' = 0$, that is $W_0(M_1) \subset M_1$;
(b) if $\mathbf{x} \in M_2$, then by $x_1 - x_2 > 0$ it is holds $x_1' - x_2' > 0$, that is $W_0(M_2) \subset M_2$;
(c) if $\mathbf{x} \in M_3$, then by $x_1 - x_2 < 0$ it is holds $x_1' - x_2' < 0$, that is $W_0(M_3) \subset M_3$.

(ii) Recall, that a fixed point of the operator W_0 is a solution of the equation $W_0(x) = x$ and this system of equations has the form

$$\begin{cases} x_1 = 4x_3^2x_2^2 + 4x_3x_1(2x_3^2 - 2x_3 + 1), \\ x_2 = 4x_3^2x_1^2 + 4x_3x_2(2x_3^2 - 2x_3 + 1), \\ x_3 = (2x_3^2 - 2x_3 + 1)^2 + 8x_3^2x_1x_2. \end{cases} \tag{10}$$

Evidently that $x_3 = 0$ is not satisfies the third equation of the above system.

Suppose $x_3 > 0$, $x_1 = 0$ then from the first equation of (10) it follows $x_2 = 0$. Then one has that $x_3 = 1$, that is we have the fixed point $\mathbf{e}_3$.

Assume $x_3 > 0$, $x_2 = 0$ then from the second equation of (10) it follows $x_1 = 0$. Then one has that $x_3 = 1$, that is we again have the fixed point $\mathbf{e}_3$. Hence if $x_3 > 0$, $x_1x_2 = 0$ the we get the fixed point $\mathbf{e}_3$.

Let $x_1x_2x_3 > 0$ then from the first and second equations of (10) one has

$$\frac{x_1}{x_2} = \frac{x_3x_2^2 + x_1\left(2x_3^2 - 2x_3 + 1\right)}{x_3x_1^2 + x_2\left(2x_3^2 - 2x_3 + 1\right)} \quad \Leftrightarrow \quad x_1^3 = x_2^3 \quad \Leftrightarrow \quad x_1 = x_2.$$

Assume that $x_1 = x_2$. Using $2x_1 + x_3 = 1$ from the third equation in the above system (10) one has

$$x_3 = (2x_3^2 - 2x_3 + 1)^2 + 8x_3^2\left(\frac{1-x_3}{2}\right)^2$$
$$x_3 = 4x_3^4 + 4x_3^2 + 1 - 8x_3^3 + 4x_3^2 - 4x_3 + 2x_3^2 - 4x_3^3 + 2x_3^4$$
$$x_3 = 6x_3^4 - 12x_3^3 + 10x_3^2 - 4x_3 + 1 \tag{11}$$

It is easy to see that (11) is the equation $f(x_3) = x_3$ (see (7)). By Theorem 3 the Eq. (11) has the solutions $x_3 = 1$ and $x_3 = x^*$. Evidently, that if $x_3 = 1$ then we get a fixed point which is the vertex $\mathbf{e}_3$. If $x_3 = x^*$ the we obtain the fixed point $\mathbf{x}^*$.

(iii) straightforward.

(iv) Let $\mathbf{x}^{(0)} \in M_1 \setminus (E_3 \cup \{\mathbf{e}_3\})$, then $x_1^{(n)} = x_2^{(n)}$ and $x_3^{(n)} = f^n\left(x_3^{(0)}\right)$. By virtue of Theorem 3 it follows that

$$\lim_{n\to\infty} f^n\left(x_3^{(0)}\right) = x^*.$$

Consequently, using $x_1^{(n)} = x_2^{(n)} = \left(1 - x_3^{(n)}\right)/2$ one has that

$$\lim_{n\to\infty} W_0^n\left(\mathbf{x}^{(0)}\right) = \left(\frac{1-x^*}{2}, \frac{1-x^*}{2}, x^*\right).$$

Thus we obtain that $\lim\limits_{n\to\infty} W_0^n\left(\mathbf{x}^{(0)}\right) = \mathbf{x}^*$ for any $\mathbf{x}^{(0)} \in M_1 \setminus (E_3 \cup \{\mathbf{e}_3\})$.

(v) Consider the function $\psi(\mathbf{x}) = |x_1 - x_2|$, $\mathbf{x} \in S^2$. We claim that the function $\psi(\mathbf{x})$ is a Lyapunov function for the nonlinear operator W_0. Indeed, the case $\mathbf{x} \in M_1$ is obvious. Let $\mathbf{x} \in S^2 \setminus M_1$ then from the (6) we have

$$\psi(W_0(\mathbf{x})) = |x_1' - x_2'| = |4x_3(3x_3^2 - 3x_3 + 1)(x_1 - x_2)| = 4x_3(3x_3^2 - 3x_3 + 1)\psi(\mathbf{x}).$$

Consequently, it follows that

$$\psi\left(\mathbf{x}^{(n+1)}\right) = 4x_3^{(n)}\left(3\left(x_3^{(n)}\right)^2 - 3x_3^{(n)} + 1\right)\psi\left(\mathbf{x}^{(n)}\right) \le \left(3\left(x_3^{(n)}\right)^2 - 2x_3^{(n)} + 1\right)^2\psi\left(\mathbf{x}^{(n)}\right),$$

where we have used well-known AM-GM relation.

From the last equation of the (6) we have

$$\left(2\left(x_3^{(n)}\right)^2 - 2x_3^{(n)} + 1\right)^2 \le x_3^{(n+1)} \le (2x_3^2 - 2x_3 + 1)^2 + 2x_3^{(n)}\left(x_1^{(n)} + x_2^{(n)}\right)^2 \Rightarrow$$

$$\frac{1}{4} \le x_3^{(n+1)} \le f\left(x_3^{(n)}\right)$$

By Theorem 3 it follows that there is a sufficiently large n_0 such that for all $n > n_0$ it is hold

$$\frac{1}{4} \le x_3^{(n+1)} \le \frac{1}{2}.$$

Consequently it follows that for all $n > n_0$ it is hold

$$\frac{11}{16} \leq 3\left(x_3^{(n)}\right)^2 - 2x_3^{(n)} + 1 \leq \frac{3}{4}.$$

Therefore there is a sufficiently large n_0 such that for any $n > n_0$ it holds

$$\psi\left(\mathbf{x}^{(n+1)}\right) \leq \frac{9}{16}\psi\left(\mathbf{x}^{(n)}\right).$$

Consequently we have that the function $\psi(\mathbf{x})$ is a Lyapunov function for the nonlinear operator W_0. Moreover one has that

$$\lim_{n\to\infty} \psi\left(\mathbf{x}^{(n)}\right) = 0. \tag{12}$$

(vi) Let $\mathbf{x}^{(0)} \in S^2 \setminus (M_1 \cup E_3)$. By assertion of the part (iv) the function $\psi(\mathbf{x})$ is a Lyapunov function. Besides, since the sets M_2 and M_3 are invariant sets, that is it follows that $\psi(\mathbf{x})$ is a decreasing Lyapunov function on the set M_2 and it is an increasing Lyapunov function on the set M_3. In other words the last means that there exists the following limit

$$\lim_{n\to\infty}\left(x_1^{(n)} - x_2^{(n)}\right) \ \left(\text{resp. } \lim_{n\to\infty}\left(x_2^{(n)} - x_1^{(n)}\right)\right) \tag{13}$$

when $x_1^{(0)} > x_2^{(0)}$ (resp. $x_1^{(0)} < x_2^{(0)}$).

Let $\mathbf{x}^{(0)} \in M_2$ then from (6) we have

$$\frac{x_2'}{x_1'} = \frac{x_2}{x_1} \cdot \frac{x_1^3 x_3 + x_1 x_2\left(2x_3^2 - 2x_3 + 1\right)}{x_2^3 x_3 + x_1 x_2\left(2x_3^2 - 2x_3 + 1\right)} > \frac{x_2}{x_1} \tag{14}$$

Iterating the inequalities we get

$$Q_n := \frac{x_2^{(n)}}{x_1^{(n)}} < 1, \quad n = 0, 1, 2, \ldots$$

By (14) we have

$$Q_{n+1} = Q_n \cdot \frac{\left(x_1^{(n)}\right)^3 x_3^{(n)} + x_1^{(n)} x_2^{(n)}\left(2\left(x_3^{(n)}\right)^2 - 2x_3^{(n)} + 1\right)}{\left(x_2^{(n)}\right)^3 x_3^{(n)} + x_1^{(n)} x_2^{(n)}\left(2\left(x_3^{(n)}\right)^2 - 2x_3^{(n)} + 1\right)} > Q_n.$$

Consequently Q_n is strictly increasing and with the upper bound 1. Hence its limit exists and equal to 1 as the supremum of Q_n in M_2 and using it we have

$$\lim_{n\to\infty}\frac{x_2^{(n)}}{x_1^{(n)}}=1\Leftrightarrow\lim_{n\to\infty}\frac{x_1^{(n)}}{x_2^{(n)}}=1\Rightarrow\lim_{n\to\infty}\left(\frac{x_1^{(n)}}{x_2^{(n)}}+\frac{x_2^{(n)}}{x_1^{(n)}}+2\right)^{-1}=\frac{1}{4}$$

Therefore for a sufficiently small $\varepsilon>0$ there is a natural n_0 such that for all $n>n_0$ it is hold

$$-\varepsilon<\frac{x_1^{(n)}x_2^{(n)}}{\left(x_1^{(n)}+x_2^{(n)}\right)^2}-\frac{1}{4}<\varepsilon\ \Leftrightarrow\ \frac{1}{4}-\varepsilon<\frac{x_1^{(n)}x_2^{(n)}}{\left(x_1^{(n)}+x_2^{(n)}\right)^2}<\frac{1}{4}+\varepsilon \tag{15}$$

The third equation of (6) we can write in form

$$x_3^{(n+1)}=\left(2\left(x_3^{(n)}\right)^2-2x_3^{(n)}+1\right)^2+8\left(x_3^{(n)}\left(1-x_3^{(n)}\right)\right)^2\frac{x_1^{(n)}x_2^{(n)}}{\left(x_1^{(n)}+x_2^{(n)}\right)^2}.$$

Using (7) and (15) from the last we have

$$f\left(x_3^{(n)}\right)-8\varepsilon\left(x_3^{(n)}\left(1-x_3^{(n)}\right)\right)^2<x_3^{(n+1)}<f\left(x_3^{(n)}\right)+8\varepsilon\left(x_3^{(n)}\left(1-x_3^{(n)}\right)\right)^2$$

By virtue of Theorem 3 it follows that

$$\lim_{n\to\infty}x_3^{(n)}=\lim_{n\to\infty}f^n\left(x_3^{(0)}\right)=x^*$$

and using $x_1+x_2=1-x_3$ one has that there is the limit

$$\lim_{n\to\infty}\left(x_1^{(n)}+x_2^{(n)}\right)=1-x^*. \tag{16}$$

Consequently, from (12), (13) and (16) we have that there are the limits

$$\lim_{n\to\infty}x_1^{(n)}=\xi,\ \lim_{n\to\infty}x_2^{(n)}=\xi,\ \lim_{n\to\infty}x_3^{(n)}=x^*\text{ and }\lim_{n\to\infty}\mathbf{x}^{(n)}=\left(\xi,\xi,x^*\right).$$

Since the limit point should be a fixed point and by assertion of item ii) we obtain that $\left(\xi,\xi,x^*\right)=\mathbf{x}^*$, that is

$$\lim_{n\to\infty}\mathbf{x}^{(n)}=\mathbf{x}^*.$$

In the case $\mathbf{x}^{(0)}\in S^2\setminus(M_2\cup E_3)$ similarly can be proved that $\lim_{n\to\infty}\mathbf{x}^{(n)}=\mathbf{x}^*$. □

Now consider another nonlinear operator $W_1:=V_1\circ V_0$ defined by superposition of the given operators and it is clear that this operator has the form

$$W_1 : \begin{cases} x_1' = 2(x_3^2 + 2x_1x_2)(x_2^2 + 2x_1x_3), \\ x_2' = 2(x_3^2 + 2x_1x_2)(x_1^2 + 2x_2x_3), \\ x_3' = 2(x_3^2 + 2x_1x_2)^2 - 2(x_3^2 + 2x_1x_2) + 1. \end{cases} \quad (17)$$

Denote $g(x) = \frac{1}{2}(9x^4 - 12x^3 + 4x^2 + 1), \quad x \in [0, 1]$. Using the above techniques one can proof the following Theorems.

Theorem 5 *For the function $g(x)$ the following statements are true:*

(i) Fix $(g) = \{1, x^{**}\}$, *where* $x^{**} = \frac{\sqrt[3]{37+\sqrt{1377}}}{9} - \frac{2}{9\sqrt[3]{37+\sqrt{1377}}} + \frac{1}{9}$;
(ii) the fixed point $x = 1$ is a repelling point and the fixed point x^{**} is attracting point;
(iii) for any value $n \geq 2$ the function $g(x)$ has no $n-$ periodic points, different from fixed points;
(iv) $\lim_{n\to\infty} g^n(x) = x^{**}$ *for any* $0 < x < 1$ *and* $g(0) = g(1) = 1$.

Similarly for the operator W_1 we have the following Theorem.

Theorem 6 *For the operator W_1 (17) the following statements are true:*

(i) the set $M_1 = \{\mathbf{x} \in S^2 : x_1 = x_2\}$ is an invariant set;
(ii) the vertex $\mathbf{e}_3$ and $\mathbf{x}^{**} = \left(\frac{1-x^{**}}{2}, \frac{1-x^{**}}{2}, x^{**}\right)$ are fixed points;
(iii) $W_1(\mathbf{x}^{(0)}) = \mathbf{e}_3$ for any $\mathbf{x}^{(0)} \in \{\mathbf{e}_1, \mathbf{e}_2, \mathbf{e}_3\}$;
(iv) if $\mathbf{x}^{(0)} \in M_1 \setminus \{\mathbf{e}_1, \mathbf{e}_2, \mathbf{e}_3\}$ then $\omega_{W_1}\left(\mathbf{x}^{(0)}\right) = \{\mathbf{x}^{**}\}$;
(v) the function $\psi(\mathbf{x}) = |x_1 - x_2|$ is a Lyapunov function;
(vi) if $\mathbf{x}^{(0)} \in S^2 \setminus (M_1 \cup \{\mathbf{e}_1, \mathbf{e}_2, \mathbf{e}_3\})$ then $\omega_{W_1}\left(\mathbf{x}^{(0)}\right) = \{\mathbf{x}^{**}\}$.

If an operator has the property of being regular, then it satisfies the ergodic hypothesis, and by Theorem 4 the SO W_0 and by Theorem 6 the SO W_1 are regular transformations, so we have the following corollary.

Corollary 1 *The SOs W_0 (6) and W_1 (17) are ergodic transformations.*

4 Notes and Comments

In the above investigated the dynamical systems generated by the non-Volterra quadratic SOs V_0 (4), V_1 (5) and by the SOs W_0 (6), W_1 (17). We know that the SO V_0 is a regular operator (Theorem 1) and the SO V_1 is non-regular operator (Theorem 2). Their superpositions defined by the SOs W_0 and W_1 are regular transformations (Theorems 4 and 6). In this case we showed that in the dynamics of superpositions of a regular and a non-regular SOs the regularity property will dominate and the non-regularity will lost.

References

1. Bernstein, S.: Solution of a mathematical problem connected with the theory of heredity. Ann. Math. Stat. **13**(1), 53–61 (1942). https://doi.org/10.1214/aoms/1177731642
2. Devaney, R.L.: An introduction to chaotic dynamical systems. In: Studies in Nonlinearity. Westview Press, Boulder, CO (2003)
3. Ganikhodjaev, N.N., Ganikhodjaev, R.N., Jamilov, U.U.: Quadratic stochastic operators and zero-sum game dynamics. Ergod. Theory Dyn. Syst. **35**, 1443–1473 (2015). https://doi.org/10.1017/etds.2013.109
4. Ganikhodjaev, N.N., Jamilov, U.U., Mukhitdinov, R.T.: Nonergodic quadratic stochastic operators for a two-sex population. Ukr. Math. J. **65**(8), 1282–1291 (2014). https://doi.org/10.1007/s11253-014-0858-2
5. Ganikhodzhaev, R.N.: A family of quadratic stochastic operators that act in S^2. Dokl. Akad. Nauk UzSSR **1**, 3–5 (1989)
6. Ganikhodzhaev, R.N.: Quadratic stochastic operators, Lyapunov functions and tournaments. Sb. Math. **76**, 489–506 (1993). https://doi.org/10.1070/SM1993v076n02ABEH003423
7. Ganikhodzhaev, R., Mukhamedov, F., Rozikov, U.: Quadratic stochastic operators and processes: results and open problems. Infin. Dimens. Anal. Quan. Probab. Relat. Top. **14**, 279–335 (2011). https://doi.org/10.1142/S0219025711004365
8. Jamilov, U.U.: On a family of strictly non-Volterra quadratic stochastic operators. J. Phys. Conf. Ser. **697**, 012013 (2016). https://doi.org/10.1088/1742-6596/697/1/012013
9. Jamilov, U.U.: On symmetric strictly non-Volterra quadratic stochastic operators. Disc. Nonlin. Comp. **5**, 263–283 (2016). https://doi.org/10.5890/DNC.2016.09.006
10. Jamilov, U.U., Ladra, M.: Non-ergodicity of uniform quadratic stochastic operators. Qual. Theory Dyn. Syst. **15**, 257–271 (2016). https://doi.org/10.1007/s12346-015-0145-0
11. Jamilov, U.U., Scheutzow, M., Wilke-Berenguer, M.: On the random dynamics of Volterra quadratic operators. Ergod. Theory Dyn. Syst. **37**, 228–243 (2017). https://doi.org/10.1017/etds.2015.30
12. Khamraev, A.Y.: On the dynamics of a quasi strictly non-Volterra quadratic stochastic operator. Ukr. Math. J. **71**, 1273–1281 (2020). https://doi.org/10.1007/s11253-019-01712-w
13. Lyubich, Y.I.: Mathematical Structures in Population Genetics. Biomathematics, vol. 22. Springer- Verlag, Berlin (1992)
14. Mukhamedov, F., Embong, A.F., Rosli, A.: Orthogonal preserving and surjective cubic stochastic operators. Ann. Funct. Anal. **8**, 490–501 (2017). https://doi.org/10.1215/20088752-2017-0013
15. Mukhamedov, F.M., Ganikhodjaev, N.N.: Quantum Quadratic Operators and Processes. Springer, Berlin (2015). https://doi.org/10.1007/978-3-319-22837-2
16. Mukhamedov, F.M., Jamilov, U.U., Pirnapasov, A.T.: On non-ergodic uniform Lotka-Volterra operators. Math. Notes **105**, 258–264 (2019). https://doi.org/10.1134/S0001434619010280
17. Mukhamedov, F., Pah, C.H., Rosli, A.: On Non-ergodic Volterra Cubic Stochastic operators. Qual. Theory Dyn. Syst. **18**, 1225–1235 (2019). https://doi.org/10.1007/s12346-019-00334-8
18. Mukhamedov, F., Saburov, M.: Stability and monotonicity of Lotka-Volterra type operators. Qual. Theory Dyn. Syst. **16**, 249–267 (2017). https://doi.org/10.1007/s12346-016-0190-3
19. Rozikov, U.A., Zhamilov, U.: F-quadratic stochastic operators. Math. Notes **83**, 554–559 (2008). https://doi.org/10.1134/S0001434608030280
20. Rozikov, U.A., Zhamilov, U.U.: Volterra quadratic stochastic operators of a two-sex population. Ukr. Math. J. **63**, 1136–1153 (2011). https://doi.org/10.1007/s11253-011-0568-y
21. Zhamilov, U.U., Rozikov, U.A.: On the dynamics of strictly non-Volterra quadratic stochastic operators on the two-dimensional simplex. Russ. Acad. Sci. Sb. Math. **200**, 1339–1351 (2009). https://doi.org/10.1070/SM2009v200n09ABEH004039

A Quadratic Worm Propagation Model

Khayotjon O. Khudoyberdiev

Abstract In the present paper, we consider a discrete time network worm propagation model generated by a non-Volterra quadratic stochastic operator. We describe the set of all fixed points of the quadratic stochastic operator and find their types. We prove that for any parameters and initial points the trajectories of such operators converge either to a fixed point from the boundary of the two-dimensional simplex or to a vertex of the simplex.

Keywords Network worms · Propagation dynamics · Modeling · Quadratic stochastic operator · Regular operator

1 Introduction

The proliferation of network worms is causing cyber attacks on network computers and various financial losses. It is not always possible to protect against them through an existing protection system. Spread worms can cause serious damage, such as huge damage to the digital economy as well, by directly stealing personal and confidential information. Network worms can spread and multiply in new appearances from different species. A powerful approach to analyzing the spread of computer worms on the Internet is analytical modeling. The advantage of analytical modeling is traditionally seen as a general solution, as well as the high speed of modeling specific processes for different initial conditions. Modeling of biological viruses was the first to start with [10]. See [19] for network worm breeding models.

There are known mathematical models of the spread of worms in computer networks. In this paper, we analyze a one-stage model, which has received the abbreviated designation SIR—model ("Susceptible—Infected—Removed"). It is assumed that in the SIR model, computers can be in the following states: $S(t)$ is the number of

K. O. Khudoyberdiev (✉)
V. I. Romanovskiy Institute of Mathematics, Uzbekistan Academy of Sciences, 9, University Str., 100174 Tashkent, Uzbekistan
e-mail: xudoyberdiyev.x@mail.ru

L. Accardi et al. (eds.), *Infinite Dimensional Analysis, Quantum Probability and Applications*, Springer Proceedings in Mathematics & Statistics 390,
https://doi.org/10.1007/978-3-031-06170-7_24

efficient computers vulnerable to infection by this worm; $I(t)$—number of infected computers; $R(t)$—number of computers immune to infection by this worm. Let's designate: t—in time units; $N = S(t) + I(t) + R(t)$ the total number of computers in the network.

Initially, the spread of worms from an infected computer will damage weak $S \to \beta \to I$ computers. The infection process is characterized by a specific parameter, for which we get the β sign. The β parameter is considered constant for a given situation. Infected computers with a constant frequency of μ at a certain point in time are treated for worms and have $I \to \mu \to R$ immunity: weak S computers μ frequency gets "immunity" from infection with. $\mu(S \to \mu \to R)$. The above SIR model of the spread of the worm is described by the following system of differential equations

$$\begin{aligned} \tfrac{ds}{dt} &= -\beta si - \mu s, \\ \tfrac{di}{dt} &= \beta si - \mu i, \\ \tfrac{dr}{dt} &= \mu i + \mu s. \end{aligned} \tag{1}$$

Note that in [3] the continuous time dynamical systems generated by operator (1) studied. In the below we consider the discrete time versions of SIR models, but generated by a quadratic stochastic operator. So here we shortly describe the essence of quadratic stochastic operator.

2 Preliminaries

Let

$$S^{m-1} = \left\{ \mathbf{x} = (x_k) \in \mathbb{R}^m : \ x_k \geq 0, \ \text{for any } k \text{ and } \sum_{k=1}^{m} x_k = 1 \right\}$$

be the $(m-1)$-dimensional simplex. A map V from S^{m-1} into itself is called a quadratic stochastic operator if

$$V : x'_k = \sum_{i,j=1}^{m} p_{ij,k} x_i x_j \tag{2}$$

for any $\mathbf{x} \in S^{m-1}$ and for all $k = 1, \ldots, m$, where

$$p_{ij,k} \geq 0, \quad p_{ij,k} = p_{ji,k}, \quad \sum_{k=1}^{m} p_{ij,k} = 1. \tag{3}$$

Such operators occur in some models of mathematical biology and describe the evolution of a population consisting of several types of individuals. (see e.g.

[1, 5–9, 12, 18, 20]) Assume that $\{\mathbf{x}^{(n)} \in S^{m-1} : n = 0, 1, 2, \ldots\}$ is the trajectory of the initial point $\mathbf{x} \in S^{m-1}$, where $\mathbf{x}^{(n+1)} = V(\mathbf{x}^{(n)})$ for all $n = 0, 1, 2, \ldots$, with $\mathbf{x}^{(0)} = \mathbf{x}$. Let $\omega(\mathbf{x}^{(0)})$ be the set of limit points of the trajectory $\{\mathbf{x}^{(n)}\}$, $n = 0, 1, \ldots$ The main problem is investigation of the asymptotical behaviour of the trajectories for a given QSO. In other words the main task is the description of the set $\omega(\mathbf{x}^{(0)})$ for any initial point $\mathbf{x}^{(0)} \in S^{m-1}$ for a given QSO. This problem is an open problem even in two-dimensional case. A point $\mathbf{x} \in S^{m-1}$ is called a *fixed point* of a QSO V if $V(\mathbf{x}) = \mathbf{x}$. A QSO V is called *regular* if for any initial point $\mathbf{x} \in S^{m-1}$, the limit $\lim_{n\to\infty} V^n(\mathbf{x})$ exists. Note that the limit point is a fixed point of a QSO that is $x^* = \lim_{n\to\infty} V(x^*) = V(x^*)$. Thus, the fixed points of a QSO describe limit or long run behavior of the trajectories for any initial point. The limit behavior of trajectories and fixed points play an important role in many applied problems.

Definition 1 ([5]) A quadratic stochastic operator is called Volterra if

$$p_{ij,k} = 0, \quad \text{for any } \; k \notin \{i, j\}, \; i, j, k = 1, \ldots, m.$$

The biological treatment of such operators is rather clear: the offspring repeats one of its parents. Such types of operators have been investigated in many papers [4, 8, 13, 14].

Fix $l \in \{1, \ldots, m\}$ and assume that coefficients $p_{ij,k}$ satisfy

$$p_{ij,k} = 0 \;\text{ if }\; k \notin \{i, j\} \;\text{ for any }\; k \in \{1, \ldots, l\}, \; i, j = 1, \ldots, m;$$

$$p_{ij,k} > 0 \;\text{ for at least one pair }\; (i, j), \; i \neq k, \, j \neq k, \forall k \in \{l+1, \ldots, m\}. \quad (4)$$

Definition 2 ([17]) For any fixed $l \in \{1, \ldots, m\}$, the QSO defined by (2), (3) and (4) is called $\ell-$ Volterra QSO.

We point out that limiting behaviors of certain classes of ℓ-Volterra QSO have been investigated in [15, 16].

3 Discrete Time SIR-models

We consider discrete time analogues of the above mentioned SIR models of worm propagation. The main problem about the asymptotical behaviour of the trajectories for any QSO is the open problem still. This is the motivation of considering the discrete time analogues of these models, presenting in a QSO form.

Although that discrete time SIR models have a form QSO, but the parameters of this operator are not related to $p_{ij,k}$. In each model to make some relations with $p_{ij,k}$ we find conditions on parameters of models (operators) rewriting them in the form (2).

The discrete version of (1) has the following form

$$V:\begin{cases} x_1' = (1-\mu)x_1 - \beta x_1x_2, \\ x_2' = (1-\mu)x_2 + \beta x_1x_2, \\ x_3' = \mu x_1 + \mu x_2 + x_3, \end{cases} \tag{5}$$

where $x_1' = s_{n+1}$, $x_2' = i_{n+1}$, $x_3' = r_{n+1}$, $x_1 = s_n$, $x_2 = i_n$, $x_3 = r_n$ for a natural n.

Using $x_1 + x_2 + x_3 = 1$ one can rewrite the operator (5) in the form

$$\begin{cases} x_1' = (1-\mu)x_1^2 + (1-\beta-\mu)x_1x_2 + (1-\mu)x_1x_3, \\ x_2' = (1-\mu)x_2^2 + (1+\beta-\mu)x_1x_2 + (1-\mu)x_2x_3, \\ x_3' = \mu x_1^2 + \mu x_2^2 + x_3^2 + 2\mu x_1x_2 + (1+\mu)x_1x_3 + (1+\mu)x_2x_3. \end{cases} \tag{6}$$

We consider QSO for the $m = 3$:

$$\begin{cases} x_1' = p_{11,1}x_1^2 + p_{22,1}x_2^2 + p_{33,1}x_3^2 + 2p_{12,1}x_1x_2 + 2p_{13,1}x_1x_3 + 2p_{23,1}x_2x_3, \\ x_2' = p_{11,2}x_1^2 + p_{22,2}x_2^2 + p_{33,2}x_3^2 + 2p_{12,2}x_1x_2 + 2p_{13,2}x_1x_3 + 2p_{23,2}x_2x_3, \\ x_3' = p_{11,3}x_1^2 + p_{22,3}x_2^2 + p_{33,3}x_3^2 + 2p_{12,3}x_1x_2 + 2p_{13,3}x_1x_3 + 2p_{23,3}x_2x_3. \end{cases} \tag{7}$$

From the Eqs. (6) and (7) we have the following relations:

$$\begin{array}{lll} p_{11,1} = 1-\mu & 2p_{12,1} = 1-\beta-\mu & 2p_{13,1} = 1-\mu \\ p_{11,2} = 0 & 2p_{12,2} = 1+\beta-\mu & 2p_{13,2} = 0 \\ p_{11,3} = \mu & 2p_{12,3} = 2\mu & 2p_{13,3} = 1+\mu \\ p_{22,1} = 0 & 2p_{23,1} = 0 & p_{33,1} = 0 \\ p_{22,2} = 1-\mu & 2p_{23,2} = 1-\mu & p_{33,2} = 0 \\ p_{22,3} = \mu & 2p_{23,3} = 1+\mu & p_{33,3} = 1 \end{array} \tag{8}$$

Proposition 1 *The operator (6) maps S^2 to itself if and only if*

$$0 \le \mu \le 1, \quad \mu - 1 \le \beta \le 1 - \mu. \tag{9}$$

Moreover, under condition (9) the operator (6) is a 2-Volterra QSO.

Proof The proof consists solving of simple inequalities which are obtained from conditions (3) for $p_{ij,k}$ given by equalities (8). □

In the below we consider (5) operators which are satisfy the conditions (9).

Let 1_Y be the characteristic (or indicator) function of a subset Y of X. For an arbitrary set T, will denote by $1_Y T$ the set

$$1_Y T = \begin{cases} T, & \text{if } x \in Y, \\ \emptyset, & \text{if } x \notin Y. \end{cases}$$

Denote by $\mathbf{e}_i = (\delta_{1i}, \delta_{2i}, \dots, \delta_{mi}) \in S^{m-1}$, $i = 1, \dots, m$, the vertices of the simplex S^{m-1}, where δ_{ij} is the Kronecker delta. Let the set $\Gamma_\tau = \{\mathbf{x} \in S^{m-1} : x_i = 0,\ i \notin \tau \subset E\}$ be a face of the simplex S^{m-1} and denote by $\mathbf{c}(\tau)$ the center of the face Γ_τ. It is evident that if $\beta = \mu = 0$ then V is the identity map.

Proposition 2 *For the operator V the following statements are true:*

(i) $\mathrm{Fix}\,(V) = \{\mathbf{e}_3\} \cup 1_{\{\mu=0,\ \beta\neq 0\}}\big(\Gamma_{\{1,3\}} \cup \Gamma_{\{2,3\}}\big)$;
(ii) if $\beta = \mu = 0$ then $\mathbf{e}_3$ is a non-hyperbolic point;
(iii) if $\mu > 0$ then $\mathbf{e}_3$ is an attracting point;
(iv) if $\mu = 0,\ \beta > 0$ then any point of the face $\Gamma_{\{2,3\}} = (0, 1-\varepsilon, \varepsilon)$, $\varepsilon \in [0, 1]$ is a non-hyperbolic point;
(v) if $\mu = 0,\ \beta < 0$ then any point of the face $\Gamma_{\{1,3\}} = (1-\varepsilon, 0, \varepsilon)$, $\varepsilon \in [0, 1]$ is a non-hyperbolic point.

Proof (i) Recall, that a fixed point of the QSO V is a solution of the equation $V(x) = x$ and this system of equations has the form

$$V : \begin{cases} x_1 = (1-\mu)x_1 - \beta x_1 x_2, \\ x_2 = (1-\mu)x_2 + \beta x_1 x_2, \\ x_3 = \mu x_1 + \mu x_2 + x_3. \end{cases} \tag{10}$$

Let $x_1 = 0$ then from the second of the system (10) we have: if $\mu = 0$ then any point from the face $\Gamma_{\{2,3\}}$ is a fixed point and if $\mu \neq 0$ then one has $x_2 = 0$, that is we obtain the fixed point $\mathbf{e}_3$.

Let $x_2 = 0$ then from the first of the system (10) one has: if $\mu = 0$ then any point from the face $\Gamma_{\{1,3\}}$ is a fixed point and if $\mu \neq 0$ then we have $x_1 = 0$, that is we get the fixed point $\mathbf{e}_3$.

Let $x_3 = 0$ then from the third of the system (10) one has: if $\mu = 0$ then any point from the face $\Gamma_{\{1,2\}}$ is a fixed point and if $\mu \neq 0$ then we have $x_1 = x_2 = 0$, that is we get the point $(0, 0, 0)$. But this point does not belong to S^2.

Let $x_3 > 0$ then from the third of the system (10) we have: if $\mu \neq 0$ then $x_1 = x_2 = 0$, that is we obtain the fixed point $\mathbf{e}_3$ and if $\mu = 0$ then any point from the set $\Gamma_{\{1,3\}} \cup \Gamma_{\{2,3\}}$ is a fixed point.

(ii) To verify the types of fixed points we rewrite the operator (5) as follows

$$\widetilde{V} : \begin{cases} x_1' = (1-\mu)x_1 - \beta x_1 x_2, \\ x_2' = (1-\mu)x_2 + \beta x_1 x_2, \end{cases} \tag{11}$$

where $(x_1, x_2) \in \widetilde{S} = \{(x_1, x_2) :\ x_1 \geq 0, x_2 \geq 0,\ x_1 + x_2 \leq 1\}$ and x_1, x_2 are the first two coordinates of a point lying in S^2.

The Jacobian of the operator (11) at the fixed point $\widetilde{\mathbf{x}} \in \widetilde{S}$ has the representation

$$D\widetilde{V}(\mathbf{x}) = \begin{pmatrix} 1-\mu-\beta x_2 & -\beta x_1 \\ \beta x_2 & 1-\mu+\beta x_1 \end{pmatrix}.$$

The equation $\det\left(D\widetilde{V}(\mathbf{x}) - \lambda E\right) = 0$ has the form

$$\lambda^2 + (\beta(x_2 - x_1) + 2(\mu - 1))\lambda + (\mu - 1)(\beta(x_2 - x_1) + \mu - 1) = 0, \tag{12}$$

where E is the identity matrix.

The solutions of the Eq. (12) (are eigenvalues of (11) matrix)

$$\lambda_1 = 1 - \mu,\ \lambda_2 = 1 - \mu + \beta(x_1 - x_2).$$

Therefore it immediately follows the statements in the proposition. □

Consider the following function $f_\mu(x) = (1 - \mu)x + \mu,\ \mu, x \in [0, 1]$.

We define $f_\mu^n = \underbrace{f_\mu \circ \cdots \circ f_\mu}_{n \text{ times}}$ to be the n-fold composition of f_μ with itself.

In the next Proposition we will give dynamics of the function $f_\mu(x)$.

Proposition 3 *For the function f_μ the following statements are true:*

(i) $\mathrm{Fix}(f_\mu) = 1_{\{\mu>0\}}\left(\{x = 1\}\right) \cup 1_{\{\mu=0\}}\left([0, 1]\right)$*;*
(ii) if $\mu \in (0, 1]$ then $x = 1$ is an attracting point and it is a non-hyperbolic point when $\mu = 0$;
(iii) if $\mu = 1$ then $f_\mu(x) = 1$ and if $\mu = 0$ then $f_\mu(x) = x$, for any $0 \le x \le 1$;
(iv) if $0 < \mu < 1$ then $\lim\limits_{n\to\infty} f_\mu^n(x) = 1$ for any $0 \le x \le 1$.

Proof (i) If $0 < \mu \le 1$, it is easy to verify that $x = 1$ is a solution of the equation $f_\mu(x) = x$ and if $\mu = 0$ then $Fix(f_\mu) = [0, 1]$.

(ii) One can easily see that $f_\mu'(x) = 1 - \mu$. Consequently we obtain that if $0 < \mu \le 1$ then $x = 1$ is an attracting point and if $\mu = 0$ then $x = 1$ is a non-hyperbolic point.

(iii) Evidently that if $\mu = 1$ for any $0 \le x \le 1$ in that case $f_\mu(x) = 1$ and if $\mu = 0$ for any $0 \le x \le 1$ in that case $f_\mu(x) = x$.

(iv) Let $0 < \mu < 1$ and $0 < x < 1$. Since the function f_μ is an increasing and bounded from the above function it follows the existence of the following limit $\lim\limits_{n\to\infty} f_\mu^n(x) = 1$. □

Theorem 1 *For any parameters (9) and for an initial $\mathbf{x}^{(0)} \in S^2 \setminus Fix(V)$ the set limit points of trajectories of QSO (5) has the form:*

$$\omega_V\left(\mathbf{x}^{(0)}\right) = \begin{cases} \{(0, 1 - x_3^{(0)}, x_3^{(0)})\}, & \text{if } \mu = 0,\ \beta > 0, \\ \{(1 - x_3^{(0)}, 0, x_3^{(0)})\}, & \text{if } \mu = 0,\ \beta < 0, \\ \{\mathbf{e}_3\}, & \text{if } \mu \ne 0,\ \mu - 1 \le \beta \le 1 - \mu. \end{cases}$$

Proof (i) Case $\mu = 0,\ \beta \ne 0$: In this case by assertion of Proposition 2 the points of the face $\Gamma_{\{1,3\}}$ and $\Gamma_{\{2,3\}}$ are fixed points of operator V.

(a) Suppose $\beta > 0$ and $\mathbf{x}^{(0)} \in S^2 \setminus (\Gamma_{\{1,3\}} \cup \Gamma_{\{2,3\}})$ be an initial point. By virtue of Proposition 3 it is follow that $\lim_{n\to\infty} f_\mu^n(x_3^{(0)}) = \lim_{n\to\infty} x_3^{(n)} = x_3^{(0)}$. Then from the (5) we have

$$x_1' = x_1(1-\beta x_2) \le x_1, \; x_2' = x_2(1+\beta x_1) \ge x_2.$$

Therefore one has

$$x_1^{(n)} \le x_1^{(n-1)}, \; x_2^{(n)} \ge x_2^{(n-1)}, \; n = 0, 1, 2, \ldots$$

Since the sequence $\{x_1^{(n)}, \; n = 0, 1, 2 \ldots\}$ (resp. $\{x_2^{(n)}, \; n = 0, 1, 2 \ldots\}$) is a monotone decreasing (resp. increasing) and bounded from the below (resp. above) sequence it is follow that there are the following limits

$$\lim_{n\to\infty} x_1^{(n)} = x_1^*, \quad \lim_{n\to\infty} x_2^{(n)} = x_2^*.$$

Therefore there is the limit $\lim_{n\to\infty} \mathbf{x}^{(n)} = \mathbf{x}^*$. Since the limit point $\mathbf{x}^*$ should be a fixed point it follows that $x_1^* = 0$.

Thus $\lim_{n\to\infty} \mathbf{x}^{(n)} = (0, 1 - x_3^{(0)}, x_3^{(0)})$ for any $\mathbf{x}^{(0)} \in S^2 \setminus (\Gamma_{\{1,3\}} \cup \Gamma_{\{2,3\}})$.

(b) Assume $\beta < 0$ and $\mathbf{x}^{(0)} \in S^2 \setminus (\Gamma_{\{1,3\}} \cup \Gamma_{\{2,3\}})$ be an initial point. Due to assertion of Proposition 3 $\lim_{n\to\infty} f_\mu^n(x_3^{(0)}) = \lim_{n\to\infty} x_3^{(n)} = x_3^{(0)}$. Then from the (5) we have

$$x_1' = x_1(1-\beta x_2) \ge x_1, \; x_2' = x_2(1+\beta x_1) \le x_2.$$

Therefore it follows that

$$x_1^{(n)} \ge x_1^{(n-1)}, \; x_2^{(n)} \le x_2^{(n-1)}, \; x_3^{(n)} = x_3^{(0)}, \quad n = 1, 2, \ldots$$

Since the sequence $\{x_1^{(n)}, \; n = 0, 1, 2 \ldots\}$ (resp. $\{x_2^{(n)}, \; n = 0, 1, 2 \ldots\}$) is a monotone increasing (resp. decreasing) and bounded from the above (resp. below) sequence it is follow that there are the following limits

$$\lim_{n\to\infty} x_1^{(n)} = x_1^*, \quad \lim_{n\to\infty} x_2^{(n)} = x_2^*.$$

Therefore there is the limit $\lim_{n\to\infty} \mathbf{x}^{(n)} = \mathbf{x}^*$. Since the limit point $\mathbf{x}^*$ should be a fixed point it follows that $x_2^* = 0$.

Thus $\lim_{n\to\infty} \mathbf{x}^{(n)} = (1 - x_3^{(0)}, 0, x_3^{(0)})$ for any $\mathbf{x}^{(0)} \in S^2 \setminus (\Gamma_{\{1,3\}} \cup \Gamma_{\{2,3\}})$.

(ii) Case $\mu \neq 0$. By virtue of Proposition 3 it is follow that
$\lim_{n\to\infty} f_\mu^n(x_3^{(0)}) = 1 \;\Rightarrow\; \lim_{n\to\infty} \mathbf{x}^{(n)} = \mathbf{e}_3$ for any $\mathbf{x}^{(0)} \in S^2 \setminus \{\mathbf{e}_3\}$. □

Corollary 1 *For any parameters (5) the QSO (5) is a regular transformation.*

References

1. Akin, E., Losert, V.: Evolutionary dynamics of zero-sum games. J. Math. Biol. **20**, 231–258 (1984). https://doi.org/10.1007/BF00275987
2. Allen, L.J.S.: Some discrete-time SI, SIR, and SIS epidemic models. Math. BioSci. **124**, 83–105 (1994)
3. Barashkov, V.M., Zadorina, N.A., Sareva, A.A.: Analysis of a one-stage mathematical models of the spread of computer viruses, The development of technical sciences in the modern world. http://izron.ru (2019) (in Russian)
4. Ganikhodjaev, N.N., Ganikhodjaev, R.N., Jamilov, U.U.: Quadratic stochastic operators and zero-sum game dynamics. Ergod. Theory Dynam. Syst. **35**, 1443–1473 (2015). https://doi.org/10.1017/etds.2013.109
5. Ganikhodzhaev, R.N.: Quadratic stochastic operators, Lyapunov functions and tournaments. Russ. Acad. Sci. Sb. Math. **76**, 489–506 (1993). https://doi.org/10.1070/SM1993v076n02ABEH003423
6. Ganikhodzhaev, R.N., Mukhamedov, F.M., Rozikov, U.A.: Quadratic stochastic operators and processes: results and open problems. Infin. Dimens. Anal. Quantum Probab. Relat. Top. **14**, 279–335 (2011). https://doi.org/10.1142/S0219025711004365
7. Jamilov, U.U.: On a family of strictly non-Volterra quadratic stochastic operators. J. Phys. Conf. Ser. **697**, 012013 (2016). https://doi.org/10.1088/1742-6596/697/1/012013
8. Jamilov, U.U., Scheutzow, M., Wilke-Berenguer, M.: On the random dynamics of Volterra quadratic operators. Ergod. Theory Dynam. Syst. **37**, 228–243 (2017). https://doi.org/10.1017/etds.2015.30
9. Jamilov, U.U., Khudoyberdiev, Kh.O., Ladra, M.: Quadratic operators corresponding to permutations. Stoch. Anal. Appl. **38**, 929–938 (2020). https://doi.org/10.1080/07362994.2020.1747494
10. Kephart, J.O., White, S.R.: Directed-graph epidemiological models of computer viruses. In: Proceedings of the 1991 IEEE Computer Society Symposium on Research in Security and Privacy, Oakland, California, pp. 343–359 (1991)
11. Kotenko, I.V., Vorontsov, V.V.: Analytical models of network worm propagation. SPIIRAS Proc. Issue 4. SPb. Nauka (2007) (in Russian)
12. Lyubich, Y.I.: Mathematical Structures in Population Genetics. Biomathematics, vol. 22. Springer-Verlag, Berlin (1992)
13. Mukhamedov, F.M., Jamilov, U.U., Pirnapasov, A.T.: On non-ergodic uniform Lotka-Volterra operators. Math. Notes **105**, 258–264 (2019). https://doi.org/10.1134/S0001434619010280
14. Mukhamedov, F., Saburov, M.: On dynamics of Lotka-Volterra type operators. Bull. Malay. Math. Sci. Soc. **37**, 59–64 (2014)
15. Mukhamedov, F., Saburov, M., Qaralleh, I.: On $\xi^{(s)}$-quadratic stochastic operators on two dimensional simplex and their behavior. Abst. Appl. Anal. **12**, 942038 (2013). https://doi.org/10.1155/2013/942038
16. Rozikov, U.A., Shoyimardonov, S.K.: Ocean ecosystem discrete time dynamics generated by l-Volterra operators. Int. J. Biomath. **12**, 1950015 (2019)
17. Rozikov, U.A., Zada, A.: On dynamics l-Volterra quadratic stochastic operators. Int. J. Biomath. **56**, 1418–1427 (2004)
18. Rozikov, U.A., Zhamilov, U.U.: Volterra quadratic stochastic operators of a two-sex population. Ukr. Math. J. **63**, 1136–1153 (2011). https://doi.org/10.1007/s11253-011-0568-y
19. Wang, Y., Wen, S., Xiang, Y., Zhou, W.: Modeling the propagation of worms in networks: a survey. IEEE Commun. Surv. Tutor. **16** (2014)
20. Zhamilov, U.U., Rozikov, U.A.: On the dynamics of strictly non-Volterra quadratic stochastic operators on the two-dimensional simplex. Russ. Acad. Sci. Sb. Math. **200**, 1339–1351 (2009). https://doi.org/10.1070/SM2009v200n09ABEH004039

Author Index

L. Accardi et al. (eds.), *Infinite Dimensional Analysis, Quantum Probability and Applications*, Springer Proceedings in Mathematics & Statistics 390,
https://doi.org/10.1007/978-3-031-06170-7